AA001072

2018 IEEE International Conference on Semiconductor Electronics (ICSE 2018)

Kuala Lumpur, Malaysia
15 – 17 August 2018

IEEE Catalog Number: CFP18421-POD
ISBN: 978-1-5386-5284-8

**Copyright © 2018 by the Institute of Electrical and Electronics Engineers, Inc.
All Rights Reserved**

Copyright and Reprint Permissions: Abstracting is permitted with credit to the source. Libraries are permitted to photocopy beyond the limit of U.S. copyright law for private use of patrons those articles in this volume that carry a code at the bottom of the first page, provided the per-copy fee indicated in the code is paid through Copyright Clearance Center, 222 Rosewood Drive, Danvers, MA 01923.

For other copying, reprint or republication permission, write to IEEE Copyrights Manager, IEEE Service Center, 445 Hoes Lane, Piscataway, NJ 08854. All rights reserved.

*** This is a print representation of what appears in the IEEE Digital Library. Some format issues inherent in the e-media version may also appear in this print version.

IEEE Catalog Number: CFP18421-POD
ISBN (Print-On-Demand): 978-1-5386-5284-8
ISBN (Online):   978-1-5386-5283-1

Additional Copies of This Publication Are Available From:

Curran Associates, Inc
57 Morehouse Lane
Red Hook, NY 12571 USA
Phone: (845) 758-0400
Fax: (845) 758-2633
E-mail: curran@proceedings.com
Web: www.proceedings.com

2018 IEEE INTERNATIONAL CONFERENCE ON SEMICONDUCTOR ELECTRONICS (ICSE2018)

Proceedings

15 – 17 August 2018
Pullman Kuala Lumpur City Centre Hotel & Residence,
Kuala Lumpur, Malaysia

Organized by
Electron Devices Chapter of IEEE Malaysia Section

Co-organized by
Institute of Microengineering & Nanoelectronics (IMEN),
Universiti Kebangsaan Malaysia

Supported by
Malaysia Convention & Exhibition Bureau (MyCEB) Ministry of Tourism,
Arts and Culture Malaysia (MoTAC) Malaysia Truly Asia

 **Message from Vice Chancellor of Universiti Kebangsaan Malaysia**

Assalamualaikum Warahmatullahi Wabarakatuh and *Selamat Datang*.

Assalamualaikum Warahmatullahi Wabarakatuh and Selamat Datang.
I would like to congratulate the Electron Devices Chapter of IEEE Malaysia Section and the Institute of Microengineering and Nanoelectronics (IMEN) of Universiti Kebangsaan Malaysia for successfully organizing the 13th IEEE International Conference on Semiconductor Electronics 2018 (ICSE2018) under the able leadership of Prof. Dato' Dr. Burhanuddin Yeop Majlis.

IMEN is one of the Centre of Excellence (CoE) of UKM and has been awarded Nano Malaysia Centre of Excellence by the Ministry of Science, Technology and Innovation (MOSTI) in 2011 and has recently been awarded the Higher Education Centre of Excellence (HiCOE) by the Malaysian Ministry of Education (MOE) in the field of micro and nanoelectronics which is one of UKM's research strength. UKM and MOE are committed in promoting microelectronics and particularly nanotechnology as a niche technology to spawn supporting and enabling technologies, to give Malaysia its leap forward into the competitive world of knowledge economy.

I very much welcome the holding of this conference which gives excellent opportunity for engineers, researchers, and educators from academia and industry to present their works and keep themselves abreast of technology development, particularly in the areas of microelectronics and nanotechnology.

I wish all of you a pleasant stay in this country and we hope that ICSE2018 will be successful and enjoyable to all participants.

Terima kasih.

Tan Sri Prof. Datuk Dr. Noor Azlan Ghazali
Vice Chancellor
Universiti Kebangsaan Malaysia

Message from IEEE Electron Devices Society (EDS) Malaysia Chapter Chair 2018

On behalf of the IEEE Electron Devices Malaysia Chapter, I would like to welcome all participants and paper presenters to the 13th IEEE International Conference on Semiconductor Electronics 2018 (ICSE2018). I am very pleased with the positive response received from various local and regional research institutions as well as institutions of higher learning.

It is a great pleasure to welcome all our plenary, keynote and tutorial speakers as well as CEOs, participants and exhibitors. We are proud to host talks from world renowned experts in nanomaterials and nanotechnology which surely will immensely benefit our participants. I would like to thank Prof Jackie Y. Ying from NanoBio Lab A*STAR, Prof Edward Yi Chang from the National Chiao Tung University as our plenary speakers; and our keynote speakers Prof. Navakantha Bhat, Assoc. Prof. Tomas Blecha, Prof. Jang Kyoo Shin and Prof. Kevin Homewood for your support. I would also like to thank IEEE Malaysia Section, IEEE EDS Distinguished Lecturers and the Malaysian Convention and Exhibition Bureau (MyCEB) for supporting ICSE2018.

This year, EDS Malaysia and ICSE2018 committee have agreed to offer some new programs to attract industrial participation in this bi-annual conference. For the first time, ICSE2018 is offering pre-conference tutorials, CEO talks and industry awards to attract more industrial participation. The Malaysian Human Resources Development Fund (HRDF) have agreed to allow ICSE2018 participation from industrial employees as a form of training which is claimable by their respective employers. ICSE2018 also offers invited talks for academicians, Best Student Paper Awards as well as Best Student Presenter Awards. We hope all these efforts will enhance the interest in electron devices research in Malaysia.

I would like to express my deepest gratitude and appreciation to the organizing committee of ICSE2018 for their relentless effort to ensure the successful organization of this conference. I would also like to acknowledge the hard work and co-operation of the staff and students of Institute of Microengineering and Nanoelectronics (IMEN), IEEE EDS Malaysia as well as IEEE EDS UKM Student Branch. Special thanks to all our plenary, keynote and tutorial speakers and especially participants who without them, this conference will not be a reality.

Thank you and *Terima Kasih*.

Assoc. Prof. Dr. P Susthitha Menon
Chapter Chair
IEEE Electron Devices Malaysia Chapter 2018
and
Technical Chair
IEEE-ICSE2018

 **Message from ICSE2018 Conference Chair**

Selamat Datang to Kuala Lumpur and ICSE2018. On behalf of the organizing committee, it is with great pleasure to welcome you to the 13th IEEE International Conference on Semiconductor Electronics 2018 (ICSE2018). Over the last twenty six years, ICSE has become the preeminent international forum on semiconductor electronics embracing all aspects of the semiconductor technology. As the initiator of ICSE conference series and person-in-charge of its very first conference way back in 1992, I am delighted to see the journey ICSE has taken to this far and the benefits we have imparted to all participants we have met.

ICSE2018 starts off with pre-tutorial sessions by Dr. Samar K. Saha who is the Junior Past President of IEEE EDS and EDS Distinguished Lecturer Region 6 (Western USA) as well as Dr. M.K. Radhakrishnan who is the Vice President (Regions & Chapters) of IEEE EDS and EDS Distinguished Lecturer Region 10 (Asia and Pacific). This time the conference offers two plenary lectures and four keynote lectures by distinguished persons in their own fields. We are honored and proud to have the eminent Prof Jackie Y. Ying from the NanoBio Lab A*STAR and Prof Edward Yi Chang from the National Chiao Tung University as our plenary speakers who are world renowned experts in nanomaterials and nanotechnology. We are proud to have prominent researchers Prof. Navakantha Bhat from Indian Institute of Sciences, Assoc. Prof. Tomas Blecha from the University of West Bohemia, Prof. Jang Kyoo Shin from the Kyungpook National University and Prof. Kevin Homewood from the Hubei University, Republic of China as our keynote speakers. It is indeed a privilege for our participants to have opportunity to meet them and listen to their great ideas. In conjunction with ICSE2018, we have an industrial session organized by EDS Malaysia and IEEE Malaysia Section where Prof. Ir. Dr. Ahmad Fadzil Mohamad Hani, CEO of SIRIM Berhad and En Ahmad Rizan Ibrahim, CEO of MIMOS Berhad will share their knowledge and insights about their respective industries.

The conference proceedings provide views into the current advances in semiconductor electronics in the region and we look forward to the presentations of our participants. We hope participants would appreciate knowledge imparted by the lectures while making new contacts with other participants.
We are grateful for all speakers and participants for your support. To participants from overseas, I wish you a pleasant stay in this country and we will endeavor to make your stay here enjoyable.

Terima kasih.

Prof. Dato' Dr. Burhanuddin Yeop Majlis, SMIEEE
Chair, 13ᵗʰ IEEE International Conference on Semiconductor Electronics

IEEE-ICSE 2018 Committee

Chair:
Prof. Dato' Dr. Burhanuddin Yeop Majlis, SMIEEE *Universiti Kebangsaan Malaysia*

Co-Chair:
Assoc. Prof. Dr. Norhayati Soin, SMIEEE *Universiti Malaya*

Hon. Secretary:
Dr. Haslina Jaafar, SMIEEE *Universiti Putra Malaysia*

Assistant Hon. Secretary:
Dr. Zubaida Yusoff, SMIEEE *Multimedia University*

Treasurer:
Ir. Dr. Hazian Mamat, SMIEEE *MIMOS Berhad*

Technical Chairs:
Prof. Dr. A.H.M Zahirul Alam, SMIEEE *International Islamic University Malaysia*
Assoc. Prof. Dr. P. Susthitha Menon, SMIEEE *Universiti Kebangsaan Malaysia*

Logistic Chair:
Assoc. Prof. Dr. Norhayati Soin, MIEEE *Universiti Malaya*

Secretariat Leader:
Dr. Dilla Duryha Berhanuddin, MIEEE *Universiti Kebangsaan Malaysia*
Dr. Rhonira Latif, MIEEE *Universiti Kebangsaan Malaysia*

Webmaster:
Dr. Aliza Aini Md Ralib, MIEEE *International Islamic University Malaysia*

Committee Member

Assoc. Prof. Dr. Badariah Bais, SMIEEE *Universiti Kebangsaan Malaysia*
Assoc. Prof. Dr. Mohd Nizar Hamidon, MIEEE *Universiti Putra Malaysia*
Assoc. Prof. Dr. Dee Chang Fu, MIEEE *Universiti Kebangsaan Malaysia*
Assoc. Prof. Dr. Azrul Azlan Hamzah, MIEEE *Universiti Kebangsaan Malaysia*
Assoc. Prof. Dr. Jumril Yunas, SMIEEE *Universiti Kebangsaan Malaysia*
Assoc. Prof. Dr. Siti Nooraya Mohd Tawil, MIEEE *National Defence University of Malaysia*
Assoc. Prof. Dr. Mohd Haris Md Khir, MIEEE *Universiti Teknologi Petronas*
Dr. Jahariah Sampe, MIEEE *Universiti Kebangsaan Malaysia*
Dr. Muhamad Ramdzan Buyong, SMIEEE *Universiti Kebangsaan Malaysia*
Dr Amrallah Mustaffa, MIEEE *Universiti Putra Malaysia*
Dr. Mohd Ambri Mohamed, MIEEE *Universiti Kebangsaan Malaysia*
Dr. Siti Noorjannah Ibrahim, MIEEE *International Islamic University Malaysia*
Dr. Rosminazuin Ab Rahim, MIEEE *International Islamic University Malaysia*
Dr. Nor Farahidah Za'bah, MIEEE *International Islamic University Malaysia*
Dr. Ahmad Rifqi Md Zain, MIEEE *Universiti Kebangsaan Malaysia*
Ir Ts. Dr. Ahmad Sabirin Zoolfakar, MIEEE *Universiti Teknologi MARA*
Dr. Afishah Alias, MIEEE *Universiti Tun Hussein Onn Malaysia*
Ir. Ts. Azrif Manut, MIEEE *Universiti Teknologi MARA*
Ir. Nor Azhadi Hj. Ngah, MIEEE *TMR&D Sdn. Bhd.*

Secretariat

Mrs. Khairul Nisha Mohd Kharuddin, MIEEE
Mrs. Aishah Fauthan
Mrs. Anezah Marsan
Mr. Abdul Hafiz Mat Sulaiman , MIEEE
Mr. Shafii Abdul Wahab

Secretariat Office

Secretariat of IEEE-ICSE2018
Electron Devices Malaysia Chapter
Institute of Microengineering & Nanoelectronics (IMEN)
Universiti Kebangsaan Malaysia, 43600 UKM Bangi, Selangor, MALAYSIA
Telephone +603 8921 6987
Fax +603 8925 0439
Email: edsmalaysia@gmail.com

Gap in pagination due to unavailable material.

Page vii

TABLE OF CONTENTS

Message from Vice Chancellor of UKM	ii
Message from IEEE EDS Malaysia Chapter Chair	iii
Message from ICSE2018 Conference Chair	iv
IEEE-ICSE 2018 Committee	v
Objective	N/A
Table of Contents	viii

Pre-Conference Tutorial

No.	Titles	Page
T1	**Advanced Silicon Devices for VLSI Circuits and Systems at Nanometer Nodes** Samar K. Saha, *Prospicient Devices, California, USA*	A1
T2	**Building in Reliability in Silicon Devices** M.K. Radhakrishnan, *NanoRel LLP Technical Consultants, Singapore*	A2

Plenary Speakers

No.	Titles	Page
P1	**Nanosystems for Food, Drug and Biomedical Applications** Jackie Y. Ying, *Institute of Bioengineering and Nanotechnology, 31 Biopolis Way, The Nanos, Singapore 138669*	B1
P2	**High-performance E-mode GaN MIS-HEMT for Power Switching Applications** Edward Yi Chang, *Department of Materials Science and Engineering, National Chiao Tung University, 1001 Ta Hsueh Road, Hsinchu 30010, Taiwan.*	B2

Keynote Speakers

No.	Titles	Page
K1	**Tunable Steep Slope MoS2 Transistor** Navakantha Bhat, *Professor and Chairperson Centre for Nano Science and Engineering (CeNSE) Indian Institute of Science, Bangalore*	C1
K2	**Band Edge Modified Rare Earths- A Route to The Mid Infrared in Silicon** Kevin Homewood, *School of Materials Science & Engineering at Hubei University, Wuhan, China*	C2

| K3 | **CMOS Pixel-Aperture and Offset Pixel-Aperture Techniques for 3-Dimensional Imaging**
Jang Kyoo Shin, *Professor, School of Electronics Engineering, Kyungpook National University, Daegu 41566, South Korea* | C3 |
| K4 | **Smart Firefighter Protective Suit – Functional Blocks And Technologies**
Tomas Blecha, Radek Soukup, Petr Kaspar, Ales Hamacek, Jan Reboun, *Department of Measurement and Technologies, Faculty of Electrical Engineering University of West Bohemia* | C4 |

Cluster 1: MEMS & Nanoelectronics

MEMS/NEMS, Device Modeling & Simulation, Nanoelectronics, Device Physics & Characterizatsion Semiconductor sensor

No.	Titles	Page
1	**Engineering the Properties of Nb2O5-ZnO Nanostructures via Dual Synthesis Techniques** Rozina Abdul Rani, Fatin Nurdini Omar, Mohd Husairi Fadzilah Suhaimi, Ahmad Sabirin Zoolfakar, Mohamad Hafiz Mamat, Salman Alrokayan, Haseeb A. Khan, Mohamad Rusop Mahmood *Universiti Teknologi MARA, MALAYSIA*	1
2	**Dielectrophoresis: Iron Dificient Anemic Red Blood Cells for Artificial Kidney Purpose** Farahdiana Wan Yunus, Azrul Azlan Hamzah, Mas Syarafina Norzin, Muhammad Ramdzan Buyong, Jumril Yunas, Burhanuddin Yeop Majlis *Universiti Kebangsaan Malaysia, MALAYSIA*	5
3	**Physical Modelling and Simulation of Au/ZnO Schottky Diode** Hira Shumail, Sadia Muniza Faraz *NED University of Engineering & Technology, PAKISTAN*	9
4	**Voltage Characterization on Dielectrophoretic Force Respond to Hematologic Cell Manipulation** Muhammad Izzuddin Abd Samad, Muhamad Ramdzan Buyong, Farahdiana Wan Yunus, Kim Shyong Siow, Azrul Azlan Hamzah, Burhanuddin Yeop Majlis *Universiti Kebangsaan Malaysia, MALAYSIA*	13
5	**Integration Design of Wide-Dynamic-Range MEMS Magnetometer and Oscillator** Chao-Hung Song, Kuei-Ann Wen *National Chiao Tung University, TAIWAN*	17

6 **Enhancing Photocatalytic Performance of Nanoporous Nb2O5 Doped Platinum** 21
Aidil Aizat Mohd Hamir, Katrul Nadia Basri, Mahzaton Aqma Abu Talip, Mohamad Hafiz Mamat, Azrif Manut, Ahmad Sabirin Zoolfakar, Rozina Abdul Rani, M.Rusop
Universiti Teknologi MARA, MALAYSIA

7 **A Better Understanding of CNTs Chemical Purification and Functionalization Processes** 25
Saman Azhari, Mohd Nizar Hamidon, Intan Helina Hasan, Muhammad Asnawi Mohd Kusaimi, Siti Amaniah Mohd Chachuli, Nur Alin Mohd Azhari, Ismayadi Ismail
Universiti Putra Malaysia, MALAYSIA

8 **Modeling and Simulation of VO2/Au Thin Film Transition Behavior** 29
Asmaa Leila Hassein-Bey, Abdelkader Hassein-Bey, Hakim Tahi, Slimane Lafane, Samira Abdelli-Messaci, Mohamed El-Amine Benamar
University Saad Dahlab Blida, ALGERIA

9 **Integration of 2.4GHz RF Transmitter with Accelerometer for Mobile Sensing Applications on 1P6M ASIC Compatible CMOS MEMS Process** 33
Pao-Min Chu, Kuei-Ann Wen
National Chiao Tung University, TAIWAN

10 **Detecting Hydrogen using TiO2-B2O3 Gas Sensor at Different Operating Temperature** 37
Siti Amaniah Mohd Chachuli, Mohd Nizar Hamidon, Md. Shuhazlly Mamat, Mehmet Ertugrul
Universiti Putra Malaysia, MALAYSIA

11 **Study on Breakdown Characteristics of AlGaN/GaN- based HFETs** 41
Tineesha Naidu, Sharifah Wan Muhamad Hatta, Norhayati Soin, Sharidya Rahman, Yasmin Abdul Wahab
University of Malaya, MALAYSIA

12 **Monolithic Integration of Digital MEMS Thermometer and Temperature Compensated RTC on 1P6M ASIC Compatible CMOS MEMS Process** 45
Ching-Wen Hsu, Kuei-Ann Wen
National Chiao Tung University, TAIWAN

13 **Influence of Titanium Oxide Coating on Mechanical Properties of Porous Nanocrystalline Silicon Membrane** 49
Muhammad Fahmi Jaafar, Rhonira Latif, Burhanuddin Yeop Majlis
Universiti Kebangsaan Malaysia, MALAYSIA

14 **Influence of Volume Variety of Waste Cooking Palm Oil as Carbon Source on Graphene Growth Through Double Thermal Chemical Vapor Deposition** 53
Robaiah Hj Mamat, Fazlena Hamzah, Azhan Hashim, Saifollah Abdullah, Salman.A.H.Alrokayan, Haseeb.A.Khan, Munirah Safiay, Salifairuz Mohamad Jafar, Asnida Asli, Zuraida Khusaimi, Mohamad Rusop
Universiti Teknologi MARA, MALAYSIA

15 **Feasability of Zinc Oxide Nanowire as a Temperature Sensor: An Analytical Study** 57
Azam Mohamad, Harzawardi Hasim, Suhana Mohamed Sultan
Universiti Teknologi Malaysia, MALAYSIA

16 **Taguchi Optimization of Graphene-Based Surface Plasmon Resonance-Kretschmann Biosensor Using FDTD** 61
Nur Akmar Jamil, P. Susthitha Menon, Sahbudin Shaari, Mohd Ambri Mohamed, Burhanuddin Yeop Majlis
Universiti Kebangsaan Malaysia, MALAYSIA

17 **Effect of Flow Rate towards Microfluidic Hollow Fiber Extraction Efficiency** N/A
Nurul Azzyaty Jayah, Badariah Bais, Burhanuddin Yeop Majlis
Universiti Kebangsaan Malaysia, MALAYSIA

18 **Separation of Micro Engineered Particle Using Dielectrophoresis Technique** 69
Nur Rabiatul Adawiyah Tajul Othamany, Norazreen Abd Aziz, Muhammad Izzuddin Abd Samad, Muhamad Ramdzan Buyong, Burhanuddin Yeop Majlis
Universiti Kebangsaan Malaysia, MALAYSIA

19 **Effect of Different Metal Contact Distance and Light on Electrical Properties of Calcium Carbonate Thin Film** 73
N.H.Sulimai, Rozina Abdul Rani, M.J.Salifairus, M.H.Mamat, M.F.Malek, A.S. Zoolfakar, Z.Khusaimi, S.Abdullah, Haseeb Khan, Salman Alrokayan, M.Rusop
Universiti Teknologi MARA, MALAYSIA

20 **Facile Synthesis of N-doped ZnO Nanorod Arrays: Towards Enhancing the UV-sensing Performance** 77
A.S. Ismail, M.H. Mamat, M.F. Malek, N.E.A. Azhar, N.H. Sulimai, R. Abdul Rani, A.S. Zoolfakar, M. Rusop
Universiti Teknologi MARA, MALAYSIA

21 **Tip Deflection of a Thermal Bimorph Cantilever Beam with Different Geometrical Structures** 81
Z. H. A. Rahman, M. H. Md. Khir, M. A. Zakariya
Universiti Teknologi PETRONAS, MALAYSIA

22 **Dielectrophoresis: Characterization of Triple-Negative Breast Cancer using Clausius-Mossotti Factor** 85
Nur Mas Ayu Jamaludin, Muhamad Ramdzan Buyong, Muhammad Khairulanwar Abdul Rahim, Azrul Azlan Hamzah, Burhanuddin Yeop Majlis, Badariah Bais
Universiti Kebangsaan Malaysia, MALAYSIA

23 **Synthesis, Properties and Humidity Detection of Anodized Nb2O5 Films** 89
Rozina Abdul Rani, Ahmad Sabirin Zoolfakar, Mohamad Fauzee Mohamad Ryeeshyam, Najwa Ezira, Ahmed Azhar, Mohamad Hafiz Mamat, Salman Alrokayan, Haseeb A. Khan, Mohamad Rusop Mahmood
Universiti Teknologi MARA, MALAYSIA

24 **Impedance based Aluminium Interdigitated Electrode (Al-IDE) Biosensor on Silicon Substrate for Salmonella Detection** 93
R.D.A.A Rajapaksha, M.N. Afnan Uda, U. Hashim, S.C.B. Gopinath, C.A.N. Fernando
Universiti Malaysia Perlis, MALAYSIA

25 **Performance Analysis of SAW Gas Sensor Based on ST- Cut Quartz for Breath Analysis** 97
Nur Fatin binti Muhammad Razali, Aliza Aini Md Ralib, Rosminazuin Ab Rahim
International Islamic University Malaysia, MALAYSIA

26 **Comparative study of the Calcium Ferrite Nanoparticles (CaFe2O4-NPs) Synthesis Process** 101
Noor Sulaiman, Jumril Yunas
Universiti Kebangsaan Malaysia, MALAYSIA

27 **Dynamic Behavior of Condenser Microphone Under the Influence of Squeeze Film Damping** 104
Siti Aisyah Zawawi, Azrul Azlan Hamzah, Reena Sri Selvarajan, Burhanuddin Yeop Majlis, Faisal Mohd-Yasin
Universiti Kebangsaan Malaysia, MALAYSIA

28 **Optimization of p-type Emitter Thickness for GaSb-Based Thermophotovoltaic Cells** 109
W. Emilin Rashid, Pin Jern Ker, M. Z. Jamaludin, N. A. Rahman, Mahdi All Khamis *Universiti Tenaga Nasional, MALAYSIA*

29 **Capacitance Effects of Ring Oscillator's Waveform Quality in Designing Physically Unclonable Functions** 113
Zulfikar Zulfikar, Norhayati Soin, Sharifah Fatmadiana Wan Muhamad Hatta
University of Malaya, MALAYSIA

| 30 | **Numerical study of Zigzag Micro Mixer with 3D Channel Dimension** | 117 |

Wan Ammar Fikri Wan Ali, Jumril Yunas, Azlan Hamzah, Kamarul 'Asyikin Mustafa, Burhanuddin Yeop Majlis
Universiti Kebangsaan Malaysia, MALAYSIA

| 31 | **A Novel Digital Etch Technique for p-GaN Gate HEMT** | 121 |

Yuan Lin, Yueh Chin Lin, Franky Lumbantoruan, Chang Fu Dee, Burhanuddin Yeop Majlis, Edward Yi Chang
Universiti Kebangsaan Malaysia, MALAYSIA

| 32 | **pH Sensing Characteristics of Silicon Nitride as Sensing Membrane based ISFET Sensor for Artificial Kidney** | 124 |

Mas Syarafina Norzin, Farahdiana Yunus, Azlan Hamzah, Jumril Yunas, Burhanuddin Yeop Majlis
University Kebangsan Malaysia, MALAYSIA

| 33 | **Characteristic Study of Doped ZnO Thin Film** | 128 |

Syafiqah Ishak, Shazlina Johari, Muhammad M. Ramli, Ashila Ashila, Azri Arbae, Muhammad Shafawi, Aidil Azlan
Universiti Malaysia Perlis, MALAYSIA

| 34 | **Characterization of Permittivity and Conductivity for ESKAPE Pathogens Detection** | 132 |

Muhammad Khairulanwar Abdul Rahim, Muhamad Ramdzan Buyong, Nur Mas Ayu Jamaludin, Hamzah Azrul Azlan, Kim S Siow, Burhanuddin Majlis
University Kebangsan Malaysia, MALAYSIA

| 35 | **Effect of ZnO Composition on the Electrical Properties of MEH-PPV: ZnO Nanocomposites Thin film via Spin Coating** | 136 |

Najwa Ezira Ahmed Azhar, Shafinaz Sobihana Shariffudin, Rozina Abdul Rani, Ahmad Sabirin Zoolfakar, Mohd Firdaus Malek, Salman Alrokayan, Haseeb A. Khan, Mohamad Rusop
Universiti Teknologi MARA, MALAYSIA

| 36 | **Synthesis of ZnO Nanoflakes by 1064 nm ND-YAG Pulsed Laser Deposition in a Horizontal Tube Furnace** | 140 |

Kong Eng Ng, Tiong Teck Yaw, Mohd Ambri Mohamed, Chang Fu Dee, Burhanuddin Yeop Majlis, Wei Sea Chang
Universiti Kebangsaan Malaysia, MALAYSIA

| 37 | **Low Transmittance of Anatase Titanium Dioxide (TiO2) Prepared via Doctor Blade Technique** | 144 |

M.S.P Sarah, N. Norizan, S.S. Shariffudin, H. Hashim
Universiti Teknologi MARA, MALAYSIA

38   **Performance Comparison on Current Consumption Between Arduino   148
Nano and Arm Coretex M3 for Portable Dialysis System**
Farrah Masyitah Mohd Shuib, Shafii A. Wahab, Abdul Hafiz Mat
Sulaiman, Hamzah Azrul Azlan, Sawal Hamid Md Ali, Mohd Nuriman
Nawi
Universiti Kebangsaan Malaysia, MALAYSIA

39   **Silver-Graphene Oxide Nanocomposite Film-based SPR Sensor for   152
Detection of Pb2+ Ions**
Wan Maisarah Mukhtar, Karsono Ahmad Dasuki, Affa Rozana Abdul
Rashid, Nur Athirah Mohd Taib, Razman Mohd Halim
Universiti Sains Islam Malaysia, MALAYSIA

40   **Optimization of Pattern Transfer in Fabrication of GFET for   156
Biosensing Applications**
Reena Sri Selvarajan, Azlan Hamzah, Siti Aisyah Zawawi, Burhanuddin
Yeop Majlis *Universiti Kebangsaan Malaysia, MALAYSIA*

41   **A Study on the Atomic Topography of Nanostructured TiO2 Thin   160
Films: Effect of Annealing**
Nur Munirah Safiay, Rozina Abdul Rani, Nur Amierah Mohd Asib, M
Robaiah, Zuraida Khusaimi, Fazlena Hamzah, Saifollah Abdullah,
Salman Alrokayan, Haseeb A. Khan, Mohamad Rusop
Universiti Teknologi MARA, MALAYSIA

42   **Platinum (Pt) Doped Nb2O5 for Enhancing Ultraviolet Photodetector   164**
M. Anas A. Basir, A. Manut, M. Hafiz Mamat, Rosmalini Abdul Kadir,
A. Sabirin Zoolfakar, Rozina Abdul Rani, M. Rusop
Universiti Teknologi MARA, MALAYSIA

43   **Activation Energy of Thermal Oxidation Germanium Oxide on   169
Germanium Substrates**
Nurul Atiqah Abdul Halim, Azrul Hamzah, Burhanuddin Yeop Majlis,
Marini Sawawi, Muhammad Kashif, Farrah Masyitah Mohd Shuib, Siti
Kudnie Sahari
Universiti Kebangsaan Malaysia, MALAYSIA

44   **Optimization of Wurtzite Gan-Based Gunn Diode as Terahertz   173
Source**
Lee Wen Zhao, Duu Sheng Ong, Kan Yeep Choo
Multimedia University, MALAYSIA

Cluster 2: Nanophotonics
Optoelectronics & Photonics Technology, Microwave Device and MMIC

No.	Titles	Page
45	**Band Gap Engineering of GaAsBi Alloy for Emission of up to 1.52 μm** Abdul Rahman Mohmad, John P. R. David *Universiti Kebangsaan Malaysia, MALAYSIA*	177
46	**Photonic Crystal Embedded Waveguide for Compact C-Band Band-Pass Filter** Mohd Nuriman Nawi, Nurulhani Diana Rashid, Dilla Duryha Berhanuddin, Burhanuddin Yeop Majlis, Mohd Adzir Mahdi, Ahmad Rifqi Md Zain *Universiti Kebangsaan Malaysia, MALAYSIA*	180
47	**Optical Properties of Multilayer Porous Silicon with Different Fabrication Conditions for Application along Telecommunication C-band** Ahmad Afif Safwan Mohd Radzi Mohamad Rusop, Saifollah Abdullah *Universiti Kebangsaan Malaysia, MALAYSIA*	184
48	**Design and Simulation of Tapered Optical Fiber by Enhancing The Evanescent Field Region for Sensing Application** Nurulhani Diana Rashid, Dilla Duryha Berhanuddin, Mohd Adzir Mahdi, Burhanuddin Yeop Majlis, Ahmad Rifqi Md Zain *Universiti Kebangsaan Malaysia, MALAYSIA*	188
49	**Jsc And Voc Optimization of Perovskite Solar Cell with Interface Defect Layer using Taguchi Method** Mohd Shaparuddin Bahrudin, Siti Fazlili Abdullah, Ibrahim Ahmad, Ahmad Wafi Mahmood Zuhdi, Azri Husni Hasani, Fazliyana Za'abar, M. Najib Harif, Mazin Malik *Universiti Tenaga Nasional, MALAYSIA*	192
50	**Flexible Photoanode on Titanium Foil for Back-Illuminated Dye Sensitized Solar Cells** Suraya Shaban, Suhaidi Shafie, Yusran Sulaiman, Noor Fadzilah binti Mohamed, Sharif Muhammad Quisar Lokman, Fauzan Ahmad *Universiti Putra Malaysia, MALAYSIA*	197
51	**Simulation Analysis on CIGS Solar Cell on Different Absorber Layer Thickness Subjectt to Temperature Change using SCAPS 1-D Software** M. Najib Harif, Siti Fazlili Abdullah, Ahmad Wafi Mahmood Zuhdi, Fazliyana Za'abar, Mohd Shaparuddin Bahrudin, Azri Husni Hasani *Universiti Tenaga Nasional, MALAYSIA*	201

| 52 | Comparative Study of the Temperature Effects on n-type and p-type Silicon Cells by Numerical Simulation
Ahmad Aizan Zulkefle, Zaihasraf Zakaria, Maslan Zainon, Zikri Abadi Baharudin, A Rahman, Mohd. Ariff Mat Hanafiah
Universiti Teknikal Malaysia Melaka, MALAYSIA | 205 |

| 53 | Optimization of Baseline Parameters and Numerical Simulation for Cu(InGa)Se2 Solar Cell
Fazliyana Za'abar, Ahmad Wafi Mahmood Zuhdi, Mohd Shaparuddin Bahrudin, Siti Fazlili Abdullah, Azri Husni Hasani, M. Najib Harif
Universiti Tenaga Nasional, MALAYSIA | 209 |

| 54 | Modelling and Simulation of Photovoltaic Solar Cell using Silvaco TCAD and Matlab Software
Azri Husni Hasani, Siti Fazlili Abdullah, Ahmad Wafi Mahmood Zuhdi, Mohd Shaparuddin Bahrudin, Fazliyana Za'abar Malaysia, M. Najib Harif
Universiti Tenaga Nasional, MALAYSIA | 214 |

Cluster 3: IC Design and Manufacturing

IC Packaging and Testing, Reliability and Failure Analysis, Semiconductor Manufacturing and Process, VLSI Design

No.	Titles	Page
55	Design and Broadband Verification of RF Amplifier for 4G/ 5G Front-End Module Basuki Rachmatul Alam *Institut Teknologi Bandung, INDONESIA*	218
56	A Compact Bidirectional Pseudo Floating Gate Front-End for Resonating Sensors based on NAND and NOR Logic Gates Luca Marchetti, Yngvar Berg, Mehdi Azadmehr *University College of Southeast Norway, NORWAY*	221
57	Design of Ultra Low Voltage Low Power DXCCII for Analog Signal Processing Mohammad Faseehuddin, Jahariah Sampe, Sawal Hamid Md Ali *Universiti Kebangsaan Malaysia, MALAYSIA*	226
58	Corner Mismatch Model for Fast Non-Monte Carlo Best and Worst Cases Simulation Philip Beow Yew Tan, Chiew Ching Tan, Mohamad Marzuki Bin Mohd Fauzi *Silterra Malaysia Sdn. Bhd., MALAYSIA*	230
59	Characterization of SOI Film Thickness, Oxide Thickness and Charges with C-V Measurement Ke Kian Seng Joseph, Tan Chan Lik, Niew Soon Huat *Infineon Technologies (Kulim) Sdn Bhd, MALAYSIA*	234

60	**Comparative Study of Si Based Micromachined Patch Antenna Operating at 5 GHz for RF Energy Harvester** Noor Hidayah Mohd Yunus, Jahariah Sampe, Jumril Yunas, Alipah Pawi *Universiti Kebangsaan Malaysia, MALAYSIA*	238
61	**Design of Phase Frequency Detector (PFD),Charge Pump (CP) and Programmable Frequency Divider for PLL in 0.18um CMOS Technology** Anim Arifah Ahmad, Sawal Hamid Md Ali, Noorfazila Kamal, Siti Raudzah Abdul Rahman, Masuri Othman *Universiti Kebangsaan Malaysia, MALAYSIA*	242
62	**The Design and Optimisation of 2.4 GHz CMOS LC Tank Voltage Controlled Oscillator (VCO) for PLL using 0.18 µm CMOS Technology** Siti Raudzah Abdul Rahman, Sawal Hamid Md Ali, Noorfazila Kamal, Anim Arifah Ahmad, Masuri Othman *Universiti Kebangsaan Malaysia, MALAYSIA*	246
63	**First Principle Study of Graphene-Carbon Nanotubes Hybrid (GCH) Structure for Advanced Nanoelectronics Devices** Lee Li Theng, Iskandar Yahya, Mohd Ambri Mohamed, Mahamad Fariz Mohamad Taib *Universiti Kebangsaan Malaysia, MALAYSIA*	250

Cluster 4: Material, Process & Product

Process Technology (CMOS, Bipolar, BiCMOS, GaAs , etc), Microelectronics Application in Product Development, Electronics Materials & Device Fabrication

No.	Titles	Page
64	**Electrokinetic Behavior and Stability of Solder Powders in Aqueous Media** Terence Lucero F. Menor, Manolo G. Mena, Herman D. Mendoza *University of the Philippines Diliman, PHILIPPINES*	254
65	**Challenges in Developing Thin Profile, Smaller Flip Chip Bump Pitch FCBGA Packaging** Shaw Fa Lim, Keith Newman *Advanced Micro Devices, Inc. (AMD), MALAYSIA*	259
66	**Ultrasonic Sensor System with a 94 Mrad Total-Ionizing-Dose Tolerance** Shinya Fujisaki, Minoru Watanabe *Shizuoka University, JAPAN*	263

67   **Effect of DRIE on the Structure of Si based Filtration Pore Arrays**   267
      Fabricated with Double Side Aluminium Coating Layer
      Kamarul 'Asyikin Mustafa, Jumril Yunas, Azrul Azlan Hamzah, Wan
      Ammar Fikri Wan Ali, Burhanuddin Yeop Majlis
      Universiti Kebangsaan Malaysia, MALAYSIA

68   **Shallow Trench Isolation Stress Effect on 45 Degree Rotated**   271
      MOSFET Layout
      Chiew Ching Tan, Philip Beow Yew Tan, M.K. Md Arshad
      Silterra Malaysia Sdn. Bhd., MALAYSIA

69   **Modeling, Simulation and Optimization of 14nm High-K/Metal Gate**   275
      NMOS with Taguchi Method
      S.K.Mah, I.Ahmad, P.J. Ker, K.P. Tan, Noor Faizah Z. A.
      Universiti Tenaga Nasional, MALAYSIA

TUTORIAL 01: Advanced Silicon Devices for VLSI Circuits and Systems at Nanometer Nodes

Samar K. Saha

Prospicient Devices, California, USA

Abstract – The silicon Integrated circuits (ICs) continues to have an unprecedented impact on improving almost every aspect of modern society including communications, military, security, healthcare, energy saving, industrial automation, transport, and entertainment. Over the last four decades, the relentless pursuit of IC device miniaturization for manufacturing high-performance and high-density IC-chips and system-on-a-chip (SoC) led to the creation of Internet and social media. The semiconductor components are used in smart cars, smart homes, smart cities, smart health, smart energy, smart security, smart appliances, and so on. The Internet enables connecting any and every smart devices or "things" creating "Internet of Things" (IoT) or Internet of everything (IoE). And, the IoT-connected smart devices constitute a smart environment and integrated ecosystem that can be accessed via personal computers, tablets, and smartphones from anywhere without human interaction. However, the performance of nanoscale-MOSFETs in the design and manufacturing of "smart" electronic products necessary to create smart networks or "smart things" to enable smart environments and integrated ecosystems is inadequate due to the fundamental physical limitations such as short channel effects (SCEs).

Shrinking conventional bulk MOSFET device dimensions in the decananometer regime degrades device performance including degradation in the subthreshold swing and decrease in device turn-on voltage. As a result, the scaled MOSFETs cannot be turned off easily by lowering the gate voltage leading to excessive leakage current. Due to SCEs, the device characteristics become increasingly sensitive to process variation that imposes a serious challenge for continued scaling of bulk-MOSFETs for the nanometer technology nodes. In addition, at gate length below 20-nm, the sub-surface leakage paths are weakly controlled by the gate irrespective of gate oxide thickness and their potential barriers can be easily lowered by drain bias through the enhanced electric field coupling to the drain. Thus, to surmount the continuous scaling challenges of conventional bulk MOSFET devices new device technologies such as FinFET has emerged as the real alternative to MOSFETs. This tutorial provides the basic features and operating principles of FinFETs required for the understanding of design and manufacturing of advanced ICs at the nanometer nodes for smart electronic products. In addition, this tutorial also discusses the emerging undoped or lightly-doped channel MOSFETs with performance comparable to FinFETs for design and manufacturing of smart IC products.

Dr. Samar K. Saha served as the President of IEEE Electron Devices Society (EDS) during 2016-2017, and is currently serving as the Junior Past President of EDS. He is an Adjunct faculty in the Electrical Engineering department at Santa Clara University and Chief Scientist at Prospicient Devices. Since 1984, he has worked at various positions for National Semiconductor, LSI Logic, Texas Instruments, Philips Semiconductors, Silicon Storage Technology, Synopsys, DSM Solutions, Silterra USA, and SuVolta. He has, also, worked as a faculty member in the Electrical Engineering departments at Southern Illinois University at Carbondale, Illinois; Auburn University, Alabama; the University of Nevada at Las Vegas, Nevada; and the University of Colorado at Colorado Springs; Colorado. He has authored over 100 research papers; one book entitled, *Compact Models for Integrated Circuit Design: Conventional Transistors and Beyond*, CRC Press, USA (2015); one book chapter on Technology Computer-Aided Design (TCAD), entitled, *Introduction to Technology Computer-Aided Design*, in Technology Computer Aided Design: Simulation for VLSI MOSFET, C.K. Sarkar (ed.): CRC Press, USA (2013), and holds 12 US patents. His research interests include exploratory device and process architectures, TCAD, compact modeling, devices for renewable energy, and TCAD and R&D management.

TUTORIAL 02: Building in Reliability in Silicon Devices
M.K. Radhakrishnan

NanoRel LLP Technical Consultants, Singapore

Abstract : As the device technology is progressing in the nanometer regime towards atomic scale, the famous comment "There is plenty of room at the bottom" by Richard Feynman about 60 years ago needs to be reviewed carefully and understood in detail. A recent comment "There is plenty of difficulty near the bottom" by device manufacturers depicts another face of the advancement in technology. The difficulty observed is in the path for building reliable devices. Device reliability is the resultant of various analyses of the design, process and product and understanding innumerable phenomenon to control the extension of even atomic level defects, especially when the dimensions are at nanometer level. At the early stages of technology development, the reliability estimation was through specific tests prediction to weed out defects. Many such tests are a routine and has helped to develop technology. However, at that time itself it was known that understanding device physics and solving the related problems at smaller dimensions are the key to success. As such, the device development at any technology level has been through the critical understanding of the failures through physical analysis and hence solving the issues. This procedure is the Building-in reliability in devices. It could be noted that many fundamental physics problems such as hot carriers, issues of dielectric breakdown and strength, electro migration and stress migration, electrostatic discharge issues, etc which paused the threat at different stages were solved fully or partially through physical failure analysis and understanding the related failure mechanisms. However, as the technology progressed with new materials at new dimensions, the problems also became more challenging. The tools which helped and being used for analysis itself have limitations due to the dimensional miniaturization. Moreover, the diversification in the manufacture from design to product through different vendors and stages invites a separate approach for building the product reliability. This tutorial discusses the fundamentals of reliability and failure analysis in its basic concepts and from where it can be understood using a physics based approach. The progression in devices from the conventional MOSFET and its dimensional shrinkage to nano scale MOSFET and then to new FinFET is discussed with a view to understand how the physical failure analysis can reveal the failure mechanisms to improve reliability. While evolving these processes to build reliability in nano scale devices, it is also understood that the fundamental problem remains the same as in early 1960s. That is where the "the plenty of room" is emerged as "plenty of difficulty".

Dr. M.K. Radhakrishnan (M82, SM94, LSM18) is the Founder Director of NanoRel LLP - Technical Consultants Singapore providing analysis based solutions to micro and nano electronic industries for improving design and process reliability of devices. As a researcher in the area of semiconductor devices, analysis and reliability physics for more than 40 years, he worked as a Senior Member Technical Staff with the Institute of Microelectronics Singapore (1993-2001), Director and Principal Consultant to Philips Electronics Singapore/Netherlands (2001-2004), Device Program Leader in ST Microelectronics (1991- 1993) and a Scientist in ISRO(1985-1991). Radhakrishnan was Adjunct Professor at National University of Singapore (1994-2007). He was a Senior Consultant to ITU, Geneva as a UN Expert in the area of Reliability and Failure Analysis in 1995-1997. Dr. Radhakrishnan is currently the Vice-President of IEEE Electron Devices Society. He was an Elected Member of the Board of Governors of IEEE Electron Devices Society (20112016), He is an IEEE EDS Distinguished Lecturer from 1997 onwards and serves as the Editor of IEEE Journal of Electron Devices (IEEE JEDS). He was Editor-in-Chief of IEEE EDS Newsletter (2013-17), Guest Editor to IEEE TDMR and the Editor of Journal of Semiconductor Technology and Science (JSTS) from 2002-2005. As a researcher, academician and technical consultant he works with various MNCs in Asia, Europe and USA. He continues to do extensive training on device analysis & reliability and ESD to various Industries, Universities and Research Centers. Dr. Radhakrishnan has given plenary and keynote talks at numerous major international conferences around the globe and has given more than 100 Distinguished Lectures. He is a Fellow of IETE, Life Senior Member of IEEE, Member of EDFAS and ESDA.

Nanosystems for Food, Drug and Biomedical Applications

JACKIE Y.YING

Institute of Bioengineering and Nanotechnology
31 Biopolis Way, The Nanos, Singapore 138669
www.ibn.a-star.edu.sg
E-mail: jyying@ibn.a-star.edu.sg

Nanotechnology allows for the unique design and functionalization of materials and devices at the nanometer scale for a variety of applications. Our laboratory has fabricated nanosystems for drug screening, *in vitro* toxicology, sample preparation, diagnostic, and food pathogen detection. The miniaturized devices allow for the rapid and automated processing of drug candidates, clinical and food samples in tiny volumes, greatly facilitating drug testing, genotyping assays, infectious disease detection, cancer diagnosis, point-of-care monitoring, and food testing.

For example, we have designed plasmonic nanocrystals for single nucleotide polymorphism (SNP) genotyping. The platform involves polymerase chain reaction (PCR) for target sequence amplification and colorimetric detection with nanoprobes for pharmacogenomics applications. We have also established polymer-based lab-on-a-cartridge for automated sample preparation and PCR detection. The integrated all-in-one system, termed MicroKit, allows for the rapid and accurate typing and subtyping of influenza and other viral infections within 2 hours. We have further developed sophisticated lab-on-a-chip system that enables us to achieve multiplexed detection of drug-resistant bacteria and food pathogens.

We have created the silicon-based Microsieve system for the rapid and selective isolation of circulating tumor cells (CTCs) from peripheral blood. This non-invasive, near real-time, inexpensive liquid biopsy approach allows for the enumeration and biomarker analysis of CTCs for cancer diagnosis, prognosis and monitoring. We have also established paper-based assays for the rapid detection of various diseases, such as Dengue, Zika, hepatitis and sexually transmitted diseases. In addition, these inexpensive test kits can be used for food pathogen detection and meat speciation.

13th IEEE International Conference on Semiconductor Electronics (IEEE-ICSE2018), Kuala Lumpur, Malaysia, August 15–17, 2018. Keynote Lecture.

High-performance E-mode GaN MIS-HEMT for Power Switching Applications

Edward Yi Chang

Department of Materials Science and Engineering, National Chiao Tung University,

1001 Ta Hsueh Road, Hsinchu 30010, Taiwan.

GaN-based wide-bandgap semiconductor devices are promising for future high power and high frequency applications. In particular, the GaN high electron mobility transistor (HEMT) grown on large-size Si substrate is ideal for high power switching device applications with low cost potential. The GaN HEMT based convertors and inverters have been demonstrated for electric vehicle (EV) applications. To achieve high switching efficiency and low current collapse GaN devices, there are many device/process issues have to be solved. These include the careful design of epitaxial structure and the suitable gate insulator for MIS (metal-insulator-gate)-HEMT device structure. Furthermore, surface passivation technique is crucial to achieve low dynamic on-resistance (Ron) and good reliability. Moreover, for safety consideration, a normally-off device is a must. Several approaches to achieve normally-off device will be presented in this talk, including gate recess, F- ion implantation, high-k interlayer MOS-HEMT and novel hybrid ferroelectric charge-trapping gate stack approaches. Finally, power modules with high-performance GaN-on-Si HEMT are demonstrated for power switching device applications.

Tunable Steep Slope MoS2 Transistor

Navakantha Bhat

Professor and Chairperson Centre for Nano Science and Engineering (CeNSE) Indian Institute of
Science, Bangalore

The CMOS scaling is now facing serious fundamental and technological challenges resulting in diminishing performance and economic returns. The concurrent reduction in power consumption, an important aspect of scaling, has become difficult, because of the inability to reduce supply voltage below 1 V. This is primarily because the fundamental nature of charge transport governed by Boltzmann's statistics restricts the Sub-threshold Swing (SS, abruptness between OFF to ON transitions) of FETs to the thermionic limit of 60 mV/dec at room temperature. Hence, as we cram in more transistors into the same footprint, energy dissipation and heat management have become fundamental bottlenecks. Clearly, the road ahead needs breakthroughs in new materials and device design. In this talk, I will present a new device architecture to beat the Boltzmann limit. We demonstrate, for the first time, sub-thermionic transport through tunable Schottky contacts in dual gated MoS2 FETs. Two device configurations the gate tunable thermionic tunnel transistor (GT3) and dynamic Vt and adaptable transport transistor (DVAT) are expounded. The GT3 transistor has the flexibility to operate either in the sub-thermionic tunnel regime, yielding steep SS<60 mV/dec OR thermionic high mobility regime. Combining the best of both tunnelling and thermionic regimes in the same operation cycle, the DVAT transistor, the closest to an 'ideal transistor', registers SS~29 mV/dec (3 dec) AND high mobility (100 cm2V-1s-1). This work is envisioned to pave a new path in the development of sub-thermionic, high performance FETs operating in the sub-0.5 V 'green computing' regime.

Band Edge Modified Rare Earths- a route to the mid infrared in Silicon

Kevin Homewood

Professor, School of Materials Science & Engineering at Hubei University, Wuhan, China

We describe a new technology, band edge modification (BEM) of rare earth (RE) optical transitions in silicon. BEM refutes previous assumptions on the interaction of REs with semiconductors and other hosts (M. A. Lourenço et al, Adv. Funct. Mater., 26, 1986-1994, 2016). This approach opens up a route to efficient, fully-silicon-based, optoelectronic devices across the near and mid-IR. Intrinsic RE transitions are internal to the RE and do not contribute directly to carrier conduction in the bands of the host. Consequently, this makes them of limited use for optical detectors. The band edge modified RE levels exampled here interact directly with the silicon bands and so offer the possibility of extrinsic photovoltaic or photoconductive detectors. Silicon detectors and cameras currently completely dominate the ultra violet, visible and very near-IR regions – however they do not work well beyond 1.1 mm, the silicon band gap. We have used ion implantation to introduce europium, ytterbium and cerium into silicon photodiodes, also formed by ion implantation. We show that BEM enables efficient silicon detectivity to be extended from 1.1 mm out to the mid-IR region. The responsivites and detectivities of these new silicon detectors offer a real challenge to existing detector materials and devices in the 2 to 6 mm range – currently dominated by more challenging, and expensive materials such as mercury cadmium telluride, indium antimonide and the arsenides. Replacing these materials with silicon would offer enormous benefits in cost, reliability and also integration with the silicon microelectronics for detection and imaging. An additional benefit is using much less toxic materials and production processes – a major concern with current technologies. Low leakage currents achievable in silicon based photodiodes mean that further development of this new mid-IR silicon technology could lead to thermoelectrically cooled or even room temperature detectors. Current commercial detectors in this area have to be cooled to liquid nitrogen temperatures (77 K) to achieve the performance needed for most applications. Higher operating temperature (HOT) detectors are an industry aim and, particularly if implemented in silicon, would be a major breakthrough. We acknowledge the Royal Society UK for the award of the 2015 Brian Mercer Award for Innovation.

CMOS pixel-aperture and offset pixel-aperture techniques for 3-dimensional imaging

Jang Kyoo Shin

Professor, School of Electronics Engineering, Kyungpook National University, Daegu 41566, South Korea

Pixel-aperture and offset pixel-aperture techniques for 3-dimensional complementary metal oxide semiconductor (CMOS) image sensors are presented in this talk. In conventional camera systems, the aperture is located between the object and the CMOS image sensor (CIS). This type of image sensor consists of a pixel array with red, green, and blue (RGB) Bayer pattern color filters. Our proposed image sensor uses red, green, blue, and white (RGBW) filters, and the aperture is located on the W pixel which is without any color filter. A sharp image can be obtained from the W pixels, and the RGB pixels produce a defocused image with blurring. The sharp image can be compared with the defocused image to obtain depth information for 3D imaging. A metal layer, such as aluminum in the conventional CIS process, is used for the aperture on the white pixel. Offset pixel-aperture structures for disparity information are also considered. We designed and simulated pixel models for the pixel-aperture and offset pixel-aperture structures using a 0.11 μm CIS process and evaluated the performance of the proposed structures using finite-difference time-domain (FDTD) analysis. The proposed structures have been fabricated and some experimental results will also be presented.

Smart firefighter protective suit – functional blocks and technologies

Tomas Blecha
Department of Measurement and Technologies, Faculty of Electrical Engineering University of West Bohemia
Pilsen, Czech Republic
tblesi@ket.zcu.cz

Radek Soukup
Department of Measurement and Technologies, Faculty of Electrical Engineering University of West Bohemia
Pilsen, Czech Republic
rsoukup@ket.zcu.cz

Petr Kaspar
Department of Measurement and Technologies, Faculty of Electrical Engineering University of West Bohemia
Pilsen, Czech Republic
petrx@ket.zcu.cz

Ales Hamacek
Department of Measurement and Technologies, Faculty of Electrical Engineering University of West Bohemia
Pilsen, Czech Republic
hamacek@ket.zcu.cz

Jan Reboun
Department of Measurement and Technologies, Faculty of Electrical Engineering University of West Bohemia
Pilsen, Czech Republic
jreboun@ket.zcu.cz

Abstract—**This paper presents a research focused on a smart textile-based protective system which is intended to bring more safety to firefighters facing hazardous conditions. The system is fully integrated into a firefighter protective suit and it is able to monitor heart rate, to detect movements of a firefighter, to detect toxic and combustible gases in the environment and to measure temperature and relative humidity inside and outside of the suit. The protective system consists of developed integrated sensor modules, e-textile wiring harnesses, suit control unit, commander control unit, body area network and wide area network as well. The protective system also include indoor and outdoor localization units. The indoor localization unit based on inertial sensors, which is placed on a protective boot, is determined for the remote tracking of firefighters in situations when GPS signal is missing. The main goal of this system is to increase more firefighter safety.**

Keywords—smart textiles, firefighter protective suit, inertial based localization system, BAN and WAN communications, sensor modules

I. INTRODUCTION

According to statistical data from 2015, based on data from 35 countries in the world, 121 firefighters lose their lives whilst saving others and more than 82,000 firefigthers were injured in fire interventions [1], [2]. Firefighter safety can be increased by introducing new smart textile-based protective suits, which not only protect the human body against the extremes of nature but also provide information about the firefighter's state of health and his environment.

Smart textile products have become a new area of dynamic R&D activities in Europe and in the USA. These products under development are based on integration of electronic modules into different types of clothing, predominantly into fashion, sport clothing and particularly into protective and health care clothing. The smart textile products aim to be user friendly and to bring more comfort and safety.

The presented smart firefighter protective suit with fully integrated electronic microsystem was developed in the close collaboration with Czech firefighter brigade and industrial partners.

The main emphasis on the development of the whole personal protective system was its functionality and long-term reliability. The system concept is based on functional building blocks that allow to create specific variants of an integrated system for on-line monitoring of life functions and environmental parameters. The whole personal protective system is designed to be open and therefore it is possible to integrate not only developed but also commercial sensors into this system.

II. PERSONAL PROTECTIVE SYSTEM

The developed firefighter personal protection system (PPS) consists of five basic parts, which are depicted in the Fig. 1. Most of the electronic components are fully integrated into the firefighter jacket. Some electronic components, such as an infrared sensor for remote temperature measurement and a motion sensor module, are integrated into a protective glove or boot in order to improve their function and usability (Fig. 2).

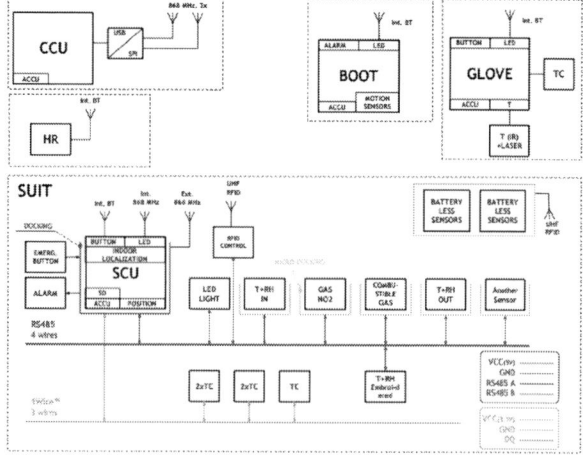

Fig. 1: The block diagram of the smart PPS for firefighters.

The personal protective system is integrated into a standard firefighter suit and it is able to monitor heart rate (HR) of firefighters, to detect firefighter movements, to monitor the concentration of toxic and explosive gases, to measure temperatures (T) and relative humidity (RH) inside and outside the suit. Fire safety of a firefighter is also increased by an integrated emergency button, acoustic and light alarm system. The protective system consists of developed integrated sensor modules, indoor and outdoor localization system (inertial + GPS), body motion capture in order to know the posture of a person during an intervention, e-textile wiring harnesses, suit control unit (SCU), commander control unit (CCU), body area network (BAN) and wide area network (WAN). All measured data are transmitted wirelessly over a WAN communication network with a dynamic mesh algorithm to the commander control unit, which is thus continuously informed about the actual state of the firefighters. If any monitored parameter exceeds the pre-set threshold, the SCU will automatically activate the acoustic and light alarm.

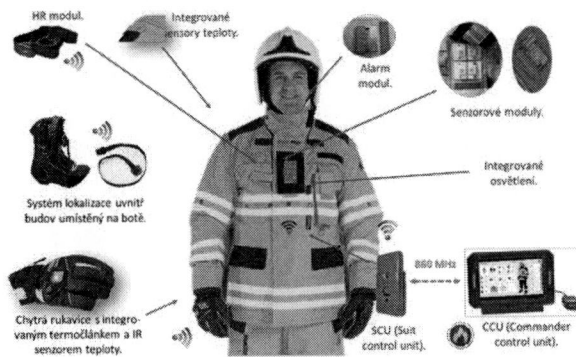

Fig. 2: Overview of PPS components.

All measured data from sensors are stored in the suit control unit on the SD card and then transmitted wirelessly to the commander control unit. The CCU allows to display data from up to 12 SCUs and also stores data on an SD card. Thanks to the integrated SD cards, the personal protection system provides the "black box" function. All data received by the CCU is also transmitted and stored on the cloud server. The personal protective system into the firefighter suit is fully automated and can work completely independently if the SCU loses its wireless connection with the CCU.

A. Commander Control Unit

The CCU allows the data visualization of up to 12 firefighters, graphical threshold visualization by traffic light approach, and announcing alarm to a particular firefighter. The CCU unit is based on a commercial tablet with added WAN communication unit for frequency of 868 MHz and charging spring probe connector (Fig. 3). The CCU is protected by specially developed robust case which is water proof with IP67 level, can resist harsh environment and complies with explosion protection standard ATEX 95 – European Directive 94/9/EG. Basic functions of commander control unit are following:

- Data processing.
- Data visualization.
- Indoor/outdoor localization data processing and visualization.

- Data storage.
- Alarm states signalization.
- Data transfer.
- Wireless communication.

Fig. 3: Commander Control Unit (CCU).

Most system functions of the commander control unit are implemented by software tools, allowing simple modification, extension, or upgrading of functions. This solution is not tied to specific hardware and can be easily transferred to a new version or other hardware. The CCU is controlled by a capacitive touchscreen with an intuitive navigation menu. The user interface is based on tiles. An overview of all monitored firefighters is displayed on the main page (Fig. 4). For each firefighter, up to 12 parameters can be selected, which can be displayed on the details screen. If any monitored parameter exceeds the set thresholds, the color of the tile changes.

Fig. 4: Basic user interface screen.

B. Suit Control Unit

All measured data from individual sensors integrated in a smart firefighter suit is stored in the suit control unit (SCU) on an SD card as well and then the data is wirelessly transmitted to the CCU. Basic functions of the suit control unit are following:

- Data evaluation.
- Data processing.
- Data storages (black box function).
- Alarm states signalization.
- Data transfer.
- Wireless and wired communication.

The SCU is integrated in each firefighter suit into the designed docking station (Fig. 5). The SCU also includes motion sensors (accelerometer, gyroscope), battery that provides system operation for 8 hours, communication modules and SD card. SCU communication with wireless modules and sensors is provided with Bluetooth version 4, communication with the CCU is at frequency of 868 MHz using the dynamic routing protocol. Some sensors and functional electronic blocks are connected to the SCU by means of textile wires that are permanently integrated into the suit. The whole suit control unit is based on the Arrietta G25 microcomputer platform. A WAN communication module and a GPS module are connected into the suit control unit and are permanently integrated in a suit in the shoulder area to increase the communication distance and minimize the shading effect of the firefighter's body.

Fig. 5: Suit Control Unit and docking station.

C. Sensor modules

The smart firefighter suit contains removable and fully integrated (no need to remove) sensor modules (Fig. 6). These sensor modules include a sensor element, pre-processing circuits, and circuits for data transmission to the SCU. The following sensors are attached to the suit:

- Thermocouple temperature sensor (6 pieces).
- Temperature and humidity sensors.
- NO_2, CO and explosive gas sensors.
- Heart rate sensors.

Fig. 6: The sensor module and its connection to the microdocking station.

The thermocouple sensor modules are encapsulated to withstand environmental conditions, wash cycles, and are permanently integrated into a suit and connected with SCU by textile wires. Data transfer from thermocouples is via a 1-wire bus. The NO_2, CO, explosion gases, temperature and humidity sensors are placed in an ergonomically rounded, planar housing and resist harsh environmental conditions. Individual removable sensor modules are placed in microdocking stations that are permanently integrated into the suit. These sensor modules are connected with the microdocking station by means of spring needles. The sensor modules communicate with the SCU via the RS-485 interface on the textile buses.

TABLE I. *SENSORS USED IN THE PERSONAL PROTECTIVE SYSTEM*

Sensor module	Description	Placement	Measurement range	Communication protocol
Thermocouple modules (6 pieces)	K-type thermocouple	Surface of the suit most exposed to the heat	-50 to 500 °C	1-wire
Humidity and Temperature sensor module	Sensirion SHT15	Outer part of the suit (chest)	-40 to 120 °C, 20 to 99 % RH	RS-485
Humidity and Temperature sensor module	Sensirion SHT15	Inner part of the suit	0 to 80 °C, 20 to 99 % RH	RS-485
NO2 module	Electrochemical sensor	Outer part of the suit (chest)	0 to 10 ppm NO2	RS-485
Combustible gas module	Pellistor Micropel 75	Outer part of the suit (chest)	0 to 100 % LEL	RS-485
CO module	Electrochemical sensor	Outer part of the suit (chest)	0 to 900 ppm	RS-485
HR module	HR belt	Directly on body	30 to 240 bpm	Bluetooth

D. Acustic alarm and active illumination system

If any measured value exceeds the set threshold, an acoustic and a light alarm is automatically activated. The intervention commander can also activate the alarm state by using button on the CCU screen. If the alarm is activated, the firefighter is warned by the flashing of the integrated LEDs and by the 90 dB$_A$ sound signal. The acoustic alarm module is integrated into the collar of a firefighter's suit. Active illumination is provided by LED modules that are permanently integrated into a suit. These modules are connected to the SCU via the RS-485 bus. This connection allows to set LED modules in different modes. LED modules can be automatically activated based on the ambient light intensity.

E. Glove with integrated temperature sensors

To detect remote hot spots, an infrared temperature sensor has been integrated into the firefighter protective gloves. The glove contains permanently integrated evaluation electronics, an infrared sensor, a thermocouple sensor, a laser pointer, visualization LEDs, and a wireless communication module (Fig. 7). Measured data from temperature sensors is transmitted via Bluetooth version 4 to SCU. Removable battery with charging circuits is placed inside the pocket of the glove.

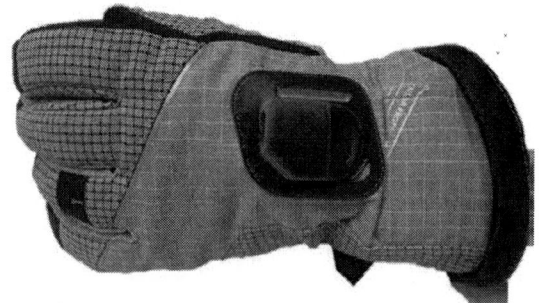

Fig. 7: Smart glove for firefigters.

III. WAN – COMMUNICATION NETWORK

The WAN communication network provides data transmission between the SCU and the CCU at frequency band of 868 MHz, which allows good signal penetration inside buildings and increases the communication distance. The CCU can communicate with up to 12 SCUs at a time. WAN communication is bi-directional and a dynamic mesh

C4

network with its own designed communication protocol has been developed to extend the communication range, especially inside buildings [3]. This communication protocol allows to send data either directly from the SCU to the CCU or sequentially between the SCUs (Fig. 8).

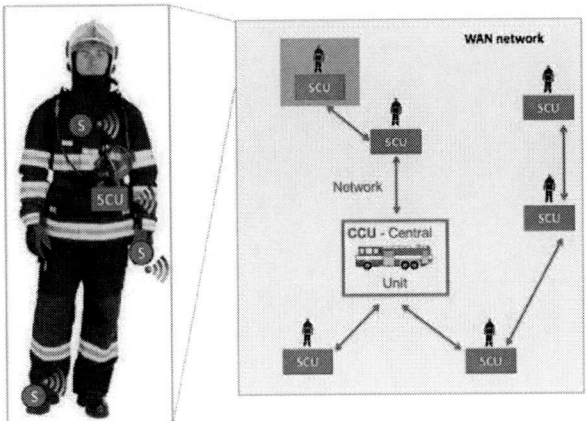

Fig. 8: Schema of wireless WAN communication system.

In order to ensure the transmission of data in particularly unfavourable spaces (cellars, reinforced concrete buildings), a signal repeater (Fig. 9) has been developed that can be distributed in the building and thus increase the communication distance between SCUs and CCUs. The WAN communication distance between the SCU and CCU can be up to 1000 m in the open space in the case of dynamic mesh networks and 300 m in the point-to-point connection. The advantage of the proposed WAN communication network compared to the TETRA / MATRA communication protocol is the higher transmission speed, robustness and their mutual independence, so it is possible to use both systems in parallel.

Fig. 9: Signal repeater to increase range.

IV. TESTING OF SMART FIREFIGHTER SUIT

A significant attention was paid to long-term reliability of the whole smart firefighter protective suit throughout the research. In the course of the system development, it was conducted a wide range of laboratory and field tests such as washing resistance tests, long term stability tests, functional tests and field tests in the flashover container. Based on the tests results the microsystem was optimized.

The high washing resistance is the crucial parameter for smart textiles products therefore the smart firefighter suit was also subjected to a washing resistance test with 30 washing cycles which was conducted according to the standards ISO 6330:2012. Temperature during the washing test was set at 60 °C and speed of spin at 500 rpm. The suit passed the test without any problem.

The research team intensively cooperated with the Czech fireman brigades during testing. Several field tests were performed in the flashover container in the firefighter testing centre in Zbiroh, Czech Republic. The flashover container simulates real conditions in a burning room. The suit passed successfully these tests.

CONCLUSIONS

Smart firefighter suit is equipped with an integrated personal protective system including WAN dynamic mesh communication network, BAN communication network, microdocking station for sensor modules, inertial localization system, interconnection system (textile wires), suit control unit with docking station, commander control unit, temperature modules, CO, NO_2, explosive gases, humidity, heart rate, motion detection, LED modules and acoustic alarm. The system includes a smart gloves for firefighters including an integrated infrared temperature sensor for detection of remote hot spots and a thermocouple for measurument of the ambient temperature. The individual sensor modules integrated in the suit communicate with the SCU via a textile bus system or Bluetooth version 4. The data from the SCU is wirelessly transmitted to the CCU, which allows to display the measured data and graphical threshold visualization by traffic light approach. If the thresholds are exceeded, a light and acoustic alarm is automatically activated. The developed smart firefighter protective suit performed well at the EMC test, washing resistance test, long-term stability and reliability testing. Very promising results were obtained in the field test in the flashover container in the firefighter testing centre. The suit was created as universal system, which can be equipped with further functions.

ACKNOWLEDGMENT

This research has been supported by the Ministry of Education, Youth and Sports of the Czech Republic under the RICE – New Technologies and Concepts for Smart Industrial Systems, project No. LO1607.

REFERENCES

[1] N. N. Brushlinsky, J. R. Hall, S. V. Sokolov, a P. Wagner, „Center of Fire Statistics", *World Fire Stat.*, 2015.

[2] F. Scandella, *Firefighters: feeling the heat.* European Trade Union Institute, 2012.

[3] R. Šalom *et al.*, „Implementation of AODV routing protocol in sensor wireless networks", in *Telecommunications Forum (TELFOR), 2012 20th*, 2012, s. 194–197.

Engineering the Properties of Nb_2O_5-ZnO Nanostructures via Dual Synthesis Techniques

Rozina Abdul Rani
[1]NANO-SciTech Centre,
Institute of Science, [2]NANO-ElecTronic Centre (NET),
Faculty of Electrical
Engineering, Universiti
Teknologi MARA
Shah Alam, Selangor, Malaysia
rozina.abdulrani@yahoo.com

Fatin Nurdini Omar
Faculty of Applied Sciences,
Universiti Teknologi MARA
Shah Alam, Selangor, Malaysia
fatinnurdini96@gmail.com

Mohd Husairi Fadzilah Suhaimi
[1]NANO-SciTech Centre, Institute
of Science, [2]Faculty of Applied
Sciences Universiti Teknologi
MARA
Shah Alam, Selangor, Malaysia
husairi5840@salam.uitm.edu.my

Ahmad Sabirin Zoolfakar
NANO-ElecTronic Centre
(NET), Faculty of Electrical
Engineering, Universiti
Teknologi MARA
Shah Alam, Selangor, Malaysia
ahmad074@ salam.uitm.edu.my

Mohamad Hafiz Mamat NANO-ElecTronic Centre (NET),
Faculty of Electrical
Engineering, Universiti
Teknologi MARA
Shah Alam, Selangor, Malaysia
mhmamat@salam.uitm.edu.my

Salman Alrokayan
Research Chair for Biomedical
Applications of Nanomaterials,
Department of Biochemistry,
College of Science, King Saud
University,
Riyadh, Saudi Arabia
salrokayan@ksu.edu.sa

Haseeb A. Khan
Research Chair for Biomedical
Applications of Nanomaterials,
Department of Biochemistry,
College of Science, King Saud
University,
Riyadh, Saudi Arabia
haseeb@ksu.edu.sa

Mohamad Rusop Mahmood
[1]NANO-SciTech Centre,
Institute of Science, [2]NANO-ElecTronic Centre (NET),
Faculty of Electrical
Engineering, Universiti
Teknologi MARA
Shah Alam, Selangor, Malaysia
nanouitm@gmail.com

Abstract—In this paper, dual synthesis techniques; anodization and hydrothermal, were introduced to synthesize the Nb_2O_5-ZnO nanostructures. The purpose of this research is to engineer the properties of the Nb_2O_5-ZnO nanostructures which can offer new optical and electrical behaviors and also lead to large surface area. Synthesis of Nb_2O_5 has been conducted by anodizing the Niobium foil in the electrolyte containing NH_4F and ethylene glycol. Meanwhile, ZnO was hydrothermalized with Zinc Nitrate precursor solution for 18 hr at different temperatures. All samples were characterized using FESEM, XRD and UV-Vis absorbance spectra. The morphological and properties changes of the Nb_2O_5-ZnO nanostructures are systematically investigated as a function of hydrothermal temperature.

Keywords— Nb_2O_5, ZnO, anodization, hydrothermal, nanostructure, dual synthesis

I. INTRODUCTION

The uniqueness of physical and chemical properties of nanostructured Nb_2O_5 metal oxides have made them eligible to incorporate in multi electronic application, for example, memory devices, sensors, batteries and solar energy devices [1-4]. Based on the advantages of nanostructures such as high surface to volume ratios, quantum confinement effects and altered surface energies, they offer the ability to develop a material with suitable properties for targeted applications and high performance of devices.

In order to engineer or tailor the properties of nanostructured metal oxide into desired specific properties, doping or an addition of coating layer in the device fabrication have been implemented. For example, Patil *et al* were developed thick ZnO with Nb_2O_5 as a dopant for gas sensor application and Ueno *et. al.* were synthesized porous

ZnO coated with a thin Nb_2O_5 thin layer for dye-sensitized solar cells [5, 6]. Both techniques are used to alter or optimize the properties of the nanostructured metal oxide. Besides that, the dual synthesis approach can be a promising technique [7]. This approach can be the combination of two synthesis technique with two discontinuous processes or one synthesis technique with two continuous processes. Thus, this technique allows the mixture of different morphologies and types of nanostructured metal oxide. Sun *et. al.* has successfully synthesized 1D-3D nanostructured TiO_2 bilayer photoanode with controlled morphology *via* hydrothermal synthesis process and $TiCl_4$ treatment [8]. The bottom layer consists of vertically aligned 1D TiO_2 nanowires, whereas the top layer consists of 3D TiO_2 dendritic nanostructures. These bilayer nanostructures possess a high specific surface area, fast electron transport, and a pronounced light-scattering effect, which demonstrated the highest energy conversion efficiency of 7.2% for rutile TiO_2. Another work related to engineer the properties of metal oxide has been conducted by Kim *et. al.*[9]. They have applied two-step processes to synthesize heterogeneous branched ZnO-SnO_2 nanostructures for chemical sensing application. Meanwhile, two-step techniques also applied in producing ZnO hollow microspheres with well-designed constituent units, in the shapes of 1D nanowire networks, 2D nanosheet stacks, and 3D mesoporous nanoball blocks for novel photosensitive materials [10].

In the present work, by implementing dual synthesis techniques we have successfully synthesize 3D Nb_2O_5-ZnO nanostructures. The bottom layer was nanoporous Nb_2O_5 which is formed *via* anodization technique and the top layer consist of ZnO nanorods structures that have been produced *via* the hydrothermal process. Additionally, the investigation of the morphological, structural, optical and electrical properties of the produced samples was also conducted.

II. MATERIALS AND CHARACTERIZATION

A. Synthesis of Nb_2O_5-ZnO nanostructures

The preparation of Nb_2O_5-ZnO nanostructures involved two step synthesis processes; anodization and hydrothermal. Firstly, to synthesize Nb_2O_5, Niobium foil was anodized in the ethylene glycol (98% anhydrous, Sigma Aldrich) consists of 4 vol % deionized water and 0.5% NH_4F (98% purity, Sigma Aldrich). During anodizing, the solution was heated at 50°C and a bias voltage of 10 V was applied for 30 min. Details on the anodization optimization are presented in the previous work [11, 12].

To synthesize the ZnO nanostructure, the solution containing Zinc Nitrate Hexahydrate ($Zn(NO_3)_2 \cdot 6H_2O$)) and Hexamethylenetetramine (HMT: $(CH_2)6N_4$) were mixed in a molar ratio of 1:1 in 150 ml distilled water and undergo pre-treatment at 50 °C for 30 minutes. The solution was put in the bottle and continuously stirred for 3 hr at 250 rpm. Before proceed for hydrothermal process, the anodized Nb_2O_5 samples were immersed in the solution and then the bottle soaked in the water bath for 18 hr at different temperatures; RT (~23 °C), 50, 65 and 80 °C. After that, the sample was rinsed with distilled water and dried at room temperature. To obtain a crystalline structure of Nb_2O_5-ZnO, the annealing process was conducted at 440 °C for 50 min.

B. Structural characterization

The morphological structures of nanoporous Nb_2O_5 were characterized using a JEOL JSM-6700F field-emission scanning electron microscope (FESEM) and the Nb_2O_5-ZnO nanostructures were characterized using scanning electron microscope (SEM). All samples were characterized by an X-ray diffraction (XRD, Panalytical X'pert PRO). Their absorbance properties were characterized using UV-visible spectra (Cary 5000 spectrophotometer, Varian) in wavelength 200 nm to 800 nm ranges.

III. RESULTS AND DISCUSSION

The FESEM image in Fig. 1 shows the morphological structure of anodized nanoporous Nb_2O_5. After 30 min anodizing the Nb foil in the NH_4F-organic based solution, nanoporous Nb_2O_5 with thickness of ~1.5 µm is obtained. It is observed that the cross-sectional of nanoporous Nb_2O_5 contain highly packed nanovein-like structures where their internal diameter size ranging from 20 to 50 nm. At the bottom part of the film, uniform pseudo-semispheres with diameters in the range of 40 to 60 are created. Meanwhile, the wall size of nanovein is about 10 to 20 nm thick. As can be seen in inset image in Fig. 1, the top view of the nanoporous Nb_2O_5 consists a highly nanoporous distribution. The pore size is almost similar to the size of the internal nanovein structure. Details synthesis process of anodized nanoporous Nb_2O_5 has been presented in the previous studies [11, 12].

The morphological structure of the Nb_2O_5-ZnO sample after hydrothermal process at different temperature is presented in Fig.2. The obtained images clearly showed that the temperature of the hydrothermal process significantly affects the ZnO morphology. From Fig. 2(a), the SEM image represents the hydrothermallized sample prepare at RT. It completely shows no obvious ZnO structure growth on the surface of nanoporous Nb_2O_5. At a temperature of 50 °C, scattered ZnO nanorod with taped hexagonal shape at the end of the structure are obtained (Fig. 2(b)). It can be observed that the diameter of the hexagonal shape, ranging from ~ 0.5 µm to 3 µm and their length also have an irregular size with the maximum length up to 8 µm. Most of the ZnO nanorods aligned parallel to the nanoporous Nb_2O_5 substrate. When increased the temperature up to 65 °C, uniform size of ZnO nanorods was produced as can be seen in Fig. 2(c). Their size is much smaller than ZnO nanorods prepared at 50 °C. The diameter of the hexagonal shape is about 1 µm, while the length is in the size of 3 µm. It was noticed that the distribution of ZnO nanorods on the surface of nanoporous Nb_2O is increased and agglomerated. The synthesis of the ZnO at 80 °C has produced a structure like flower bunch made of ZnO nanorods. The structure composed of many hexagonal shaped nanorods, which are combined in the centre of main ZnO nanorods. The size of the whole structure of the flower bunch ZnO is approximately 5 µm. The SEM image in Fig. 2(d) clearly shows that the flower bunch ZnO nanorods almost covered the whole surface of nanoporous Nb_2O. According to these results, the formation of different morphologies of ZnO nanorods can be related to the different reaction rates which is contributed by the different hydrothermal temperatures [13]. The compactness of the ZnO growth is greater at higher reaction temperature due to higher molecular mobility which changed the nucleation and growth of the ZnO nanorods.

The structural properties of the samples were further characterized using X-ray diffraction analysis (XRD). In Fig. 3, the XRD patterns of the nanoporous Nb_2O_5 and the Nb_2O_5-ZnO nanostructures obviously shows a dominant orthorhombic phase, as distinguished by peaks appearing at 22.9, 28.4, 34.6, 36.4, 46.2, 46.4, 47.7, 49.9, 55.5, 56.7, 58.4, 63.2, 66.5 and 69.2° [11]. It is also notified that the XRD spectra of Nb_2O_5-ZnO nanostructures also containing ZnO phases corresponding to the hexagonal wurzite structure at 2θ values of 31.9, 34.6, 36.4, 47.7,56.7, 63.2, 66.5, 68.1 and 69.2° [14, 15]. From the obtained XRD spectra, the temperature of hydrothermal process can influence the crystalline peak intensity of the ZnO nanorods. For sample prepared at RT and 50 °C, low intensity of wurzite crystalline peak was observed in the XRD spectra. Meanwhile, the peak intensity of the ZnO nanorods at (100), (002) and (101) diffraction peaks (which is at 31.9, 34.6 and 36.4°, respectively) enhanced significantly for sample prepare at 65 and 80 °C. The enhanced and narrow width of the ZnO diffraction peak indicated high quality of crystalline ZnO nanorods [16].

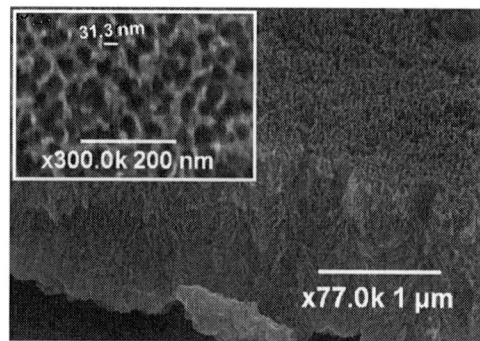

Fig. 1. FESEM image of cross-sectional view of the whole nanoporous Nb_2O_5 film. Inset image shows top view of nanoporous Nb_2O_5.

Fig. 2. SEM images of crystalline Nb_2O_5-ZnO nanostructures prepared using 0.1 M zinc nitrate at different hydrothermal temperatures; RT (a), 50 °C (b), 65 °C (c) and 80 °C (d).

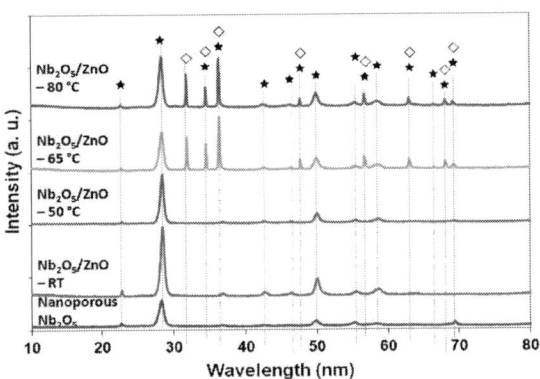

Fig. 3. X-ray diffraction spectra for the annealed nanoporous Nb_2O_5 and Nb_2O_5-ZnO nanostructures prepared at different hydrothermal temperatures; RT, 50, 65 and 80 °C.

In order to provide more information on the relation between the optical band gap, E_g and hydrothermal temperature, the absorption spectra of all samples were conducted by UV-Vis spectra and presented in Fig. 4. In term of absorption intensity, Nb_2O_5-ZnO nanostructures prepared at a temperature of 80 °C showed the highest among other samples. From the obtained results, it obviously showed that the optical band gap of nanoporous Nb_2O_5 has been engineered by depositing ZnO nanorods in different hydrothermal temperatures. The E_g value of nanoporous Nb_2O_5 is about 3.307 eV, while synthesize the Nb_2O_5-ZnO at RT has reduced the E_g value to 2.995 eV. However, it has noticeable when further increased the hydrothermal temperature the E_g value of Nb_2O_5-ZnO has increased again. As presented in Fig. 4(b-e), the E_g value of Nb_2O_5-ZnO nanostructures synthesize at RT, 50, 65 and 80 °C are 2.995, 3.054, 3.085 and 3.123 eV, respectively. In this characterization, the E_g value is determined according to Planck's Law and the following equation [17]:

$$Eg = hc/\lambda = 1240 \text{ eV. nm}/\lambda \qquad (1)$$

where h is Planck's constant ($4.13566733 \times 10^{-15}$ eV s); c is the speed of light ($2.99792458 \times 10^{17}$ nm/s) and λ is the wavelength.

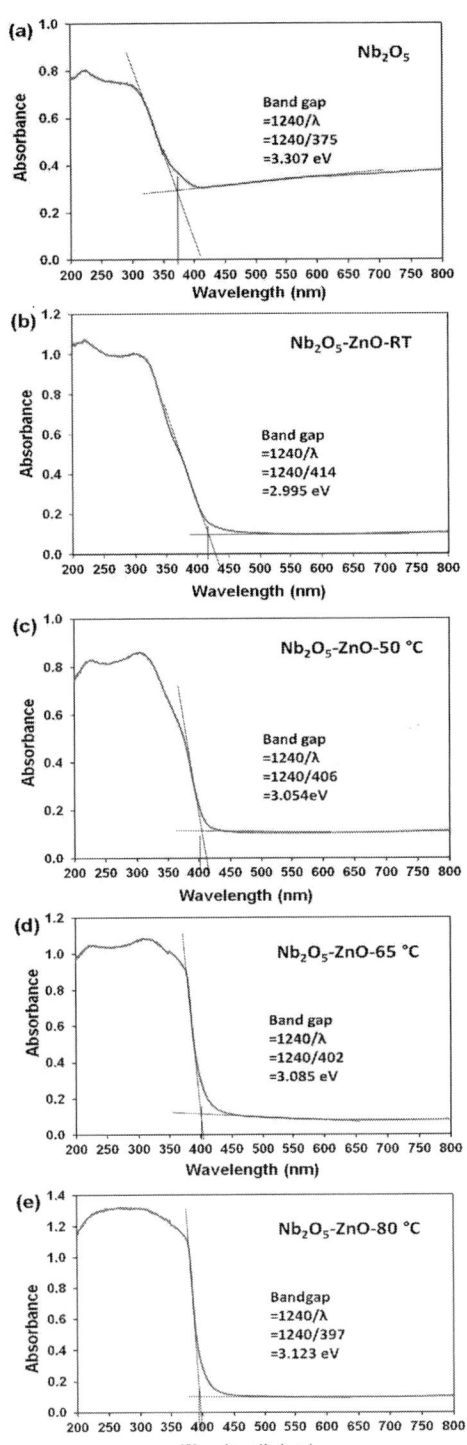

Fig. 4. UV-Vis absorption spectra of nanoporous Nb_2O_5 and Nb_2O_5-ZnO nanostructures prepared at different hydrothermal temperatures; RT, 50, 65 and 80 °C. The band gap value for each sample was determined by utilize the energy-wavelength relationship; E=hc/λ.

IV. CONCLUSIONS

Novel Nb_2O_5-ZnO nanostructures were synthesized *via* dual synthesis techniques which are anodization and hydrothermal. Their properties were engineered by implementing dual synthesis and vary the hydrothermal temperature. It is concluded that by increasing the hydrothermal temperature, the compactness of the ZnO growth is greater and the intensity of wurtzite peak is increased. Meanwhile, the E_g of the initial nanoporous Nb_2O_5 has been reduced from 3.307 eV to 2.995 eV when undergoes the hydrothermal ZnO at RT. The E_g of Nb_2O_5-ZnO nanostructures were increased to 3.054, 3.085 and 3.123 eV when the hydrothermal temperature was increased to 50, 65 and 80 °C, respectively. These unique nanostructures also possess a high surface area to volume ratio structure. Thus, it is believed that the Nb_2O_5-ZnO nanostructures can promote important applications such as sensing layer, photocatalyst and photoanodes, with good selectivity and sensitivity behaviour and remarkable enhancement device performance.

ACKNOWLEDGMENT

This work is supported by Ministry of Education Malaysia (MOE) under the Bestari Perdana Grant Scheme, Project code: 600-IRMI/PERDANA 5/3 BESTARI (101/2018). Special thanks to Research Chair for Biomedical Applications of Nanomaterials, Department of Biochemistry, College of Science, King Saud University, Riyadh 11451, Saudi Arabia.

REFERENCES

[1] M. M. Rahman, R. Abdul Rani, A. Z. Sadek, A. S. Zoolfakar, M. R. Field, T. Ramireddy, K. Kalantar-zadeh, and Y. Chen, "A vein-like nanoporous network of Nb_2O_5 with a higher lithium intercalation discharge cut-off voltage," *J. Mater. Chem. A,* vol. 1, pp. 11019-11025, 2013.

[2] R. Abdul Rani, A. S. Zoolfakar, J. Subbiah, J. Z. Ou, and K. Kalantar-zadeh, "Highly ordered anodized Nb_2O_5 nanochannels for dye-sensitized solar cells," *Electrochem. commun.,* vol. 40, pp. 20-23, 2014.

[3] R. A. Rani, A. S. Zoolfakar, A. P. O'Mullane, M. W. Austin, and K. Kalantar-Zadeh, "Thin films and nanostructures of niobium pentoxide: fundamental properties, synthesis methods and applications," *Journal of Materials Chemistry A,* vol. 2, pp. 15683-15703, 2014.

[4] W. Helge, R. Elena, S. Stefan, and M. Thomas, "Integration of niobium oxide-based resistive switching cells with different select properties into nanostructured cross-bar arrays," *Semiconductor Science and Technology,* vol. 30, p. 115014, 2015.

[5] S. Ueno and S. Fujihara, "Effect of an Nb2O5 nanolayer coating on ZnO electrodes in dye-sensitized solar cells," *Electrochimica Acta,* vol. 56, pp. 2906-2913, 2011.

[6] A. Patil, C. G Dighavkar, S. K Sonawane, S. Patil, and R. Borse, *Influence of Nb₂O₅ doping on ZnO thick film gas sensors* vol. 12, 2010.

[7] Q. Liu, Z. Sun, Y. Dou, J. H. Kim, and S. X. Dou, "Two-step self-assembly of hierarchically-ordered nanostructures," *Journal of Materials Chemistry A,* vol. 3, pp. 11688-11699, 2015.

[8] Z. Sun, J. H. Kim, Y. Zhao, D. Attard, and S. X. Dou, "Morphology-controllable 1D-3D nanostructured TiO_2 bilayer photoanodes for dye-sensitized solar cells," *Chemical Communications,* vol. 49, pp. 966-968, 2013.

[9] S. S. Kim, S.-W. Choi, H. G. Na, D. S. Kwak, Y. J. Kwon, and H. W. Kim, "ZnO–SnO2 branch–stem nanowires based on a two-step process: Synthesis and sensing capability," *Current Applied Physics,* vol. 13, pp. 526-532, 2013.

[10] Z. Sun, T. Liao, J.-G. Kim, K. Liu, L. Jiang, J. H. Kim, and S. X. Dou, "Architecture designed ZnO hollow microspheres with wide-range visible-light photoresponses," *Journal of Materials Chemistry C,* vol. 1, pp. 6924-6929, 2013.

[11] R. A. Rani, A. S. Zoolfakar, J. Z. Ou, M. R. Field, M. Austin, and K. Kalantar-zadeh, "Nanoporous Nb_2O_5 hydrogen gas sensor," *Sens. Actuators, B,* vol. 176, pp. 149-156, Jan 2013.

[12] J. Z. Ou, R. A. Rani, M. H. Ham, M. R. Field, Y. Zhang, H. Zheng, P. Reece, S. Zhuiykov, S. Sriram, M. Bhaskaran, R. B. Kaner, and K. Kalantar-Zadeh, "Elevated temperature anodized Nb_2O_5: A photoanode material with exceptionally large photoconversion efficiencies," *ACS Nano,* vol. 6, pp. 4045-4053, 2012.

[13] V. V. Burungale, V. V. Satale, A. J. More, K. K. K. Sharma, A. S. Kamble, J. H. Kim, and P. S. Patil, "Studies on effect of temperature on synthesis of hierarchical TiO_2 nanostructures by surfactant free single step hydrothermal route and its photoelectrochemical characterizations," *Journal of Colloid and Interface Science,* vol. 470, pp. 108-116, 2016.

[14] R. Wahab, S. G. Ansari, Y. S. Kim, M. Song, and H.-S. Shin, "The role of pH variation on the growth of zinc oxide nanostructures," *Applied Surface Science,* vol. 255, pp. 4891-4896, 2009.

[15] S. Amizam, N. Abdullah, H. A. Rafaie, and M. Rusop, "SEM and XRD Characterization of ZnO Nanostructured Thin Films Prepared by Sol - Gel Method with Various Annealing Temperatures," *AIP Conference Proceedings,* vol. 1217, pp. 37-41, 2010.

[16] K. A. Eswar, J. Rouhi, F. S. Husairi, R. Dalvand, S. A. H. Alrokayan, H. A. Khan, M. Rusop Mahmood, and S. Abdullah, "Hydrothermal growth of flower-like ZnO nanostructures on porous silicon substrate," *Journal of Molecular Structure,* vol. 1074, pp. 140-143, 2014.

[17] E. Abdelkader, L. Nadjia, and B. Ahmed, "Preparation and characterization of novel $CuBi_2O_4$/SnO_2 p–n heterojunction with enhanced photocatalytic performance under UVA light irradiation," *Journal of King Saud University - Science,* vol. 27, pp. 76-91, 2015.

Dielectrophoresis: Iron Dificient Anemic Red Blood Cells for Artificial Kidney Purposes

Farahdiana Wan Yunus, Azrul Azlan Hamzah, Mas Syarafina Norzin, Muhammad Ramdzan Buyong, Jumril Yunas, Burhanuddin Yeop Majlis

Institute of Microengineering and Nanoelectronic (IMEN),

Universiti Kebangsaan Malaysia (UKM)

46300 Bangi, Selangor, Malaysia

Email: wydianaaa@siswa.ukm.edu.my

Abstract— The main objective of this research is to address certain medical problems faced by kidney renal disease patients. A sample of iron deficient blood was studied using dielectrophoresis to determine any differences between the properties of anemic red blood cells and normal red blood cells. The crossover frequencies show that anemic red blood cells and normal red blood cells offer different results due to their differing properties. The results for crossover frequencies successfully demonstrate the differences between the red blood cell types. Crossover frequencies for normal red blood cells is at 380kHz, while the frequency for iron deficient red blood cells is at 7MHz. These differences in particle properties were proven by the analysis of crossover frequency differences using MATLAB.

Keywords—dielectrophoresis, iron deficient red blood cells, anemic, renal kidney disease, artificial kidney.

I. Introduction

Dielectrophoresis (DEP) was first introduced by Herbert A. Polh in the 1950s [1]. Over the course of 40 years, dielectrophoresis has been used to study bio particles [2][3][4]. Studies on matters pertaining to bio particles by dielectrophoresis has been widely used such as the study of diverse phenomena such as blood cells [5], yeast cells [6][7], leukaemia cells [8], stem cells [9], breast cancer [10] and so on. Therefore, it is likely that dielectrophoresis is broadly known because this technique has helped in many ways, including in solving medical problems. A previous paper by M.R. Buyong [11] stated that it is possible for the dielectrophoresis concept to be implemented on artificial kidneys. In this research, the dielectrophoresis method is used to improve the condition of patients with kidney renal problem as stated in the previous paper [12]. To address this problem, the properties of cells need to be known. Patients with kidney problems are commonly known to have anaemic disease. Thus, iron deficient anaemic red blood cells (RBCs) were tested as a reference to determine the change in the electrical properties of cells. Similar research has also strengthened the evidence that anaemic blood properties may differ [13]. [13] also demonstrated differences in the values and properties between healthy red blood cells and anaemic red blood cells. Therefore, the crossover frequency has been measured as shown in the results to prove the differences in properties. A tapered electrode per M. R. Buyong [14] was used in this research.

II. Theory of Dielectrophoresis Force (FDEP)

Dielectrophoresis is the movement of particles in a non-uniform electric field subject to varying time (AC) [15]. Dielectrophoresis depends strictly on the medium and particles' electrical properties, particle shapes and sizes and also the frequency of the electric field. In an electric field E, dielectric particles behave as an effective dipole with dipole moment p proportional to the electric field, which is [15],

$$\rho \, \alpha \, E \qquad (1)$$

The constant of the proportional depends, in general, on the geometry of the dielectric particle. In the presence of an electric field gradient, the force on a dipole is given by [15],

$$F = (\rho . \, \mathrm{d} \,) \, E \qquad (2)$$

Where ρ is the constant of a dipole moment vector, d is del vector and E is the external electric field. As the two equations are combined, the (F_{DEP}) for a homogenous particle suspended in an electric field gradient is given by [15],

$$F_{DEP} = 2\pi\varepsilon_{medium} R^3 \, CMF \, (\mathrm{d}E^2) \qquad (3)$$

where r is the radius of particle, ε_m is the permittivity of the suspending medium, and dE is the gradient of the rms electric field. CMF is the real part of the Clausius-Mossotti factor given by [16],

$$CMF = (\varepsilon_{particle} - \varepsilon_{medium}) / (\varepsilon_{particle} + 2\varepsilon_{medium}) \qquad (4)$$

$\varepsilon_{particle}$ and ε_{medium} are the complex permittivity of the particles of cells and medium [16].

Therefore, in this research, the CMF will be calculated using MATLAB to find out the crossover frequencies of RBCs. As stated in Eq. (4), CMF is dependent on the electrical properties of both the particles and medium.

III. Apparatus and Materials

A. Apparatus

An Olympus STM6 microscope was used to view the RBCs, with a prober to connect electrodes, one-touch diabetic pen to pin out the blood, and function generator to supply the AC field.

B. Materials

Iron deficient RBC, RinsCap NS 0.9% sodium chloride (NaCl), and heparin were used in this research.

IV. METHODS

A ratio of 1:2 (RBC: NaCl) was used on the electrodes. Heparins are considered neglected because the usage is around 1000ml: 1ml (RBC: heparin). The electrodes were connected to the function generator to supply the alternating current (AC). Fig. 1. is an illustration of the setup for the experiment.

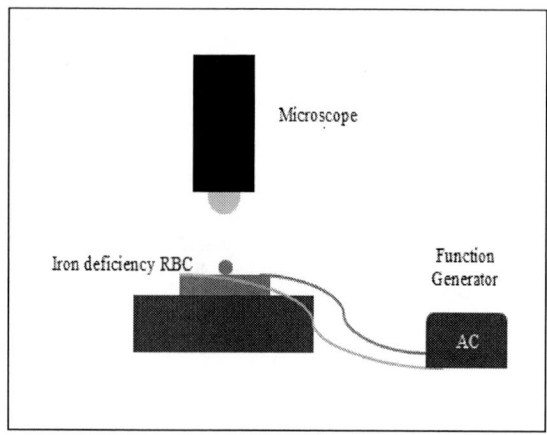

Fig. 1. Illustration of experiment setup

V. PROPERTIES OF PARTICLES

Table I below shows the comparison particle properties of healthy and anemic RBC. The values were fed into MATLAB to find the CMF as shown in Eq. 4.

TABLE I.   COMPARISON BETWEEN TWO PROPERTIES OF RBC

Properties	Healthy RBC	Anemic RBC
Permittivity of Vacuum, $\varepsilon_\circ$	8.85×10^{-12}	8.85×10^{-12}
Relative Permittivity of Medium, ε_m	$78\,\varepsilon_\circ$	$120\,\varepsilon_\circ$
Conductivity of Medium, σ_m	0.055	0.0013
Shell Relative permittivity of Particles	$4.4\,\varepsilon_\circ$	$5\,\varepsilon_\circ$
Shell electrical conductivity of Particles	1×10^{-6}	1×10^{-6}
Particle relative permittivity of RBC	$59 \times \varepsilon_\circ$	$65 \times \varepsilon_\circ$
Particle conductivity of RBCs, σ_p	0.31	50×10^{-3}
Particle diameter RBC	R3 $=7 \times 10^{-6}$	R3 $= 5.6 \times 10^{-6}$
Shell thickness	R4 = R3 – (9×10^{-9})	R4 = R3 – (9×10^{-9})

VI. RESULTS AND DISCUSSIONS

In previous research [17], M. R. Buyong studied normal RBC using the same electrode sputtered [18] by titanium. Those results came out differently compared to iron deficient RBC. This is possible because the shapes of anemic RBC are different from those of normal RBC. Anemic RBC are considered to be sickle-shaped and not donut-shaped, as they are longer and thinner. This may affect all properties of the particles.

The results in the table below show the response of iron deficiency RBC from 30kHz to 15000kHz. The RBC responded to positive dielectrophoresis (P_{DEP}) and the particles are attracted to the electrodes during 30kHz until 6000kHz. The RBs are in a uniform state at 7000kHz where there is no mobility in terms of direction, $P_{DEP} = N_{DEP}$. Particles start to repel at 9000kHz which gives negative dielectrophoresis (N_{DEP}).

TABLE II.   RESPONSE OF IRON DEFICIENCY RBC AT CERTAIN FREQUENCIES

No	Frequency, kHz	Mobility		Type of DEP
		Mobility	*Direction mobility*	
1	30	Available	Attract	P_{DEP}
2	60	Available	Attract	P_{DEP}
3	90	Available	Attract	P_{DEP}
4	200	Available	Attract	P_{DEP}
5	300	Available	Attract	P_{DEP}
6	500	Available	Attract	P_{DEP}
7	1000	Available	Attract	P_{DEP}
8	3000	Available	Attract	P_{DEP}
9	4000	Available	Attract	P_{DEP}
10	5000	Available	Attract	P_{DEP}
11	6000	Available	Attract	P_{DEP}
12	**7000**	**None**	**Constant**	$P_{DEP}= N_{DEP}$
13	9000	Available	Repel	N_{DEP}
14	12000	Available	Repel	N_{DEP}
15	15000	Available	Repel	N_{DEP}

Each RBC at different frequencies was recorded for 60 seconds. The timing was recorded to see how long it takes for the RBCs to be cleared from the region of interest (ROI) during P_{DEP} and to see if they will flock together at the center of ROI during N_{DEP}. During the experiment, RBCs are focused on ROI. The RBCs can be confirmed as P_{DEP} or N_{DEP} if they are able to respond when AC is supplied.

Fig. 2. shows the movement of RBCs during P_{DEP}. RBCs are trying to clear up the ROI and move away from the ROI as they are attracted to the electrodes.

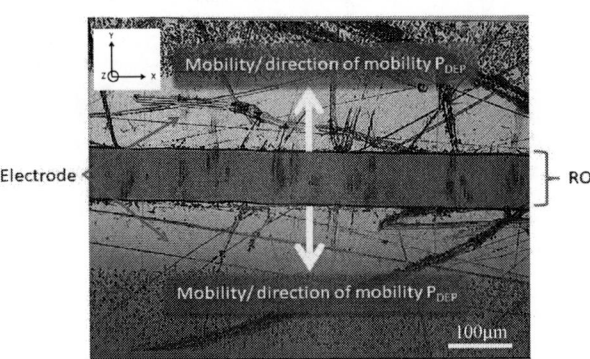

Fig. 2. Direction of RBCs during P_{DEP}

Fig. 3. shows the results at 7000kHz when RBCs are in a uniform state, $P_{DEP} = N_{DEP}$. RBCs did not respond to the AC. This is where the crossover frequencies appear, f_{ox}.

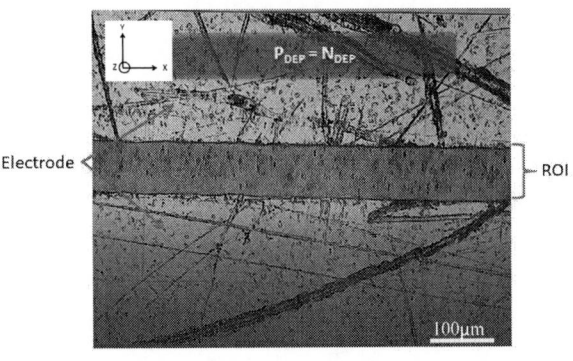

Fig. 3. Particles at uniform state $P_{DEP} = N_{DEP}$

Fig. 4. shows the direction of RBCs during N_{DEP} response. The RBCs flock to the centre of ROI and try to repel the electrodes.

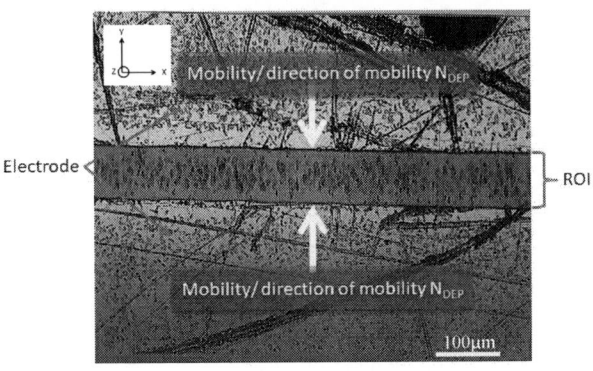

Fig. 4. Direction of RBCs during N_{DEP}

The results in Fig. 5. were obtained using MATLAB. Equation 4 was applied to find out the crossover frequencies,

f_{ox}. The graphs show that f_{ox} for anemic RBC is proven to be 6.87MHz $\simeq$ 7MHz. Normal RBC f_{ox} is found to be at 380kHz [17].

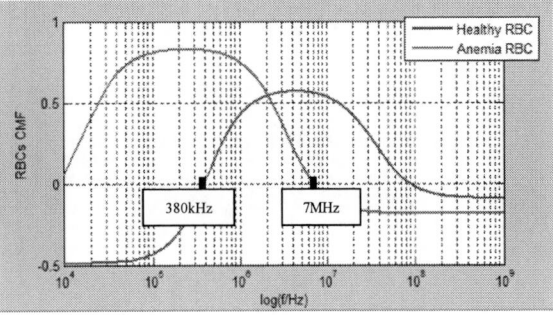

Fig. 5. Cross over frequencies (f_{ox}) of normal RBC [17] vs a nemic RBC

Based on the graph in Fig. 5., the curves shows the movement of RBC from P_{DEP} to N_{DEP} for the anemic RBC while N_{DEP} to P_{DEP} and back to N_{DEP} for healthy RBC. The movements of the particles can be related using Fig. 2., Fig. 3, and Fig. 4. The speed of movement of the particles depends on their frequency.

Anemic RBCs begin to respond faster to AC at 20kHz, and remained at maximum speed from 60kHz to 670kHz. This is the fastest movement when AC was applied during P_{DEP} (Refer Fig. 2.). Later, RBC began to slow down until reaching CMF=0, 7MHz where $P_{DEP} = N_{DEP}$ (refer Fig. 3.). RBCs start to switch from P_{DEP} to N_{DEP} at frequencies of 7MHz and above (Refer Fig. 4.). The movement of particles responds to AC during the change to N_{DEP}, starting slow and beginning to move faster at 39MHz.

Thus, based on Fig. 5. above, this research proves that anemic RBCs yield different results due to differences in particle properties.

VII. CONCLUSIONS

It is important to prove that different diseases on RBCs will result in different properties. This is one of the precautions that should be taken for artificial kidneys. The artificial kidneys should be matched to the type of disease that the patients carry. Further research will be done to determine whether different diseases in RBCs will affect other biological properties of the body.

ACKNOWLEDGMENT

The author would like to acknowledge with gratitude the sponsor HICOE-AKU-95 grant funded by The Ministry of Higher Education, DPP-2018-006 Peranti bioperubatan with IOT funded by Dana Pembangunan Penyelidikan PTJ and PRGS/1/2017/TK05/UKM/02/1 Design and Fabrication of Silicon Membrane Filtration System for Artificial Kidney funded by SKIM GERAN PENYELIDIKAN PEMBANGUNAN PROTOTAIP (PRGS).

REFERENCES

[1] R. Pethig, Y. Huang, X. Wang, and J. P. H. Burt, "Positive and negative dielectrophoretic collection of colloidal particles using interdigitated castellated microelectrodes.pdf," vol. 881, 1992.

[2] M. P. Hughes, "Fifty years of dielectrophoretic cell separation technology," *Biomicrofluidics*, vol. 10, no. 3, 2016.

[3] T. Z. Jubery, S. K. Srivastava, and P. Dutta, "Dielectrophoretic separation of bioparticles in microdevices: A review," *Electrophoresis*, vol. 35, no. 5, pp. 691–713, 2014.

[4] J. Kadaksham, P. Singh, and N. Aubry, "IMECE2003-4 3950 PARTICLE SEPARATION USING DIELECTROPHORESIS," no. 2, pp. 1–5, 2016.

[5] N. Piacentini, G. Mernier, R. Tornay, and P. Renaud, "Separation of platelets from other blood cells in continuous-flow by dielectrophoresis field-flow-fractionation," *Biomicrofluidics*, vol. 5, no. 3, pp. 1–8, 2011.

[6] F. H. F. Morales, J. E. Duarte, and J. S. Marti, "Non-uniform electric field-induced yeast cell electrokinetic behavior," *Rev. Ing. E Investig.*, vol. 28, no. 3, pp. 116–121, 2008.

[7] W. M. Arnold, "Dielectrophoretic Cell Separation : Some Hints and Kinks," *Proc. ESA Annu. Meet. Electrost.*, vol. 2, no. 1, pp. 1–11, 2010.

[8] F. F. Becker, X.-B. Wang, Y. Huang, R. Pethig, J. V Vykoukal, and P. R. C. Gascoyne, "The removal of human leukemia cells from blood using interdigitated microelectrodes," *J. Phys. D. Appl. Phys.*, vol. 27, pp. 2659–2662, 1994.

[9] R. Pethig, A. Menachery, S. Pells, and P. De Sousa, "Dielectrophoresis: A review of applications for stem cell research," *J. Biomed. Biotechnol.*, vol. 2010, 2010.

[10] F. F. Becker, X. B. Wang, Y. Huang, R. Pethig, J. Vykoukal, and P. R. Gascoyne, "Separation of human breast cancer cells from blood by differential dielectric affinity.," *Proc. Natl. Acad. Sci.*, vol. 92, no. 3, pp. 860–864, 1995.

[11] M. R. Buyong *et al.*, "Implementing the concept of dielectrophoresis in glomerular filtration of human kidneys," *IEEE Int. Conf. Semicond. Electron. Proceedings, ICSE*, vol. 2016–Septe, pp. 33–37, 2016.

[12] F. W. Yunus, A. A. Hamzah, M. R. Buyong, J. Yunas, and B. Y. Majlis, "Negative charge dielectrophoresis by using different radius of electrodes for biological particles," *Proc. 2017 IEEE Reg. Symp. Micro Nanoelectron. RSM 2017*, pp. 84–87, 2017.

[13] S. Sang *et al.*, "Portable microsystem integrates multifunctional dielectrophoresis manipulations and a surface stress biosensor to detect red blood cells for hemolytic anemia," *Sci. Rep.*, vol. 6, no. May, pp. 1–8, 2016.

[14] M. R. Buyong, F. Larki, M. S. Faiz, A. A. Hamzah, J. Yunas, and B. Y. Majlis, "A tapered aluminium microelectrode array for improvement of dielectrophoresis-based particle manipulation," *Sensors (Switzerland)*, vol. 15, no. 5, pp. 10973–10990, 2015.

[15] M. R. Buyong, N. A. Aziz, A. A. Hamzah, M. F. M. R. Wee, and B. Y. Majlis, "Finite element modeling of dielectrophoretic microelectrodes based on a array and ratchet type," *IEEE Int. Conf. Semicond. Electron. Proceedings, ICSE*, no. 3, pp. 236–239, 2014.

[16] M. R. Buyong, N. A. Aziz, A. A. Hamzah, and B. Y. Majlis, "Dielectrophoretic characterization of array type microelectrodes," *IEEE Int. Conf. Semicond. Electron. Proceedings, ICSE*, no. Dc, pp. 240–243, 2014.

[17] M. R. Buyong, F. Larki, C. E. Caille, Y. Takamura, A. A. Hamzah, and B. Y. Majlis, "Determination of lateral and vertical dielectrophoresis forces using tapered microelectrode array," *Micro Nano Lett.*, no. 2, pp. 1–6, 2017.

[18] A. A. Hamzah, J. Yunas, B. Y. Majlis, and I. Ahmad, "Sputtered encapsulation as wafer level packaging for isolatable MEMS devices: A technique demonstrated on a capacitive accelerometer," *Sensors*, vol. 8, no. 11, pp. 7438–7452, 2008.

Physical Modelling and Simulation of Au/ZnO Schottky diode

Hira Shumail

Department of Electronic Engineering, NED University of
Engineering & Technology, 75270
Karachi, Pakistan
hira.oct56@hotmail.com

Dr. Sadia Muniza Faraz

Department of Electronic Engineering, NED University of
Engineering & Technology, 75270
Karachi, Pakistan
smuniza@neduet.edu.pk

Abstract—A Schottky diode of Gold (Au)/ ZnO is simulated using an open source TCAD device simulator (Minimos-NT) and the effect of different device models on the electrical characteristics of ZnO Schottky diode have been studied. The simulated results have been compared with electrical characterization experimental results reported in [1] and the reasons for discrepancy in the results have been addressed. For effective simulation, material parameters of ZnO collected from literature published in various research papers and magazine articles were supplied to the software. Hence, development of a complete database of ZnO parameters was also aimed which will assist in future simulations of the said material.

Keywords—Zinc Oxide, Schottky diode, TCAD device simulator, Simulation

I. INTRODUCTION

Wide bandgap semiconductors such as zinc oxide (ZnO), silicon carbide (SiC) and gallium nitride (GaN) mark the beginning of a new era in electronics. Compared to the conventional semiconductor materials such as silicon (Si), such materials have higher operating temperatures and ensure durable and reliable performance at higher voltages and frequencies. Amongst the various wide bandgap semiconductors, the global research interest has been attracted towards zinc oxide (ZnO) due to its numerous properties as a semiconductor material and in 2007 it has been reported to be second only to Silicon in popularity [2]-[4].

Zinc oxide belongs to the II-VI group of the periodic table. It has a direct and wide bandgap (3.37ev), and possesses a large exciton binding energy (60meV) at room temperature leading to effective emission and indicating its suitability for use in optical devices [5-6]. It has the richest family of nanostructures. Nanowires, nanorods, nanorings, nanoneedles, nanotubes, nanosprings are to name a few [7].

Various growth approaches for ZnO nanostructures are reported in literature including Aqueous Chemical Growth (ACG) technique [8]. The grown ZnO commonly results in n-type semiconductor material. Due to the lack of a reliable and reproducible p-type doping in ZnO, ZnO heterostructures with other p-type semiconductors are more common than ZnO homostructures [8]. Hence good quality heterojunctions have been fabricated using various p-type materials e.g. Si, GaN, AlGaN, NiO, CdTe etc. [9]. ZnO heterostructures possess many deep radiative levels which emit radiations of various colours covering the whole visible region [9][10]. N-ZnO/p-Si heterojunctions are also being explored to determine their potential applications in solar cells as can be seen in [11] and [12].

Schottky diodes are the basis of a large number of electronic devices such as field effect transistors, microwave diodes and photovoltaics. Schottky diodes generate a low noise level which makes them particularly useful for use in detectors, microwave and RF receivers, as well as mixers [13]. High quality Schottky contacts are critical for realizing ZnO based devices [14].

Modeling of electronic and photonic devices using physical drift-diffusion/hydrodynamic simulation is an efficient and powerful tool for optimization in terms of its physical structure, for applications ranging from DC switching to RF signals and opto-electronics. This paper is therefore aimed at physical modeling of Gold (Au)/ Zinc Oxide Schottky diode using an open source TCAD device simulator (Minimos-NT) to investigate its electrical characteristics (current–voltage characteristics etc). The simulated results were compared with the experimental results and the mismatch in the results has been addressed. Material parameters of ZnO were collected from published literature and the development of a complete database of ZnO parameters was aimed which will assist in future simulations of the said material.

II. DEVICE DESCRIPTION & DRAWING

A. Device Model

The device model as depicted in [1] is shown in Fig. 1. ACG method has been used to grow ZnO nanorods on a glass substrate. A layer of silver (Ag) has been deposited to behave as the ohmic contact on the glass substrate, on top of which ZnO nanorods have been grown. The thickness of Ag layer is 139 nm. Finally circular gold (Au) contacts of thickness 56 nm and diameter 1.5 mm have been deposited on the nanorods which will behave as the Schottky contacts owing to the work function difference of the metal Au and semiconductor ZnO. The fabrication has been performed at room temperature (300 K). The fabricated nanorods were hexagonal shaped with lengths varying from 1.3 - 1.9 µm and diameters varying from 300 - 450 nm [1].

Due to various constraints, we were restricted to the performance of 2-D Simulations only. Hence, following assumptions were made before the device model was entered in the Minimos-NT simulator.

1. Instead of nanorods, a continuous layer of ZnO is present

2. Simulations have been performed for the front face of the diode while the third dimension is assumed to be 1 μm. The obtained readings have then been multiplied by a factor of $2\pi R$ to determine results for the complete circular Schottky diode.

Fig. 2(a) demonstrates the assumptions while Fig 2(b) shows the 2-D Front face of the Schottky diode simulated in Minimos-NT. The part shaded in black indicates the third dimension, assumed to be 1 μm. Amongst the varying lengths of the nanorods, a dimension of 1.5 μm was selected for simulation. Fig. 5.4 shows the final device dimensions of the diode simulated in Minimos-NT.

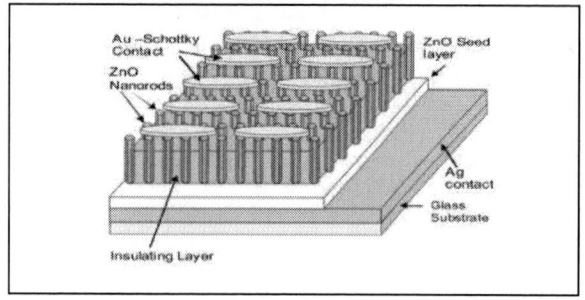

Fig. 1. Schematic diagram of Au/ZnO nanorods Schottky diodes given in [1]

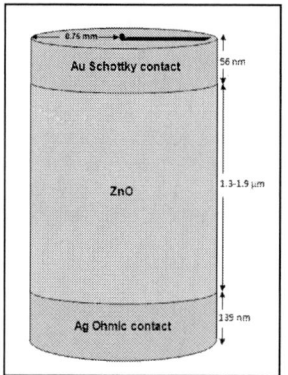
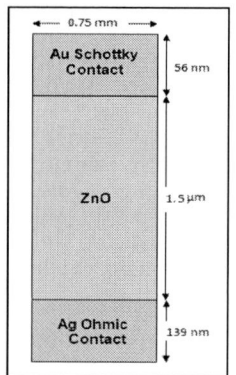

Fig. 2. (a) Au/ZnO Schottky diode model assumed for Simulations (b) 2-D Front-face of Schottky diode model

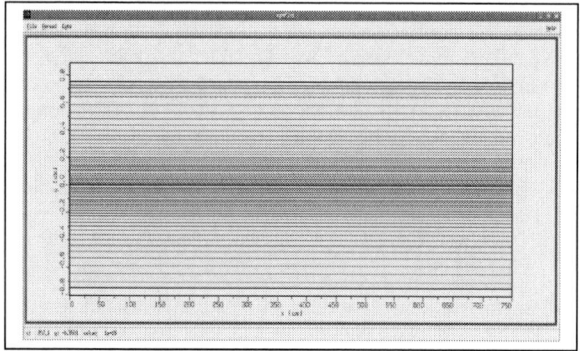

Fig. 3. Final dimensions of 2D-simulated Schottky diode model

III. ZINC OXIDE MODEL PARAMETERS

ZnO is a fairly new addition to the family of semiconductor materials and its modeling parameters are not clearly specified and fixed as yet. Also, many of the available properties are the basic physical properties of bulk zinc oxide which are notalways accurate since ZnO undergoes a phenomenon called the "quantum size effects" in which as the dimensions of ZnO continues to shrink down to nanometer, their physical properties change [15]. The properties of ZnO are also highly dependent on the type of nanostructure fabricated and the growth technique [16].

Table 1 shows the various design parameters used in the modeling of Au/ZnO Schottky diode collected from literature. Models for which ZnO parameters were not available, parameters for Si or other wide bandgap semiconductors (e.g. GaN, SiC etc.) were used as suitable. GaN is a wide bandgap semiconductor (3.4 eV) [17] whereas Si is the standard semiconductor with widely available parameters. It is taken as the reference semiconductor and is assumed to predict the general behaviour of semiconductors. These parameters were included in the Minimos-NT material database under specially created section for ZnO, through programming.

IV. RESULTS AND DISCUSSION

Fig. 4 shows a comparison between the simulated and measured Current-Voltage (I-V) Characteristics for Au/ZnO Schottky diode. The variation between the two curves in the forward characteristics can be attributed to a number of reasons. The simulated model is a 2D model and various assumptions have been made to incorporate the 3D effects to match the simulated device with the fabricated device. Moreover, a continuous slab of ZnO has been assumed as opposed to ZnO nanorods in the fabricated device. Naturally, a continuous slab will be more conductive than a group of nanorods spread over the same area.

Also, the ZnO parameter model was not available in the simulator. Various parameters were collected from literature and used to create the ZnO parameter model which cannot be an exact representation of the ZnO material used in the device, since the properties of ZnO are highly dependent on the type of nanostructure fabricated and the growth technique.

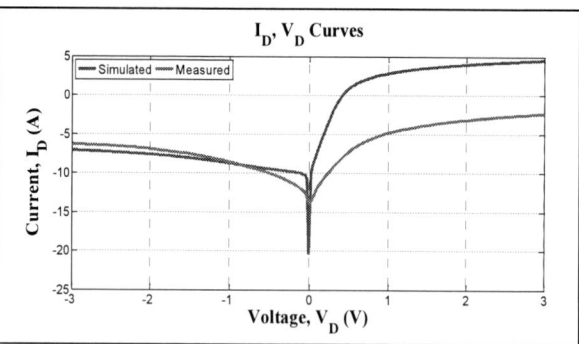

Fig. 4. Current-Voltage (I-V) Characteristics for Au/ZnO Schottky diode (Measured [1] & Simulated)

Moreover, model parameters of Si and other wide bandgap semiconductors have been used for those models where ZnO parameters were not available. Lastly, a number of secondary effects occur during the experimental process of device fabrication, the provision for which is not given in any software and those variations can only be incorporated in the simulations by adjustment of other model parameters.

Figure 5, 6, 7 & 8 show the mass density distribution profile, electron mobility distribution profile, hole mobility distribution profile and the permittivity distribution profile for Schottky diode respectively, indicating that the entered ZnO model in the simulation software has been accepted and is operational during simulation.

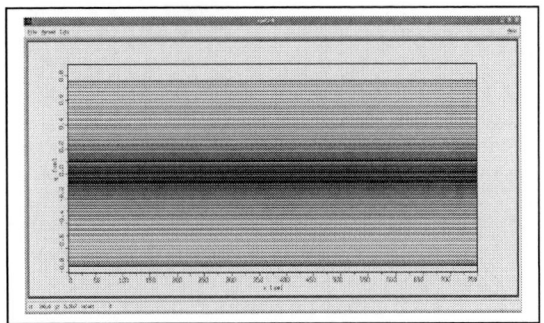

Fig. 8. Permittivity distribution profile for Schottky diode

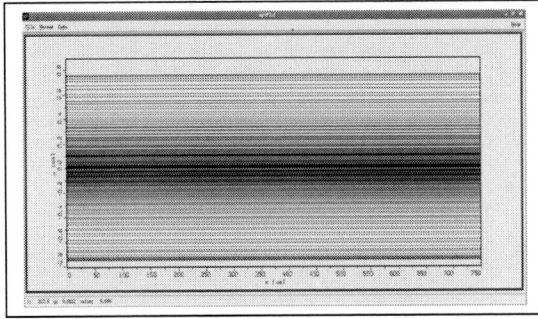

Fig. 5. Mass density distribution profile for Schottky diode

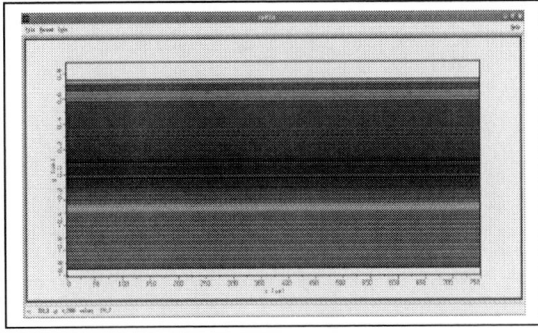

Fig. 6. Electron mobility distribution profile for Schottky diode

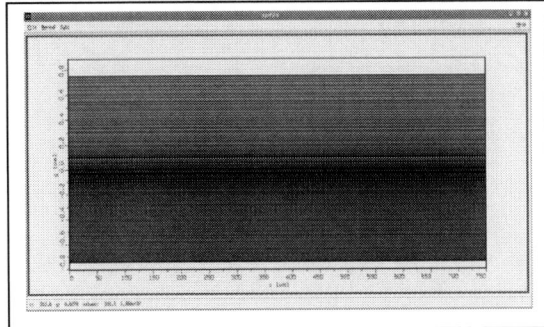

Fig. 7. Hole mobility distribution profile for Schottky diode

TABLE I.   PARAMETER SET FOR SIMULATION OF AU/ZNO SCHOTTKY DIODE

Model	Parameter	Keyword(Minimos- NT)	Value
Generation & Recombination			
Shockley-Read-Hall & Surface Recombination (Ltm)	NT	Nt	1×10^{13} cm-3 [18]
	ε_T	Et	0.0 eV [18]
		type	Donor [18]
	$\sigma_{T,n}$	ste	1.0×10^{-19} m^2 [18]
	$\sigma_{T,p}$	sth	1.0×10^{-19} m^2 [18]
Auger Recombination	CnAU	Ce	2.8×10^{-31} cm^6/s [18]
	CpAU	Ch	9.9×10^{-32} cm^6/s [18]
Direct Recombination	C^{DIR}	Cdir	1×10^{20} cm^3/s [18]
Impact Ionization (Simple)	C_{n1}	Cn1	3×10^{12} s^{-1} [18]
	C_{n2}	Cn2	1.2 [18]
	C_{h1}	Ch1	3×10^{12} s^{-1} [18]
	C_{h2}	Ch2	1.2 [18]
Carrier Mobility			
Mobility (Constant Mobility Model)	μ_e	Const	250 cm^2/V s [19]
	μ_h	Const	180 cm^2/V s [19]
Velocity Saturation (SiC)	$v^{sat}_{v,300,\parallel}$	vsat300	2.5×10^7 cm/s [18]
	δ^v_{sat}	delta	-1 [18]
Energy Relaxation Time (Const)	$\tau_{\varepsilon,n}$	twe	0.35 ps [18]
	$\tau_{\varepsilon,p}$	twh	0.4 ps [18]
Band Structure Parameters			
Bandgap Energy (Ltm_Pure)	E0	Eg0	64 ± 7 meV [20]
	α	Alpha	0.00020 ± 0.00002 eV/K [20]
	β	Beta	325 ± 20 K [20]
	ε_{off}	Eoffs	- Eg0/2 [18]
Bandgap Narrowing (Si_Slotboom)	ε_{ref}	Eref	18.7×10^{-3} ev [21]
	N_{ref}	Nref	7×10^7 cm^3 [21]
	$\Delta\varepsilon_c / \Delta\varepsilon_g$	dEcdEg	0.5 [18]

Model	Parameter	Keyword(Minimos-NT)	Value
Effective Carrier Mass (Ltm_Pure)	$m_{0,n}$	me0	0.7 [18]
	$m_{1,n}$	me1	0 [18]
	$m_{0,p}$	mh0	0.82 [18]
	$m_{1,p}$	mh1	0 [18]
	$m_{2,p}$	mh2	0 [18]
Effective Density of States (Ltm_Pure)	Equivalent Energy Minima	Mc	6 [22]
Lattice & Thermal Properties			
Permitivity	ε_r	epse	9 [19]
Mass Density	ρ	rho	5.606 g/cm³ [23]
Thermal Heat Flux	K300	kappa	148 W/K m [18]
	α	alpha	-1.65 [18]
Lattice Heat Capacity (Ltm_Pure)	heat	heat	711 J K⁻¹ kg⁻¹ [18]
	dheat	dheat	255 J K⁻¹ kg⁻¹ [18]
	alpha	alpha	1.85 [18]
Boundary Conditions			
Au/ZnO Schottky Contact	Barrier height (ϕ_B/ε_w)	Ew	0.62 eV [1]
	Electron recombination velocity (v_n)	rve	1x10⁶ cm/s [24]
	Hole recombination velocity (v_n)	rvh	1x10⁴ cm/s [24]
	Series resistance (Rs)	R	20.5 Ohm [1]
Ag/ZnO Schottky Contact	Thermal Contact Resistivity	Rth	4.0e⁻⁴ K cm²/W [18]
	Global Thermal Contact Resistance	Rg	0.017 K/W [15]
	Series resistance (Rs)	R	28 Ohm [1]

V. CONCLUSION

A Schottky diode comprising of Gold (Au)/ ZnO heterojunction is modeled & simulated using an open source TCAD device simulator (Minimos-NT, release 2.1). The electrical characteristics of ZnO Schottky diode have been studied and the results compared with electrical characterization experimental results reported in [1]. The reasons for discrepancy in the results have been addressed. Also, the database of ZnO parameters has been developed.

REFERENCES

[1] Faraz, S. M., M. Willander, and Q. Wahab. "Interface state density distribution in Au/n-ZnO nanorods Schottky diodes." IOP Conference Series: Materials Science and Engineering. Vol. 34. No. 1. IOP Publishing, 2012.

[2] Duy-Thach Phan, Gwiy-Sang Chung, Appl. Surf. Sci., 257 ,p.p: 328, (2011).

[3] N. H. Alvi, Linköping Studies in Science and Technology Dissertation No. 1378, (2011).

[4] R. M. S.Jarrah, "Fabrication and Characteristics Study of ZnO/Si Heterojunction by DC Magnetron Sputtering," Ph.D. dissertation, Dept. Phy., University of Baghdad, 2013.

[5] J. Aranovich, A. Ortiz, and R. H. Bube, "Optical and electrical properties of ZnO films prepared by spray pyrolysis for solar cell applications," J. Vac. Sci. Technol. 16(4), 994 (1979).

[6] X. W. Sun, J. Z. Huang, J. X. Wang, and Z. Xu, "A ZnO nanorod inorganic/organic heterostructure lightemitting diode emitting at 342 nm," Nano Lett. 8(4), 1219–1223 (2008).

[7] Xiang-Dong Gao, Xiao-Min Li, Sam Zhang, Wei-Dong Yu, Ji-Jun Qiu. "ZnO submicron structures of controlled morphology synthesized in zinc-hexamethylenetetramine-ethylenediamine aqueous system." Journal of materials 22, no.7(2007): 1815-1823.

[8] Muniza Faraz, Sadia, Arne Henry, Omer Nur, and Magnus Willander. "Post fabrication annealing effects on electrical and optical characteristics of n-ZnO nanorods/p-Si heterojunction diodes."

[9] Bano, Nargis, Siama Zaman, A. Zainelabdin, S. Hussain, I. Hussain, O. Nur, and Magnus Willander. "ZnO-organic hybrid white light emitting diodes grown on flexible plastic using low temperature aqueous chemical method." Journal of Applied Physics 108, no. 4 (2010): 043103.

[10] Willander, Magnus, O. Nur, Q. X. Zhao, L. L. Yang, M. Lorenz, B. Q. Cao, J. Zúñiga Pérez et al. "Zinc oxide nanorod based photonic devices: recent progress in growth, light emitting diodes and lasers." Nanotechnology 20, no. 33 (2009): 332001.

[11] Chen, Cheng-Pin, Pei-Hsuan Lin, Liang-Yi Chen, Min-Yung Ke, Yun-Wei Cheng, and JianJang Huang. "Nanoparticle-coated n-ZnO/p-Si photodiodes with improved photoresponsivities and acceptance angles for potential solar cell applications." Nanotechnology 20, no. 24 (2009): 245204.

[12] Shih, Jeanne-Louise.2007. Zinc Oxide Silicon Heterojunction Solar Cells by Sputtering.Master thesis, McGill University, Canada (ISBN: 978-0-494-51474-0).

[13] Sheng, S. Li. "Semiconductor physical electronics." (1993).

[14] H. L. Mosbacker, Y. M. Strzhemechny, B. D. White, P. E. Smith D. C. Look, D. C. Reynolds, C. W. Litton, and L. J. Brillson, "Role of near-surface states in ohmic-Schottky conversion of Au contacts to ZnO", Appl. Phy. Lett., vol. 87, no.1, pp.2102, Jul. 2005.

[15] Liu, Fei, et al. "Morphology study by using scanning electron microscopy."Microscopy: Science, Technology, Applications and Education (2010): 1781-1792.

[16] B. G. Streetman and S. Banerjee, "Solid State Electronic Devices", Chapter – 5, Prentice Hall, p220-227 (2000)

[17] Minimos-NT, Device & Circuit Simulator, release 2.1 (manual).

[18] R. M. S.Jarrah, "Fabrication and Characteristics Study of ZnO/Si Heterojunction by DC Magnetron Sputtering," Ph.D. dissertation, Dept. Phy., University of Baghdad, 2013.

[19] Rai, R. C. "Analysis of the Urbach tails in absorption spectra of undoped ZnO thin films." Journal of Applied Physics 113.15 (2013): 153508.

[20] Li, L. M., et al. "Bandgap narrowing and ethanol sensing properties of In-doped ZnO nanowires." Nanotechnology 18.22 (2007): 225504.

[21] Razeghi, Manijeh. Fundamentals of solid state engineering. New York: Springer, 2006.

[22] Norton, David P., et al. "ZnO: growth, doping & processing." Materials today7.6 (2004): 34-40.

[23] Tripathi, Brijesh, et al. "Influence of optical properties of ZnO thin-films deposited by spray pyrolysis and RF magnetron sputtering on the output performance of silicon solar cell." IOP Conference Series: Materials Science and Engineering. Vol. 43. No. 1. IOP Publishing, 2013.

[24] Madhusudana, Chakravarti V., and C. V. Madhusudana. Thermal contact conductance. New York: Springer-Verlag, 1996.

978-1-5386-5284-8/18 $31.00 © 2018 IEEE

Voltage Characterization on Dielectrophoretic Force Response to Hematologic Cell Manipulation

Muhammad Izzuddin Abd Samad[1], Muhamad Ramdzan Buyong[1], Farahdiana Wan Yunus[1], Kim Shyong Siow[1], Azrul Azlan Hamzah[1], Burhanuddin Yeop Majlis[1]

[1]Institute of Microengineering and Nanoelectronic (IMEN), Universiti Kebangsaan Malaysia (UKM)
43600, Bangi, Selangor, Malaysia
Email: muhdramdzan@.ukm.edu.my

Abstract—**This paper presents a voltage characterization of dielectrophoretic force response on hematologic cell manipulation. In this case, red blood cells (RBC) were used as the target cell to quantify voltage DEP response, which is manipulate on tapered DEP microelectrode. In this research, RBC have been attracted to the higher electric density region are classified are positive DEP (pDEP). The magnitude of the particle moment was characterized with different peak-peak input voltage (V p-p), namely 5 and 10 V p-p for 400 kHz in 40 seconds. The magnitude of the cell dipole moment at 10 V p-p for 40 s is a significant clearance of the region of interest compared to 5 V p-p after 40 s. Therefore, our findings make it possible to optimization for isolate RBC in blood plasma for medical sciences research.**

Keywords—Dielectrophoretic, Tapered Microelectrode, Red Blood cell, Voltage Characterization.

I. INTRODUCTION

The DEP is an alternative technique to manipulate a dielectric particle by experience on a non-linear electric fields as illustrate in Fig 1, it have observer by Herbert Pohl in 1950[1]. Basically, DEP concept have used in a multiple research field such as electromagnetic [2], biomechanics [3-4] and bio-sensing [5-6]. It able been operation on biosensor through separating [7], trapping [8] and isolating [9] bioparticle on lab on a chip (LOC) platform. If particle polarization less than medium polarization, it have called positive DEP (pDEP). On contrary of pDEP, the particle polarization will higher than medium polarization. A particle will repelling from higher electric field area, it is called negative DEP (nDEP). The magnitude of particle movement can represent based on particle DEP respond due to electrode configuration [10].

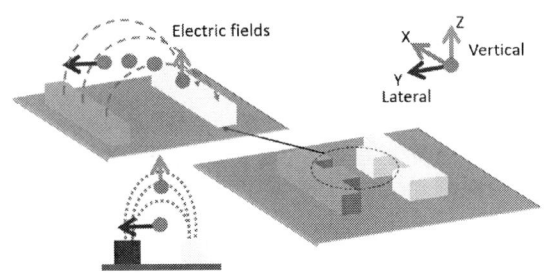

Fig.1 The spatial non-uniformity of electric field magnitude induce classical DEP force.

Recently, the electric field able to influence based on electric spot intensity on microelectrode profile [11-12]. In previous research, a tapered microelectrode profile are generating two spots intensity at top and bottom edge electrode [12]. A higher input voltage will induce high strength electric fields at spots intensity, causing high DEP forces on target cell [13].Based on that phenomena, particle trajectory of DEP mechanism at high magnitude of dipole moment.

In this research, an alternative current DEP application which is able enhance RBC manipulation due magnitude of electric fields density through characterize voltage DEP respond. Increment input voltage was enhance electrical charge density will influence DEP force response. The integration of DEP force response able enhance effectiveness in sensitivity and selectivity on discriminate target cell. The implementation of current DEP method in biosensing science research, which enhance high selectivity and sensitivity on particle manipulation.

II. MAGNITUDE OF DIPOLE MOMENT

It is relationship electrical polarization of surface charge conductance due non-uniform electric fields [14]. The target cell will accelerating dipole moment, m depending on their magnitude and polarity. Otherwise, it was expressed as cell polarization factor or CMF on electrical field, E. so that:

$$m \propto CMF.E \qquad (1)$$

CMF is a technical analysis indicator to represent the polarity of particle polarization either pDEP or nDEP [15] as shown in Fig 2. If there are pDEP, particle attraction is occurred due to CMF value greater than 0 to maximum of 1.0. Otherwise, the nDEP case, particle repulsion is occurred due to CMF value lower than 0 to minimum -0.5. When polarization of particle and medium are equilibrium each other, CMF value will represent 0 value. It is called crossover frequency.

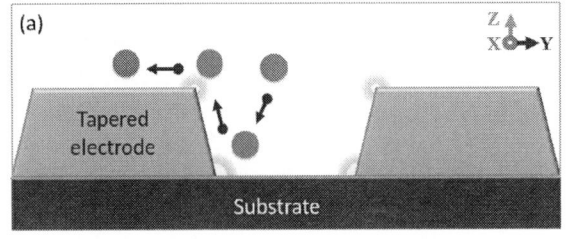

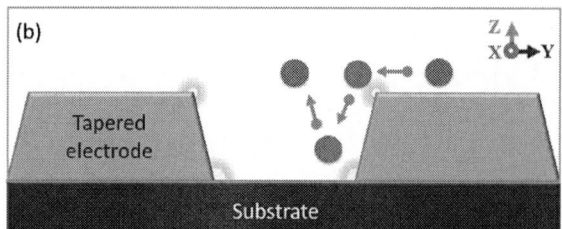

Fig 2. The particle manipulation mechanism on (a) lateral and (b) vertical in tapered microelectrode.

The dieletrophoresis forces (F_{DEP}) was described due single particle expose to translational electro-kinetic forces.

$$F_{DEP} = (m, \nabla E) \qquad (2)$$

The DEP force is directly proportional to particle volume in ∇E, vector electric field. Where R is spherical radius of cell.

$$F_{DEP} = 4\pi R^3 CMF(E, \nabla)E \qquad (3)$$

If the cases is electrostatic or quasi-static applied field, $E. (\nabla E) = 1/2 \nabla E^2$. Therefore F_{DEP} is

$$F_{DEP} = 2\pi R^3 CMF \nabla E^2 \qquad (4)$$

III. METHODOLOGY

In our research, RBC have used as target cell with diameter of 6 um to 8 um with blood cell conductivity, σ_p 0.055 S/m and relative permittivity particle, ε_p 2.5 x 10^{-12}. It was suspended in sodium chloride (NaCl), 0.9 % and deionizes water with ratio 1:3 ratio. The suspension medium consist medium conductivity, σ_m 0.6 mS/m and medium permittivity, ε_m 79 x 10^{-12}. The Maxwell-Wagner theoretical model used in calculate polarization factor or CMF for determine direction target cell either it attractive or repulsive DEP forces using MATLAB R2014a.

This paper is about positive DEP manipulation at 400 kHz with applied current voltage at 5 and 10 V p-p. The objective is to characterize magnitude of dipole moment on target cell manipulation. The amount of sample manipulation were conducted with 100 uL using droplet technique within 40 seconds. The dipole movement of target cell was verify under Olympus STM6 microscope with 20 objective lens as shown in Fig 3. The magnitude of dipole moment were verify based on area RBC distribution on ROI at inter-electrode using Image J software.

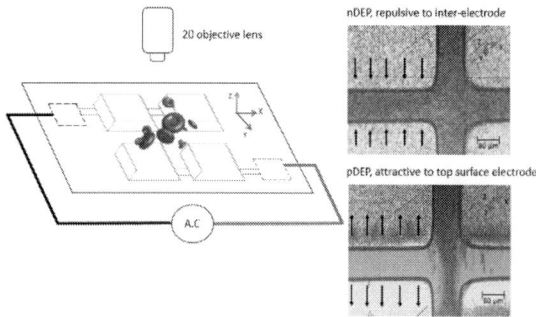

Fig. 3. Experimental setup for capturing magnitude effect.

IV. RESULT AND DISCUSSION

The real part of CMF for RBC is plotted due a function of the frequency as illustrated in Fig 4. The plotted is start from negative polarization of RBC in the range of 10 kHz to 280 kHz with polarization factor of -0.5 to 0. Further of 280 MHz is positive polarization of RBC with polarization factor of 0 to -0.4.

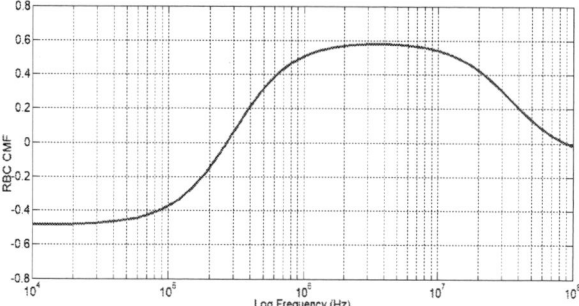

Fig. 4. Analytical of RBC CMF on 10 kHz to 10 GHz using MATLAB.

The RBC are induce positive DEP at 400 kHz with attractive due high electric field. It is based on calculated CMF with their dielectric properties and particle size. Therefore, the RBC are attractive to top surface electrode with different magnitude of cell dipole moment. The increment of supplied voltage are able to generating different electric fields density on top and bottom edge of tapered electrode. The Fig 5 as shown the verification area of RBC distribution at region of interest (ROI) which is inter-electrode.

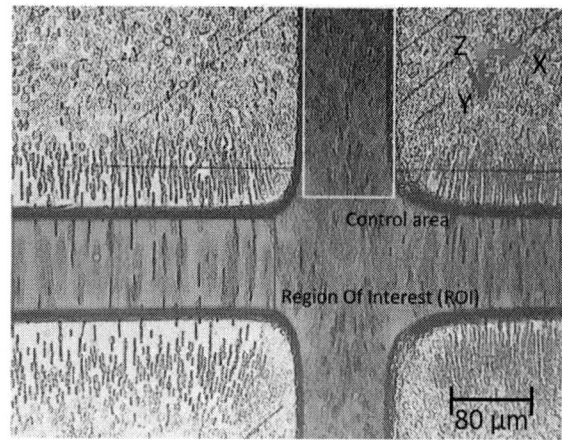

Fig. 5. The verification of RBC magnitude dipole moment at region of interest on tapered electrode.

In order to determine percentage of RBC distribution via magnitude RBC dipole moment, the area of RBC localization at ROI have are being calculate referring to Fig. 2. The higher electric field density on edge tapered electrode are lowest percentage of RBC distribution on ROI. In pDEP, the RBC have are being attractive to top surface electrode. This effected to the percentage of RBC distribution at ROI which is reduced as increasing time interval manipulation.

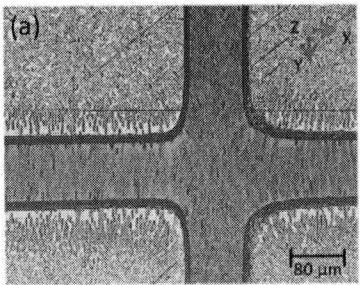 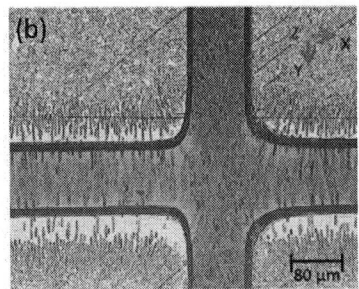 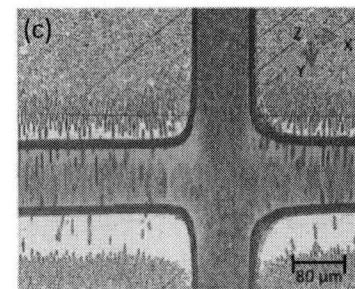

Fig 6. The RBC voltage dielectrophoresis response on 5 V p-p: (a) 10 s, (b) 20 s and (c) 40 s.

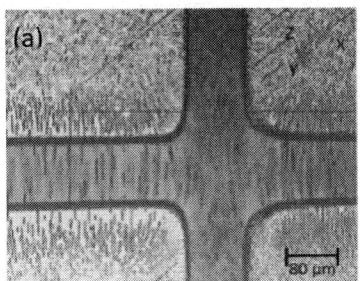 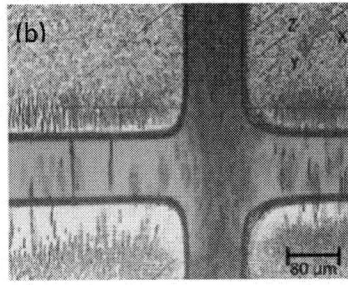 

Fig 7. The RBC voltage dielectrophoresis response on 10 V p-p: (a) 10 s, (b) 20 s and (c) 40 s.

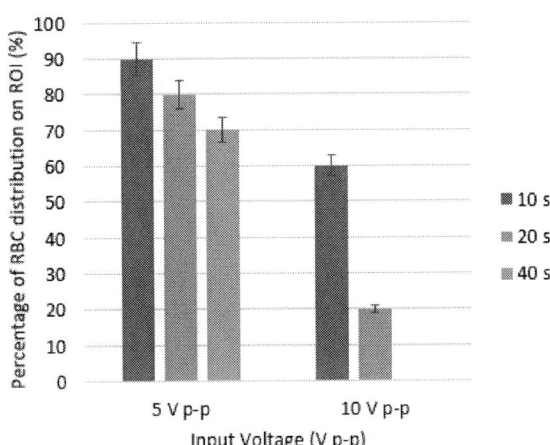

Fig 8. Percentage RBC distribution on ROI on 5 Vp-p and 10 Vp-p for 10 s, 20s and 40 s.

In Fig 6 and 7 as shows the RBC manipulation at 5 V p-p and 10 V p-p respectively, the percentage of RBC for each photo have been analyses RBC distribution area as illustrate in Fig 8 at 10 s, 20 s and 40 s were illustrate different manipulation pattern at different time interval. The area of RBC distribution at ROI for each manipulation period was indicating a different DEP force to attracting RBC via top surface electrode. At 10 V p-p, the magnitude of electric field density was the highest, causing high DEP forces for enhance RBC trajectory to top surface electrode.

V. CONCLUSION

Based on voltage characterization of RBC manipulation is related to cell dipole moment under electrical field application. It occurred when particle surface charge is induce external polarization charge density on cell. The input voltage are induce external electro-kinetic force due dipole moment of cell. At 10 V p-p, the RBC was induce higher DEP forces causing 0% RBC distribution at ROI within 40 s manipulation time. This is due to RBC, blood cell dipole moment trajectory effect on magnitude voltage applied. As conclusion, voltage characterization are to enhance RBC, blood cell manipulation.

ACKNOWLEDGMENT

This work was financially supported by the (GGPM-2017-028) research fund from the University Grant and Fundamental Research Grant Scheme (FRGS/1/2017/TK04/UKM/02/14) research fund from the Ministry of Education, Malaysia.

REFERENCES

[1] Pethig, Ronald. "Dielectrophoresis: An assessment of its potential to aid the research and practice of drug discovery and delivery." *Advanced drug delivery reviews* 65.11-12 (2013).

[2] Martinez-Duarte, Rodrigo, et al. "Fluido-dynamic and electromagnetic characterization of 3D carbon dielectrophoresis with finite element analysis." *Proceedings of NSTI Nanotech.* 2008.

[3] Polniak, Danielle V., et al. "Separating large microscale particles by exploiting charge differences with dielectrophoresis." *Journal of Chromatography A* (2018).

[4] Perez-Gonzalez, Victor H., et al. "A simple approach to reducing particle trapping voltage in insulator-based dielectrophoretic systems." *Analytical chemistry* (2018).

[5] Mavrogiannis, Nicholas, Francesca Crivellari, and Zachary R. Gagnon. "Label-free biomolecular detection at electrically displaced liquid interfaces using interfacial electrokinetic transduction (IET)." *Biosensors and Bioelectronics* 77 (2016): 790-798.

[6] Suzuki, Masato, et al. "Negative dielectrophoretic patterning with different cell types." *Biosensors and bioelectronics* 24.4 (2008): 1043-1047.

[7] D'Amico, Lorenzo, et al. "Isolation and concentration of bacteria from blood using microfluidic membraneless dialysis and dielectrophoresis." *Lab on a Chip* 17.7 2017.

[8] Kim, Soo Hyeon, and Teruo Fujii. "Efficient analysis of a small number of cancer cells at the single-cell level using an electroactive double-well array." *Lab on a Chip* 16.13 (2016): 2440-2449.

[9] Cheng, I-Fang, et al. "Antibody-free isolation of rare cancer cells from blood based on 3D lateral dielectrophoresis." *Lab on a Chip* 15.14 (2015): 2950-2959.

[10] Perez-Gonzalez, Victor H., et al. "A simple approach to reducing particle trapping voltage in insulator-based dielectrophoretic systems." *Analytical chemistry* (2018).

[11] Buyong, Muhamad Ramdzan, et al. "Tapered microelectrode array system for dielectrophoretically filtration: fabrication, characterization, and simulation study." *Journal of Micro/Nanolithography, MEMS, and MOEMS* 16.4 (2017).

[12] Buyong, Muhamad Ramdzan, et al. "Determination of lateral and vertical dielectrophoresis forces using tapered microelectrode array." *Micro & Nano Letters* 13.2 (2018).

[13] Buyong, Muhamad Ramdzan, et al. "A tapered aluminium microelectrode array for improvement of dielectrophoresis-based particle manipulation." *Sensors* 15.5 (2015): 10973-10990.

[14] Khoshmanesh, Khashayar, et al. "Dielectrophoretic platforms for bio-microfluidic systems." *Biosensors and Bioelectronics* 26.5 2011

978-1-5386-5284-8/18 $31.00 © 2018 IEEE

Integration Design of Wide-Dynamic-Range MEMS Magnetometer and Oscillator

Chao-Hung Song
Department of Electronic Engineering
National Chiao Tung University
Hsinchu 300, Taiwan
Email:nonox1257@gmail.com

Kuei-Ann Wen
Department of Electronic Engineering
National Chiao Tung University
Hsinchu 300, Taiwan
Email:stellawen@mail.nctu.edu.tw

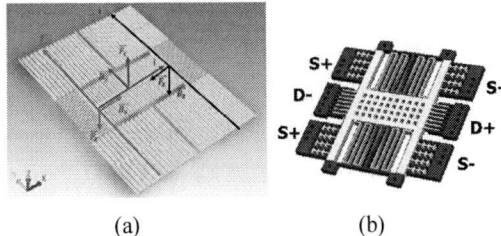

(a)   (b)

Fig. 1. Structure of proposed MEMS (a) magnetometer (b) oscillator.

Abstract—This paper presents a Lorentz current-based MEMS magnetometer integrated with MEMS oscillator. The auto capacitance offset calibration magnetometer can achieve a tunable sensitivity from 12 to 2.94×10^6 T/V and eliminate the capacitance mismatch caused by process variation under 0.084fF by SAR calibrator. The natural frequency of proposed differential oscillator is around 18.8 kHz, Q-factor of 183 under the bias voltage 20 V and at 760Torr ambient pressure. The proposed biasing method of oscillator obtains a frequency tuning range of 58510 ppm, resulting in a wide-dynamic-range which is about 107.8dB. FOM of the oscillator design is 4.75×10^{20} Hz$^2\Omega^2$. The integrated magnetometer and oscillator is implemented in 1P6M ASIC compatible CMOS MEMS process.

Keywords—MEMS, magnetometer, differential oscillator, wide-dynamic-range

I. INTRODUCTION

In past few years, magnetometer has become one of the key element among Inertial Measurement Unit (IMU).Since accelerometers and gyroscopes only sense the movement in 3D with an unknown position, magnetometer can be used to fix the relative movements, helping in detecting the absolute orientation of users. Hence, it is widely used for navigation, alignment, and height detection in wearable devices. For MEMS magnetometer, some of researches have been implemented by permanent magnet, and high resolution is achieved [1]. However, the require of specialized materials increase the cost and difficulties in integrated with other sensors. For Lorentz-based magnetometer, compared with others, which has zero hysteresis or crosstalk behavior [2], can be monolithic integrated with other sensors. Besides, research has shown that the signal of angular velocity can be obtained by the magnetometer sensor with the aim of software [3]. By this, the conventional 9-axis IMU can be implemented without the requirement of gyroscope. Moreover, the received signal also comes from different magnitude and frequency range. Therefore, how to design a wide range magnetometer becomes more prevailing in wearable device application.

In this paper, a Lorentz-based magnetometer integrated with MEMS oscillator is presented. Dependent on the relative difference of resonance frequency between magnetometer and oscillator, the proposed oscillator which has large tuning range in resonance frequency can not only offer the Lorentz current but also adjust the sensing range of magnetometer from micro-Tesla to Tesla, which can fit in many applications.

II. SYSTEM DESCRIPTION

A. Mechanical Structure

Proposed MEMS magnetometer is implemented in standard 0.18-μm CMOS 1P6M process. Fig. 1(a) shows that

under the presence of external magnetic field perpendicular to structure, the Lorentz force will be induced an in-plane motion. Relatively, the in-plane magnetic field will induce an out-plane motion. The area of magnetometer is 499.8×481.4 μm^2 and gaps between the structure are 2.5 μm. The direction of the current flow is set symmetric to avoid the magnetic couple.

MEMS oscillator is illustrated in Fig. 1(b). S and D represent sensing parts and driving parts of the oscillator. Under the operation, driving parts activate the shuttle along x-axis and the induced current will flow from sensing parts to the following sustaining amplifier. The dimension of structure is 348.4×346.1 μm^2. The natural frequency and quality factor of structure is 18.8 kHz and 183 under 760 Torr ambient pressure. The interdigital combs are designed to make the unwanted feedthrough current caused by the substrate or combs can be ideally cancelled and prevent the resonance frequency from shifting away.

B. Linear Model of Oscillator

Under linear operation, the RLC model shown in Fig. 2 is usually utilized to model the mechanical behavior of the MEMS oscillator. The electrical parameters Rm, Lm, Cm are motional impedance, inductance and capacitance as given below,

$$R_m = d_0^2 \, (mk)^{1/2} / (Q \zeta^2 N_{gn}^2 \varepsilon_0^2 h_r^2 V_{dc}^2) \tag{1}$$

$$L_m = d_0^2 \, m / (\zeta^2 N_{gn}^2 \varepsilon_0^2 h_r^2 V_{dc}^2) \tag{2}$$

$$C_m = \zeta^2 N_{gn}^2 \varepsilon_0^2 h_r^2 V_{dc}^2 / d_0^2 \tag{3}$$

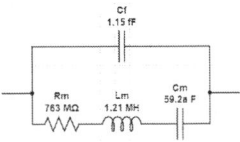

Fig. 2. RLC equivalent model of MEMS oscillator.

where d_0 is the gap between the comb, ε_0 is permittivity, ζ is a constant that models the fringing effect of capacitance, m is the mass of structure, k is spring constant, Q is quality factor, N_{gn} is the number of gap sets, h_r is thickness of the structure, and V_{dc} is the bias voltage on the movable part of MEMS. R_m results in a power loss in oscillation, which the power loss must be compensated in the following amplifier to sustain the oscillation. R_m is 763 MΩ, L_m is 1.21 MH, C_m is 0.0592 fF for this design. In addition, the feedthrough parasitic capacitance C_f is also introduced in the model. The influence of C_f would cause a phase shift on the resonance frequency, which is unessential for the design. Normally, the gap of the comb dominates C_f and C_m the most. In this process, the effect of feedthrough current cannot be negligible. From simulation and calculation, C_f is 1.15 fF, and it is quite larger compared to C_m. According to the model in Fig. 2, it will result in a large phase shift and make the oscillator unable to work. Therefore, the proposed MEMS oscillator is presented to resolve the problem and will be mentioned in section C.

C. Oscillator Design

To realize a robust oscillator, the non-ideality of MEMS oscillator like the effect of feedthrough capacitance should be resolved. The fully-differential oscillator is presented to eliminate the problem. As shown in Fig. 3, if the feedthrough current flows through the positive port of the oscillator, the presence of the elimination current in the negative port will absorb the current as soon as C_{feed1} is equal to C_{feed2}. To make sure that C_{feed1} and C_{feed2} are equal, sensing and driving parts are separate with equal distance.

In addition, to realize high tuning range oscillator, tuning combs are usually used to change the effective spring constant and vary the resonance frequency. The resonance frequency of capacitive transducer is given by

$$f_0= [(k_m-k_e)/m_m]^{1/2}/2\pi \qquad (4)$$

where k_m is mechanical stiffness, k_e is electrical stiffness, and m_m is the mass of moving part. Under different bias on tuning comb, the electrical stiffness changes, and resonance frequency decreases. However, this tuning method requires additional bias voltage of tuning comb and the tuning range constraint when mechanical stiffness is large. The proposed oscillator is free of tuning combs. The tuning method in this work is to vary the both k_m and k_e. The sensing combs of sides are set anti-symmetric along x axis as shown in Fig. 4(a). With this design, there is a slight displacement along x-axis which causes a variation in k_m before AC signals involved. Under different bias voltages on moving shuttle, the frequency range will expand by about 1.5 times from FEM simulation as shown in Fig. 4(b)

D. Sustaining Amplifier

To overcome the large loss caused by oscillator, the TIA amplifier is utilized to maintain the oscillation. The architecture of sustaining amplifier and MEMS oscillator is shown in Fig 5. The gain of the amplifier should be large and can be expressed as

$$R_{AMP} \geq R_{in} + R_{out} + R_m \qquad (5)$$

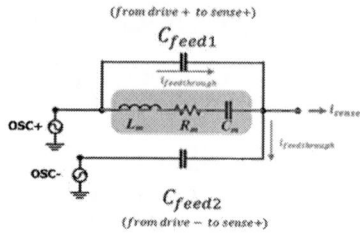

Fig. 3. The differential oscillator to eliminate the feedthrough effect.

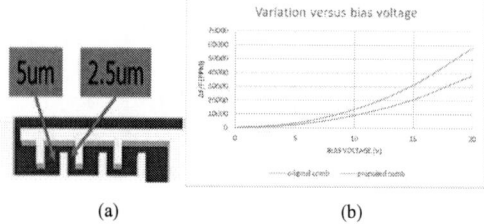

(a)   (b)

Fig. 4. (a) The proposed interdigital comb structure. (b) Frequency variation versus bias voltage with different comb structures.

where R_{in} and R_{out} is the input and output resistance of the amplifier. An integrator combined with differentiator typology is used for TIA since it offers large gain and small phase shift at lower frequency in kHz order. Also, the differentiator offers a feedback control for gain control. The auto gain control (AGC) circuit is adopted to maintain the operation in linear region. A reference voltage is applied to avoid the oscillation signal from containing higher harmonics, and make loop gain equal to unity. The simulation result shows that the loop gain at the 18.8kHz is 195dB and phase shift is about 0.5 degree.

E. Readout System

The readout system is illustrated in Fig 6. The system combines magnetometer readout circuit with Lorentz current generator made by MEMS oscillator. Since the signal band of natural signals such as gravity, magnetic field are usually

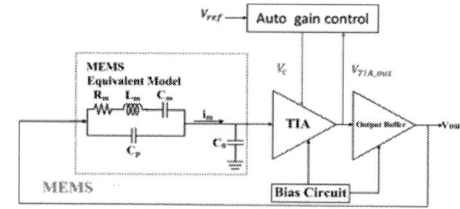

Fig. 5. The integration of sustaining amplifier and MEMS oscillator.

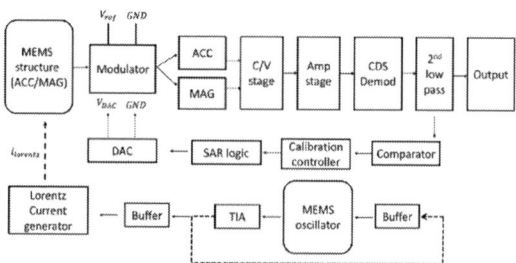

Fig. 6. Function blocks implemented in magnetometer readout system.

below 1000Hz, the magnetic variation is detected by a capacitive interface which is composed of chopper-amplifier and CDS circuit. Chopper amplifier reduces the flicker noise in low frequency band. The architecture of readout circuit is shown in Fig. 6. Some of the front-end interface implement programmable gain amplifier (PGA) to change the sensitivity of the magnetometer. However, it may suffer from the mismatch of the passive element and result in a common mode offset. Also, the noise like kT/C would restrict the performance of circuit. In this paper, the tunable sensitivity is achieved by neither PGA nor any electrical way, and the details will be shown in other section.

F. Calibrator

Due to the process variation, the stress relief will cause curvature. In the result, the mismatch of the capacitance will either decrease the signal swing or DC level. Therefore, the requirement of the calibrator is significant. The successive-approximation-register (SAR) calibrator based on [4] and [5] is illustrated in Fig 7(a) and (b). It is composed of shift D-flip-flops, R-2R DAC, comparator, and logic controller. The calibration process can be divided into two parts, fine and coarse tune parts. By separating two parts, 10-bit resolution is achieved by a 5-bit DAC. In coarse part, the control logic controls UP or DN switch. Next, in fine part, UP1 and DN1 switch are controlled. By connecting the capacitance to V_{DAC} or ground, the DC level of input can be pull high or low and eventually to a desired reference voltage. By setting C_2 32 times smaller than C_1, the other 5-bit accuracy is obtained. With C_1 and C_2 equal to 86 and 2.68fF, the capacitance mismatch caused by process variation under 0.084fF by SAR calibrator. The function of the calibrator is simulated and shown in Fig 7(b). After ten comparison cycles, the initial offset between positive (pink line) and negative (blue line) can be decreased.

G. High Dynamic Range Design

As mention above, for different applications, magnitude of received magnetic field can be from micro-Tesla to Tesla order. The proposed way to achieve high dynamic range is implemented by changing the bias voltage of MEMS oscillator. Due to softened frequency effect, the resonance frequency of the oscillator is varied. The frequency of Lorentz current is then changed. Although the resonance frequency of MEMS magnetometer is fixed, the variation in Lorentz current can make magnetometer to operate in different frequency. From finite element method (FEM) simulation, the resonance frequency of oscillator varies from 18.8 to 17.7 kHz. The tunable range is about 58510 ppm under the bias voltage from 0 to 20 V. When operating on resonance, the resolution of the magnetometer is 12 μT. As bias voltage varied, the sensing range is expanded and achieves 2.94 T, and the wide-range design is obtained.

III. SIMULATION RESULTS

Fig. 8 shows the co-simulation result of MEMS oscillator. The MEMS is replaced by Verilog-A model to model mechanical behavior. The oscillation is start-up within about 2 ms and the amplitude of the signal is restricted to 1.3-Vpp. It allows the loop gain of oscillator is nearly unity and removes high order harmonics as well. The control voltage $V_{control}$ controls gain of the loop and maintain when output voltage equals to reference voltage. The phase noise at 10kHz offset is -92.9(dBc/Hz). Monte Carlos simulation is run for fifty times under ±3 standard deviation. The result shows that the average of oscillation frequency is 18.801kHz.

The readout circuit is then simulated with the oscillator. Fig. 9(a) shows that the frequency response of magnetometer and oscillator. The in-plane and out-plane motion have different resonance frequency. By controlling the natural frequency of oscillator with different bias voltage (blue line),

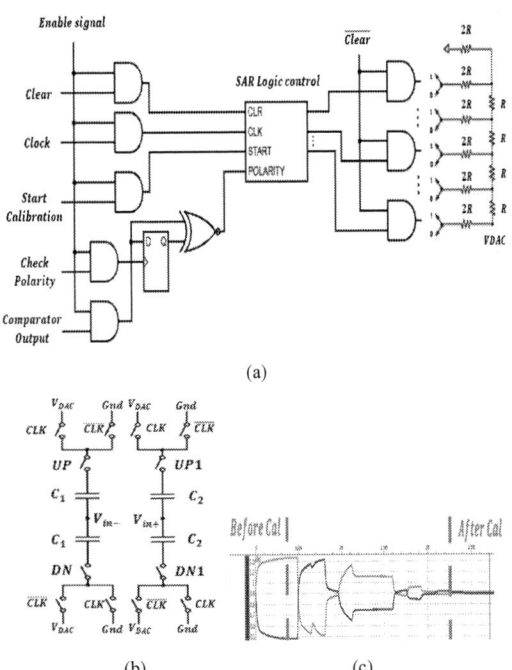

(a)

(b)   (c)

Fig.7. The circuit of (a) SAR-based calibrator and (b) control switch. (c) The simulation result of the calibration circuit.

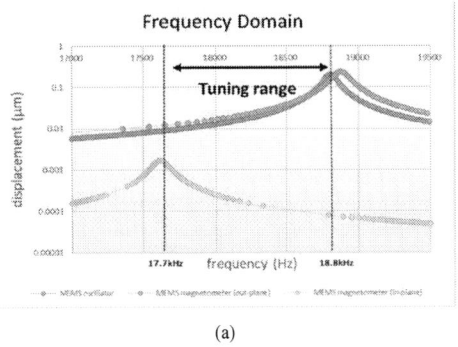

(a)

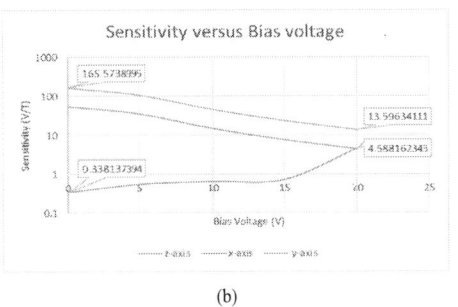

(b)

Fig.8. (a) Frequency response of proposed MEMS. (b) Sensitivity versus bias voltage under an atmospheric presuure.

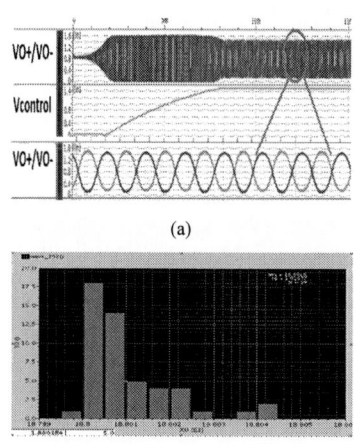

(a)

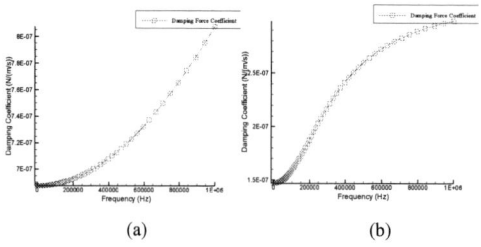

(b)

Fig. 9. (a) Transient response of output voltage. (b) Monte Carlos analysis of the oscillation frequency.

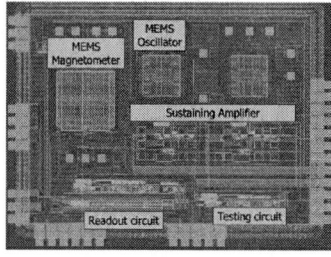

(a)   (b)

Fig. 10. Damping coefficient analysis under (a) 760 Torr and (b) 10Torr ambient pressure.

Fig. 11. Layout of the proposed CMOS MEMS integrated circuit

a tunable scheme is achieved and the oscillation frequency can be varied from 17.7 to 18.8 kHz. It allows the frequency response of MEMS oscillator to shift between the frequency tone of in-plane (gray line) and out-plane (orange line) motion. Therefore, high dynamic range magnetometer is obtained and achieve the sensitivity from 0.34 to 4.59 (V/T) and 13.59 to 165.57 (V/T) under 760 Torr ambient pressure for out-plane and in-plane detection, respectively. The result is shown in Fig. 9(b). Moreover, the range could be even expanded under lower damping environment. From Fig. 10, as damping coefficient decrease from 6.87×10^{-7} to 1.47×10^{-7} (N/(m/s)), Q factor which is inversely proportional to damping coefficient will be enlarged from 183 to 903. As maximum displacement increases, the sensing range can be expanded by about 5 times compared to the magnetometer under 760 Torr air pressure. The layout view of the proposed chip and comparison of MEMS magnetometer are shown in Fig. 11 and Table I.

TABLE I. COMPARISON OF MAGNETOMETER

	[6]	[7]	[8]	This work	
Technology	0.35-μm CMOS	AIN-on-SOI		0.18-μm CMOS	
Axis	X,Y,Z	Out-plane	Out-plane	X,Y	Z
Sensitivity (V/T)	250	0.28	0.0109	247,81 [a]	7.2 [a]
Resolution (nT/√Hz)	4	N/A		64.2,87.5	44.6
Driving current (mA)	1.08	5	N/A	2	
Sensing range(μT)	±2000	±3000	$\pm 1.18 \times 10^5$	$\pm 12 \sim 2.94 \times 10^6$	
Sensitivity range(V/T)	N/A			13.6~166	0.34 ~4.6

a. The value is maximum sensitivity when operating at resonance frequency

IV. CONCLUSIONS

In this paper, an integration of MEMS oscillator and magnetometer implemented using UMC 0.18-μm 1P6M CMOS process is presented. The sensitivity is in a range from 12 μT to 2.94 T, and dynamic range is about 107.8dB. The offset cancellation circuit is implemented by SAR-based logic circuit and mismatch capacitance can be cancelled with the resolution of 0.084fF.

ACKNOWLEDGMENT

This design was sponsored in part by the National Science Council of Taiwan under a grant of MOST 104-3115-E-009-022. The authors appreciate the National Chip Implementation Center (CIC), Taiwan, for supporting the chip manufacturing.

REFERENCES

[1] H.H. Yang, N.V. Myung, J. Yee, D.-Y. Park, B.-Y. Yoo, M. Schwartz, K. Nobe, J.W. Judy, "Ferromagnetic micromechanical magnetometer, " Sensors and Actuators A: Physical, vol. 97-98, pp. 88-97,4/1/2002.

[2] D. Ettelt, P. Rey, G. Jourdan, A. Walther, P. Robert and J. Delamare, "3D Magnetic Field Sensor Concept for Use in Inertial Measurement Units (IMUs)," in Journal of Microelectromechanical Systems, vol. 23, no. 2, pp. 324-333, April 2014.

[3] K. Kunze, G. Bahle, P. Lukowicz and K. Partridge, "Can magnetic field sensors replace gyroscopes in wearable sensing applications?," International Symposium on Wearable Computers (ISWC) 2010, Seoul, 2010, pp. 1-4.

[4] C. J. Leo, B. Zhao and Y. Gao, "A fully integrated capacitance boosting offset calibration circuit for capacitive pressure sensor," *2016 International Symposium on Integrated Circuits (ISIC)*, Singapore, 2016, pp. 1-4.

[5] D.Y. Shin, H.Lee and S. Kim, "A delta-sigma interface circuit for capacitive sensors with an automatically calibrated zero point," IEEE Trans. Circuits Syst. II, vol. 58, no. 2, pp. 90-94, February 2011.

[6] Martijn F. Snoeij, , Viola Schaffer, Sudarshan Udayashankar, and Mikhail V. Ivanov, "Integrated Fluxgate Magnetometer for Use in Isolated Current Sensing", IEEE JOURNAL OF SOLID-STATE CIRCUITS, VOL. 51, NO. 7, JULY 2016.

[7] S. Ghosh and J. E. Y. Lee, "An ultra-sensitive piezoelectric-on-silicon flapping mode MEMS lateral field magnetometer," 2017 Joint Conference of the European Frequency and Time Forum and IEEE International Frequency Control Symposium (EFTF/IFCS), Besancon, 2017, pp. 502-505.

[8] S. Ghosh and J. E. Y. Lee, "A lorentz force magnetometer based on a piezoelectric-on-silicon radial-contour mode disk," 2017 19th International Conference on Solid-State Sensors, Actuators and Microsystems (TRANSDUCERS), Kaohsiung, 2017, pp. 830-83

[9] F. Y. Kuo, C. Y. Lin, P. C. Chuang, C. L. Chien, Y. L. Yeh and S. K. A. Wen, "Monolithic Multi-Sensor Design With Resonator-Based MEMS Structures," in IEEE Journal of the Electron Devices Society, vol. 5, no. 3, pp. 214-218, May 2017.

978-1-5386-5284-8/18 $31.00 © 2018 IEEE

Enhancing Photocatalytic Performance of Nanoporous Nb_2O_5 Doped Platinum

Aidil Aizat Mohd Hamir, Katrul Nadia Basri, Mahzaton
Aqma Abu Talip, Mohamad Hafiz Mamat, Azrif Manut,
Ahmad Sabirin Zoolfakar*
Faculty of Electrical Engineering, Universiti Teknologi
MARA, 40450 Shah Alam, Selangor,
Malaysia
Email : ahmad074@salam.uitm.edu.my

Rozina Abdul Rani, M.Rusop
Nano-SciTech Centre, Institute of Science, Universiti
Teknologi MARA, 40450, Shah Alam, Selangor Darul Ehsan,
Malaysia.

Abstract — **Highly ordered Nb_2O_5 nanoporous network *via* anodization for photocatalytic activity of methyl orange (MO) is presented. Persistent organic pollutants (POPs) are organic combinations that are unaffected to environmental degradation through chemical, biological, and photocatalytic processes. POPs have significant effects on human health and environment. This study will help reducing POPs. Anodization method is used to synthesis Nb_2O_5 nanoporous structure. In this research, anodization duration is varies from 30 to 90 minutes to study the effect of durations on photocatalytic performance. To determine the rate of absorption of MO, UV-VIS Spectrophotometer is used to measure the absorption. The enhancement of photocatalytic property is desired in reducing contamination of POP in water produce by textile industries. The degradation of MO were influenced by the dimension of Nanoporous Nb_2O_5, showing that highly ordered nanoporous network as well as larger size of nanoporous Nb_2O_5 have better performance of photocatalytic activity.**

Keywords-Nb_2O_5, Anodization, POPs, Photocatalytic activity

I. INTRODUCTION

Persistent organic pollutants (POPs) are organic (carbon-based) chemicals that remain in the environment and POPs can be found in certain pesticides and industrial chemicals, and as by-products of manufacturing processes and waste incineration [1]. In 2016, Department of Environment Malaysia has reported 35% of dam water in Malaysia was slightly polluted and 5% was totally polluted and the scenario is now deteriorating [2].

However, in some cases, the high concentrations of contamination could not be decontaminated fully due to the presence of highly complex and abundant mixture of pollution [3]. To overcome these difficulties, several POPs sample need to be studied and pre-treatment is required to minimize dam water being fully contaminated in the future. Scientist have discovered several techniques to eliminate or minimize POP. The most effective method to eliminate POPs is *via* photocatalytic process which has proven with higher efficiencies when incorporated with catalyst. In order to enhance the photocatalytic activity, Copper (Cu) doped Zinc Oxide (ZnO), p-NiO/n-Nb_2O_5[4] nanocomposites, Nb_2O_5/ZnO nanorod [5] composites have been studies which it can create a better and faster as well

as economical solution to eradicate pollution. Metal oxide have numerous applications in the petroleum, semiconductor, optoelectronics and chemical industries [6]. Several nanostructured metal oxides including Nb_2O_5, TiO_2, ZnO, CuO, WO_3, MoO_3, and binary metal oxides such as $BiVO_4$, $BiWO_4$, $SnWO_4$, and $CoWO_4$ have been examined for photocatalytic applications [7]. Scientist has taken initiative to examine and focused toward enhancing the performance of the photocatalyst either *via* doping or hybrid material.

Huiyao Wang *et al.* had studied the photocatalytic properties of rational designed Nb_2O_5 doped with thin film of WO_3 hybrid nanostructures, which were fabricated by the reactive magnetron sputtering method. Compared to pure components of Nb_2O_5, these nanoporous has a higher photocatalytic activity. Niobium film become more amorphous with the present of WO_3. High doping is not recommended where it could detrimental at high content. The advantage of doping in little quantity will improve trapping of electron recombination during irradiation. Additionally, research conducted by Yuanzhi indicate that pure Nb_2O_5 will not produce a high photocatalytic effect other than combination with catalyst. In his study, the experiment combines Nb_2O_5 with C_3N_4 where the content of heterojunction showed highest Photocatalytic efficiency.

Studies conducted by Andre *et. al* on self-cleaning products shows that Nb_2O_5 doped with TiO_2 is frequently used. It is showed that niobium has a possibility to segregate on the top of TiO_2 under oxidation condition. There is some dopant material is being used to delay the anatase-to-rutile transformation (ART) concurrently changing physical chemistry to the surface morphologies of TiO_2 in order to enhance photocatalytic performance. The presence of anatase and rutile crystallites also induces the high level of photocatalytic activity [8].

In this work, the photocatalytic property were investigated by evaluating the degradation of methyl orange (MO) and Nb_2O_5 doped Pt were characterized using field emission scanning electron microscopy (FESEM). The degradation of MO were influenced by the dimension of Nanoporous Nb_2O_5, showing that highly ordered nanoporous network as well as larger size of nanoporous Nb_2O_5 have better performance of photocatalytic activity. The developed nanoporous on surface of Nb_2O_5 and the performance of Nb_2O_5 doped Pt is measured in absorbance of sunlight through phtotocatalytic activity degradation of Methyl Orange. The

---

The authors would like to acknowledge with gratitude to Faculty of Electrical Engineering UiTM for the facilities and funding. This work is fully supported by Universiti Teknologi MARA under the Geran Inisiatif Penyelian (GIP) (600-IRMI/MYRA 5/3/GIP (070/2017))

result will determine the best anodization time with presence of Pt in Nb_2O_5 towards photocatalytic activity.

II. METHODOLOGY

A. Anodization of Niobium

The niobium foil with a thickness of 0.25 mm was cut into smaller pieces with dimensional of 1.0 cm X 2.5 cm (Sigma-Aldrich with 99.9% purity). Niobium foil was cleaned ultrasonically with acetone, ethanol, distilled water and blow with nitrogen gas to avoid any contaminates and grease. Electrolyte solution for anodization was a mixture of 50 ml of ethylene glycol, 0.25g of ammonium fluoride and 4 vol % of DI water. Then the mixture is stirred at 300 rpm and heated at constant 60 °C using hot plate magnetic stirrer until the powder completely dissolved.

The anodization voltage was fixed at 10V and the electrolyte temperature was fixed at 60 °C as well as the duration of the anodization duration was varies for 30, 60 and 90 min. The anodization process was carried out in fume cupboard in order to avoid others exposed or inhale hazardous gases released during the process. After desired duration of anodization, the sample is then dried in ambient environment. Finally, the anodized niobium foils were annealed in furnace at 440 °C for 30 minutes. The furnace is set with slow ramp up and down temperature of 2 °C per minutes to increase the oxide layer. After annealed is done, the next step is to dope Nb_2O_5 with Pt (thickness = 10nm) *via* thermal evaporator.

B. Structural Characterization

The surface morphologies of the nanoporous Nb_2O_5 were characterized *via* Field Emission Scanning Electron Microscope (FESEM, JSM-7600F). Platinum doping on the surface of nanoporous Nb_2O_5 is carried out by using thermal evaporator. The absorbance of the dye solution was recorded using UV-VISS (JASCO V-670 spectrophotometer) with wavelength in the range of 400nm to 800nm.

C. Photocatalytic

In order to determine the photocatalytic performance of Nb_2O_5 doped Pt, Methylene Orange (MO) dye solution is used with concentration of 100mg/L. Solar Simulator (Buko Keiki-EP 2000-xenon lamp) with wavelength from 400nm to 800nm is used to simulate/mimic sunlight irradiation. In order to ensure the absorption-desorption equilibrium of MO solution, the Nb_2O_5 sample was placed in the beaker containing 25ml of MO and placed in the dark room for 24 hrs. Then, the solar simulator is turned on to expose the mixture and the process will take placed for 4 hours, but every half hour, the solution was taken out for UV-Vis to observe the change of its decolorization. The absorbance is investigated by measuring the ability of Nb_2O_5 thin film to decolorize MO dye solution using UV-VISS Spectrophotometer.

III. RESULT AND DISCUSSION

A. Niobium Oxide Film Characterization

Niobium Oxide is generated *via* a processed called anodization. During the anodization process, several parameters affect the behaviour of film thickness and surface morphology such as duration of anodization, electrolyte solution and temperature of electrolyte. Fixed voltage supply is applied where positive terminal was connected to Niobium and negative terminal was applied to platinum mesh electrode. When both electrode is immersed in electrolyte solution and voltage is applied, tiny bubble is started to appear at Niobium foil where oxidation process occurred. While at cathode, hydrogen is formed where the reduction process occurred. Niobium absorb the oxygen thus Nb_2O_5 films formed on the surface of Niobium foil.

As the anodization duration is varied throughout the experiments, thus nanoporous structured produced highly depends on anodization duration which related closely on the electrochemical reaction happen on the surface of niobium. Although other study had found that anodization process need high voltage supply in order to achieve more growth of anodic oxide. However, in this study, 10V is used as constant voltage. The surface morphologies of the Nb_2O_5 are shown in Fig. 1. The variation of pore dimensional on the surface of Nb_2O_5 formed is due to variation of anodization duration.

Average porous hole diameter for Fig. 1(a) is 10-20 nm. It was the smallest diameter compared to other samples due to smallest duration (30 min). Fig. 1(b) shows hihghly ordered nanoporous network of Nb_2O_5. Effect of duration during anodization contributed huge impact on the structure created. Generally, when there is optimum time for nanoporous to form, it will have built with highly organized and well pore size. But with longer time of synthesizing, the nanoporous will start to collapse and formed a wide diameter hole and formed mesh-net structured as shown in Fig. 1(c). Excessive voltage flow through time ruptured a well-structured nanoporous as seen in Fig. 1(c). The cross-sectional image is shown in Fig. 1(d). Layer of oxide shows is at 4µm thickness to prove that anodization is successfully occur on surface of niobium.

B. Photocatalytic Performance

Fig. 2 illustrated photocatalytic activity to observe the variations in absorption of simulated sunlight source by Nb_2O_5 doped Pt. Fig. 2(a - c) show absorption rate for 30 min, 60 min and 90 min of anodization time in 4 hours under irradiation of simulated sunlight. Fig. 2(b) clearly shows that the highest efficiency of absorption of sunlight followed by 90 minutes sample and 30 minutes sample. Anodization time is the key to create a well-generated nanoporous followed by high surface area with larger pore size formed. In this study, Pt act as catalyst in the photocatalytic reaction to enhance the absorption of sunlight into the nanoporous of niobium. Fig. 1(a) clearly show the size of nanoporous of Nb_2O_5 is smaller than 60 min sample. It is due to insufficient time for the voltage to construct highly ordered nanoporous wall structure. The opening of the hole is determining by a steady amount of voltage and time.

By increasing the duration of anodization to more than 60 minutes will result in collapsing of nanoporous wall structure. This is due to expose to the voltage and oxidation occur not in optimum pace. Sample for 90 minutes in Fig. 1(c) shows a slight decreasing in absorption of light. The tiny

drop is due to unable to absorb light due to collapse structure. The presence of Pt helps the photocatalytic activity although the structure is not well build. This indicate the presence of metal catalyst boost the photocatalytic activity.

Fig 2(d) compare the highest peak of absorption at every 30 minutes of sun irradiation for 30 min, 60 min and 90 min sample. It can be clearly observed 60 min' sample has a highest and the best performance of photocatalytic activity due to highest absorption of light in every 30 minutes exposed to sunlight radiation. Addition Pt catalyst to boost the absorption of light is accepted and can be used widely in reducing POP's in the river. A lower performance of photocatalytic occur for 90 minutes sample. A slower and little absorption of MO by the Nb_2O_5 is because deformation of nanoporous and collapse of the wall of the structure prevent a high absorption of MO into Nb_2O_5.

$$Nb_2O_5 + hv\ (UV) \longrightarrow Nb_2O_5\ (e_{CB}^- + h_{VB}^+) \qquad [1]$$

$$Nb_2O_5\ (h_{VB}^+) + H_2O \longrightarrow Nb_2O_5 + H^+ + OH \qquad [2]$$

$$Nb_2O_5\ (h_{VB}^+) + OH^- \longrightarrow Nb_2O_5 + OH \qquad [3]$$

$$Nb_2O_5\ (e_{CB}^-) + O_2 \longrightarrow Nb_2O_5 + O_2^- \qquad [4]$$

$$O_2^- + H^+ \longrightarrow HO_2 \qquad [5]$$

$$Dye + OH \longrightarrow degradation\ products \qquad [6]$$

$$Dye + h_{VB}^+ \longrightarrow oxidation\ products \qquad [7]$$

$$Dye + e_{CB}^- \longrightarrow reduction\ products \qquad [8]$$

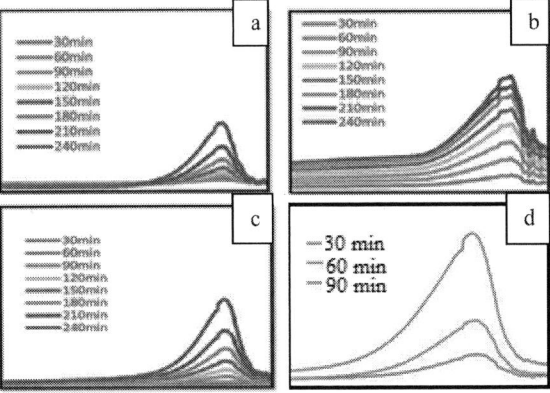

Fig. 2: Absorption of MO for variation anodization durations a) 30 min, b) 60 min, c) 90 min and d) comparison of 3 anodization time of Nb_2O_5 doped Pt. y-axes represent arb units and x-axis represent wavelength (nm)

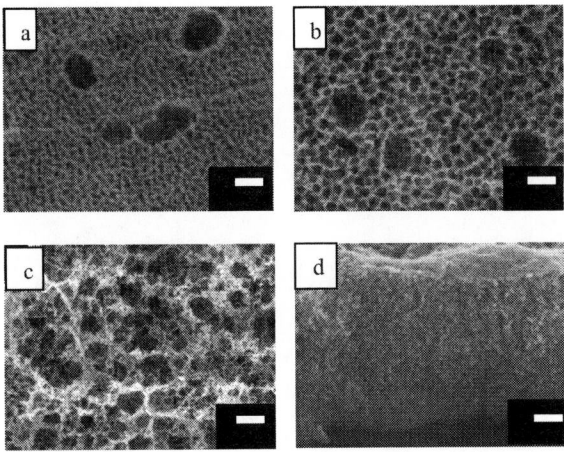

Fig. 1: Top view of image of nanoporous annealed Nb_2O_5 film prepared in different anodization time: (a) 30 minute, (b) 60 minute, (c) 90 minute and (d) cross sectional view of 60 minutes. All scale bars represent 100nm.

C. Photocatalytic activity

Fig. 3 illustrate the mechanism of photocatalytic activity irradiated by sunlight. The absorption of photon with sufficient equivalent energy or higher than bandgap energy is involved in this reaction. When photon has high absorption energy electron from valence band will excited to conduction band while it leaves a positive hole in valence band. In order to ensure photocatalytic action is happening, recombination of electron and hole must be prevented. The photo catalytic action occurred when activated electron with an oxidant agent produced a reduced product and the generated hole with the reductant agent produced an oxidized product. The reaction is expressed as shown in these equations:

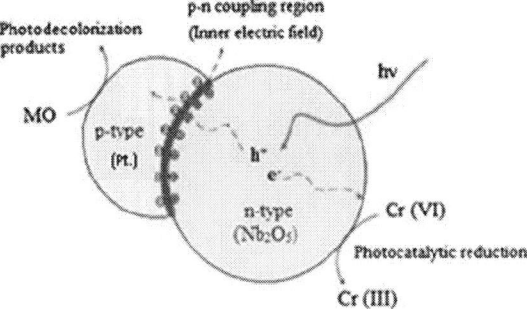

Fig. 3: The Schematic mechanism of charge transfer of photogenerated electrons and holes of Nb_2O_5 doped Pt Photocatalytic activity.

Based on the oxidation and reduction process, the Nb_2O_5 (h_{VB}^+) is oxidize the organic compounds by reacting with water to generate OH Eq. 2. The photo generated positive holes will oxidise the organic molecule to react with H_2O and produced OH radicals Eq. 3. The electron acceptors will react with the photogenerated electrons as Eq. 4, the O_2 absorbed by Nb_2O_5 surfaced and reduced it to superoxide radical anion O_2^-. Thus, surface morphology of metal oxide semiconductor will act as the main factor of the performance of photocatalytic activity. Fig. 3 illustrate the schematic mechanism of Nb_2O_5 doped Pt photocatalytic activity.

CONCLUSION

The effects of anodized duration toward photocatalytic performance of Nb_2O_5 doped Pt have been successfully investigated. The best result obtained was 60 minutes anodize time where the film has a highly ordered organized nanoporous network structure and contributed to the highest of photocatalytic activity. The enhancement of photocatalytic activity is due to high surface areas well as highly ordered nanoporous network of Nb_2O_5 exposed to the MO.

REFERENCES

[1] WHO, "Persistent Organic Pollutants: Impact on Child Health," *World Heal. Organ,* pp. 1–67, 2010.

[2] O. Indicators and E. E. Indicators, "Asean Iwrm Performance Reports & Monitoring Indicators Malaysia 2013 Report (Water Pollution Management) Outcome Indicators," no. 5, pp. 2013–2016, 2015.

[3] C. E. Manyi-Loh, S. N. Mamphweli, E. L. Meyer, A. I. Okoh, G. Makaka, and M. Simon, "Microbial anaerobic digestion (bio-digesters) as an approach to the decontamination of animal wastes in pollution control and the generation of renewable energy," *Int. J. Environ. Res. Public Health*, vol. 10, no. 9, pp. 4390–4417, 2013.

[4] F. Hashemzadeh, A. Gaffarinejad, and R. Rahimi, "Porous p-NiO/n-Nb2O5 nanocomposites prepared by an EISA route with enhanced photocatalytic activity in simultaneous Cr(VI) reduction and methyl orange decolorization under visible light irradiation," *J. Hazard. Mater,* vol. 286, pp. 64–74, 2015.

[5] S. M. Lam, J. C. Sin, I. Satoshi, A. Z. Abdullah, and A. R. Mohamed, "Enhanced sunlight photocatalytic performance over Nb2O 5/ZnO nanorod composites and the mechanism study," *Appl. Catal. A Gen.*, vol. 471, pp. 126–135, 2014.

[6] Y. Chen and I. E. Wachs, "Tantalum oxide-supported metal oxide (Re2O7, CrO 3, MoO3, WO3, V2O5, and Nb2O5) catalysts: Synthesis, Raman characterization and chemically probed by methanol oxidation," *J. Catal.*, vol. 217, no. 2, pp. 468–477, 2003.

[7] S. Thangavel, K. Krishnamoorthy, V. Krishnaswamy, N. Raju, S. J. Kim, and G. Venugopal, "Graphdiyne-ZnO Nanohybrids as an Advanced Photocatalytic Material," *J. Phys. Chem. C*, vol. 119, no. 38, pp. 22057–22065, 2015.

[8] F. F. de Brites-Nóbrega, A. N. B. Polo, A. M. Benedetti, Mô. M. D. Leão, V. Slusarski-Santana, and N. R. C. Fernandes-Machado, "Evaluation of photocatalytic activities of supported catalysts on NaX zeolite or activated charcoal," *J. Hazard. Mater.* vol. 263, pp. 61–66, 2013.

[9] C. Cheng, A. Amini, C. Zhu, Z. Xu, H. Song, and N. Wang, "Enhanced photocatalytic performance of TiO2-ZnO hybrid nanostructures.," *Sci. Rep.*, vol. 4, p. 4181, 2014.

[10] A. Pal, T. K. Jana, and K. Chatterjee, "Silica photocatalytic application under visible light irradiation," *Mater, Res. Bull.*, vol. 76, pp. 353–357, 2016.

[11] R. Abdul Rani, A. S. Zoolfakar, J. Subbiah, J. Z. Ou, and K. Kalantar-Zadeh, "Highly ordered anodized Nb2O5 nanochannels for dye-sensitized solar cells," *Electrochem. commun.*, vol. 40, pp. 20–23, 2014.

A better understanding of CNTs chemical purification and functionalization processes

Saman Azhari
Institute of Advanced Technology (ITMA)
University Putra Malaysia (UPM)
Kuala Lumpur, Malaysia
Samanar@gmail.com

Mohd Nizar Hamidon
Institute of Advanced Technology (ITMA)
University Putra Malaysia (UPM)
Kuala Lumpur, Malaysia
mnh@upm.edu.my

Intan Halina Hasan
Institute of Advanced Technology (ITMA)
University Putra Malaysia (UPM)
Kuala Lumpur, Malaysia
i_helina@upm.edu.my

Muhammad Asnawi
Institute of Advanced Technology (ITMA)
University Putra Malaysia (UPM)
Kuala Lumpur, Malaysia
muhdasnawi92@gmail.com

Siti Amaniah Mohd Chachuli
Institute of Advanced Technology (ITMA)
University Putra Malaysia (UPM)
Kuala Lumpur, Malaysia
sitiamaniah@utem.edu.my

Alin Azhari
Institute of Advanced Technology (ITMA)
University Putra Malaysia (UPM)
Kuala Lumpur, Malaysia
Alinazhari94@gmail.com

Ismayadi Ismail
Institute of Advanced Technology (ITMA)
University Putra Malaysia (UPM)
Kuala Lumpur, Malaysia
ismayadi@upm.edu.my

Abstract— Purification and functionalization of carbon nanotubes have been examined using different acid treatments. CNTs treatment may vary depending on the desired outcome. The results show that HCl as a strong acid, oxidizes all the materials present in the sample. HNO_3 acts as an oxidizing agent which is useful to produce functional CNTs. H_2O_2 behaves as a regulator in presence of both acids. The results suggest that H_2O_2 may be a critical element to control the outcome of CNTs purification and functionalization process.

Keywords—Acid Treatment, CNTs, Purification, Functionalization

I. INTRODUCTION

In the past couple of decades; since discovery of carbon nanotubes (CNTs) by Ijimma [1], CNTs have attracted attention from research communities due to their extraordinary physical and chemical characteristics. CNTs astonishing behavior in correlation with their morphology, for instance the sp2 hybridization of carbon atoms besides CNTs chiral angle and diameter; makes them a peculiar member of carbon-based material. As a result, CNTs have been utilized for enhancement of polymeric, ceramic and other composite materials [2], [3]; in addition, characteristics such as ballistic conductance in single walled carbon nanotubes (SWCNT) have been theorized.

First reported CNTs were produced via the arc discharge process [1]. The nanotubes produced via this technique were high in purity and had few numbers of walls; however, this method was costly and had low production yield. On the other hand, Chemical vapor deposition (CVD) was found to be worthy of the cost due to its high production yield; although the low purity of the produced CNTs and complexity in controlling the morphology of the product could be listed as disadvantages of this method. Nonetheless CNTs produced via CVD process are easily obtainable by researchers due to their low cost.

Transition metals such as Nickel (Ni), Iron (Fe) and Cobalt (Co) are used as seed material to synthesize CNTs in CVD. These metal particles are frequently observed to remain either at the base, tip or in the middle of the nanotubes. This is due to the mechanism of CNTs formation. The presence of catalyst impurities in CNTs would drastically affect nanotubes properties such as electrical and thermal conduction. These effects reduce the likability of CNTs to be utilized for engineering applications such as electronic devices [4].

There have been extensive studies on removal of these metal impurities from CNTs via physical and chemical methods. The physical methods mainly revolve around high temperature treatment of CNTs to remove any impurities; on the other hand, chemical methods revolve around acid treatment and chemical etching of metal impurities. These treatments each affect CNTs greatly and change their properties drastically. In one hand purification of CNTs seems necessary because impurities affect various properties of CNTs and on the other hand acid treatment of CNTs would change the morphology and structure of nanotubes while oxidizing the impurities [5].

There are numerous chemical etching methods for metal impurities. Usually these methods are selected depending on chemical composition of the metal catalyst and preferred outcome. Many studies have compared the effects of various acids and their concentration, duration of the treatment as well as temperature to determine the optimum condition for purification of CNTs. Although there have been extensive studies on removal of metal impurities from CNTs; the effect of each acid treatment on CNTs have not been fully understood. Herein the effects of different etchants on the purity and functionality of CNTs are examined.

978-1-5386-5284-8/18 $31.00 © 2018 IEEE

II. MATERIALS AND METHODS

A. Sample Preparation

The material used for this work were as follow: Commercial multiwall carbon nanotubes (MWCNTs) produced by catalytic decomposition (L.MWNTs-2040, length=5-15 μm, diameter=20-40 nm, Shenzhen Nanotechnologies Port Co. Ltd., Shenzhen, China); 65% nitric acid, 37% hydrochloric acid and 30% hydrogen peroxide (R&M Chemicals). For this work 4 samples were prepared in addition to the purchased CNTs. For each sample 20mg of CNTs were magnetically stirred in a reflux for 6 hours; table 1 presents the treatment performed on each sample. To collect MWCNTs, final mixture for each sample was centrifuged followed by washing twice with distilled water and deionized water respectively until pH 7 was obtained. MWCNTs were dried in the oven for 12 hours at 80°C. These method, materials and temperature were preferred as they were reported to be highly effective based on previous studies and reviews.

Samples	Material	Temperature (°C)	References
A	HCl	110	[6]
B	HNO_3	90	[7]
C	HCl/H_2O_2	125	[8]
D	HNO_3/H_2O_2	120	[9]

Table. 1. Treatments performed

B. Characterization

Raman spectroscopy (Witec Alpha 300R) with laser excitation wavelength of 532 nm to validate the degree of graphitization, Energy Dispersive Spectroscopy (EDS) (FEI Nova Nanosem EDS) for elemental studies, Thermogravimetric Analysis (TGA) (Mettler Toledo TGA/DSC) to determine the degree of oxidation were utilized. Last but not least Fourier transform infrared spectroscopy (FTIR) (Thermo Nicolet AEM FTIR) was employed to determine the functionality of pristine and treated CNTs.

III. RESULTS AND DISCUSSION

A. FTIR

Qualitative investigation of the functional groups attached to surface of CNTs could contribute to better understanding the chemical reaction which occurs during each acid treatment. The result confirms the presence of carboxyl and hydroxyl groups on the surface of as-purchased sample due to the purification treatment done by the supplier prior to shipping.

This allows us to witness the etching effects of HCl in sample A and C, as the O-H stretch, C-H stretch and C=O stretch observed at 3200-3600 cm⁻¹ (phenol, alcohol), 2850-2925 cm⁻¹ (aliphatic) and 1680-1740 cm⁻¹ (aldehyde, lactone, carbonyl) respectively, weaken and disappear [10]–[13]. These peaks get noticeably stronger in samples B and D.

Similarly samples B and D display various weak peaks which indicate the presence of CO_2 (2348 cm⁻¹) and anhydride (1030-1270 cm⁻¹) functional group [14] as a result

of HNO_3 treatment. These peaks are either very weak or completely omitted in samples treated with HCl.

The peaks observed at 1631, 1490, 1384 cm⁻¹ correspond to the C=C vibrations of aromatic carbon structure [15]. These peaks as well seem to disappear mildly in samples A and strongly in C. Figure 1 displays the FTIR spectrum of all samples. These results confirm the effectiveness of HNO_3 for attachment of carbonyl, carboxyl and hydroxyl functional groups to the surface of CNTs. We may also conclude that the presence of H_2O_2 improves the effectiveness of each acid for the purpose they may be utilized.

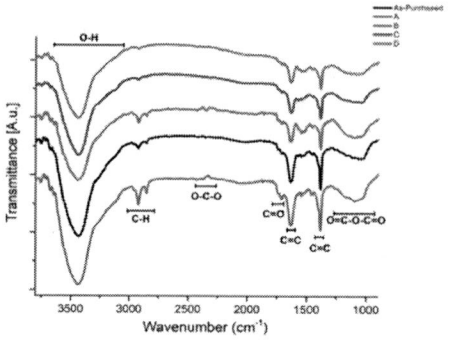

Figure. 1. FTIR of treated samples

B. Raman

The Raman result of as-purchased CNTs indicate the three prominent peaks at 1338 cm⁻¹, 1566 cm⁻¹ and 2673 which represent D band (Defect), G band (Graphitized) and 2D band (second order scattering of D band) respectively [16]. The ratio of (I_D/I_G) which is calculated to determine the quality of CNTs, for as-purchased sample is 0.9441. This indicates the presence of slightly more graphitized carbon structures in comparison with amorphous carbons. The ratio of (I_{2D}/I_G) for as-purchased CNTs is 0.4546 which confirms the presence of MWCNTs.

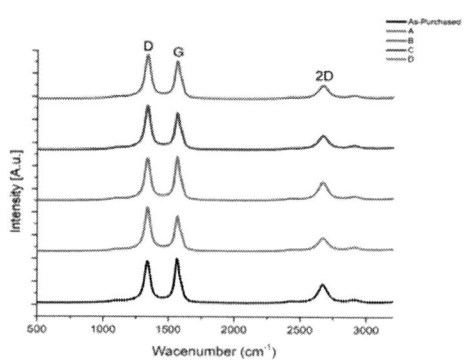

Figure 2 Raman spectra of treated samples

Raman results obtained from sample A, C and D display a strong peak at 1338 cm⁻¹ (D band) and relatively a weaker peak at 1566 cm⁻¹ (G band); on the other hand, sample B displays strong peak at 1566 cm⁻¹ (G band) and relatively weaker peak at 1338 cm⁻¹ (D band). All samples display a 2D peak at 2673 although the intensity of this peak varies as a result of different treatment. These result are shown in Figure 2.

978-1-5386-5284-8/18 $31.00 © 2018 IEEE

The results from sample A and C clearly display the etching strength of HCl in comparison with HNO₃. CNTs tend to oxidize much faster in presence of HCl which induces the disordered graphitic materials (stronger D band) and deteriorated the CNTs [17]. Although the H_2O_2 in presence of acids tend to behave as a strong oxidizer, the mixture of H_2O_2 and HCl (sample C) display a lower ratio of I_D/I_G in comparison with HCl (less disorder); on the other hand, the presence of H_2O_2 in HNO₃ (sample D) increases the I_D/I_G ratio in comparison with HNO₃.

The 2D band does not correlate with defects of the CNTs [18]. The ratio of I_{2D}/I_G corresponds to the variation of layers in graphitic materials and at the same time could indicate the band structural defects as a result of sp^3 formation. In short the decrease in I_{2D}/I_G ratio indicates structural damage to the honeycomb lattice. From the result we may conclude that the highest oxidation and functionalization takes place in sample D, on the other hand HNO₃ treatment may enhance the structure of CNTs. Table 2 displays the percentage changes of I_D/I_G and I_{2D}/I_G in relation to as-purchased sample.

Sample	I_D/I_G	I_{2D}/I_G	I_D/I_G ($\Delta\%$)	I_{2D}/I_G ($\Delta\%$)
As-Purchased	0.94412	0.4546	0	0
A	1.230966	0.43932	30.3824	-3.3613
B	0.97013	0.46501	2.75495	2.28993
C	1.183334	0.443301	25.3372	-2.4856
D	1.14137	0.409969	20.8925	-9.8177

Table. 2. Ratio of graphitized, defective and structural alteration

C. EDS

The EDS analysis of the CNTs provides us with useful information regarding composition of the materials present in the CNTs sample. EDS analysis indicate the presence of Nickel (Ni) nanoparticles which are used as catalytic material to synthesize CNTs. The EDS results obtained from sample A to D indicate the presence of the same elemental elements (C, O and Ni) in all samples. Since EDS provides data regarding a small area in the sample it cannot be assumed that the result would be consistent all over the samples. Figure 3 displays the EDS data regarding the presence of each element (Carbon, Oxygen and Nickel) in each sample are presented.

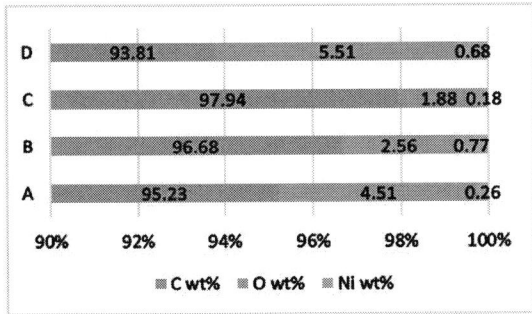

Fig. 3. EDS result of treated samples

As it could be seen from the provided charts the Nickel content in sample A and C are relatively lower than B and D. Oxygen content is higher in samples A and D while the carbon content is the highest in sample B and C. Figure 3 displays the elemental analysis of treated samples.

D. TGA

To confirm the working principle of each oxidative treatment the TG curve for each sample was recorded in oxygen rich atmosphere. Thermal oxidation of each sample is displayed from 30 – 610 ºC and not further due to buoyancy error which interfere with final residual value.

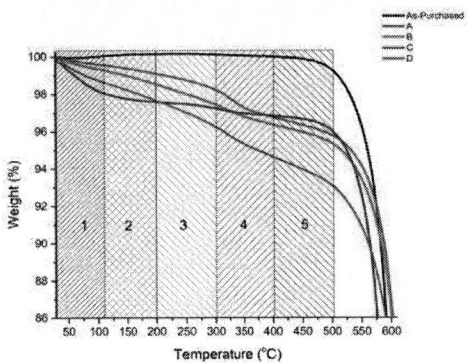

Fig. 4. TG curves of treated samples

The five shaded segments in the Figure 4 indicate oxidation and conversion range for different oxygen containing functional groups, it also shows that CNTs immediate oxidation begins at 510 ºC and ends at 680 ºC (not shown in Fig. 4.). Segment 1 corresponds to evaporation of moisture, while conversion of hydroxyl to carboxyl group because of oxidation reaction, takes place in segment 2 [11], [19]. Sample C displays 2% weight loss which indicates evaporation of water vapor. In contrast sample A had the minimum weight loss compared to other samples in segment 1. Segment 3 is decomposition region for aldehyde and carboxyl groups [20]. It is clear that sample C does not experience much weight loss unlike sample A, B and D. Decomposition of lactone and anhydride groups takes place in segment 4 while new ether group is formed due to oxidation of graphitized carbon structure [11]. Sample A displays greater weight loss in segment 4 in comparison to segment 3 which indicates higher percentage of C=O bond. With that in mind, sample C displays the smallest weight loss in this segment. Segment 5 corresponds to decomposition of remaining phenol and carbonyl groups. Sample D displays the highest weight loss in this segment unlike sample C which displayed the least weight loss. Table 3 displays the weight loss percentage of each sample in each segment.

Segment	Samples			
	A	B	C	D
1	0.45%	0.70%	2%	1.40%
2	0.55%	0.80%	0.40%	1.00%
3	0.80%	1.10%	0.20%	1.40%
4	1.50%	1.10%	0.40%	1.60%
5	0.90%	0.90%	0.80%	1.50%

Table. 3. Weight loss of treated sample in oxygen rich, thermal oxidative environment

IV. CONCLUSION

In this work we have demonstrated the effects of each treatment on the CNTs. This results confirm that HCl protonates the catalyst particles in the sample and reduces the hydrocarbons which as result increases the oxygen content. HNO_3 as a weaker acid does not deteriorate CNTs or catalyst particles as brutally as HCl, while oxidizing the CNTs, from the result obtained it may be concluded that by donating a proton, HNO_3 reduces the oxygen containing functional groups and alters them to produce carboxylic acids and aldehydes. In sample C the reaction is not like sample A, The presence of H_2O_2 appears to assist with the removal of catalyst particles. The Cl⁻ and OH⁻ ions are produced as a result of HCl and H_2O_2 reaction which further reduces the hydrocarbons. Hence, higher moisture content and milder deterioration of graphitized carbon is witnessed in compared to its counterpart. In sample D the carbon content reduces while oxygen content increases and the amount of catalyst particles appear to be close to that of sample B. Unlike sample C the presence of H_2O_2 seems to increase the oxygen containing functional groups which result in increase of oxygen content. This could be explained by considering the mixture of HNO3 and H_2O_2 which results in production of HNO4 that assists in further oxidation of the sample and deterioration of graphitized structure.

Based on the presented result it is clear that HCl should be utilized when purification of CNTs is desired, although oxidation of graphitized carbon structure is inevitable. Further addition of H_2O_2 increases the moisture content in the sample and assists with the purification process while regulating structural damage. Although there are number of works that suggest HNO_3 is useful for purification of CNTs [21], this result suggest that may not be the case in all types of CNTs. It is though that HNO_3 improves the structure of graphitized CNTs considering oxidation that takes place in the process [22]. Various oxygen containing functional groups are introduced as a result of HNO_3 oxidation. The addition of H_2O_2 though enhances the functionalization process, it damages the graphitized structure despite the presence of HNO_3.

REFERENCES

[1] S. Iijima, "Helical microtubules of graphitic carbon," *Nature*, vol. 354. pp. 56–58, 1991.

[2] N. Gopal, S. Rana, J. Whan, L. Li, and S. Hwa, "Polymer nanocomposites based on functionalized carbon nanotubes," *Prog. Polym. Sci.*, vol. 35, pp. 837–867, 2010.

[3] A. TermehYousefi, S. Azhari, A. Khajeh, M. N. M. N. Hamidon, and H. Tanaka, "Development of haptic based piezoresistive artificial fingertip: Toward efficient tactile sensing systems for humanoids," *Mater. Sci. Eng. C*, vol. 77, pp. 1098–1103, 2017.

[4] R. Brukh and S. Mitra, "Mechanism of carbon nanotube growth by CVD," *Chem. Phys. Lett.*, vol. 424, no. 1–3, pp. 126–132, 2006.

[5] P. Chaturvedi, P. Verma, A. Singh, P. K. Chaudhary, Harsh, and P. K. Basu, "Carbon nanotube - Purification and sorting protocols," *Def. Sci. J.*, vol. 58, no. 5, pp. 591–599, 2008.

[6] A. R. Harutyunyan, B. K. Pradhan, J. Chang, G. Chen, and P. C. Eklund, "Purification of single-wall carbon nanotubes by selective microwave heating of catalyst particles," *J. Phys. Chem. B*, vol. 106, no. 34, pp. 8671–8675, 2002.

[7] P. Paunović *et al.*, "Effect of activation/purification of multiwalled carbon nanotubes (MWCNTs) on the activity of non-platinum based hypo-hyper d-electrocatalysts for hydrogen evolution," *Mater. Res. Bull.*, vol. 44, no. 9, pp. 1816–1821, 2009.

[8] Y. Wang, H. Shan, R. H. Hauge, M. Pasquali, and R. E. Smalley, "A highly selective, one-pot purification method for single-walled carbon nanotubes," *J. Phys. Chem. B*, vol. 111, pp. 1249–1252, Feb. 2007.

[9] F. Avilés, J. V Cauich-Rodríguez, L. Moo-Tah, A. May-Pat, and R. Vargas-Coronado, "Evaluation of mild acid oxidation treatments for MWCNT functionalization," *Carbon N. Y.*, vol. 47, no. 13, pp. 2970–2975, 2009.

[10] S. S. S. M. Zebarjad and J. Khaki, "ORIGINAL RESEARCH A study on the dependence of structure of multi-walled carbon nanotubes on acid treatment," *J. Nanostructure Chem.*, vol. 5, no. 3, pp. 287–293, 2015.

[11] L. Li, X. Yao, H. Li, Z. Liu, and W. Ma, "Thermal Stability of Oxygen-Containing Functional Groups on Activated Carbon Surfaces in a Thermal Oxidative Environment Thermal Stability of Oxygen-Containing Functional Groups on Activated Carbon Surfaces in a Thermal Oxidative Environment," no. July, 2014.

[12] A. Zadehnazari, "An investigation on the effects of functionalized multi-walled carbon nanotube on mechanical and thermal properties of dopamine-bearing poly (amide – imide) composite films," no. September 2017, 2014.

[13] J. Coates, "Interpretation of Infrared Spectra , A Practical Approach," pp. 1–23.

[14] J. L. Figueiredo, M. F. R. Pereira, M. M. A. Freitas, and J. J. M. Orfao, "Modification of the surface chemistry of activated carbons," vol. 37, pp. 1379–1389, 1999.

[15] S. A. Miners, G. A. Rance, L. Torre, S. M. Kenny, and N. Khlobystov, "Controlled Oxidative Cutting of Carbon Nanotubes Catalysed by Silver Nanoparticles," no. c, pp. 1–11, 2014.

[16] M. G. S. Bernd, S. R. Braganc, N. Heck, C. P. Luiz, and S. Filho, "Synthesis of carbon nanostructures by the pyrolysis of wood sawdust in a tubular reactor," vol. 6, no. 2, pp. 171–177, 2016.

[17] M. S. Dresselhaus, A. Jorio, A. G. S. Filho, R. Saito, and P. T. R. S. A, "Defect characterization in graphene and carbon nanotubes using Raman spectroscopy Defect characterization in graphene and carbon nanotubes using Raman spectroscopy," pp. 5355–5377, 2010.

[18] S. Azhari *et al.*, "Sensors and Actuators A : Physical Fabrication of piezoresistive based pressure sensor via purified and functionalized CNTs / PDMS nanocomposite : Toward development of haptic sensors," *Sensors Actuators A. Phys.*, vol. 266, pp. 158–165, 2017.

[19] S. A. Chernyak, A. S. Ivanov, N. E. Strokova, K. I. Maslakov, S. V Savilov, and V. V Lunin, "Mechanism of Thermal Defunctionalization of Oxidized Carbon Nanotubes," 2016.

[20] M. S. Shafeeyan, W. Mohd, A. Wan, A. Houshmand, and A. Shamiri, "A review on surface modification of activated carbon for carbon dioxide adsorption Journal of Analytical and Applied Pyrolysis A review on surface modification of activated carbon for carbon dioxide adsorption," *J. Anal. Appl. Pyrolysis*, vol. 89, no. 2, pp. 143–151, 2017.

[21] H. Hu, B. Zhao, M. E. Itkis, and R. C. Haddon, "Nitric Acid Purification of Single-Walled Carbon Nanotubes," *Society*, vol. 107, no. 50, pp. 13838–13842, 2003.

[22] V. Datsyuk *et al.*, "Chemical oxidation of multiwalled carbon nanotubes," *Carbon N. Y.*, vol. 46, pp. 833–840, 2008.

2018 IEEE International Conference on Semiconductor Electronics (ICSE)

Modeling and Simulation of VO₂/Au Thin Film Transition Behavior.

Asmaa Leila Hassein-Bey
FUNDAPL, Physics Department, Faculty of Sciences, University Saad Dahlab Blida 1(USDB 1), P.O. Box 270, 09000
Blida, Algeria
asmaahasseinbey@univ-blida.dz

Abdelkader Hassein-Bey
Micro & Nano Physics Group, Physics Department, Faculty of Sciences, University Saad Dahlab Blida 1 (USDB 1), P.O. Box 270, 09000 Blida, Algeria
Blida, Algeria
a_hassein@univ-blida.dz

Hakim Tahi
Microelectronics and Nanotechnology Division (CDTA) Cité 20 Août 1956, Baba Hassen, BP: 17, DZ-16303
Algiers, Algeria
htahi@cdta.dz

Slimane Lafane
Ionized Medium and Laser Division (CDTA)Cité 20 Août 1956, Baba Hassen, BP: 17, DZ-16303
Algiers, Algeria

Samira Abdelli-Messaci
Ionized Medium and Laser Division (CDTA)Cité 20 Août 1956, Baba Hassen, BP: 17, DZ-16303
Algiers, Algeria

Mohamed El-Amine Benamar
FUNDAPL, Physics department, Faculty of Sciences, University Saad Dahlab Blida 1(USDB 1), P.O. Box 270, 09000
Blida, Algeria

Abstract—**Vanadium dioxide (VO₂) exhibits a metal-insulator transition (MIT) near 68°C with a unique sharp resistivity change. Below this temperature, it behaves as a semiconductor with a high electrical resistivity, above it the material behaves as a metal with a low electrical resistivity. In this paper, modeling and FEM simulation using Comsol Multiphysics of metal–insulator transition behavior in VO₂/Au thin film have been carried out in order to evaluate the electrical resistance behavior as a function of applied voltage and temperature sweeps. The simulation results present a similar transition behavior as in experiments owning a good sensitivity even at room temperature under a specific voltage. In addition, these simulations could be of a great interest to design potential sensors based on this transition metal oxide materials thin layer. The simulation results could be used as a powerful design tool for several MIT-based sensors and predict the feasibility of a variety of device with a high sensitivity at ambient temperature. Our simulations predict a transition at room-temperature under polarization of about 4.25V. Practically, in term of operating temperature range many applications are targeted going from data storage device to biosensors.**

Keywords—vanadium dioxide, metal-insulator transition, electrical resistance, FEM simulation, modeling

I. INTRODUCTION

Vanadium dioxide is one of the most studied metal-insulator transition (MIT) close to room temperature near 68°C with a unique sharp resistivity change [1,2] reaching ratios in the order of 10^5 over 0.1 °C [3]. The physical mechanisms of the MIT switching in VO₂ are still under debate, with two possible explanations being considered for this transition: (i) a structural phase transition (from monoclinic (M1) to tetragonal (rutile, R)) and (ii) electron–electron correlations (Mott transition) [4-7]. Indeed, the high sensitivity to temperature makes VO₂ interesting

[8,9]. The most attracting aspect in VO₂ is that the transition temperature of the first-order MIT [10,11] could be modified by doping [12,13], mechanical stress [14] and electric field [15]. Thus transition could be modulated by external parameters making VO₂ extremely interesting for a broad range of applications such as: energy efficient and high-speed switches [16], ultra-steep transition slope transistors [17], gas sensors [18] and thermal biosensors [19].

In this work, modeling and FEM simulation of VO₂/Au thin film switching behavior of thermally-induced metal–insulator transition (MIT) have been carried out based on experimental results in order to obtain a more accurate model close to actual behavior as a function of external applied voltage sweep.

II. EXPERIMENTAL BEHAVIOR OF ELECTRICAL RESISTANCE IN VANADIUM DIOXIDE THIN FILM

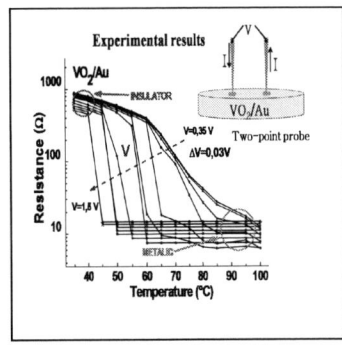

Fig. 1. Experimental behavior of electrical resistance in vanadium dioxide thin film under various voltage.

978-1-5386-5284-8/18 $31.00 © 2018 IEEE

29

For the studied structure, geometry of two superposed thin layers (VO₂/Au) is used. The simulation parameters are directly extracted from the experimental characterization data done for a 500 nm VO_2 thin film deposited by pulsed laser deposition (PLD) technique on a 200 nm thick gold buffer layer (VO₂/Au) [20].

Experimental results were obtained from a series of Agilent HP 4156C characterizations current-voltage (I-V characteristic) for different temperature stage Fig. 1 detailed in [20]. The electrical connections are performed by the deposition of a micro-probe (two tips with a radius of 20μm) on the top surface of VO₂ thin film. For each temperature rise from 35°C to 100°C with a step of 5°C, the resistance value R was deduced from the I-V characteristics. A sweep voltage range was applied. This process was performed automatically with an application developed under (Labview).The figure 1 shows the change in electrical resistance R as a function of the temperature T of VO₂ thin layer for different polarization voltage value V in the range 0.35-1.5V.

III. MODELING OF THIN FILM STRUCTURE

Fig. 2 shows the proposed structure used for simulation which consists of two thin layers. The top layer is set to be VO₂ with a lateral dimensions sufficient to avoid edge effects. The bottom layer is set to have crystalline gold (Au) (111) properties. The electric current is injected and recovers through two probe tips. The shape of these two electrical terminals is assumed to be circular with radius equals to 20 μm.

In order to study the electrical behaviour of the MIT material through its electrical resistance and its interaction with the temperature and the voltage, several simulations were carried out. The resistive response was simulated as a function of temperature and polarization voltage using the two-point method for a thin layer of VO₂/Au. The two point probe spacing distance is about 4000 μm as in the experimental conditions [20].

An imposed current is sent through one probe and exits through the second probe. The resulting voltage between the two probes is calculated by simulation. By combining both the electrical voltage and the current values into the two surface probes, it is possible to calculate material surface resistance. The electric currents interface in Comsol is used to compute electric field, current, and potential distributions in conducting media thin layers.

The modeling of the electric current flow in a conductive medium in the stationary case is governed by the continuity equation [21] as:

$$\nabla (J\text{-}J_e+\partial D/\partial t)=Q_i \qquad (1)$$

Where J is the induced current density vector, J_e is the external current density, D is the electric displacement field and finally Q_i is the source current term. In general, Q_i and J_e are supposed to be null in our material [21]. In the stationary mode, we could link the expression of the current density to the electric field E using the fundamental equation:

$$J=\sigma E \qquad (2)$$

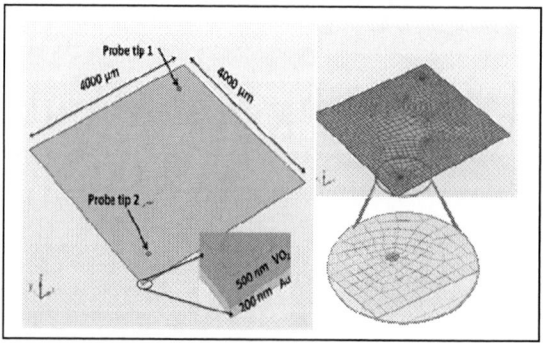

Fig. 2. Proposed structure for simulation: The top layer is set to be VO₂, bottom layer is set to have crystalline gold Au (111).

Where σ is the electrical conductivity (the reverse of electrical resistivity ρ). Thus, the model used to simulate the electric resistance currents interface in Comsol is used to compute electric field, current, and potential distributions in conducting media thin layers. Comsol software solves a current conservation equation based on Ohm's law using the scalar electric potential as the dependent variable. The structure is meshed with a fixed layer swept mesh.

In order to obtain the closest model to the actual behavior, the parameters describing the electrical properties through the laws of variation of the resistivity as a function of the temperature are, in fact, extracted from experimental measurements [20] and introduced manually into the resistivity function of the Comsol Multiphysics simulation software interface [21].

The main parameters in our model for the upper layer (VO₂) and bottom layer (Au) are given in Table 1 .

TABLE I.   MAIN SIMULATION PARAMETERS

Parameters	Materials	
	VO₂ (unit)	Au (unit)
Resistivity	10^{-2}-10^{+2} (Ω.cm) [22,23]	4-15 (μΩ.cm) [24,25]
Layer thickness	500 (nm)	200 (nm)
Lateral dimension	4000-6000 (μm)	4000-6000 (μm)

IV. FEM SIMULATION AND RESULTS

This section describes the simulations performed to estimate the electrical behavior of the VO₂/Au thin layer for various temperatures T and voltage V, using Comsol Multiphysics software.

The simulation results of the structure (Fig. 3) restore a similar behavior of the temperature dependence of the electrical properties (resistance) with polarization voltage bias. This transition temperature controlled by voltage seems very interesting in view of developing a potential sensing application. These multiphysic simulations of VO₂ on Au could be an important design tool for VO₂ functionalization in several sensing devices.

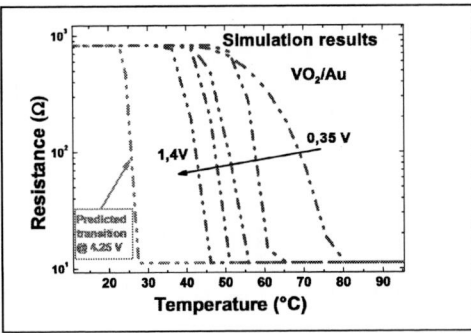

Fig. 3. Change in electrical resistance R as a function of the temperature T for different polarization voltage by simulation.

The simulations results of the structure (Fig. 3) are in agreement with experimental results, a similar metal-insulator transition behavior in dependence with polarization voltage bias was found. A comparison between experimental and simulation results are given in Table 2. We focus only on the main parameters *i.e.* the transition temperature value and the slope value in the transition zone. Based on the importance of a room temperature transition for targeted applications, we estimate by simulation the needed voltage polarization. Our simulations predict a transition at room-temperature under polarization of about 4.25 V. The MIT under ambient temperature is very interesting for biosensing application. Therefore simulation could be useful to design a new VO_2 based biosensors.

TABLE II.   A COMPARISON BETWEEN SIMULATION AND EXPERIMENT RESULTS

Polarization Voltage (V)	Average error ratio on slope in transition zone (%)	Error ratio on transition temperature (%)
0.51	15.9	3.17
0.75	7.26	1.72
1.11	5.7	2.94
0.95	7.53	<0.02
1.35	3.26	<0.02

V. CONCLUSION

In this paper, a 3D finite element analysis is conducted using Comsol Multiphysics to simulate the switching behavior and to compute the electrical resistance variation for various temperature and voltage values. Several parameters effects are studied: VO_2 and Au layer thickness, structure dimension, probe tips distance, thin film electrical resistivity value for both VO_2 and Au layers. The electrical response of the VO_2 layer against temperature has been studied by introducing different polarization voltage values into between the two terminal tips.

The main conclusions obtained are: (i) Simulation results of the structure restore a similar behavior of the temperature dependence of the electrical resistance with polarization voltage. (ii) Simulations predict a transition at room-temperature under a voltage of 4.25 V.

These simulations could be of a great interest to design potential sensors based on this transition metal oxide materials thin layer.

The simulation results could be used as a powerful design tool for several MIT-based sensors and predict the feasibility of a variety of device with a high sensitivity at ambient temperature. Practically, in term of operating temperature range many applications are targeted going from data storage device to biosensors.

We are currently investigating how to improve adapted devices through the design of high sensitivity biosensor based on ambient temperature transition.

ACKNOWLEDGMENT

The authors would like to thank the faculty of Sciences and FUNDAPL Laboratory. The authors would like to thank Nano Physics Group (university of Blida 1). We thank the Semiconductor Devices Reliability group from DMN division also Laser Matter Interaction group from the DMIL division of CDTA. Great thanks to Ahmed Tahroui.

REFERENCES

[1] M. B. Sahana, G. N. Subbanna and S. A. Shivashankar, "Phase transformation and semiconductor-metal transition in thin films of VO_2 deposited by low-pressure metalorganic chemical vapor deposition," Journal of Applied Physics, vol. 92, no. 11, pp. 6495–6504, 2002.

[2] Zheng Yang, Changhyun Ko, and Shriram Ramanathan, "Oxide Electronics Utilizing Ultrafast Metal-Insulator Transitions" Annual Review of Materials Research 41:1, 337-367, 2011.

[3] F. J. Morin, "Oxides which show a metal-to-insulator transition at the neel temperature," Phys. Rev. Lett., vol. 3, no. 34, pp. 34–36, Jul. 1959.

[4] Shadrin, E.B. & Il'inskii, "On the Nature of Metal–Semiconductor Phase Transition in Vanadium Dioxide" A.V. Phys. Solid State 42: 1126. https://doi.org/10.1134/1.1131328, 2000.

[5] R. M. Wentzcovitch, W. W. Schulz, and P. B. Allen, "VO2: Peierls or Mott-Hubbard? A view from band theory," Phys. Rev. Lett. 72, 3389, 1994.

[6] A. Pergament, "Metal-insulator transition: the Mott criterion and coherence length", J. Phys.: Condensed Matter 15, 3217, 2003.

[7] Hyun-Tak Kim, Yong Wook Lee, Bong-Jun Kim, Byung-Gyu Chae, Sun Jin Yun, Kwang-Yong Kang, Kang-Jeon Han, Ki-Ju Yee, and Yong-Sik Lim, "Monoclinic and Correlated Metal Phase in VO_2 as Evidence of the Mott Transition: Coherent Phonon Analysis" Physical Review Letters, vol. 97, Issue 26, id. 266401, 12/2006.

[8] Hyun-Tak Kim, Bong-Jun Kim, Sungyoul Choi, Byung-Gyu Chae, Yong Wook Lee, T. Driscoll, M. M. Qazilbash, and D. N. Basov. "Electrical oscillations induced by the metal-insulator transition in VO_2". Journal of Applied Physics 107, 023702, 2010; doi: 10.1063/1.3275575.

[9] Liu, M. K. et al. Terahertz-field-induced insulator-to-metal transition in vanadium dioxide metamaterial. Nature 487, 345–348, 2012.

[10] Decoupling of Structural and Electronic Phase Transitions in VO_2 2012 Zhensheng Tao, Tzong-Ru T. Han, Subhendra D. Mahanti, Phillip M. Duxbury, Fei Yuan, and Chong-Yu Ruan PRL 109 166406.

[11] Hyun-Tak Kim, Yong Wook Lee, Bong-Jun Kim, Byung-Gyu Chae,1 Sun Jin Yun, Kwang-Yong Kang, Kang-Jeon Han, Ki-Ju Yee, and Yong-Sik Lim 2006 Monoclinic and Correlated Metal Phase in VO_2 as Evidence of the Mott Transition: Coherent Phonon Analysis PRL 97 266401.

[12] H. Futaki and M. Aoki 1969 Effects of Various Doping Elements on the Transition Temperature of Vanadium Oxide SemiconductorsJpn. J. Appl. Phys. 8 1008.

[13] J.C.Rakotoniaina, R.Mokrani-Tamellin, J.R.Gavarri, G.Vacquier, A.Casalot, G.Calvarin "The Thermochromic Vanadium Dioxide: I. Role of Stresses and Substitution on Switching Properties" J. Solid State Chem. 103 81-94, 1993.

[14] Y. Muraoka and Z. Hiroi "Metal–insulator transition of VO_2 thin films grown on TiO2 (001) and (110) substrates" Appl. Phys. Lett. 80 583, 2002.

[15] P. P. Boriskov, A. A. Velichko, A. L. Pergament, G. B. Stefanovich, and D.G. Stefanovich "The effect of electric field on metal-insulator phase transition in vanadium dioxide" Tech. Phys. Lett. 28 406, 2002.

[16] You Zhou, Xiaonan Chen, ChanghyunKo, Zheng Yang, Chandra Mouli and ShriramRamanathan "Voltage-Triggered Ultrafast Phase Transitionin Vanadium Dioxide Switches" IEEE ELECTRON DEVICE LETTERS 34 2 2013.

Nikhil Shukla, Arun V. Thathachary, AshishAgrawal, Hanjong Paik, Ahmedullah Aziz, Darrell G. Schlom, Sumeet Kumar Gupta, Roman Engel-Herbert &SumanDatta "A steep-slope transistor based on abrupt electronic phase transition" NATURE COMMUNICATIONS 6 7812, 2015.

[17] Haitao Fu, Xuchuan Jiang, Xiaohong Yang, Aibing Yu, Dawei Su, Guoxiu Wang "Glycothermal synthesis of assembled vanadium oxide nanostructures for gas sensing" J Nanopart Res 14 871, 2012.

[18] Naoki Inomata, Libao Pan, Zhuqing Wang, Mitsuteru Kimura, Takahito Ono "Vanadium oxide thermal microsensor integrated in a microfluidic chip for detecting cholesterol and glucose concentrations" Microsyst Technol, Issue 7/2017.

[19] Asmaa Leila Sabeha Hassein-Bey, Hakim Tahi, Slimane Lafane, Amina Zouina AitDjafer, Abdelkader Hassein-Bey and N. Belgroune "Substrate Effect on Electrical Properties of Vanadium Oxide Thin Film for Memristive Device Applications" 2016 IEEE International Conference on Semiconductor Electronics (ICSE) Kuala Lumpur, Malaysia 10.1109/SMELEC.2016.7573636.

[20] https://www.comsol.com/documentation.

[21] N. F. Mott, Metal-Insulator Transitions, 2nd ed. London: Taylor and Francis, 1990.

[22] A. Pergament, G. Stefanovich, and A. Velichko, "Oxide Electronics and Vanadium Dioxide Perspective: A Review," J. Sel. Top. Nano Electron. Comput. 1, 24, 2013.

[23] G. Kästle,H.-G. Boyen, A. Schröder, A. Plettl, and P. Ziemann, "Size effect of the resistivity of thin epitaxial gold films", PHYSICAL REVIEW B 70, 165414, 2004.

[24] f. avilès, o. ceh and a. i. oliva, "Physical properties of au and al thin films measured by resistive heating" surface review and letters, vol. 12, no. 1 101–106, 2005.

Integration of 2.4GHz RF Transmitter with Accelerometer for mobile sensing applications on 1P6M ASIC compatible CMOS MEMS process

Pao-Min Chu

Department of Electronics Engineering

National Chiao Tung University

Hsinchu 300, Taiwan

chupaomin.ee05g@nctu.edu.tw

Kuei-Ann Wen

Department of Electronics Engineering, NCTU

National Chiao Tung University

Hsinchu 300, Taiwan

twtstella@gmail.com

Abstract—A novel design to combine RF transmitter as well as sensor monolithically has been proposed. A 2.4GHz RF transmitter is integrated with a single axis accelerometer. The two port spiral transformer of which the secondary inductor is designed to be suspended with MEMS structure. The mutual inductance of the transformer will be changed due to displacement of the proof mass. The magnetic flux variation can be sensed by demodulation signal and the measured sensitivity is 11.8 Hz/g.

Keywords—Accelerometer, Inductive, Transformer couple, Impedance Matching, CMOS-MEMS, Transmitter

I. INTRODUCTION

With development of internet of things, it brings in a large number of mobile sensing applications and hence inspired the design of sensor integrated transceiver. Conventionally, accelerometers will be designed with comb-finger to provide capacitive sensing structure. In this work, coil structure of the transformer is fabricated with MEMS process to make an inductive accelerometer. The inductance of the sensing coil structure can be designed process more immune to variation [1]. With hybrid integration process of MEMS and ASIC, a novel design of integrated transmitter and accelerometer can be implemented. The inductive accelerometers [2] [3] using s-type spring as the inductance of the quality factor is not high enough for matching circuit, and it will be more power consumptive. This paper proposed a spiral inductor based design which has better quality factor in inductive

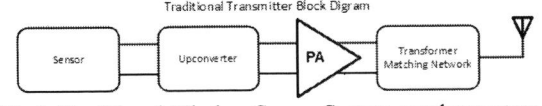

Fig 1. Traditional Wireless Sensor System accelerometers.

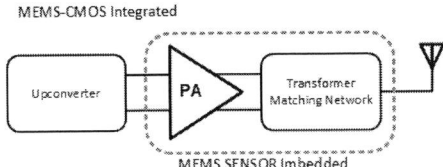

Fig 2. Proposed sensor integrated Transmitter

II. SENSOR INTEGRATED TRANSMITTER

The comparison of conventional wireless sensor module and the proposed sensor integrated transmitter building blocks are illustrated in Fig.1 and Fig.2. Conventionally, capacitive accelerometer will be designed with a chopper amplifier for readout. Since conventional amplifiers are designed in frequencies within MHz, and the demodulated will be low-frequency signal. There always will be an up-converter for RF transmission. If the inductor in the transmitter can be designed as series S-type spring, the displacement of the spring owing to outside acceleration will produce spring compression and cause mutual inductance value changes, and result frequency changes. The advantage is that high-frequency signal can increase the sampling rate to giga-hertz and effectively improve sensitivity, and the modulation signal is exactly an FM signal which can be used for transmitter using impedance matching antenna transmission. Thus the inductance of the matching circuit can be innovatively transformed to be sensor with MEMS structure which can response to outside force with displacement and cause mutual inductance change and thus cause frequency which can be demodulated with FM demodulation. Because of the S-type spring as the inductance of the quality factor is not high enough for matching circuit, and it will be more power consumptive. The quality factor of the whole inductance structure can be greatly improved if the spiral inductor is in series with S-type spring. In addition to effectively reducing the system power consumption, spiral inductor also contribute to the concentration of magnetic flux. It can obtain good mutual inductance, and improve the sensitivity of the single-axis induction. The inductance of the sensing coil structure can be designed more immune to variation in fabrication process .

III. STRUCTURE DESIGN

The acceleration is presented by the contrast displacement motion between two inductors, resulting in the change of mutual inductance to affect the coupling signal frequency change. The lowest layer of polycrystalline silicon will be etched in 1P6M ASIC compatible CMOS MEMS process, and the quality factor of the inductor can be enhanced. Different coil numbers and the mutual inductance change under different displacements are simulated. In TABLE I, coupling factor (K) is defined in Eq. (2). Coupling factor means that the amount of flux linkage as a fraction of the total possible flux linkage between the coils, and expressed as a fractional number between 0 and 1 instead of a percentage (%) value. The L1 is the inductance of primary inductor, and L2 is secondary inductor. The secondary inductor was embedded in proof mass. In Fig 3. (a) , the vertical direction, the mass will move down and up as showed in Fig. 3.(b),it makes the secondary inductor far away from primary coil, and let mutual inductance smaller/larger. If mutual inductance changes, the instance of each inductor will change and the response frequency will change accordingly. The variation of frequency was related to the outside acceleration. By using frequency demodulation, acceleration information can be sensed.

$$\text{Mutual Inductance (M)} = \textit{Im}\,(\mathbf{Z12})/\,(2\pi f) \qquad (1)$$

$$K = M/\sqrt{L1L2} \qquad (2)$$

TABLE I.　　DISPLACEMENT VERSUS K

Displacement at 0.68(um)	
Turns of Primary inductor	*K*
1	0.168
2	0.167
3	0.156
4	0.144
5	0.135

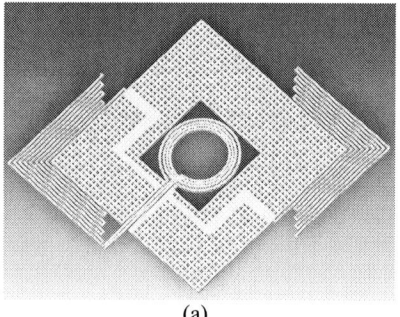

(a)

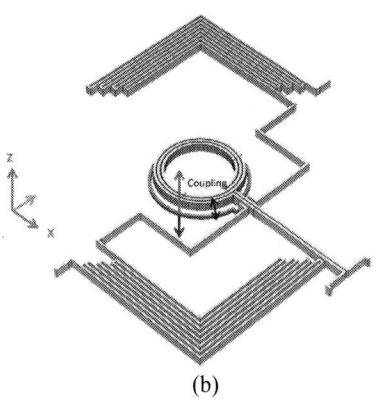

(b)

Fig 3. (a)Sensing Structure integrated with Inductor and the displacement direction (b) 3D Structure integrated with Inductor embedded in proof mass

For optimizing of the induced voltage, the variety turns of coil number was simulated. In simulation result as shown in Fig.4 and Fig.5, more turns of the coils results in higher quality factor. Quality factor is defined as the ratio of inductor's imaginary part and real part impedance.

Mutual Inductance in Eq. (3) indicate the two port impedance variety to frequency. Mutual Inductance is the interaction of one coil magnetic field on another coil. However, more turns of the coils cause lower the magnetic coupling, Fig.6. Because of the inductance will be increased fast according to Eq. (2).[5], [6].

The coupling factor -K is the strength of the magnetic coupling between winds. At the equation the mutual inductance (M) is influenced by magnetic flux directly, then the self-inductance (L) will be the magnetic force against coupling magnetic field direction. The bigger self-inductance means counter magnetic field stronger.

It can be observed that three turns is the best number ratio after the signal through the structure of the transformer as shown in Fig.7.

$$\text{Quality factor} = \textit{Im}(1/\mathbf{Y11})/\textit{Re}(1/\mathbf{Y22}) \qquad (3)$$

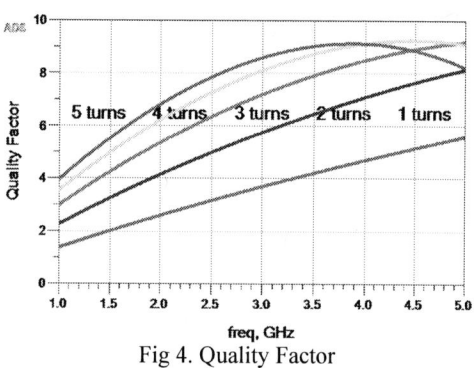

Fig 4. Quality Factor

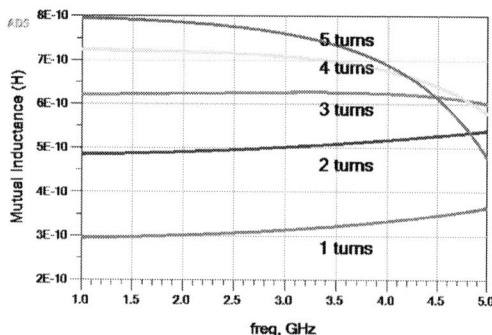

Fig 5. Mutual inductance

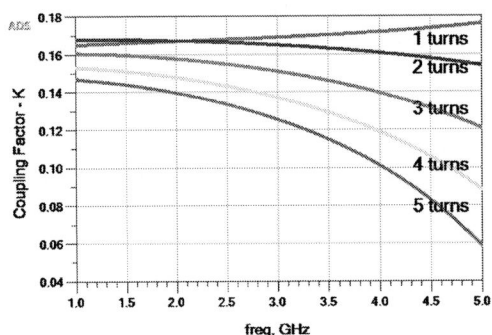

Fig 6. Coupling inductance

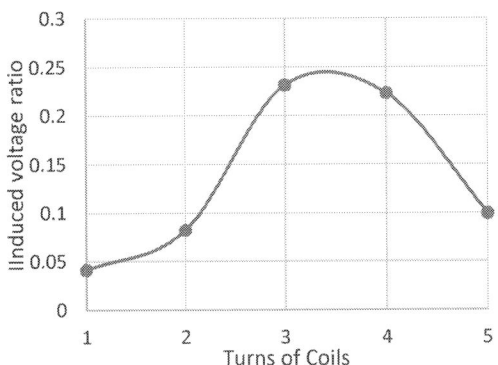

Fig 7. Induced voltage ratio versus Turns of coils

TABLE II.   INDUCED VOLTAGE VERSUS TURNS

| Turns | Inductor induced voltage (V) | | Voltage ratio |
	Secondary	Primary	
1	0.294	7.149	0.04112
2	0.566	6.899	0.08204
3	1.172	5.065	0.23139
4	1.061	4.756	0.22309
5	0.620	6.250	0.09920

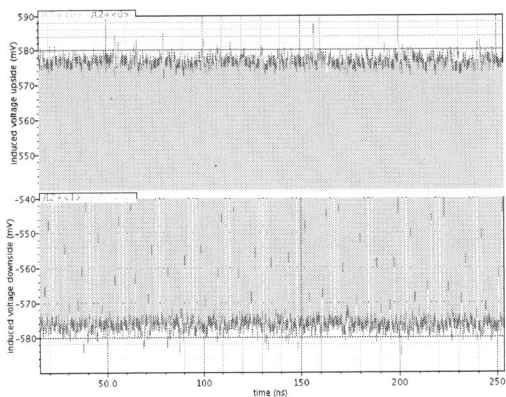

Fig 8. Difference of induced average voltage

By apply high-frequency circuit simulation with 2.4GHz, 1.8V carry wave, we can measure the secondary induced amplitude, and compare the difference about variety numbers of turns. See TABLE II. We did electromagnet simulation about induced voltage by fixed displacement with 0.68 μm. In Fig. 8. , the simulation of three turns transformer induced average voltage difference is 4mV.

IV. MEASUREMENT

The sensor structure was implemented by 1P6M 0.18-um CMOS process with dry-etching release process. Figure Fig. 9. , shows the scanning electron micrographs (SEM) of a fabricated and released structure. In vibration tests, the accelerometer was mounted on the shaker which is shown in Fig 11.

For a vibration with frequency of 1.5 KHz and variation amplitude form 1-g to 9g, the spectrum of single axe oscillation output is shown in Fig. 12. The signal input power with 2.4 GHz sinusoid wave is 5-dBm. From Fig. 13. ,it shows the sensitivity is 11.8 Hz/g. The sensing functionality of the proposed MEMS structure is successfully demonstrated.

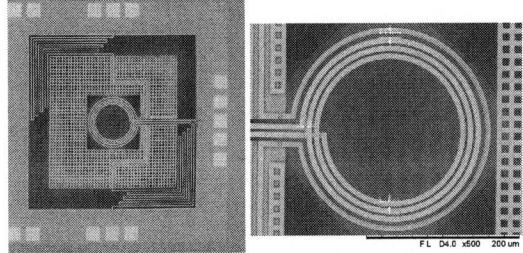

Fig. 9. SEM of fabricated structure

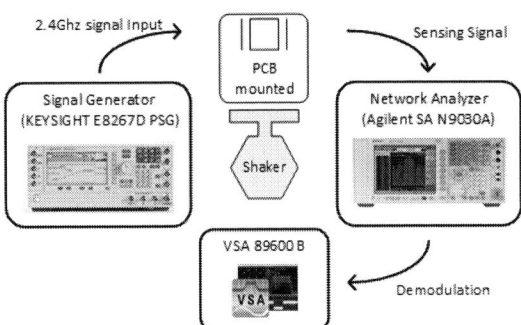

Fig. 10. Measurement Environment Setting

Fig. 11. Fabricated on PCB mounted at Shaker

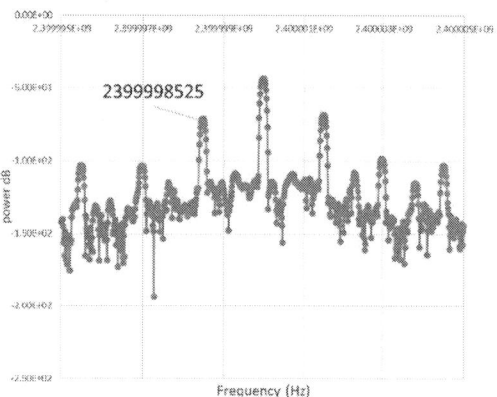

Fig. 12. Output Spectrum which 2nd harmonic was Sensed signal

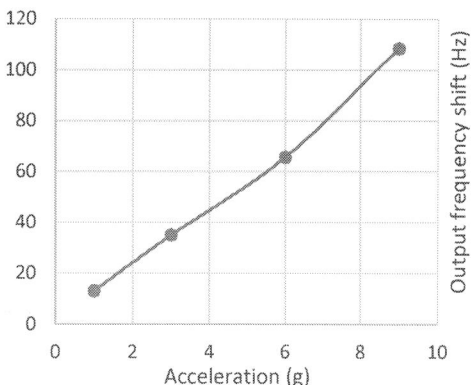

Fig. 13. Sensitivity of proposed accelerometer

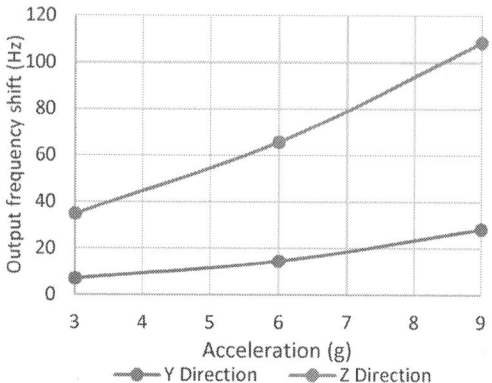

Fig. 14. Sensitivity of proposed accelerometer with cross-direction

V. CONCLUSIONS

A single axe CMOS MEMS inductive accelerometer integrated with matching scheme for transmitter was proposed and demonstrated. Measurement showed that sensitivity is 11.8 Hz/g and the sensed signal can be modulated exactly with the transmitter 2.4 GHz sinusoid wave without any other circuit integrated. In Fig. 14, Cross-axis sensitivity was also measured by position the sensor along different directions in the dynamic tests. The z axis has a relatively high cross-axis sensitivity than cross-direction.

ACKNOWLEDGMENT

This design was sponsored in part by the National Science Council of Taiwan under a grant of MOST 104-3115-E-009-022. The authors would like to acknowledge chip fabrication support provided by National Chip Implementation Center (CIC), Taiwan, R. O. C

REFERENCES

[1] Yi Chiu, Hao-Chiao Hong, Chi-Ming Chang, "Three-axis CMOS MEMS inductive accelerometer with novel Z-axis sensing scheme" Transducers 2017, Kaohsiung, TAIWAN, June 18-22, 2017, 410-413

[2] Yi Chiu, Hao-Chiao Hong, and Chia-Wei Lin," Inductive CMOS MEMS accelerometer with integrated variable inductors" IEEE MEMS 2016, Shanghai, China 2016, 974-977

[3] LI Hila, SHI Geng-chen , "The Stiffness Analysis of MEMS Planar Microspring", PIEZOELECTECTRICS & ACOUSTOOPTICS, VOL. 29 NO. 2 Apr. 2007.

[4] Behzad Razavi, RF Microelectronics (2nd Edition) Chapter 2

[5] J. R. Long," Monolithic transformers for silicon RF IC design" IEEE JSSC 2000, 1368 – 1382 Sept. 2000

[6] A. Lindblom, J. Isberg, H. Bernhoff, " Calculating the coupling factor in a multilayer coaxial transformer with air core" IEEE Transactions on Magnetics, Volume: 40, Issue: 5, 3244 – 3248,Sept. 2004

Detecting Hydrogen Using TiO₂-B₂O₃ at Different Operating Temperature

Siti Amaniah Mohd Chachuli
Institute of Advanced Technology
Universiti Putra Malaysia
Serdang, Selangor, Malaysia
sitiamaniah@utem.edu.my

Mohd Nizar Hamidom
Institute of Advanced Technology
Universiti Putra Malaysia
Serdang, Selangor, Malaysia
mnh@upm.edu.my

Md. Shuhazlly Mamat
Faculty of Science
Universiti Putra Malaysia
Serdang, Selangor, Malaysia
shuhazlly@upm.edu.my

Mehmet Ertugrul
Engineering Faculty
Ataturk University
Erzurum, Turkey
ertugrul@atauni.edu.tr

Abstract— Performance of TiO_2-B_2O_3 gas sensor that annealed using nitrogen at 650°C for 30 minutes was observed and analyzed. The sensing film of the gas sensor was prepared by mixing TiO_2-B_2O_3 with an organic binder. The sensing film was characterized by field emission scanning electron microscopy (FESEM) and X-ray diffraction (XRD). The gas sensor was exposed to hydrogen at a concentration of 100 - 1000 ppm with operating temperatures of 100°C and 200°C. However, no response was detected for 100 ppm at 100°C. But, as the operating temperature was increased to 200°C, the gas sensor indicated a good response for 100 ppm of hydrogen. The gas sensor exhibited p-type response based on decreased current when exposed to hydrogen. The sensitivity of gas sensor was calculated at 1.00, 2.18 and 3.58 for 100 ppm, 500 ppm and 1000 ppm respectively, at an operating temperature of 200°C.

Keywords— gas sensor; TiO2-B2O3; hydrogen; nitrogen; organic binder

I. Introduction

Among the many different types of gases, hydrogen is one that needs rapid detection due to its colourless, odourless, and highly explosive in a wide range [1]. Various materials have been employed to detect hydrogen such as palladium, [2]–[4], metal-oxides [5]–[11] and carbon-based materials [12]. Of which, metal-oxides are the most common materials used for hydrogen detection due to its high sensitivity to most gases, low cost and simple fabrication techniques [13]. The common metal-oxides gas sensor used to sense hydrogen are SnO2 [5], [6], ZnO [7] and TiO2 [8]–[11].

Metal-oxide gas sensor based on Titanium dioxide (TiO_2) are extensively studied because it is chemically stable, nontoxic, biocompatible, inexpensive and wide band gap semiconducting material [14]. Besides, it is also stable at higher temperatures and possesses good sensing characteristics at operating temperatures below 400°C [15]. Structurally, TiO_2 consists of three different phases: rutile (tetragonal), anatase (tetragonal) and brookite (orthorhombic) [16]. The most common applications of anatese and rutile phases are gas sensing, dye-sensitized solar cell, photo catalyst and environmental applications. However, Titania weakness is high resistance, which can be improved by the addition of proper dopant to the TiO_2 [17].

The recent advent of nanotechnology increased the development of TiO_2 nanostructures. TiO2 nanocrystalline [8], nanoparticles [9], and nanotube [10], [11] have shown a good response to low concentrations of hydrogen and are also capable to operate at low operating temperature, to as low as room temperature.

Therefore, in this current work, the hydrogen gas sensor based on TiO_2 nanoparticles was fabricated on alumina substrate and was exposed to hydrogen at concentrations of 100 - 1000 ppm. Two different operating temperatures of 100°C and 200°C were tested to examine the sensing capability.

II. Experimental Work

A. Preparation and Fabrication of Gas Sensor

The sensing film of the gas sensor was prepared by mixing TiO_2-B_2O_3 with an organic binder. Initially, 90 wt. % of TiO_2 (P25) was mixed with 10 wt. % of glass powder, Boron Oxide (B_2O_3). The glass powder was added to the paste to produce good adhesion between the paste and alumina substrate. The 96% Alumina was chosen as a substrate for the gas sensor because it can withstand high temperature up to 1000°C. Also, the paste can be easily deposited atop alumina using the screen-printed method. Hence, TiO_2 was mixed with B_2O_3 using m-xylene as a medium in an ultrasonic bath. Then, it was dried in an oven to make it into a powder form. The produced powder was then grinded in a mortar to ensure uniformity. Next, in order to make a paste, TiO_2-B_2O_3 was mixed with an organic binder. The organic binder was prepared by mixing linseed oil with m-xylene and α-terpineol. For measurement of gas sensor, a silver paste was used as an interdigitated electrode at the bottom of sensing film. The sensing film and interdigitated electrode were deposited on alumina substrate using screen-printed method.

For firing process, TiO_2-B_2O_3 sample was calcined in the furnace with air as carrier gas using temperature of 500°C for 30 minutes. Then, the sample was annealed with nitrogen at 650°C for 30 minutes.

B. Experimental Setup for Gas Sensor

The measurement of the TiO_2-B_2O_3 gas sensor was conducted in a gas chamber with connection to the mass flow controller, temperature controller and Kiethley 487

Picoammeter / Voltage source. Three different hydrogen concentrations were flowed into the gas chamber: 100, 500 and 1000 ppm. During the measurement, 500 sccm of nitrogen was used as carrier gas and two different operating temperatures which are 100°C and 200°C were tested to the gas sensor. For measurement, 10 V voltage source was applied to the electrode of gas sensor and current was observed as a response of gas sensor.

III. RESULTS AND DISCUSSION

A. Characterization of TiO₂-B₂O₃ thick film

TiO_2-B_2O_3 thick film on alumina substrate was characterized using field emission scanning electron microscopy (FESEM) and X-ray diffraction (XRD). Fig. 1 displays the nanoparticles structure in the TiO_2-B_2O_3 thick film. Based on observation, the structure of the nanoparticles was spherical and it was clearly seen in the sample. This also indicated the prepared paste is homogenous. The diameters of TiO_2 nanoparticles range from $40 - 70$ nm.

Fig. 1. FESEM image of TiO₂-B₂O₃ thick film

On the other hand, the XRD pattern of TiO_2-B_2O_3 thick film after heat treatment for $2\theta = 20°$ to $80°$ as demonstrated in Fig. 2. The analysis shown that TiO_2-B_2O_3 thick film consists of TiO_2 (anatase and rutile), B_2O_3 (glass powder) and Al_2O_3 (substrate) phases. Anatase peaks were observed at $2\theta = 25.36°$ (101), $37.05°$ (103), $37.90°$ (004), $48.15°$ (200), $62.86°$ (204), $68.97°$ (116), $70.47°$ (220) and $75.27°$ (215). The sharp peak of anatase phase at $2\theta = 25.36°$ (101) indicated that the sample might have a high degree of crystallization. Whereas, the peak of B_2O_3 was observed at $2\theta = 27.76°$. This result strongly postulate that the sensing film consists of B_2O_3. In addition, alumina (Al_2O_3) was the dominant phases in this analysis based on high peaks of Al_2O_3. The high prevalence of Al_2O_3 peaks has overshadowed small peaks of rutile phase in the XRD analysis. For instance, one of rutile peak observed at $2\theta = 35.48°$ (101) was located at the same position with the highest peak of Al_2O_3. This analysis showed that Al_2O_3 substrate has a high crystallinity and covered some of the other elements present in the sample.

B. Resistance of gas sensor

The resistance of gas sensor was measured using I-V testing. The measured resistance at different operating temperatures is illustrated in Fig. 3. At room temperature (22°C), the resistance of the gas sensor was very high with approximately 1.45T Ω. Meanwhile, at 100°C, the resistance decreased rapidly to 7.49 MΩ. There was also a reduction in the resistance at 200°C with approximately value of 0.17 MΩ. It can be observed that, the resistance of the gas sensor decreases with operating temperature increasing. When resistance decreases, more current is able to flow through the gas sensor and better conductivity is achievable with elevated temperatures.

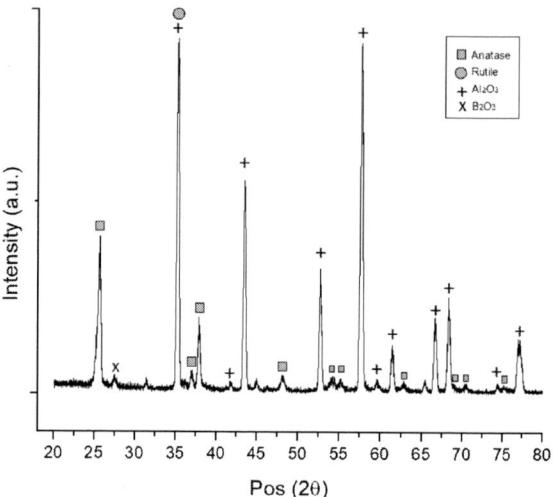

Fig. 2. XRD Pattern of TiO₂-B₂O₃ thick film

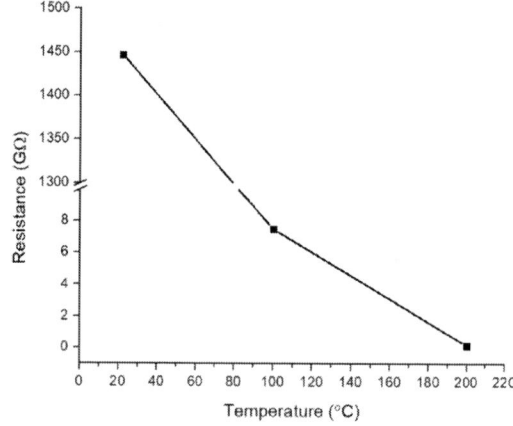

Fig. 3. Resistance of gas sensor at different operating temperature

C. Response of TiO₂-B₂O₃ gas sensor to hydrogen at different operating temperature

Fig. 4 exhibits the response of gas sensor to hydrogen at different operating temperatures. No response was observed at room temperature. Thus, the experiment was conducted at operating temperatures of 100°C and 200°C. The measurement was recorded once the current was stabilized at respective operating temperature. A slight increase in current was observed when nitrogen was flowed through the gas sensor at both operating temperatures. It also can be observed that the increment of current at operating

978-1-5386-5284-8/18 $31.00 © 2018 IEEE

temperature of 200°C was not too high as operating temperature at 100°C. This phenomenon occurred might because of behaviour of gas sensor where it will increase current when exposed to the nitrogen. From the response, it showed that the gas sensor exhibited a p-type behaviour based on the decreased current flow when exposed to a reducing gas such as hydrogen. Hence, it can be conclude that the resistance of gas sensor increases with the exposure hydrogen in the gas chamber.

sensitivity after the gas sensor was exposed to a low concentration of hydrogen. However, the response was unable to reach the original value about 1 when exposed to 500 ppm of H_2 at 100°C as indicated in Fig. 5(a). Meanwhile, a better sensor response was obtained at an operating temperature of 200°C. Therefore, it can be seen that the response and recovery of the gas sensor are quite stable over different concentrations of hydrogen at this temperature.

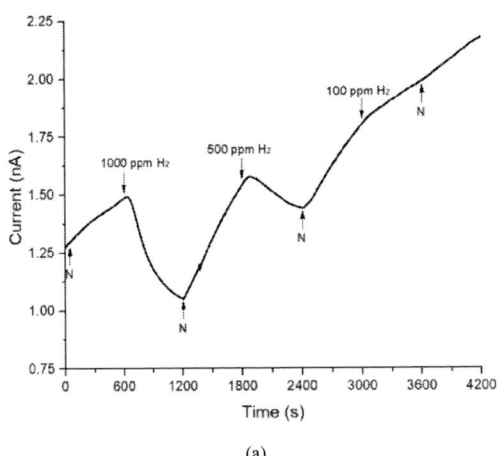

(a)

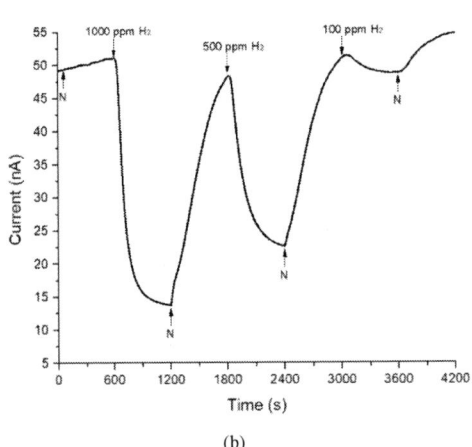

(b)

Fig. 4. Response of the gas sensor to hydrogen at different operating temperatures (a) T = 100°C and (b) T = 200°C

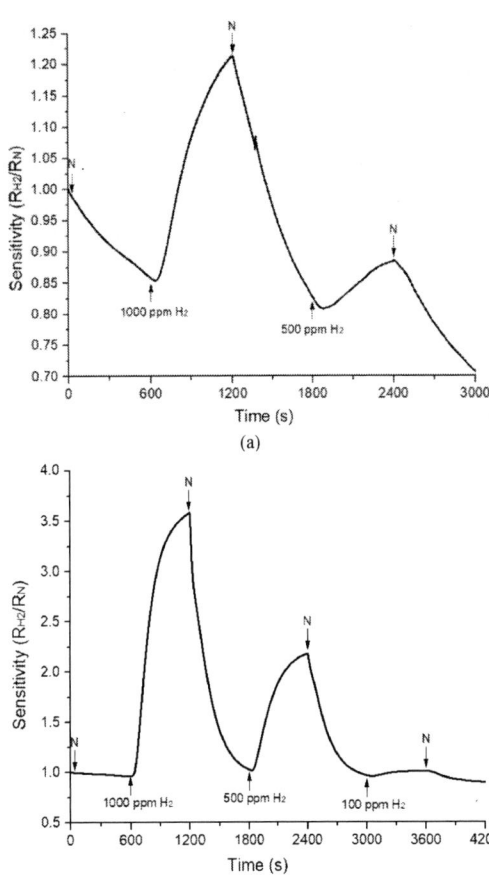

(a)

(b)

Fig. 5 Response-recovery characteristics of the gas sensor to hydrogen at different operating temperature (a) T = 100°C and (b) T = 200°C

D. Properties of the TiO₂-B₂O₃ gas sensor to hydrogen at different operating temperature

Fig. 5 presents the response-recovery characteristics of the gas sensor at operating temperatures of 100°C and 200°C. The sensitivity of the gas sensor can be calculated as follows:

$$Sensitivity = \frac{R_{H2}}{R_N} \qquad (1)$$

Where R_N is the initial resistance during nitrogen flow and R_{H2} is resistance during hydrogen flow. All the responses were greater than 1 and exhibit a decreased tendency of

Next, the sensitivity of TiO₂-B₂O₃ gas sensor to various concentrations of hydrogen at operating temperatures of 100°C and 200°C is displayed in Fig. 6. At the operating temperature of 200°C, a linear graph was generated with the highest sensitivity achieved. The calculated sensitivity values are 1.00, 2,18 and 3.58 for 100 ppm, 500 ppm and 1000 ppm of H_2 respectively. On the contrary, the difference in sensitivity is almost two folds between the two operating temperatures. This analysis showed that the sensitivity of gas sensor can be improved by tested it at higher operating temperature.

IV. CONCLUSIONS

In conclusion, the TiO₂-B₂O₃ gas sensor was fabricated using screen-printed method demonstrated its ability to detect low concentrations of hydrogen to as low as 100 ppm at an operating temperature of 200°C. Importantly, the gas sensor exhibited p-type response based on decreased current when exposed to the hydrogen and increased current when exposed to nitrogen. In addition, 200°C was the optimal operating temperature to produce highest sensitivity which approximately 1.00, 2,18 and 3.58 at 100 ppm, 500 ppm and 1000 ppm of H_2 respectively.

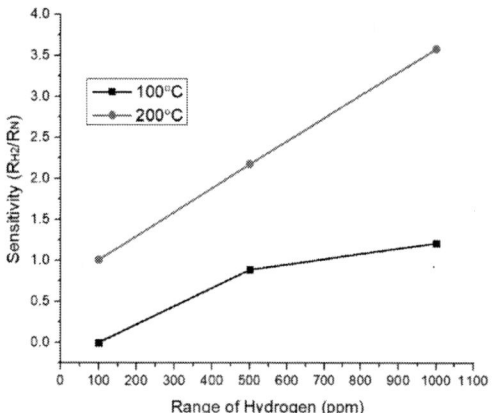

Fig. 6. Sensitivity of TiO_2-B_2O_3 gas sensor at different operating temperature

ACKNOWLEDGMENT

The author would like to thank the members of the Physics Laboratory, Science Faculty, Ataturk University, Turkey for providing facilities to conduct gas sensor measurement for this work.

REFERENCES

[1] Z. Zhang, X. Zou, L. Xu, L. Liao, W. Liu, J. Ho, X. Xiao, C. Jiang, and J. Li, "Hydrogen gas sensor based on metal oxide nanoparticles decorated graphene transistor," Nanoscale, vol. 7, no. 22, pp. 10078–10084, 2015.

[2] S. Öztürk and N. Kılınç, "Pd thin films on flexible substrate for hydrogen sensor," J. Alloys Compd., vol. 674, pp. 179–184, 2016.

[3] T. K. Kyun, J. S. Sang, and M. C. Sung, "Hydrogen gas sensor using Pd nanowires electro-deposited into anodized alumina template," IEEE Sens. J., vol. 6, no. 3, pp. 509–513, 2006.

[4] M. Z. Atashbar, D. Banerji, and S. Singamaneni, "Room-temperature hydrogen sensor based on palladium nanowires," IEEE Sens. J., vol. 5, no. 5, pp. 792–797, 2005.

[5] G. Tournier and C. Pijolat, "Selective filter for SnO2-based gas sensor: Application to hydrogen trace detection," Sensors Actuators, B Chem., vol. 106, no. 2, pp. 553–562, 2005.

[6] S. Shukla, L. Ludwig, C. Parrish, and S. Seal, "Inverse-catalyst-effect observed for nanocrystalline-doped tin oxide sensor at lower operating temperatures," Sensors Actuators, B Chem., vol. 104, no. 2, pp. 223–231, 2005.

[7] D.-T. Phan and G.-S. Chung, "Effects of different morphologies of ZnO films on hydrogen sensing properties," J. Electroceramics, vol. 32, no. 4, pp. 353–360, 2014.

[8] O. Krško, T. Plecenik, T. Roch, B. Grančič, L. Satrapinskyy, M. Truchlý, P. Ďurina, M. Gregor, P. Kúš, and A. Plecenik, "Flexible highly sensitive hydrogen gas sensor based on a TiO2 thin film on polyimide foil," Sensors Actuators, B Chem., vol. 240, pp. 1058–1065, 2017.

[9] X. Peng, Z. Wang, P. Huang, X. Chen, X. Fu, and W. Dai, "Comparative study of two different TiO2 film sensors on response to H2 under UV light and room temperature," Sensors (Switzerland), vol. 16, no. 8, 2016.

[10] E. Şennik, Z. Çolak, N. Kilinç, and Z. Z. Öztürk, "Synthesis of highly-ordered TiO2 nanotubes for a hydrogen sensor," Int. J. Hydrogen Energy, vol. 35, no. 9, pp. 4420–4427, 2010.

[11] G. K. Mor, M. A. Carvalho, and C. A. Grimes, "A room-temperature TiO2 -nanotube hydrogen sensor able to self-clean photoactively from environmental contamination," 2004.

[12] S. Dhall, N. Jaggi, and R. Nathawat, "Functionalized multiwalled carbon nanotubes based hydrogen gas sensor," Sensors Actuators, A Phys., vol. 201, pp. 321–327, 2013.

[13] S. A. Y. S. Lee, O. S. Kwon, S. J. Park, E. Y. Park and H. Y. and J. Jang, "Fabrication of ultrafine metal-oxide-decorated carbon nanofibers for DMMP sensor application," ACS Nano, vol. 5, pp. 7992–8001, 2011.

[14] A. N. Banerjee, "The design, fabrication, and photocatalytic utility of nanostructured semiconductors: Focus on TiO2-based nanostructures," Nanotechnol. Sci. Appl., vol. 4, no. 1, pp. 35–65, 2011.

[15] A. Wisitsoraat, E. Comini, G. Sberveglieri, W. Wlodarski, and A. Tuantranont, "TiO2 based nanocrystalline thin film gas sensors prepared by ion-assisted electron beam evaporation," Proc. 2nd IEEE Int. Conf. Nano/Micro Eng. Mol. Syst. IEEE NEMS 2007, pp. 109–112, 2007.

[16] M. de la L. O. Morales-Bautista, J., A. Maldonado, "Gas Sensing Performance of TiO2-Al2O3 Pellets," pp. 320–324, 2009.

[17] E. Comini, G. Faglia, and G. Sberveglieri, "Characterization of p-type Cr : TiO2 gas sensor," pp. 1320–1322, 2005.

[18] Y. Tang, O. Byl, Y. Yoon, B. E. Tien, and S. Bishop, "Ion Implanter Performance Improvement for Boron Doping by Using Boron Trifluoride (BF3) and Hydrogen (H2) Mixture Gases," vol. m, pp. 2–5.

Study on breakdown characteristics of AlGaN/GaN-based HFETs

Tineesha Naidu[1], Sharifah Wan Muhamad Hatta[1,2], Norhayati Soin[1,2], Sharidya Rahman[1], Yasmin Abdul Wahab[3]

[1]Department of Electrical Engineering, Faculty of Engineering, University of Malaya, 50603, Kuala Lumpur, Malaysia
[2]Centre of Printable Electronics, University of Malaya, 50603, Kuala Lumpur, Malaysia
[3]Nanotechnology & Catalysis Research Centre, University of Malaya, 50603, Kuala Lumpur, Malaysia

Abstract— **GaN-based devices are gaining popularity due to its high electrical performance and are widely used in high frequency and high power microwave applications. Unfortunately, there are limitations faced by HFETs which are current leakage and breakdown characteristics which affect the performance of the device. This work focuses on the investigation of geometrical and process variation impact on the electrical characteristics and breakdown voltage of the AlGaN/GaN based High Field Effect Transistor (HFET). The Sentaurus Technology Computer Aided Design (TCAD) by Synopsys was used to facilitate this study on the electronics properties of the HFET specifically the current density in the two-dimensional electron gas (2DEG) channel, carrier mobility, band energy and transfer characteristics. It has been observed that process variations specifically substrate material selection have significant impact on the performance of the HFET as compared to geometric variation. HFETs with the Silicon Carbide substrate demonstrate higher breakdown voltage as compared to the conventional silicon substrate. Optimization of the conventional structure by means of introducing silicon carbide as the substrate demonstrates an increase of 183.33% in terms of breakdown voltage.**

Keywords- breakdown voltage; 2DEG; carrier mobility; current density

I. INTRODUCTION

The semiconductor industry has been rapidly transforming whereby the technology advancement has been formulated for multiple applications that contributes to the society. Semiconductors are highly essential for cloud integration, connectivity which comprises of computing sensors, communications and interactivity which makes up the Internet of Things (IoT) [1]. Radio frequency (RF) and wireless technology is a crucial element in the IOT ecosystem. [2] The conventional Silicon-based transistors used in most integrated circuits (ICs) chips comprise of critical bottlenecks which limit the RF performance. The low energy bandgap and low breakdown strength of the silicon material results in a poor choice for high power applications. Compound semiconductors, specifically the Gallium Nitride (GaN), are more favourable for high power and high frequency applications [3]. The High field effect transistors (HFETs) or also known as High Mobility Effect Transistors (HEMTs) were developed for high speed applications in a high power and high frequency environment while maintaining low noise [4]. HFETs consist of a heterojunction of two materials, for instance the GaN and AlGaN, with different bandgaps and the conducting channel which is formed at the epitaxial interface [5]. The GaN displays the capability to be the

displacement technology for silicon semiconductors due to it being a stable material with very large heat capacity, wide bandgap material that is suitable for high temperature and higher power switching applications as well as high electrical breakdown voltage thus making it ideal for high bias voltages in application. Bandgap of the GaN is 3.4eV and it has a breakdown electric field of 3.3MV/cm, a high electron saturation velocity of 2.5×10^7cm/s and a high carrier density in the two dimensional electron gas(2DEG) channel [6]. However, HFET devices suffer from high current leakage which affects the device performance. Besides that, the parasitic and reliability issues limit the static and dynamic performance and thus affecting the maximum switching frequency of the device. High switching loss also affects the device performance, in addition to self-heating, current collapse and lattice mismatch [7]. This work acknowledges the many limitations of the HFETs and the lack of research and investigations to improve the performance of this highly potential device. Little knowledge surrounding the trapping mechanisms and electrical behavior of the devices makes up the motivation of this work. To understand better the behaviour of HFETs, this work investigates on the scaling effects and process variations on off-state breakdown voltages and electrical characteristics of the AlGaN/GaN HFETs as well as influences of current leakages by studying the effect of semi-insulating substrate material and process variations on the performance of AlGaN/GaN HFETs for high power applications.

II. DEVICE AND SIMULATON CONDITIONS

This simulation work is conducted using the Synopsys Sentaurus TCAD tool. The simulated conventional HFET structure is depicted in Fig.1. The structure below is analyzed from various aspects in terms of scaling effects and geometrical variations in order to optimize the HFET structure and study its effect on breakdown voltage performance as well as other electrical characteristics which alleviate the efficiency of the device. The device is made up of a 20nm $Al_{0.2}Ga_{0.8}N$ barrier layer grown on top of GaN buffer layer. The device is assumed to be grown on a lightly n-doped silicon substrate. The thickness of the buffer layer is varied in this work. A 50nm thick passivation layer of nitride is grown on the $Al_{0.2}Ga_{0.8}N$ barrier layer. The gate contact was set to the Schottky type with metal workfunction of 4.3V. The initial mesh interface of AlGaN/GaN is specified at 0.5nm [8].

In this work, the buffer thickness was varied along with changes in channel length (Lg), length of gate to drain (Lgd) and length of source to gate (Lsg). When one parameter was

978-1-5386-5284-8/18 $31.00 © 2018 IEEE

being varied, the others were kept constant. Table I shows the parameters taken into account in the geometric variation. In addition to these geometrical variations, the performance of HFET device was studied for different substrate materials and Aluminium composition.

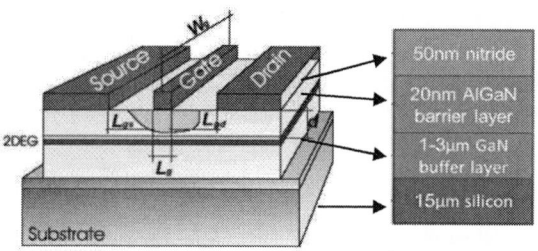

Fig.1. The HFET structure used in this work

TABLE I.   PARAMETERS FOR GEOMETRIC VARIATIONS

Geometric Variations		
Parameter	*Constant dimensions*	*Varied dimensions*
Gate- drain length, Lgd	3μm	3μm-24μm [11]
Gate length, Lg	1.5μm	1.5μm-20μm[15]
Source-gate length, Lsg	1.5μm	0.2μm-2.4μm [16]
GaN buffer thickness	1.0μm	1.0μm &3.0μm[9]
$Al_{0.2}Ga_{0.8}N$ barrier layer	20nm	20nm
Silicon substrate	15μm	15μm

III. RESULTS AND DISCUSSIONS

A. Geometrical Variations

Fig.2 and 3 represent the Id-Vd characteristics near to breakdown for different Lgd and buffer thicknesses. It can be observed when the Lgd increases, the device requires more voltage to achieve breakdown state. Device experiencing breakdown shows a sudden increase in drain current when high voltage is applied as shown in Fig.2. An upward trend for breakdown voltage is observed in Fig. 3. For breakdown simulation, resistance of $10^7\Omega\mu m$ was connected in series to the drain metal electrode and the device was biased with Vgs = -7V at off state. It can be suggested that Lgd is a crucial factor in determining the breakdown voltage. As the distance of Lgd is increased, the concentrated electric field reduces and the devices breakdown at higher voltages thus making it more suitable for higher power applications. In addition, as the buffer thickness increases, the device has a higher breakdown voltage and performs better as Lgd increases. As the buffer layer thickens, the device will have lower dislocation density and lower leakage current through the buffer layer which results in higher breakdown voltage [9]. Thinner buffer layers have less dependency on Lgd as observed in the attained breakdown voltage due to the fact that carrier accumulation at the GaN buffer layer is less which in turn causes the drain voltage to drop vertically across the buffer layer. It is understood that the device breakdown at the drain-sided end of the channel at low drain voltage and majority of the current flow during breakdown

condition is across the silicon substrate [10]. For thicker buffer layer the current flows across GaN buffer layer while only a small part flows through silicon substrate.

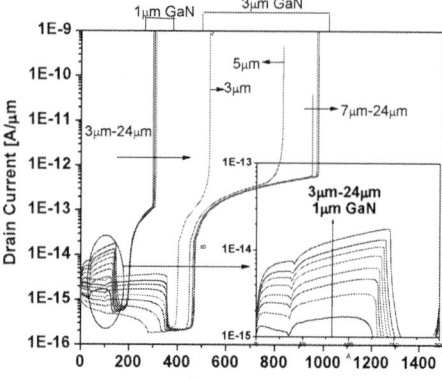

Fig. 2. Id-Vd near breakdown voltage at buffer thickness of 1μm and 3 μm

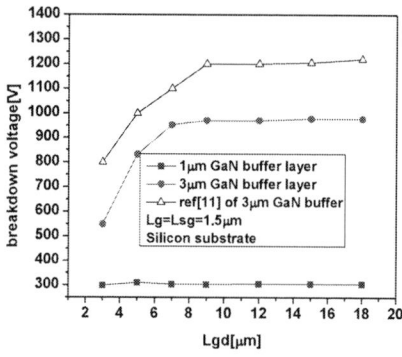

Fig. 3. Breakdown voltage range for varying Lgd

It can also be deduced that for larger Lgd, the electron mobility is substantially higher. Higher electron mobility is an advantage for the device to perform in high power application as device allows more conductivity.

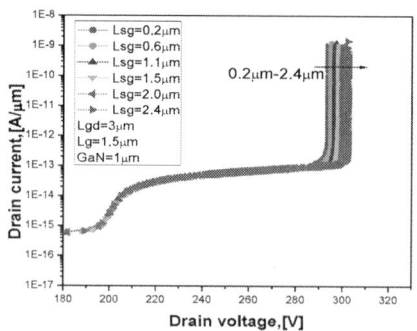

Fig. 4. Breakdown voltage range for varying Lsg

Fig. 4 and Fig. 5 depict the effect in the variation of Lsg and Lg respectively to the electrical properties of the device. It was observed that the downscaling of the source -gate distance improves the device performance but there are limits to the breakdown voltage. At smaller distances, the device experiences hard breakdown while at larger

978-1-5386-5284-8/18 $31.00 © 2018 IEEE

distances, the device suffers from soft breakdown. The breakdown through the scaling of Lsg is affected by two factors which are electric field and carrier concentration. Besides improving the device performance, scaling Lsg has positive consequences on output current and transconductances. Fig. 4 shows that as the distance between source and gate is increased, the higher the breakdown voltage is. Fig.5 suggests that the longer the distance of the gate length, the higher the on-state resistance. As a result, maximum drain current decreases with the increase of gate length. The breakdown voltage increases when the gate length decreases. Device breakdown voltage was induced by buffer leakage current. The breakdown at Lg=3µm is induced by gate leakage current and the increase of breakdown voltage from Lg=3µm to Lg=1.5µm is due to the suppression of buffer leakage current and impact ionization For gate length larger than 3µm, the high electric field is completely depleted and it is difficult for electrons to be injected from the source The type of breakdown after 3µm are induced breakdown voltage. The thicker the buffer layer the higher breakdown voltage of the device.

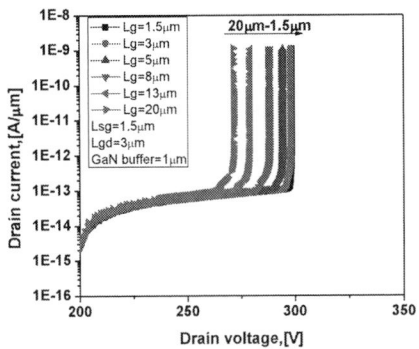

Fig. 5. Id-Vd near breakdown voltage for different Lg

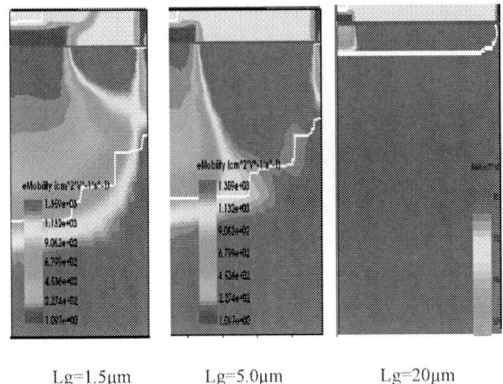

Lg=1.5µm   Lg=5.0µm   Lg=20µm

Fig. 6. Electron mobility countours for different Lg

Fig. 6 presents the electron mobility of the HFET across different gate length. The maximum electron mobility near the gate region is at 1.32e+03cm²V⁻¹S⁻¹. As Lg increases, the electron mobility decreases.

B. Process Variations

The aluminium composition in the AlGaN barrier layer was varied to study its effect on the device performance. It was observed that higher mole fraction will increase the defect density which results in higher contact resistance. This is in turn results in higher breakdown voltages. However, there is a limit of aluminium composition that can influence the performances of device [12]. Higher aluminium composition causes more dislocations and surface defects. In this research, 25% of aluminium composition shows the highest breakdown voltage which is overall a decent performance for other electrical properties. From Fig.7 below can be concluded that 25% of aluminium composition requires the highest voltage for the device to reach breakdown. Lower than 25% of aluminium composition, the device has no definite effect on the quality of the crystalline. The optimum aluminium composition results in high Ion/Ioff ratio and reduces current leakage. Aluminium composition in AlGaN barrier layer affects the 2DEG channel created between barrier layer and the buffer layer. The electron mobility in the 2DEG channel was affected by the variations of the aluminium composition in AlGaN barrier layer. Electron mobility is limited by the scattering due to the Al composition.

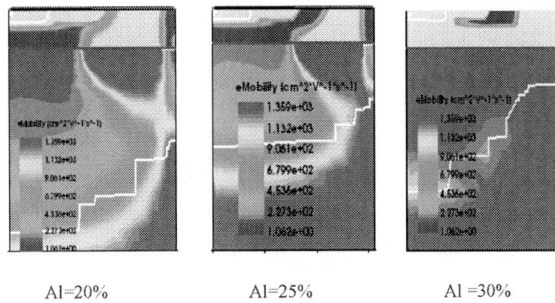

Al=20%   Al=25%   Al =30%

Fig. 7. Electron mobility contours for different Aluminium composition

Fig. 7 suggests that when the Al composition in the AlGaN barrier layer increases, the electron mobility under the gate region which is in the 2DEG channel increases until the composition reaches the optimal value. Al composition is represented by spontaneous polarization. Higher Al composition causes high piezoelectric that results in more electric field and more energy bending. Polarization density was also affected because of lattice mismatch with AlGaN barrier layer and GaN buffer layer which yields high electron density at the interface. Higher aluminium composition increases the electron density in the channel. Total current density increases when aluminium composition increases as shown in in Fig. 8. Gate leakage current also reduces when composition increases, thus improving the device performance [13]. The ideal aluminum composition is around 25% with electron density of 8.23e+09 $^{cm-3}$. High Al composition causes large conduction band discontinuity and wider bandgap [13]. In addition to that, 2DEG sheet density in the channel also improves. This explains why AlGaN barrier layer has higher conduction energy than GaN buffer layer and increasing of aluminium composition in barrier layer electrons requires more energy and in turn making the device perform better in high power application.

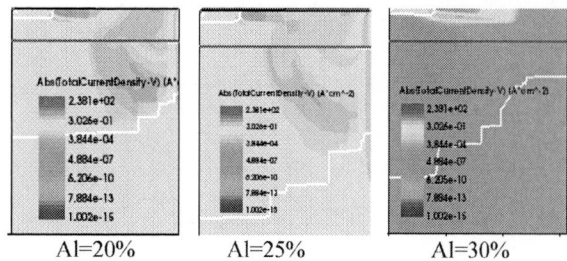

Fig. 8. Total current density contours for different Aluminium composition

This work had also looked into the potentiality of using SiC as the substrate material. Material compatibility was observed through the combination of the GaN and SiC materials. Furthermore, it was reported that the GaN- silicon carbide HFET reduces the crystal lattice difference and lattice mismatch of the device [14]. Based on Fig. 9, the substrate material of SiC reaches breakdown at a higher voltage compared to silicon. The breakdown voltage is 850V when the substrate used was silicon carbide.

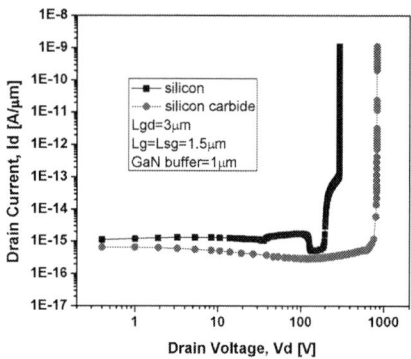

Fig. 9. Id-Vd characteristics for different the HFET device with Si substrate compared to the HFET device with SiC substrate

IV. CONCLUSION

This work investigates the variations conducted on HFET structures for high power and high frequency applications. to study the breakdown voltage and electrical characteristics of the device. Both geometric and process variation conducted was to increase the breakdown voltage. Process variation results in higher breakdown voltage compared to geometric variation. Type of substrate and buffer thickness improved the breakdown voltage. The breakdown voltage of the conventional structure with buffer thickness of 1μm is 300V and when buffer thickness increases to 3μm is 500V. When the substrate material was changed from silicon to silicon carbide at buffer thickness the breakdown voltage is 850V. Besides breakdown voltage, conduction band energy of the device is high that aids HFET to perform better in high voltage application. Current leakage through the substrate triggers the device to breakdown at a faster rate. Besides that, process variation played an important role in reducing current leakage. The aluminium composition in barrier layer reduces current leakage at the gate. The 2DEG channel and the polarization and piezoelectric effect helps reduces current leakage. Silicon carbide as substrate reduces

current leakage through substrate than silicon. This work provides substantial recommendations in increasing the performance of the HFET.

ACKNOWLEDGMENT

The authors are grateful for the financial support provided by the RU Grant (UM.0000482/HRU.OP.RF).

REFERENCES

[1] Zeinab, Kamal Aldein Mohammed, and Sayed Ali Ahmed Elmustafa. "Internet of Things applications, challenges and related future technologies." *World Scientific News* 2.67 (2017): 126-148

[2] Terán, Marco, et al. "IoT-based system for indoor location using bluetooth low energy." *Communications and Computing (COLCOM), 2017 IEEE Colombian Conference on.* IEEE, 2017.

[3] Riel, Heike, Lars-Erik Wernersson, Minghwei Hong, and Jesus A. Del Alamo. "III–V compound semiconductor transistors—From planar to nanowire structures." *Mrs Bulletin* 39, no. 8 (2014): 668-677.

[4] Andrei, C., Doerner, R., Chevtchenko, S. A., Heinrich, W., & Rudolph, M. (2017, October). On the optimization of GaN HEMT layout for highly rugged low-noise amplifier design. In *Microwave Integrated Circuits Conference (EuMIC), 2017 12th European* (pp. 244-247). IEEE.

[5] Rahman, Sharidya, Nurul Aida Farhana Othman, Sharifah Wan Muhamad Hatta, and Norhayati Soin. "Optimization of Graded AlInN/AlN/GaN HEMT Device Performance Based on Quaternary Back Barrier for High Power Application." *ECS Journal of Solid State Science and Technology* 6, no. 12 (2017): P805-P812.

[6] Kaplar, R. J., Andrew A. Allerman, A. M. Armstrong, Mary H. Crawford, Jeramy Ray Dickerson, Arthur J. Fischer, A. G. Baca, and E. A. Douglas. "Ultra-wide-bandgap AlGaN power electronic devices." *ECS Journal of Solid State Science and Technology* 6, no. 2 (2017): Q3061-Q3066.

[7] Du, Jiangfeng, Ruonan Li, Zhiyuan Bai, Yong Liu, and Qi Yu. "High breakdown voltage GaN-on-insulator based heterojunction field effect transistor with a partial back barrier layer." *Superlattices and Microstructures* 111 (2017): 760-766.

[8] Sentaurus Technology Template, "Simulation of Silicon Substrate Effect on AlGaN/GaN HFET offstate Breakdown Voltage", Synopsys Inc. 2016.

[9] Selvaraj, Susai Lawrence, Takaaki Suzue, and Takashi Egawa. "Breakdown enhancement of AlGaN/GaN HEMTs on 4-in silicon by improving the GaN quality on thick buffer layers." *IEEE electron device letters* 30, no. 6 (2009): 587-589.

[10] del Alamo, Jesús A., Alex Guo, and Shireen Warnock. "Gate dielectric reliability and instability in GaN metal-insulator-semiconductor high-electron-mobility transistors for power electronics." *Journal of Materials Research* 32, no. 18 (2017): 3458-3468.

[11] Visalli, Domenica, Marleen Van Hove, Puneet Srivastava, Joff Derluyn, Johan Das, Maarten Leys, Stefan Degroote, Kai Cheng, Marianne Germain, and Gustaaf Borghs. "Experimental and simulation study of breakdown voltage enhancement of AlGaN/GaN heterostructures by Si substrate removal." *Applied Physics Letters* 97, no. 11 (2010): 113501.

[12] Chakraborty, Apurba. "Trapping Characteristics and Reverse Leakage Current Mechanism of Algan/Gan and Algan/Ingan/Gan Heterostructures." PhD diss., IIT, Kharagpur, 2016.

[13] Suh, Chang Soo. "Lateral GaN HEMT Structures." In *Gallium Nitride-enabled High Frequency and High Efficiency Power Conversion*, pp. 29-49. Springer, Cham, 2018.

[14] Kovac, J., A. Kósa, R. Szobolovszky, Aleš Chvála, and Juraj Marek. "GaN/SiC based High Electron Mobility Transistors for integrated microwave and power circuits." *eLearn central, Institute of Electronics and Potonics* (2015).

[15] L.Jun,"Effect of gate length on breakdown voltage in AlGaN/GaN HEMT," *Chinese Physics b*,vol.25,no. 2, p.5 2016

[16] R. Stefano, "Influence of source-gate distance on the AlGaN/GaN HEMT,"IEEE Transaction on elecrtonic devices, vol.54, no. 5, p. 5, 2007

Monolithic Integration of Digital MEMS Thermometer and temperature compensated RTC on 1P6M ASIC compatible CMOS MEMS process

Ching-Wen Hsu
Department of Electronic Engineering
National Chiao Tung University
Hsinchu 300, Taiwan
Email: s952111.ee05g@g2.nctu.edu.tw

Kuei-Ann Wen
Department of Electronic Engineering
National Chiao Tung University
Hsinchu 300, Taiwan
Email: stellawen@mail.nctu.edu.tw

Abstract—**A monolithic resonator based digital MEMS thermometer has been proposed. It is further integrated with Real time clock (RTC) to demonstrate multifunction design as thermometer and temperature compensation. The MEMS/ASIC combined SOC is fabricated on UMC 0.18μm 1P6M ASIC compatible CMOS MEMS process. The resonant frequency of thermometer has been simulated with sensitivity -18Hz/℃. The temperature dependency of RTC frequency is 13.15(ppm/℃) in a temperature range from 20℃ to 80℃. Power consumption of the readout circuit is 440.3μW.**

Keywords—thermometer, RTC, calibration, MEMS resonator

I. INTRODUCTION

Nowadays, the MEMS resonators are widely used for many applications, such as thermometer, pressure sensor and gyroscope. Due to wearable device and IoT become more popular, the demand of sensors grows up.

MEMS resonator has a strong temperature coefficient of frequency [1]. This characteristic is an advantage for temperature sensing. However, it is a drawback for most of the in-chip circuitry which needs stable frequency. In this work, we tried to use the sensitivity of MEMS resonator to sense temperature and to provide compensation as well. RTC is chosen as the sample circuit for our MEMS ASIC integrated SOC design.

The conventional thermometer has been shown in Fig. 1., with phase lock loop (PLL) being integrated. It can track the resonant frequency shifts due to the temperature changes. The VCONT will change when temperature change, and the temperature can be found by VCONT. As we want the digital data, the additional analog to digital converter (ADC) should be integrated.

The comparison of common oscillator has been listed in Table I [1]- [5], we can find that the crystal oscillator has the highest performance among all the oscillators. However, the packing of the crystal oscillator is complex, and it is unable to be total integration. MEMS-based oscillator has high quality and high integration. These characteristics make MEMS- based oscillator be competitive to crystal oscillator. MEMS- based oscillator can replace crystal oscillator to be a novel design of RTC.

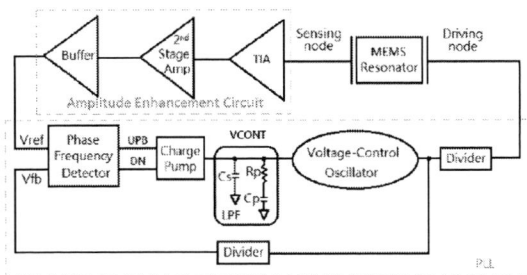

Fig. 1. System Archiecture of conventional thermometer [6]

TABLE I.   COMPARISON OF THE OSCILLATOR

	Oscillator type			
	Ring oscillator	**LC oscillator**	**Crystal oscillator**	**MEMS-based oscillator**
Quality factor	20.3	20.4	2.5x106	>100
TCF(ppm/℃)	800	15	±3	-30

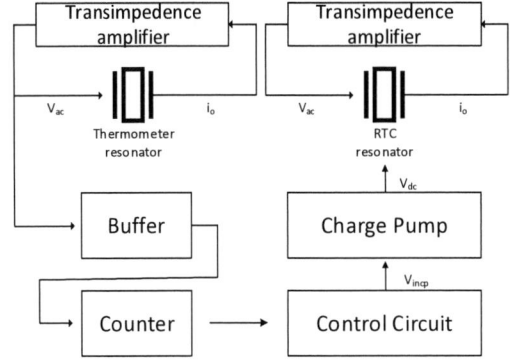

Fig. 2. Architecture of proposed thermometer and RTC.

978-1-5386-5284-8/18 $31.00 © 2018 IEEE

II. PROPOSED THERMOMETER AND RTC

A. Architecture

The architecture of system is shown in Fig. 2 and The top-view of the proposed resonator is shown in Fig. 4. The frequency of thermometer MEMS resonator would decrease while temperature grows up. The Coventorware simulation result of temperature-frequency characteristics is shown in Table II [7]. Using counter to count what the resonator frequency it is, the temperature can be convert to digital data with no need for additional ADC. After having the digital data, the control circuit can convert these digital data to voltage for charge pump input. The charge pump can provide an ideal V_{dc} voltage to resonator. According to the soften effect, this characteristic can provide the calibration method for RTC resonator. The detail of the electrostatic force actuation mechanism, soften effect and circuit design will be discussed below.

B. Electrostatic Force Actuation Mechanism

As illustrated in Fig. 3, when the driving electrode is given an AC signal and the shutter was given a V_{dc} voltage, the shutter would be actuated by electrostatic induction. When the frequency of the AC signal reaches the natural frequency of the resonator, the shutter would produce the maximum displacement. The capacitance C_{sense} would vary, and the variation of the capacitance C_{sense} convert to current signal.

C. MEMS Resonator-based Thermometer Design

The mass-spring-damper model is shown in Fig. 5. This model assumed that the mass of resonator was lumped into an effective mass m_e and it displacement was denoted by x. Similarly, the mechanical stiffness of spring was lumped into an effective spring constant k_e and all energy loss mechanisms were lumped into a single damper with an effective coefficient b_e. For any external actuation Force F_{act}, we have the classical differential equation of mechanical equilibrium as (1).

$$m_e x'' + b_e x' + k_e x = F_{act} \qquad (1)$$

By taking Laplace transform of (1), we obtain the transfer function as (2)

$$x(s)/F_{act}(s) = k^{-1}/(1 + s/(Q*w_n) + s^2/w_n^2) \qquad (2)$$

From (2), we have some important parameters as (3) and (4)

$$w_n = (k_e/m_e)^{1/2} \qquad (3)$$

$$Q = ((k_e*m_e)^{1/2})/b_e \qquad (4)$$

w_n and Q are natural frequency and quality factor. m_e is not a temperature dependent parameter, but k_e is. As the result, the natural frequency is a temperature dependent parameter.

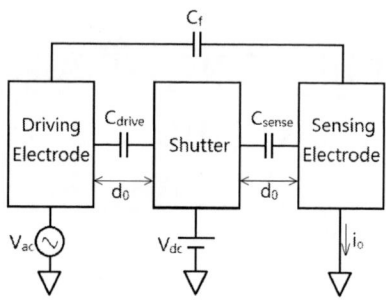

Fig. 3. Electrostatic force actuation diagram.

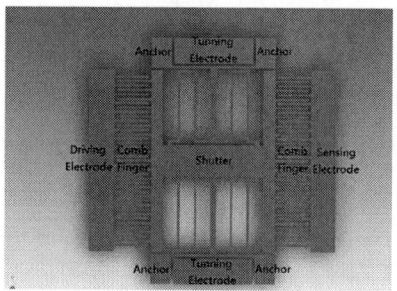

Fig. 4. The top-view of the proposed resonator.

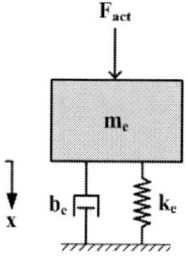

Fig. 5. The mass-spring-damper model.

D. Soften Effect

The voltage V_{dc} of Shutter affects the viscous force between fingers and thus affects the frequency. The higher the voltage, the greater the viscosity and the lower the frequency. We use this characteristic to make oscillators whose frequency do not change with temperature. When the temperature rises, the voltage is reduced so that the real-time clock frequency does not decrease with increasing temperature. Conversely, when the temperature drops, the voltage rises, so that the real-time clock frequency does not rise with the temperature drop. As the result, the soften effect can provide RTC a stable frequency even under temperature changes.

E. Trans-impedance Amplifier

The trans-impedance amplifier (TIA) is used to convert current input signal, i_o, into voltage output signal, V_{ac}. The trans-impedance gain is defined as $A_{TIA} = \partial V_{ac}/\partial I_o$ and has

978-1-5386-5284-8/18 $31.00 © 2018 IEEE   46

the unit of Ω(V/A). The conventional TIA is resistive feedback TIA, where the output voltage will be sensed and return current signal to input through the feedback resistor.

F. Control Circuit

When the temperature is low, the frequency of the thermometer is higher and the digital output value which generated by the counter is larger. After the control circuit, refer to Fig. 2, a higher V_{incp} voltage can be generated for the subsequent charge pump circuit. Conversely, if the temperature is high, a lower V_{incp} voltage is generated for the subsequent charge pump circuit and the real-time clock calibration is completed. The control circuit is completed by resistive voltage divider and MOS switch.

G. Charge Pump

Taking [8] as reference, the diode of conventional charge pump is replaced by PMOS. Two auxiliary MOSFET can control the body voltage, and the charge transfer MOSFET can avoid the influence of body effect. Charge pump can convert V_{incp} to V_{dc}, and calibrates the real-time clock oscillator to eliminate the influences of temperature. The V_{incp} to V_{dc} correspondence diagram is shown in Table III.

III. EXPERIMENT RESULTS

The simulation result of thermometer is shown in Table II. The temperature coefficient of frequency (TCF) can achieve 419 ppm/℃. The effect of calibration is shown in Fig. 6. and Fig. 7. Between 20°C and 80°C, the non-calibrated frequency difference is 678.04Hz and can be reduced to 25Hz after calibration. The overall layout diagram of the chip is shown in Fig. 8. The performance summary and conclusion of RTC part is shown in Table IV. Due to MEMS resonators and test circuit, this paper consumes lots of area. In order to achieve the Barkhausen criterion for the oscillator, high gain-bandwidth product of TIA is needed. As the result, the TIA circuit consumes lots of power. Reducing the natural frequency of resonator and lowering the motional impedance can lessen the requirement of TIA's power. It can be an improvement for power performance. Although this paper consumes higher area and power, but it can achieve 13.15ppm/℃. The stress analysis of thermometer and RTC are shown in Fig. 9 and Fig. 10. The measurement result of thermometer resonator is shown in Fig. 11. The natural frequency is 42.6kHz.

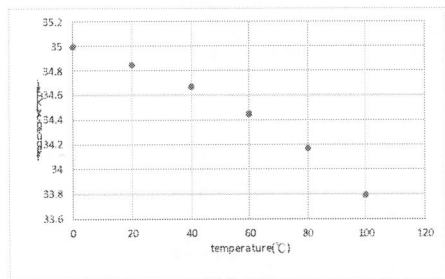

Fig. 6. RTC's before calibration diagram.

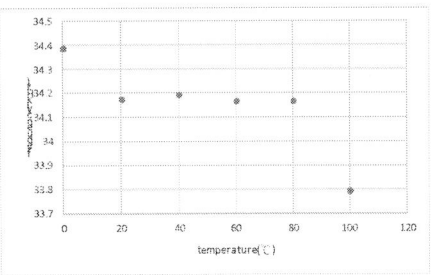

Fig. 7. RTC's after calibration diagram.

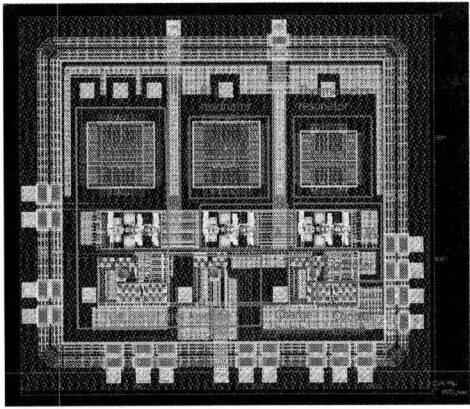

Fig. 8. The overall layout diagram of the chip

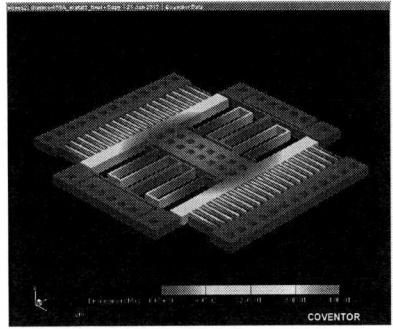

Fig. 9. The stress analysis of thermometer.

TABLE II. COVENTORWARE SIMULATION RESULT OF TEMPERATURE-FREQUENCY

Temperature(℃)	Frequency(Hz)
0	45078.44
20	45026.51
40	44728.79
60	44351.66
80	43859.07
100	43188.49

TABLE III. THE V_{INCP} TO V_{DC} CORRESPONDENCE DIAGRAM

Temperature(℃)	V_{incp}	V_{dc}
0~20	1.8	26.6
~40	1.5	21.6
~60	1.2	16.4
80~100	0.6	6.47

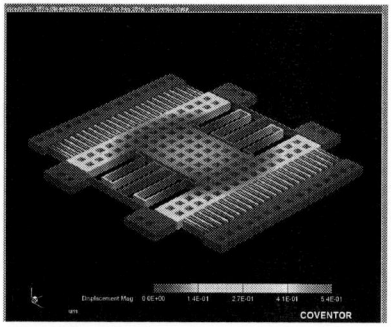

Fig. 10. The stress analysis of RTC.

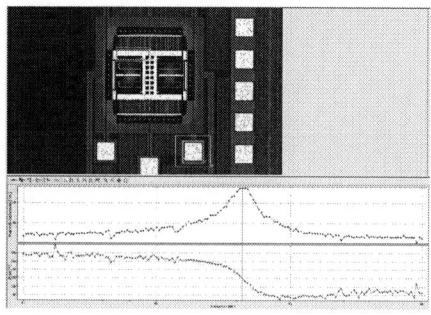

Fig. 11. The measurement result of thermometer resonator.

TABLE IV.   PERFORMANCE SUMMARY AND COMPARISON

Reference	[10]	[11]	[12]	This work
Tech[nm]	350	180	65	180
Area[mm^2]	0.1	0.11	0.022	2.677
V_{DD}	1~2.5	1.0~1.8	1.1~1.3	1.8
T[$^{\circ}$C]	-20~80	-40~100	-40~80	20~80
F_{out}[kHz]	3.3	32.6	32.5	34.2
Power	11nW	472nW	271nW	440.2uW
$\Delta f/f\Delta T$ [ppm/$^{\circ}$C]	<500	120	138	13.15

IV. CONCLUSIONS

This paper proposed two novel designs. The one is to convert the temperature into a digital output without going through the PLL and the ADC circuit. The other is to complete a RTC whose frequency has stable temperature response. Although the Q value of the MEMS resonator is worse than that of the quartz oscillator, it is much better than the LC oscillator and the ring oscillator. And it can be highly integrated with the circuit.

ACKNOWLEDGMENT

This design is sponsored in part by the National Science Council of Taiwan under grant of MOST 104-3115-E-009-022. The authors appreciate the UMC and the National Chip Implementation Center (CIC), Taiwan, for supporting the CMOS chip manufacturing.

REFERENCES

[1] J. Gronicz et al., "A 2µA temperature compensated mems-based real time clock with ±4 ppm timekeeping accuracy " In 2014 IEEE international symposium on circuits and systems (ISCAS) , pp. 514–517

[2] D. Salt, Hy-Q Handbook of Quartz Crystal Devices, Van Nostrand Reinhold UK Co. Ltd.

[3] J. Imbaud et al., "Analyzes of very high-q quartz crystal aimed to high quality 5 MHz resonators achievement" , 24th European Frequency and Time Forum, Noordwijk, NL, 13-16 April, 01_03_Imbaud.pdf, (2010).

[4] F. Brandonisio and M. P. Kennedy, "Comparison of ring and LC oscillator-based ILFDs in terms of phase noise, locking range, power consumption and quality factor," in Proc. Ph.D. Res. Microelectron. Electron., Jul. 2009, pp. 292–295.

[5] J.-C. Chiou et al., "Novel application of ring-oscillator-based sensor for 3D IC temperature and IR drop monitoring" IEEE Electrical Design of Advanced Packaging and Systems Symposium, 14-16 Dec. 2017

[6] C. Y. Lin and K. A. Wen, "MEMS resonator based thermometer SoC design in CMOS 0.18µm standard process," Soc Design Conference (ISOCC). Jeju, South Korea, pp. 13-14, Dec 2016.

[7] Coventorware-https://www.coventor.com/

[8] J. Shin, I.-Y. Chung, Y. J. Park, and H. S. Min, "A new charge pump without degradation in threshold voltage due to body effect[memory applications]," IEEE J. Solid-State Circuits, vol. 35, no. 8,pp. 1227–1230, Aug. 2000.U.

[9] T. Tanzawa and T. Tanaka, "A dynamic analysis of the dickson charge pump circuit," Solid-State Circuits, IEEE Journal of, vol. 32, pp. 1231– 1240, Aug 1997.

[10] Denier et al., "Analysis and design of an ultralow-power CMOS relaxation oscillator," IEEE Tran. Circuits Syst. I, vol. 57, no. 8, pp. 1973-1982, 2010.

[11] K. Tsubaki et al., "A 32.55-kHz, 472-nW, 120ppm/°C, fully on-chip, variation tolerant CMOS relaxation oscillator for a real-time clock application," in Proc. ESSCIRC, 2013, pp.315-318.

[12] H. Asano et al., "An area-efficient, 0.022-mm2, fully integrated resistor-less relaxation oscillator for ultra-low power real-time clock applications" in ISCAS, Sept 2017.

Influence of Titanium Oxide Coating on Mechanical Properties of Porous Nanocrystalline Silicon Membrane

Muhammad Fahmi bin Jaafar, Rhonira Latif, Burhanuddin Yeop Majlis
Institute of Microengineering and Nanoelectronics (IMEN)
Universiti Kebangsaan Malaysia
Bangi, Malaysia
p87404@siswa.ukm.edu.my

Abstract—**Porous nanocrystalline silicon (pnc-Si) membrane is a promising material to be used as the filtration membrane in an artificial kidney for its physical characteristics such as having a very small thickness and pores. However, hemocompatibility issue needs to be addressed to ensure the reliability of the materials used for the filtration membrane. Titanium oxide (TiO) has been demonstrated to be a hemocompatible material. In this paper, the pnc-Si membrane is coated with TiO layer and the change in membrane's mechanical strength is simulated. Simulation result has shown positive outcome in which the Von Mises stress acted upon the membrane has been seen decreasing with the addition of TiO layer on top and underneath the pnc-Si membrane. Besides, the displacement of the membrane has also been observed to be smaller. Thus, it is feasible to fabricate membrane with larger dimensions.**

Keywords — pnc-Si membrane, TiO coating, hemocompatibility, mechanical strength.

I. INTRODUCTION

In the development of an artificial kidney, filtration membrane takes the spotlight for being the most important site on the complex system, that is for the filtration of blood, which will mimic the main function of a healthy kidney. In our work, porous nanocrystalline silicon (pnc-Si) is considered to be used as the filtration membrane. Bare silicon that is brought into contact with blood will cause adverse reactions. The human blood is capable of sensing foreign material in its flow and activates the platelet adhesion, protein adsorption and coagulation (thrombin-anti thrombin complex, TAT generation). One of the most problematic factor that reduces the performance of the conventional hollow fiber in dialysis machine is its material which only have the ability to be used for less than a hundred hour due to the accumulation of protein fouling and thrombosis around the membrane [1-2].

To overcome this problem, surface modifications or hemocompatible coating applications should suffice. Polyethylene glycol (PEG), polysulfobetaine methacrylate (pSBMA) and diethylaminoethyl (DEAE) cellulose are types of materials used for surface modifications of silicon surface that have been proven to show hemocompatibility and hydrophobicity factor [3]. In addition, titanium oxide (TiO) thin film is a well-known biocompatible material coating in biomedical applications. Deposition of titanium and titanium oxide layer on top of silicon has been demonstrated to significantly reduce the accumulation of thrombosis and

platelet on the membrane surface [4-5]. The downside of coating extra layers of materials is that the total thickness of the membrane will increase, reducing transport of blood due to steric hindrance at the pore entrance and frictional interactions between the molecules and the pore walls as the blood is passing through. This will affect the filtration rate of blood passing through the membrane and will require calculation/simulation to compare with the performance specifications of a proposed artificial kidney.

However, thicker membrane will improve the mechanical strength of the whole structure and make it possible to increase the surface area of the membrane which will be fabricated free-standing on the silicon substrate [6-7]. Mechanical strength of a pnc-Si membrane with dimension 200 μm × 200 μm and thickness of 15 nm has been demonstrated in the previous work that it can stand up to 760 mmHg [8]. According to study, the membrane for artificial kidney should be able to withstand an applied pressure of 10 – 55 mmHg which is the blood pressure around the kidney area [9]. Thus, in this paper we have studied the limitation of membrane size in regards to the membrane thickness and the influence of TiO coating on the mechanical strength of pnc-Si membrane.

II. SIMULATION MODEL

In our study, COMSOL Multiphysics 5.1 has been utilized to simulate von Mises stress and membrane displacement with respect to the applied pressure on the membrane. For the model geometry, 3D structure has been built with varying width dimension from 1 μm to 10 μm and total thickness from 15 nm to 40 nm. The influence of surface area and membrane thickness onto the resulting von Mises stress and the maximum displacement of the membrane are simulated. First, a bare pnc-Si membrane geometry is simulated as reference before thin layer of TiO will be included in the simulation. The TiO layer will be varied from 5 nm to 20 nm on top and below of the pnc-Si membrane in order to find the influence of adding this hemocompatible coating on the mechanical strength of the membrane. For this study, the applied pressure on the top of the membrane is set as 55 mmHg, the maximum assumptions for the blood pressure surrounding the kidney area. Four sides of the membrane are set as fixed constraint, which will hold the free standing membrane at the center. The strength calculation is set based on the yield strength of silicon and titanium for the von Mises stress limit, which are 1.1 GPa and 0.434 GPa, respectively. The structural mechanics study is done with tetrahedral meshing under stationary condition in which time is independent from the calculation.

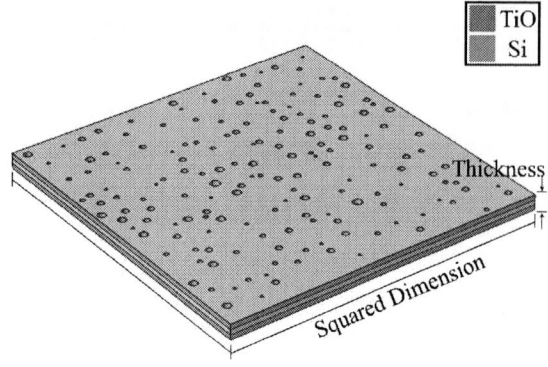

Fig. 1. 3D-Geometry structure of the pnc-Si membrane sandwiched between thin layers of TiO coating.

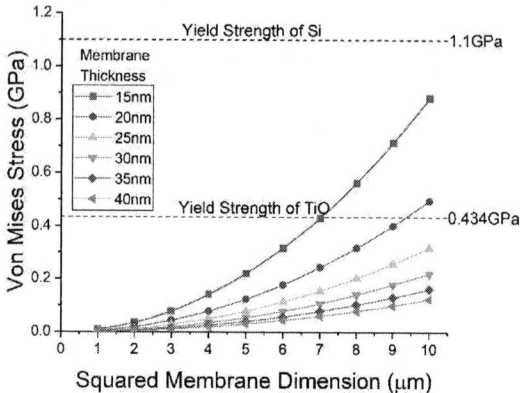

Fig. 2. Von Mises stress acting on a bare pnc-Si membrane at different membrane thickness.

III. RESULTS AND DISCUSSION

A. Mechanical strength of bare pnc-Si membrane

Von Mises stress resulting from 55 mmHg of applied pressure on the bare silicon membrane is shown in Fig. 2. The stress has been simulated to increase with respect to the increase of membrane size and decrease as the membrane thickness increases. Up to 10 μm squared membrane, the von Mises stress exerted is still lower than the value of yield strength for silicon. Yield strength of silicon is around 1.1 GPa while the maximum stress simulated on the 10 μm squared membrane with thickness of 15 nm is 0.9 GPa. Based on our simulation, the von Mises stress exerted by 55 mmHg of pressure on top of the membrane will not affect the elasticity of the pnc-Si membrane. The stress will definitely increase along with the membrane size area and consequently reaches its limit, however, in this study, we are trying to add the influence of titanium oxide coating which hypothetically will further increase its mechanical strength.

In Fig. 3, the displacement value increases drastically in thinner membranes as the membrane area increases. On the other hand, the membrane displacement decreases with the increase of membrane thickness. In thin film, the maximum displacement or deflection that can act on the membrane should be the same with its thickness value to prevent breaking [10]. From the simulation for 15 nm membrane thickness, squared dimensions which can handle the displacement due to the applied pressure is only up to 2 μm before the displacement value turns out to be higher than the thin film thickness. Even for a membrane with 40 nm thickness, the value of squared dimension cannot exceed 8 μm to withstand 55 mmHg pressure. Increasing the square dimensions of the membrane will further increase the maximum displacement value at the center of the membrane. This situation can be explained by looking at the fixed constraint acting on the four sides of the membrane. The maximum vibration of the membrane which is located at the center becomes further apart from its holder as the membrane size area increases.

By increasing the thickness of the membrane, the maximum displacement value that can be hold without breaking will also be increased. Thicker membrane has a higher flexural strength, which can indirectly be equated with yield strength [11].

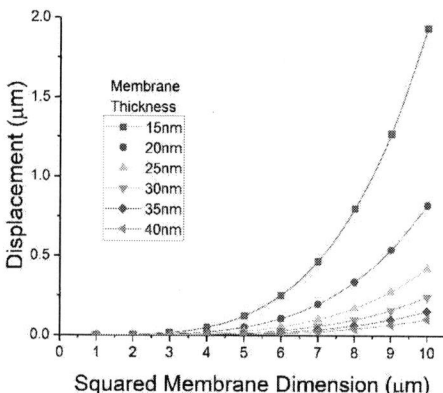

Fig. 3. Maximum displacement of a bare pnc-Si membrane at different membrane thickness.

B. Influence of TiO layer on the mechanical properties of pnc-Si membrane

We have found out that the coating of titanium oxide layer on top and underneath the pnc-Si membrane has increased the mechanical strength of the membrane. From the simulation result shown in Fig. 4, we found out that the increase of TiO thickness increases the membrane's ability to withstand higher pressure. The addition of TiO layer onto pnc-Si membrane reduces the von Mises stress acting upon it. Further increase in the TiO layer thickness has reduced the stress significantly.

978-1-5386-5284-8/18 $31.00 © 2018 IEEE

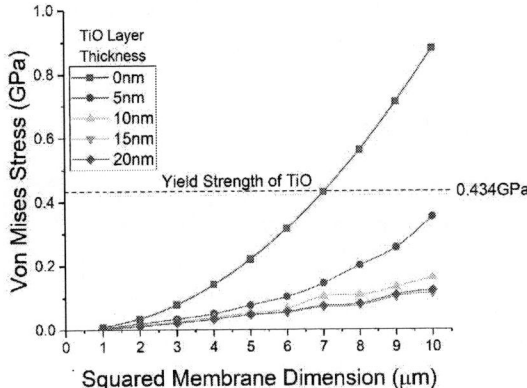

Fig. 4. Von Mises stress acting on a 15 nm pnc-Si membrane with different TiO layer thickness.

We found out that adding a 15 nm layer of TiO will be enough to set the von Mises stress at a lower value and at the same time provide sufficient pore size on the membrane that is needed for transport of molecules. This value is assumed as we tried to halve the value from the previous study which shows the Gaussian distribution of pore size across a pnc-Si membrane [6]. The hypothetically pores distribution and density changes is shown in Fig. 5.

With an additional layer of 15 nm TiO, we can assume that the mean pore diameter will be 10 nm in size, which is good enough to be implemented in the artificial kidney. That is because the size of the pores on the membrane should be made small to only allow small molecules such as urea and toxin to pass through. It is noted that the pore density is also reduced.

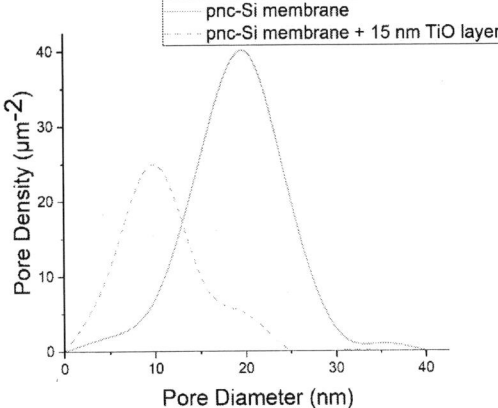

Fig. 5. Average pore density and diameter for pnc-Si membrane.

Fig. 6 shows the influence of pnc-Si membrane thickness at a constant 15 nm TiO layer on stress. We can see that the highest value of stress exerted on the membrane is only 0.11 GPa for a squared dimension of 10 μm. This amount equals to approximately 88% reduction from the maximum stress exerted on the bare pnc-Si membrane of the same size.

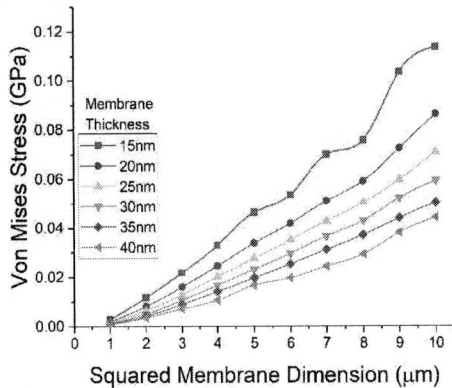

Fig. 6. Von Mises stress acting on pnc-Si membrane at different thickness with 15 nm TiO layer.

The displacement value for the membrane is also increased directly with the increase of pnc-Si membrane thickness at a constant TiO thin layer thickness of 15 nm as shown in Fig. 7. In comparison to bare pnc-Si membrane, the additional layer of TiO helps to reduce the maximum membrane displacement enormously. Thus with additional of 15 nm TiO, a 40 nm pnc-Si membrane can be fabricated to more than 10 μm × 10 μm without breaking.

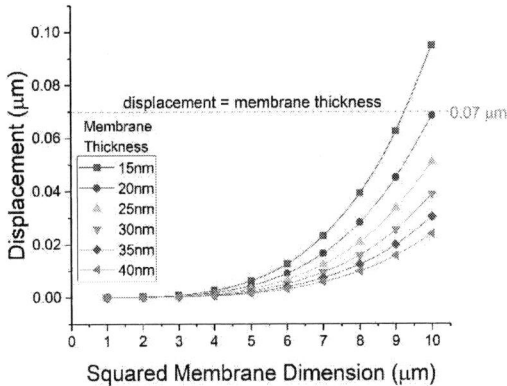

Fig. 7. Maximum displacement of pnc-Si membrane at different thickness with 15 nm TiO layer.

IV. CONCLUSION

From the simulation result, it shows that the addition of TiO thin film layer for pnc-Si membrane will not only be useful for solving the hemocompatibility issue but at the same time helps in reducing the total residual stress within the filtration membrane layer. However, the downside of this additional layer on the membrane is that the frictional force will become higher, and the pores on the membrane will also reduce in size due to the coating thickness. Thus, further study needs to be done in order to optimize the thickness of TiO coating on the membrane so that the flow of molecules passing through can match the performance specifications for

978-1-5386-5284-8/18 $31.00 © 2018 IEEE

an artificial kidney. Exact size of pores which is needed to be fabricated for the application as a filtration membrane will be studied. Here, we found a way to further reduce the pore size by adding a hemocompatible layer on top and underneath the pnc-Si membrane.

ACKNOWLEDGMENT

The work was supported by GGPM-2016-032 research grant funded by Universiti Kebangsaan Malaysia (UKM).

REFERENCES

[1] S. T. Kelly, A. L. Zydney, "Protein fouling during microfiltration: comparative behavior of different model protein," in Biotechnology Bioengineering. Vol 55, pp 91-100, 1997.

[2] K. Vernon, J. Peasegood, A. Riddell, A. Davenport, "Dialyzers designed to increase internal filtration do not result in significantly increased platelet activation and thrombin generation," in Nephron Clinical Practice. London, vol. 117(4) pp c403–c408, 2011.

[3] L. Muthusubramaniam, R. Lowe, W. H. Fisell, L. Li, R. E. Marchant, T. A. Desai, S. Roy, "Hemocompatibility of Silicon-Based Substrates for Biomedical," in Annals of Biomedical Engineering. Vol. 39, No. 4 pp. 1296-1305, April 2011.

[4] N. Huang, P. Yang, Y. X. Leng, J. Y. Chen, H. Sun, J. Wang, G. J. Wang, P. D. Ding, T. F. Xi, Y. Leng, "Hemocompatibility of titanium oxide films," in Biomaterials. Vol. 24, pp 2177-2187, 2003.

[5] C. He, A. Qin, J. Diao, and Y. Gao, "Progress in the research of hollow fiber membrane," in Materials China. Vol. 32, pp. 354–361, 2013.

[6] N. Burham, A. A. Hamzah, B. Y. Majlis, "Mechanical characteristics of porous silicon membrane for filtration in artificial kidney" in Semiconductor Electronics (ICSE), 2014 IEEE International Conference, Kuala Lumpur, pp 119-122, 2014.

[7] M. F. Jaafar, R. Latif, B. Y. Majlis, "Mechanical Properties of Porous Nanocrystalline Silicon Membrane for Artificial Kidney," in IEEE Regional Symposium on Micro and Nanoelectronics (RSM), pp 26-29, 2017.

[8] C. C. Striemer, T. R. Gaborski, J. L. McGrath, P. M. Fauchet, "Charge- and size-based separation of macromolecules using ultrathin silicon membranes," in Nature. Vol. 445, No.7129, pp 749-753, 2007.

[9] N. A. Brunzel, Fundamentals of Urine & Body Fluid Analysis, Elsevier Third Edit, 2013.

[10] J. M. Hodgkinson, Mechanical Testing of Advanced Fibre Composites. Woodhead Publishing, Ltd.: Cambridge, 2000, pp. 132–133.

[11] D. Z. Fang, "Fabrication, Characterization, and Functionalization of Porous Nanocrystalline Silicon Membranes," PhD thesis, Edmund A. Hajim School of Engineering and Applied Sciences, University of Rochester, 2010.

Influence of volume variety of waste cooking palm oil as carbon source on graphene growth through double thermal chemical vapor deposition

Robaiah Hj Mamat
NANO-SciTech Centre, Institute of Science, Faculty of Applied Science, UniversitiTeknologi MARA, 40450 Shah Alam, Selangor, Malaysia
robaiah_hjmamat@yahoo.com

Fazlena Hamzah
Faculty of Chemical Engineering, UniversitiTeknologi MARA, 40450 Shah Alam, Selangor,Malaysia
fazlena@salam.uitm.edu.my

Azhan Hashim
Faculty of Applied Science, Universiti Teknologi MARA, Kampus Jengka,26400 Bandar Tun Abdul Razak Jengka, Pahang, Malaysia
dazhan@pahang.uitm.edu.my

Saifollah Abdullah
NANO-SciTech Centre, Institute of Science, Faculty of Applied Science, UniversitiTeknologi MARA, 40450 Shah Alam, Selangor, Malaysia
saifollah@salam.uitm.edu.my

Salman.A.H.Alrokayan, Haseeb.A.Khan
Research Chair for Biomedical Applications of Nanomaterials Department of Biochemistry College of Science King Saud University Riyadh
salrokayan@ksu.edu.sa
haseeb@ksu.edu.sa

Munirah Safiay, Salifairus Mohamad Jafar
NANO-SciTech Centre, Institute of Science, Faculty of Applied Science, UniversitiTeknologi MARA, 40450 Shah Alam, Selangor,Malaysia
Munirahsafiay@gmail.com
salifairuz@salam.uitm.edu.my

Asnida Asli, Zuraida Khusaimi
NANO-SciTech Centre, Institute of Science, Faculty of Applied Science, UniversitiTeknologi MARA, 40450 Shah Alam, Selangor, Malaysia
asnidasli@yahoo.com
zurai142@salam.uitm.edu.my

Mohamad Rusop
NANO-SciTech Centre, Institute of Science, NANO-ElecTronic Centre, Faculty of Electrical Engineerin, Universiti Teknologi MARA, 40450 Shah Alam, Selangor, Malaysia
nanouitm@gmail.com

Abstract—In this paper, the growth of graphene on a nickel (Ni) substrate using waste cooking palm oil (WCPO) as a carbon precursor by double thermal chemical vapor deposition (DTCVD) has been reported. Various amounts of WCPO which comprised of 5μl, 10 μl, 50 μl and 100 μl are employed for the growth of graphene. At 10 μl, graphene effectively formed due to segregation and precipitation more appreciable amount of carbon on the Ni surface during cooling process. Using an ample amount of carbon source resulted in the formation of graphene oxide and thick amorphous carbon. The prepared samples were characterized using FESEM, XRD and Raman spectroscopy. FESEM analysis reveals prominent structural changes upon applying various amount of waste cooking palm oil. From Raman spectroscopy, 10 μl samples suggesting multilayer of the graphene derived from the ratio peak intensity (I_{2D}/I_G equal to ~0.30).

Keywords—graphene, nickel, palm oil, amount, thermal chemical vapor deposition

I. INTRODUCTION

Graphene is a single two-dimensional (2D) layer or a monolayer of sp^2-bonded carbon atoms in a hexagonal lattice structure [1]. Graphene is an incredible material with unique properties due to their superior electrical, excellent high charge-carrier mobilities, exceptional mechanical stiffness and chemical properties for electrical device application. The mass production of high quality graphene sheets at a reduced cost is vital to encourage multinationals industry such as electronics and composite key player to adopt graphene-based technology.

Therefore, the developments of environmental friendly methods paired with cost effective is crucial for large scale production of enormous industrial applications [2]. Several methods have been developed to synthesis graphene such as mechanical cleavage of graphite crystals, exfoliation of graphite through its intercalation compound, organic synthesis processes, chemical method, and chemical vapor deposition (CVD) on different substrate crystals. Nevertheless, synthesis of graphene through double thermal chemical vapor deposition method (DTCVD) is favorably adopted due to its mechanism process, inexpensive precursor, natural and eco-friendly approach, with effective and reliable method [3].

Out of the ordinary, Suriani et al. has noticed that refined palm oil is natural carbon source which results in more efficiency compared to the non-renewbale carbon sources such as ethanol, methane, acetylene and ethylene in CNT synthesing process [4]. Therefore, utilization of waste cooking palm oil as carbon source for producing graphene is well suitable for mass production due to the low-cost procedure. The palmitic acid is a saturated long chain fatty acid with a 16 carbon backbone ($CH_3(CH_2)_{14}COOH$) and it is a major component of palm oil. In fact, the carbon content acts as main substituent atom supplier to synthesis graphene [5]. According to Meesuk & Sami (2010), the palmitic acid of used cooking palm oil is higher compared to unused cooking palm oil with 41% and 35%, accordingly. Therefore, execution of waste cooking palm oil as carbon source might eliminates environmental problem and significantly minimizes the manufacturing cost to acquire carbon source in graphite synthesis [6].

In this paper, we reported the growth of graphene at different volume of waste cooking palm oil as carbon source via double thermal chemical vapor deposition on Ni

978-1-5386-5284-8/18 $31.00 © 2018 IEEE

substrate. Influence of temperature on various waste cooking palm oil as precursor has been reported in our previous report [6 – 7]. All the synthesized samples undergo characterization of FESEM, EDX, and Raman spectroscopy.

II. EXPERIMENTAL DETAILS

Waste cooking palm oil (WCPO) was obtained from refined domestic cooking palm oil that has been used to fry chickens. Each frying cycle lasted about 18 mins with temperature not exceeding 180 °C. The oil keeps constantly used until three-time frying cycles at the same amount of chicken, time and temperatures. The WCPO was filtered using the filter funnel whereas the filter paper is used to remove solid objects and precipitate. The substrate was prepared by cutting the nickel film into an area of 1 cm² and cleaned with acetone, ethanol, methanol, and de-ionized water, respectively before dried them. Double thermal chemical vapor deposition (DTCVD) was utilized to synthesize graphene. An alumina boat containing different amount of waste cooking palm oil (WCPO), carbon source, was placed at a distance of 15 cm from the center point of the tube in furnace 1. In this work, various precursor concentrations were employed at 5 µl, 10 µl, 50 µl and 100 µl., Separately, Ni substrate was placed on alumina boat at a distance of 15 cm from the center point of the tube in furnace 2. Argon gas was flown for 10 minutes at 200 sccm to remove any impurities inside the quartz tube and to establish an inert atmosphere for the deposition purpose [7]. The temperature of precursor and deposition were fixed at 350 °C and 900 °C, respectively. Upon reaching the desired temperature; the synthesis process took 15 mins and the sample was cooled down to room temperature before being extracted for further analysis [7]. Based on graphene growth mechanism the formation of graphene by the segregation and precipitation of carbon atoms on the nickel surface during the cooling process.

III. RESULTS AND DISCUSSIONS

A. Surface Morphology

The surface morphology of graphene film was observed by the field emission scanning electron microscope (FESEM) at 5000x of magnification. Fig. 1(a) has shows the surface morphology of blank nickel without introducing of the carbon source. Fig. 1(b)-(e) exhibits the samples grown at 5 µl, 10 µl, 50 µl, and 100 µl amount of precursor, respectively. The observation verifies that the amount of carbon source has the significant effect on the surface morphologies of graphene. The amount of carbon source is correlated directly to the rate of decomposition of carbon atoms on nickel surface. At 5 µl amount of carbon source, a small trace of hexagonal structure can be observed. Apparently, the Ni surface seems to be covered due to the formation of porous structures, dark fragments and wrinkles which attributed to low segregation of carbon particles [8]. In contrast, the hexagonal shapes with bright particles were observed on the surface of graphene films as depicted in Fig 1(c). In brief, graphene successfully grown on the nickel substrate with utilization of 5 µl amount of carbon source as confirmed by Raman analysis. It might be postulated that the amount of waste cooking palm oil has undergo thermal decomposition with subsequent diffusion onto Ni substrate. In fact, increment of amount carbon source up to 50 µl seems sufficient to produce graphene oxide on transition metals.

From Fig. 1(d), it reveals a densely compact of porous graphene structure with wrinkles-like and coarse appearance during the formation of graphene oxide. In fact, a non-uniform/irregular hexagonal shape was discovered. This phenomenon happened due to (i) suffcient supply of carbon and incomplete diffusion into Ni substrate by the decomposed carbon source. As a result, accumulation of particles occur on the surface of Ni substrate [7]. According to the graphene growth mechanism on Ni, graphene is established on Ni surface as an outcome of the segregation and precipitation of carbon atoms on Ni substrate during cooling process [9]. High deficiency is observed with further increment for the amount of carbon source as evidenced in FESEM image of Fig 1(e). It exposed that carbon atoms preferably assembles along the grain-boundaries which indicates a high curvature and superior density of atomic steps at the boundaries [10]. Previous study have shown that the abundant of amount carbon source produced amorphous carbon [11]. It can be speculated that the amorphous carbon is acquired due to excessive amount carbon source used which exceeds the synthesis time and incomplete diffusion diffuse into nickel substrate. Consequently, the amount of carbon source is greatly influenced the quality of generated graphene [11].

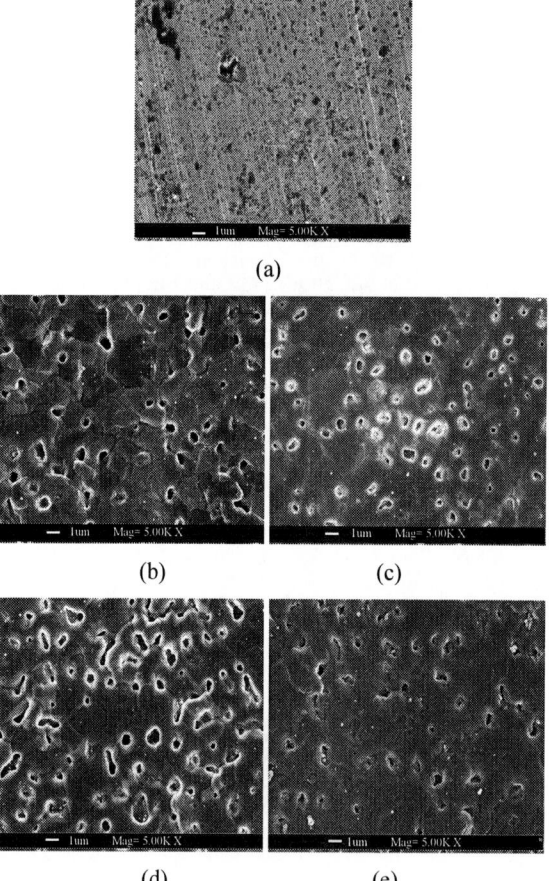

Fig. 1. FESEM images of graphene at different amount of precursor, (b) 5 µl, (c) 10 µl, (d) 50 µl, (e) 100 µl.

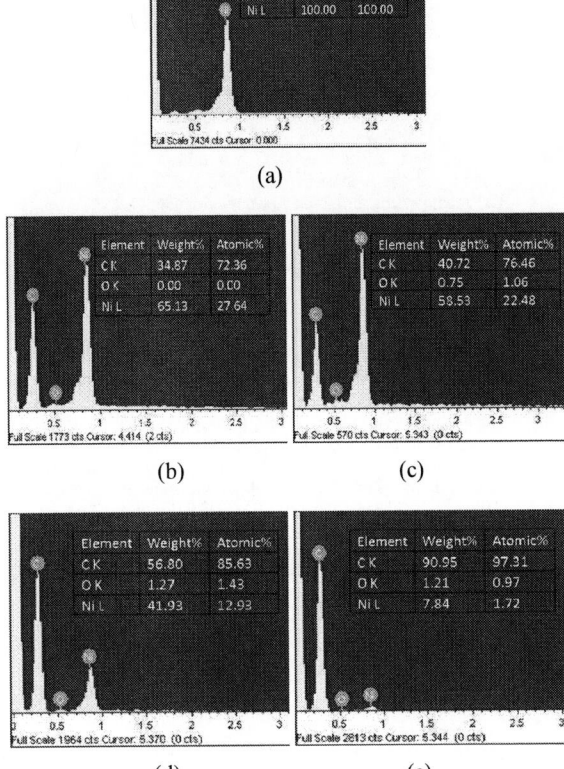

(a)

(b)   (c)

(d)   (e)

Fig. 2. EDX analysis of graphene at different amount of precursor, (b) 5 µl, (c) 10 µl, (d) 50 µl, and (e) 100 µl.

B. EDX Analysis

From Fig. 2, the composition of a specimen and the composition of individual components can be determined using Energy Dispersive X-ray spectroscopy (EDX). In EDX process, a focused beam of electrons solid samples will be bombarded onto samples which consequently emitted the X-Ray spectrum thus produced a localized chemical analysis. Additionally, EDX method is capable to detect all elements from atomic number 4 (Be) until 92 (U). In fact, X-ray have a sufficient energy to escape from the material surface. Consequently, a spectrum showing peaks associated with characteristic energies belong to certain existent elements are recorded. Hence, elements as light as oxygen and carbon can be detected in this technique. All the fabricated samples undergo Energy Dispersive X-ray spectroscopy (EDX) as illustrated in Fig. 2 to determine the composition of species in the as-synthesized samples. From Fig. 2(a), there is only nickel, Ni element detected at 100 wt.%. From Fig. 2(b), the as-synthesized graphene shows the presence of carbon (C), oxygen (O) and nickel with weight percentages of 34.87 %, 0.00 % and 65.13 %, respectively. The appearance of Ni and C peaks indicates an incorporation of graphene in the nickel matrix has occurred [12]. Notably, increment in carbon, C content was recorded at 40.72 % of weight percent. Based on FESEM images, the amount of carbon source seems greatly influenced the segregation of carbon atoms on the nickel surface and subsequently forms the six combinations of

carbon atoms. In order to verify the formation of graphene, Raman analysis was performed on the obtained 2D peak. It is noteworthy that increasing the amount of carbon source lead to high presence of carbon weight percentage as evidenced/proven in Fig 2(d). It is believes that an increment of carbon might contributed to high carbon content in an optimum of waste cooking palm oil as mentioned by Shaharin et al. [11]. The composition of graphene oxide, GO prominently detected in sample 50 µl of carbon source as revealed by EDX and Raman's result. It obviously contains carbon, oxygen and nickel with an approximation of 56.80 wt.%, 1.27 wt.%, and 41.93 wt.%, respectively. Meanwhile, increasing the carbon source attain to 100 µl (Fig. 2(e)) favourably increased the weight percent of carbon atoms leading to high defect occurrence. A high carbon content has tendency to give out carbon as well as susceptible to establish single layer of graphene on nickel surface [13]. Nevertheless, abundance of carbon atoms on nickel surface might induce formation of amorphous carbon.

C. Raman Analysis

Different volume of carbon source has been executed on the graphene film on nickel substrate and their Raman spectra is illustrated in Fig. 3 and Table 1. The characterization of graphene spectrum and its structural elucidation of graphitic materials have been conducted using Raman spectroscopy. The Raman spectrum of graphene is knowledgeably highly-sensitive to the quantity of atomic layers and degree of structural disorder or defects [14]. All the spectra notably exhibit three prominent Raman peaks for graphene which dedicates to the D, G and 2D peaks. From Fig. 3(b), the D band is observed at 1357 cm-1 and the G band recorded at 1581 cm-1 whereby the 2D band is detected at 2711 cm-1. The Raman spectrum at this volume reveals a low value of I_{2D}/I_G ratio (0.28) and I_D/I_G ratio (0.04) as summarized in Tab. 1. The values of I_{2D}/I_G peak ratio is obtained at lower than 1, which indicates that the sample is multi-layer graphene as suggested by G. Gutierrez et al. [15]. Increasing the volume of carbon source from 5 µl to 10 µl has evidently enhances the I_{2D}/I_G ratio in range of 0.28 to 0.30. This phenomenon occurs as the crystallite of graphene undergo quality improvement [16]. Since the carbon atoms are supplied by waste cooking palm oil decomposition, at an optimum volume, a large portion of carbon presence eventually will diffuse into Ni surface. In addition, the segregation and precipitation of carbon atoms could be more effective at an optimum volume of carbon source [11]. In contrast, increasing the volume of carbon source has led to decrement of I_{2D}/I_G ratio at merely 0.13 as depicted in Fig. 3(d). Apparently, the quality and crystallite of graphene seems deteriorated at this growth condition.

Both the G and D band demonstrate/exhibit the shifting peak towards higher wave number in comparison to the pristine graphite. It happened due to the oxygenation of graphite leading to the formation/establishment of sp3 carbon atoms [17]. In addition, the presence of the broadened defect is attributed to the formation of graphene oxide, GO [18]. The intensity ratio of I_D/I_G is 0.8 upon execution of high volume of carbon source as displayed/manifested in Fig. 3(e). The D band displays stronger and broader peak due to a higher level of disorder of the amorphous carbon [11]. It is in agreement with Negishi et al. whereby they have employed CVD process on a non-catalytic substrate to synthesize

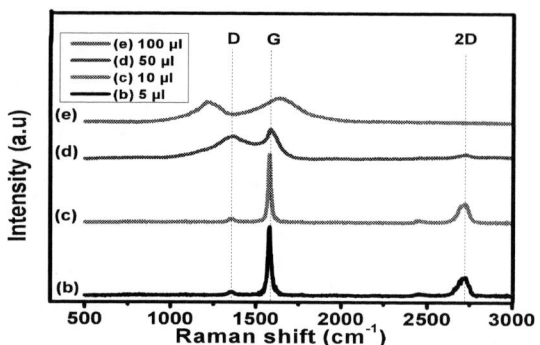

Fig. 3. Raman analysis of graphene at different amount of precursor, (b) 5 μl, (c) 10 μl, (d) 50 μl, (e) 100 μl.

amorphous carbon. An amorphous state is reported as temperature of CVD reactor has passed beyond thermal decomposition temperature of the carbon source [19]. Therefore, it is proven that quality of the graphene heavily depends on the amount of carbon source.

TABLE I.   RAMAN PEAKS OF THE GRAPHENE LAYER AT DIFFERENT VOLUME OF CARBON SOURCE

Sample	D band Position (cm^{-1})	G band Position (cm^{-1})	2D band Position (cm^{-1})	I_D/I_G	I_{2D}/I_G
(b)	1357	1581	2711	0.04	0.28
(c)	1357	1580	2712	0.04	0.30
(d)	1376	1596	2760	0.97	0.13
(e)	1219	1633	2716	0.80	-

IV. CONCLUSION

In this paper, we have compared the growth of graphene on nickel substrate using various amounts of waste cooking palm oil (WCPO). Sufficient carbon source (50 μl and 100 μl) has successfully produced GO and amorphous carbon on the nickel surface. By reducing the volume of carbon source, the structure and quality of graphene has been improved. Further decreasing the amount of carbon source supply suppressed the graphite sedimentation and low ratio of I_{2D}/I_G. Consequently, utilization of waste cooking palm oil as the carbon source for the formation of multilayer graphene has been successfully performed. This study has fixed the time constant at 15 mins but quality of graphene might further improves with execution of different synthesis times. As a result, a decrement pattern in the nucleation of multilayer graphene is observed [11].

ACKNOWLEDGMENT

This research is supported by Ministry of Higher Education Malaysia and Research Management Institute of Universiti Teknologi MARA (UiTM) under Research Acculturation Grant Scheme [600-RMIS/1/RAGS 5/3] and Research Bestari Perdana Grant Scheme [600-IRMI/PERDANA 5/3 BESTARI (100/2018). The authors extend grateful acknowledge to Research Chair for Biomedical Applications of Nanomaterials, Deanship of Scientific Research, King Saud University, Riyadh Saudi Arabia. Finally, a sincere gratitude to Aslam Mahmud for proofreading the manuscript.

REFERENCES

[1] D. R. Cooper, B. D'Anjou, N. Ghattamaneni, B. Harack, M. Hilke, A. Horth, N. Majlis, M. Massicotte, L. Vandsburger, E. Whiteway, and V. Yu, "Experimental review of graphene," *Condens. Matter Phys.*, vol. 2012, pp. 1–56, 2011.

[2] M. K. Kumar, N. S. Jha, S. Mohan, and S. K. Jha, "Reduced graphene oxide-supported nickel oxide catalyst with improved CO tolerance for formic acid electrooxidation," *Int. J. Hydrogen Energy*, vol. 39, no. 24, pp. 12572–12577, 2014.

[3] P. Taylor, W. Choi, I. Lahiri, and R. Seelaboyina, "Synthesis of Graphene and Its Applications," *Solid State Mater. Sci.*, vol. 35, no. 1, pp. 52–71, 2010.

[4] A. B. Suriani, R. Nor, and M. Rusop, "Vertically aligned carbon nanotubes synthesized from waste cooking palm oil," *J. Ceram. Soc. Japan*, vol. 118, no. 10, pp. 963–968, 2010.

[5] M. J. Salifairus, "Surface Topography of Synthesized Graphene from Green Carbon Source using Thermal Chemical Vapor Deposition," *Inst. Electr. Electron. Eng. Inc.*, vol. DOI: 10.11, pp. 522–526, 2015.

[6] M. Robaiah, M. Rusop, S. Abdullah, and H. Azhan, "Green Approach of Graphene Layer Produced from Waste Cooking Palm Oil at Different Precursor Temperatures," *AIP Conf. Proc.*, vol. 30008, no. Cvd, pp. 2337–2341, 2017.

[7] M. Robaiah, M. Rusop, S. Abdullah, Z. Khusaimi, H. Azhan, and N. A. Asli, "Synthesis graphene layer at different waste cooking palm oil temperatures," *AIP Conf. Proc.*, vol. 1877, no. 1, pp. 1–7, 2017.

[8] G. Gutierrez, F. Le Normand, D. Muller, F. Aweke, C. Speisser, F. Antoni, Y. Le Gall, C. S. Lee, and C. S. Cojocaru, "Multi-layer graphene obtained by high temperature carbon implantation into nickel films," *Carbon N. Y.*, vol. 66, pp. 1–10, 2013.

[9] K. . Al-Shurman and H. Naseem, "CVD Graphene Growth Mechanism on Nickel Thin Films," in *COMSOL Conference in Boston*, 2014, pp. 1–7.

[10] X. Song, L. Song, X. Chen, and T. Zhang, "The characterization of graphene prepared using a nickel film catalyst pre-deposited to fused silica," *RSC Adv.*, vol. 6, no. 27, pp. 22244–22249, 2016.

[11] M. R. M. & A. M. H. Shaharin Fadzli Abd Rahman, "Growth of Graphene on Nickel using a Natural Carbon Source by Thermal Chemical Vapor Deposition," *Sains Malaysiana*, vol. 43, no. 8, pp. 1205–1211, 2012.

[12] A. Jabbar, G. Yasin, W. Q. Khan, M. Y. Anwar, R. M. Korai, M. N. Nizam, and G. Muhyodin, "Electrochemical deposition of nickel graphene composite coatings: effect of deposition temperature on its surface morphology and corrosion resistance," *RSC Adv.*, vol. 7, no. 49, pp. 31100–31109, 2017.

[13] M. Lapuerta, J. M. Herreros, L. L. Lyons, R. García-contreras, and Y. Briceño, "Effect of the alcohol type used in the production of waste cooking oil biodiesel on diesel performance and emissions," *Fuel*, vol. 87, pp. 3161–3169, 2008.

[14] L. Fang, W. Yuan, B. Wang, and Y. Xiong, "Growth of graphene on Cu foils by microwave plasma chemical vapor deposition: The effect of in-situ hydrogen plasma post-treatment," *Appl. Surf. Sci.*, vol. 383, pp. 28–32, 2016.

[15] U. Narula, C. M. Tan, and C. S. Lai, "Growth mechanism for low temperature PVD graphene synthesis on copper using amorphous carbon," *Sci. Rep.*, vol. 7, pp. 1–13, 2017.

[16] A. Dathbun and S. Chaisitsak, "Effects of three parameters on graphene synthesis by chemical vapor deposition," *8th Annu. IEEE Int. Conf. Nano/Micro Eng. Mol. Syst.*, vol. 1, pp. 1018–1021, 2013.

[17] M. T. H. Aunkor, I. M. Mahbubul, R. Saidur, and H. S. C. Metselaar, "The green reduction of graphene oxide," *RSC Adv.*, vol. 6, no. 33, pp. 27807–27828, 2016.

[18] S. Perumbilavil, P. Sankar, T. Priya Rose, and R. Philip, "White light Z-scan measurements of ultrafast optical nonlinearity in reduced graphene oxide nanosheets in the 400-700 nm region," *Appl. Phys. Lett.*, vol. 107, no. 5, pp. 1–5, 2015.

[19] K. M. Negishi, Ryota, Hiroki Hirano, Yasuhide Ohno, "Layer-by-layer growth of graphene layers on graphene substrates by chemical vapor deposition.," *Thin Solid Film*, vol. 519, no. 19, pp. 6447–6452, 2011.

978-1-5386-5284-8/18 $31.00 © 2018 IEEE

Feasability of Zinc Oxide Nanowire as a Temperature Sensor: An Analytical Study

Azam Mohamad
Department of Electronic and Computer Engineering
School of Electrical Engineering,
Universiti Teknologi Malaysia,
Johor Bahru, Malaysia
azam.mohd88@gmail.com

Harzawardi Hasim
Malaysia Nuclear Agency,
Bangi,Malaysia
harzawardi.hasim@gmail.com

Suhana Mohamed Sultan
Department of Electronic and Computer Engineering
School of Electrical Engineering,
Universiti Teknologi Malaysia,
Johor Bahru, Malaysia
suhanasultan@utm.my

Abstract— The feasibility of Zinc Oxide (ZnO) nanowire as a temperature sensor was demonstrated by analytical study. A good quality I(V) model had been fitted with the experimental data on a single ZnO nanowire. It was found that the carrier concentration and mobility measured were 2.95×10^{18} cm^{-3} and 1.72 cm^2 V^{-1} s^{-1}, respectively. The I(V) model suit with the three-dimensional structure because their de Broglie wavelength smaller than the sample size. The current was observed to increase when the temperature applied increased from 27 °C to 277 °C. It was found that the carrier (electron) play an important part on current change. It was also found that the nanowire structure is more sensitive by a factor of 2 compared to nanowire film although the performances of the nanowire film was enhanced by the piezotronic effect.

Keywords— ZnO, nanowire, temperature sensor, Ohmic contact

I. INTRODUCTION

Good response in Ultra Violet (UV) light, humidity, biomolecules and a variety of gas makes Zinc Oxide (ZnO) nanostructure is a promising candidate for nanosensors [1-7]. In addition, the material is non-toxic and enables for low cost production which also required low temperature during deposition [8]. Further review reported a single nanowire can actually exhibit strong photoconduction gain as high as 10^{10} in the UV region and capability to detect single photons or at least to detect very low intensity light [9]. This semiconductor also capable to be a temperature sensor through measuring and monitoring the temperature in an ambient atmosphere [10]. This is very important characteristics to enable the ZnO nanowire for a temperature sensor applications. In 2013, Bercu *et al.* successfully fabricated a single ZnO nanowire using metalorganic chemical vapour deposition (MOCVD) and an Ohmic behaviour was demonstrated by the nanowire [11]. In the same year, they extended their research on the ZnO nanowire by investigating the effect of the nanowire diameter between head to tail towards I(V) characteristic [9]. An Ohmic I(V) characteristic was found when the nanowire tail diameter is similar size with the head. On the other hand a Schottky I(V) characteristic was dominant when the tail diameter smaller than head. However, there is lack of research on I(V)

modelling of ZnO Ohmic behaviour and the effect of this behaviour changes with temperature.

In this article, we present an Ohmic I(V) model for measured results and extend the research in investigating the response of the current with different temperatures. This work also explored the physical parameters which give an impact towards the current modulation when temperature was increased. In addition, the sensitivity of nanowire and nanowire film was studied and compared with inclusion of the piezoelectronic effect. This is important to identify which sample is the most sensitive towards the temperature components, incorporating the applicable criteria that follow.

II. ANALYTICAL METHOD

A. Current-voltage model

The nature of carrier transport in a semiconductor in the form of the rate of charge flow has been explored by Arora [12]. It had been proposed that the movement of the carriers across the length of a semiconductor can be represented by a series of bulk carriers transiting the semiconductor from one end to the other end. The current, I, which is equal to the rate of charge flow is thus given by [12]

$$I = -q n_3 v_D A_c \qquad (1)$$

where the q, v_D, n_3, and A_c are the electronic charge, linear drift velocity, carrier concentration and the cross-section area of the sample that is perpendicular to the direction of charge flow (current). The current in (1) is valid for the case of carrier concentration in a volume and only considered in homogenous sample with no concentration gradient. The current expression in (1) can also be expressed by

$$I(V) = \frac{q n_3 \mu A_c}{L} V \qquad (2)$$

978-1-5386-5284-8/18 $31.00 © 2018 IEEE

where new parameters are introduced such as the carrier mobility, μ, sample length, L, and bias voltage, V. Equation (2) is in response to the applied voltage and this equation is produced by replacing the v_D which is equal to $-\mu\xi$ showing that the drift voltage is in response to the carrier mobility, μ and electric field, $\xi = V/L$. In order to fit with the ZnO nanowire structure, (2) needs to be modified. The resulting current, I_{ZnO}, in response to the applied voltage, V, is written as

$$I_{ZnO}(V) = \frac{q\mu n_3 \pi r^2}{L} V \qquad (3)$$

where the r is the ZnO radius. This modification occur when A_c is replaced with πr^2 known as the nanowire area assumed to be a circular structure. The thermal de Broglie wavelength, λ_D is given by [12]

$$\lambda_D = \frac{1.23\ \text{nm} \cdot (\text{eV})^{1/2}}{\sqrt{(m^*/m_o)E(\text{eV})}} \qquad (4)$$

where the m^*, m_o, and E are the effective mass, electron rest mass, and thermal energy of ZnO material. This wavelength is employed to determine if the semiconductor is in three-dimensional structure (bulk), two-dimensional structure (sheet), or one-dimensional structure (line).

B. Current-temperature model

In section A, the current is shown to change with voltage. However, the current also in response to the temperature. Current response to the applied temperature is expressed in (5)

$$I(T) = \frac{q\mu\pi r^2 V \cdot N_c \exp\left(-E_g/2k_B T\right)}{L} \qquad (5)$$

Here, the ZnO carrier concentration is a main factor for the current being modulated. The carrier of nanowire can be expressed in (6) [9] which indicated that the carrier is a function of temperature.

$$n_3(T) = N_c \cdot \exp\left(-E_g/2k_B T\right) \qquad (6)$$

where the E_g and h are the material band gap energy and Planck's constant, respectively. Here we consider the carrier concentration is in non-degenerate regime. The effective density of states, N_c, is given by [12]

$$N_c = 2\left(\frac{2\pi m_e^* k_B T}{h^2}\right)^{\frac{3}{2}}. \qquad (7)$$

III. RESULTS AND DISCUSSIONS

Table I shows a complete parameters of ZnO nanowire with their respective values taken from experimental data. These values are crucial in order to validate the I-V model. Figure 1 shows a comparison of I(V) measurement between experiment and model. It is shown that the model is capable to fit well with experimental data. The carrier concentration and the mobility are measured to be 2.95×10^{18} cm^{-3} and 1.72 cm^2 V^{-1} s^{-1}, respectively. The nanowire exhibits low mobility due to large nanowire resistance of 1.57 MΩ at fixed temperature of 300 K. This value of resistance is consistent with past studies by Bercu et al., which found their resistance of nanowire ranged from 0.25 to 2.4MΩ. In addition, results in Fig.1 indicates that the ZnO nanowire conforms well with the three-dimensional structure (bulk) compared to one-dimensional structure (line) because the dimension exceeds the thermal de Broglie wavelength. The calculated de Broglie wavelength of ZnO nanowire is 15.6 nm. This value is smaller than the ZnO nanowire structure used in this work thus confirming that the bulk structure is more relevant in describing the ZnO I(V) model.

TABLE 1   ZINC OXIDE NANOWIRE PARAMETERS

Parameters	Value/Units
ZnO nanowire radius, r	100 nm [9]
ZnO nanowire length, L	4 μm [9]
Electronic charge, q	1.6×10^{-19} C [12]
Temperature, T	300 K
ZnO effective mass, m^*	0.24m$_O$ [13]
Boltzmann constant, k_B	1.38×10^{-23} J/K [12]
Electron rest mass, m_o	9.1×10^{-31} kg [12]

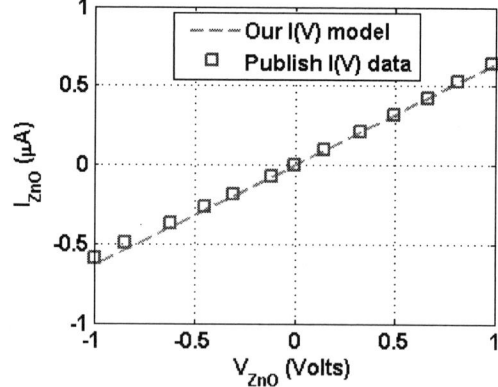

Fig. 1. I–V plot of a single ZnO Nanowire with Ohmic contact of our suggested I(V) model and published I(V) data [9].

Figure 2 shows the current response with various temperatures. The temperature of ZnO nanowire was increased from 27 ℃ to 277 ℃. The result shows an increase of current on ZnO nanowire with an increase of the temperature. The investigation on current response toward temperature is extended by investigating the response with the temperature (I-T) at a fixed bias of 2 V as displayed in Figure 3. It was found that the current increased linearly as the temperature increased from 10 ℃ to 20 ℃. This result conforms with Xue *et al.* research on the temperature variation [10]. We identify the carriers (electron) play an important part on current change. The inset in Figure 3 shows the carriers (electron) become denser as the temperature increases resulting the current on ZnO nanowire to be increased significantly.

In addition, the carrier sensitivity between ZnO nanowire and nanowire thin film are compared against temperature change. Results of nanowire thin film are taken from experimental data from Xue *et al.* sensor device [10]. Table 2 presents a comparison of the sensitivity between the ZnO nanowire and ZnO nanowire film with inclusion of the piezotronic effect. The sensitivity of nanowire is 2 times bigger than the nanowire film. This finding confirms that the nanowire is the best candidate for temperature sensor.

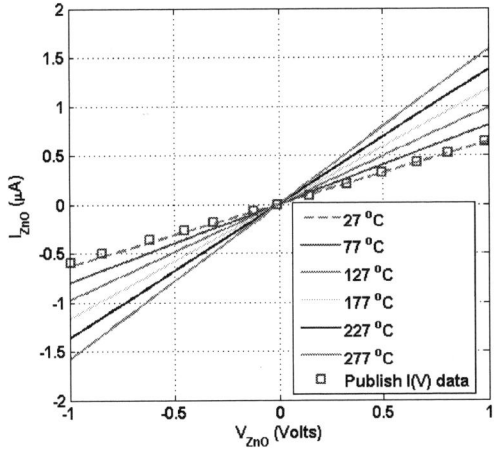

Fig. 2. I–V curves of a single ZnO Nanowire with Ohmic contact at different temperatures.

TABLE 2   SENSITIVITY OF ZINC OXIDE NANOWIRE AND NANOWIRE FILM

Type of ZnO	Sensitivity
Nanowire	6.2 nA℃⁻¹
Nanowire film	3.4 nA℃⁻¹

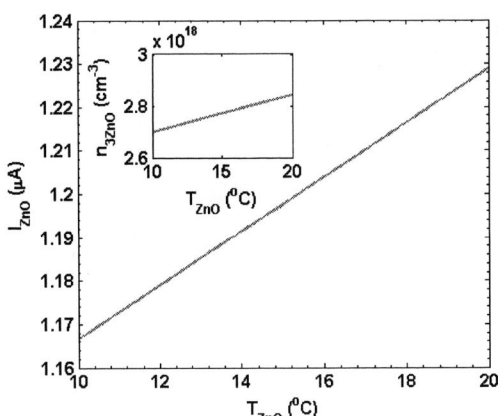

Fig. 3. Response current of the ZnO nanowire to the temperature. Inset shows the carrier (electron) response of a ZnO nanowire-based sensor to the temperature, varying from 10 ℃ to 20 ℃.

IV. CONCLUSION

In summary, a feasibility study of Zinc Oxide (ZnO) nanowire as a temperature sensor is demonstrated through analytical study. By comparing the I(V) model with experiment from Bercu *et al.*, a good quality I(V) model have been fitted for a single ZnO nanowire. It is found that the carrier concentration and mobility are measured to be 2.95×10^{18} cm^{-3} and 1.72 cm^2 V^{-1} s^{-1}, respectively. The I(V) model follows on the three-dimensional structure because their de Broglie wavelength smaller than the device dimension. The current is increased as the temperature increased from 27 ℃ to 277 ℃. It is found that the carrier (electron) play an important part on current change. The carrier become denser at high temperatures that make the ZnO current increased significantly. It is also found that the ZnO nanowire is more sensitive than ZnO nanowire film although the performances of nanowire film is enhanced by the piezotronic effect. The finding shows that the sensitivity of the nanowire is 2 times bigger than the nanowire film.

ACKNOWLEDGMENT

Authors would like to acknowledge the financial support from Research University grant of Universiti Teknologi Malaysia under Projects Q.J130000.2523.14H63, Q.J130000.2523.19H33 and Ministry of Higher Education of Malaysia.

REFERENCES

[1] S. Thongma, A. Boontan, T. Boonkoom, and K. Tantisantisom, "Electrical Studies of ZnO Nanowires and Metal Contacts," in Advanced Materials Research, 2016, pp. 3-7.

[2] N. M. Kiasari, S. Soltanian, B. Gholamkhass, and P. Servati, "Room temperature ultra-sensitive resistive humidity sensor based on single zinc oxide nanowire," Sensors and Actuators A: Physical, vol. 182, pp. 101-105, 2012.

[3] J. Zhou, Y. Gu, P. Fei, W. Mai, Y. Gao, R. Yang, et al., "Flexible piezotronic strain sensor," Nano letters, vol. 8, pp. 3035-3040, 2008.

[4] W. Khoo, S. Sultan, M. Sahdan, S. Pu, M. Shamsudin, and J. McBride, "Carbothermal Reduction Chemical Vapour Deposition Growth of Amorphous ZnO Nanostructures Towards Humidity

Sensing," Journal of Nanoelectronics and Optoelectronics, vol. 13, pp. 319-323, 2018.

[5] S. Sultan, M. de Planque, P. Ashburn, and H. Chong, "Effect of Phosphate Buffered Saline Solutions on Top-Down Fabricated ZnO Nanowire Field Effect Transistor," Journal of Nanomaterials, vol. 2017, 2017.

[6] W. Khoo and S. Sultan, "A study on the gas sensing effect on current-voltage characteristics of ZnO nanostructures," in Semiconductor Electronics (ICSE), 2014 IEEE International Conference on, 2014, pp. 221-224.

[7] W. Khoo, S. Sultan, and M. Sahdan, "The effects of gas flow rate and annealing on the morphological properties of zinc oxide nanostructures thin film using chemical vapour deposition process," 2014.

[8] S. Chirakkara and S. Krupanidhi, "Study of n-ZnO/p-Si (100) thin film heterojunctions by pulsed laser deposition without buffer layer," Thin Solid Films, vol. 520, pp. 5894-5899, 2012.

[9] B. Bercu, W. Geng, O. Simonetti, S. Kostcheev, C. Sartel, V. Sallet, et al., "Characterizations of Ohmic and Schottky-behaving contacts of a single ZnO nanowire," Nanotechnology, vol. 24, Oct 18 2013.

[10] F. Xue, L. Zhang, W. Tang, C. Zhang, W. Du, and Z. L. Wang, "Piezotronic effect on ZnO nanowire film based temperature sensor," ACS applied materials & interfaces, vol. 6, pp. 5955-5961, 2014.

[11] W. Geng, S. Kostcheev, C. Sartel, V. Sallet, M. Molinari, O. Simonetti, et al., "Ohmic contact on single ZnO nanowires grown by MOCVD," physica status solidi (c), vol. 10, pp. 1292-1296, 2013.

[12] V. K. Arora, Nanoelectronics: Quantum engineering of low-dimensional nanoensembles: CRC Press, 2015.

[13] B. Enright and D. Fitzmaurice, "Spectroscopic determination of electron and hole effective masses in a nanocrystalline semiconductor film," The Journal of Physical Chemistry, vol. 100, pp. 1027-1035, 1996.

2018 IEEE International Conference on Semiconductor Electronics (ICSE)

Taguchi optimization of Surface Plasmon Resonance-Kretschmann biosensor using FDTD

Nur Akmar Jamil, P. Susthitha Menon, Sahbudin Shaari, Mohd Ambri Mohamed, Burhanuddin Yeop Majlis

Institute of Microengineering and Nanoelectronics (IMEN),
Universiti Kebangsaan Malaysia (UKM)
43600 UKM Bangi, Selangor, Malaysia
susi@ukm.edu.my

Abstract— **This paper reports on the optimization of full-width-at-half-maximum (FWHM) of the surface plasmon resonance (SPR) curve from a graphene-based SPR biosensor. The biosensor was designed in the Kretschmann configuration which is commonly known as the most effective technique for plasmon excitation using Lumerical's finite-difference-time-domain (FDTD) analysis. The performance of the SPR biosensor can be monitored by analysing the FWHM of the SPR curves where a smaller value is desired due to a narrower resonance peak which generally corresponds to a higher transmission power resulting in a more sensitive bio-detection. To optimise the biosensor's FWHM, Taguchi's L9 orthogonal array method was utilized involving four control factors at three level values using the objective function of Smaller-the-Better (STB) signal-to-noise ratio (SNR) characteristic. The results show that the FWHM improved to 1.1261° with best settings of A3B2C3D1. The Taguchi-optimized biosensor achieved 36.98% increase in sensitivity for urea detection.**

Keywords— *Taguchi orthogonal array method; Surface plasmon resonance; Kretschmann; graphene; numerical modeling; FWHM;*

I. INTRODUCTION

Over the past years, surface plasmon resonance (SPR)–based biosensors have gained wide attention because of their unique capabilities in real time and label-free detection. The SPR biosensor based on the Kretschmann configuration for plasmon excitation and analyte-ligand interaction on functionalized surfaces can be used for various applications such as antibody-antigen detection, protein bindings and DNA hybridization [1].

In the conventional SPR biosensor, the Kretschmann setup has been widely used which includes a prism, glass slide, metal thin film and a dielectric layer where the plasmon field is excited. The transverse magnetic (TM) incident light source is passed through the prism and reflected off the back side of the chip sensor and then the reflected light is captured by a detector [2]. At two mediums of altered refractive index, light coming from the side of a higher refractive index is partly reflected and partly refracted. Above a certain critical of incidence angle, no light is refracted and total internal reflection (TIR) is observed. The absorbed light interacts with the electron gas in the metal film of the sensor causing a resonance and sharp decrease in the intensity of the incident light. The SPR biosensor has high sensitivity to changes in the refractive index of the sensing medium and the surrounding environment. A slight change of refractive index at the dielectric layer will affect surface plasmon polaritons (SPP).

This alteration is measured optically through the attenuated total internal reflection (ATR) technique [3]. The modification of refractive index changes the SPR characteristic such as the reflective minimum angle and the full-width-at-half maximum (FWHM) of the SPR signal. FWHM is the spectral width of two extreme values at the half of reflectance curve. A narrower FWHM is desired due to a deeper resonance peak for an effective detection. Fig. 1 shows the FWHM of SPR reflectance curve at 670 nm and 785 nm for experimental and simulation [4].

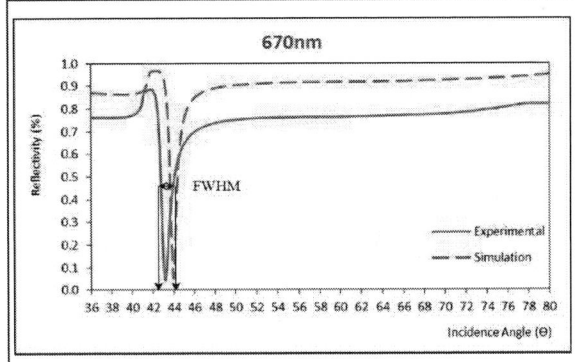

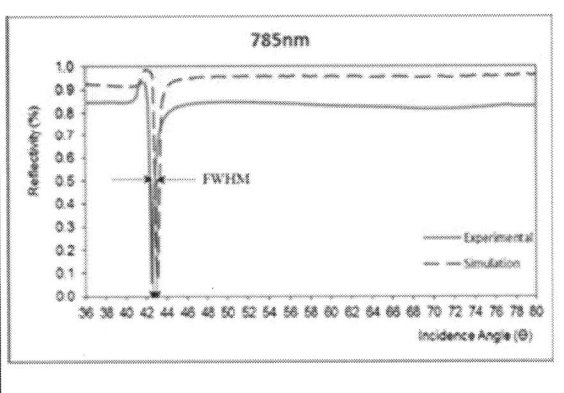

Fig. 1. Full-width-at-half-maximum (FWHM) at wavelength 670 nm and 785 nm

In SPR applications, it is useful and critical to select the metal film which is capable to resonate the light at a suitable wavelength. Owing to superior optical performances and good resistance to oxidation and corrosion, gold (Au) and silver (Ag) are considered as the most suitable material for

978-1-5386-5284-8/18 $31.00 © 2018 IEEE

biosensing applications and favored to be deposited on the SPR prism. However, the ability of biomolecules to bind on gold is poor, which limits the sensitivity of the conventional SPR biosensor. To overcome these issues, another promising material to be used is a biomolecular recognition element (BRE) such as graphene on gold film which enhances the performance of the SPR biosensor. In fact, the specific control over the geometry and optical properties of nanostructures such as graphene are challenging [5-6].

Over the last decade, numerous methods have been applied to the configuration to improve sensing performances including nanoholes, metallic nanoslits and colloidal gold nanoparticles in buffered solution [7]. Graphene is a supreme candidate as a biosensor due to its good optical absorbance, great surface-to-volume ratio, carbon single atomic layer, two-dimensional honeycomb lattice and zero band gap semi-metal with electromagnetic properties. Graphene sensor chip is predictable to extend the range of analyte that can be used to recognize biomolecules on metal coating. Graphene-based sensor exploits the surface plasmon wave (SPW) resonance state, metal film dielectric coefficients, liquid or gas sensed material and multi-layered graphene deposited on thin films [8].

In order to design a robust quality of SPR biosensor, process parameters have to be optimized to make the output features insensitive to variants in environmental conditions and other noise factors. Design of experiment using Taguchi's orthogonal array (OA) methodology is a well-established practice which is invented for designing experiments and effectively obtain the optimization of control parameters. Furthermore, Taguchi method is an execution process of centering or fine-tuning when the levels of control factors have some variations around its central values where repeatable performance are determined at the central values of control factors [9-11]. Many works in the past have utilized the Taguchi method to optimize numerous product and process issues.

Device simulants based physics offer a more accurate and inclusive analysis and are low cost and faster in delivering comprehensive device performances as compared to a laboratory measurement. The numerical simulation is designed using Lumerical finite-domain-time-domain analysis (FDTD)-based simulators [12]. It provides more insight into the numerous physical phenomenon and device characteristics by observing trend plots over device structural parameters rather than the analytic model. Hence, the SPR biosensor-based graphene layers were developed using Lumerical using a modified design from the experimental device in the past [13].

This work combines the use of numerical modeling and FDTD simulations based on Taguchi OA to obtain the optimized FWHM of a graphene-based SPR biosensor operating in the 632.8 µm, 670 µm and 785 µm wavelength region. We designed a Taguchi L9 orthogonal array which deals with four critical parameters in the process including wavelength, the thickness of chromium layer, the thickness of gold layer and thickness of graphene layers and each parameter is varied at three different level values.

This work was supported by the Ministry of Education using the Higher Institution Centre of Excellence (HiCOE) Grant of AKU-95, UKM grant DIP-2017-022 and GUP-2016-062. Authors would like to thank IMEN, National University of Malaysia (UKM) and MyBrain for the support.

II. METHODOLOGY

A. Numerical modeling

The SPR-based biosensor was modelled using the FDTD Solutions software from Lumerical, Inc. The design of the graphene-based SPR biosensor is performed in the high optical solver for capturing light interaction with the desired wavelength scale. The ultra-fine setting and conformal meshing are required to simulate the metal geometries of plasmonic structures. Several optical parameters are set for designing the model. A plane wave source is set as Bloch or periodic type at desired wavelengths and a sweep parameter is fixed at source incident angle from 36° to 80° for seeking the source angle which excites the SPR mode [14,16].

Fig. 1 shows the simulated design of graphene-based SPR biosensor. In this paper, the well-known Kretschmann prism configuration is used as a basis for SPR sensing. The designed SPR biosensor formation involves four components, namely BK7 glass, chromium layer (Cr), gold metal layer (Au) and the graphene layer. All values of the refractive index and extinction coefficient of the chromium and gold layer are obtained from Bionavis Navi SPR 200 at three optical wavelengths of 632.8 nm, 670 nm and 785 nm that excite the SPR mode [10]. The thickness of graphene is given as $L \times 0.34$ nm where L is the number of graphene layers. The complex refractive index of graphene in visible light is given by the following equation:

$$n = 3 + i\,\frac{C}{3} \tag{1}$$

where $C = 5.446$ µm^{-1} is the value of the constant implied by the opacity measurement and λ is the optical wavelength.

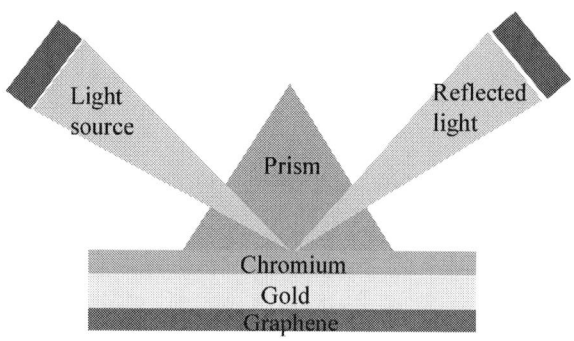

Fig. 2. Kretschmann configuration SPR biosensor based graphene.

B. Taguchi L9 Orthogonal Array Method

In Taguchi method, the optimization of the control factors for the best combinations is used to achieve high performance with a less number of experiments. There are a total of 9 simulation experiments which consist of four control factors namely wavelength, chromium thickness, gold thickness and graphene layer thickness at three different levels in order to achieve the best FWHM. The control factors and the respective level values are listed in Table I.

Meanwhile, the experiment target is to get a minimum value of FWHM, thus the FWHM is optimized using signal-to-noise ratio (SNR) of Smaller-the-Better. The SNR

(Smaller-the-Better), η_{STB} is defined by the following equation [15]

$$\eta_{STB} = -10 \log\left(\frac{1}{n}\sum_{i=1}^{n} y_i^2\right) \tag{2}$$

where y is the value of the i experiment in each group and n is the number of simulation experiments.

TABLE I.   CONTROL FACTORS AND THEIR LEVELS

Control factor	Label	Level 1	Level 2	Level 3
Wavelength (nm)	A	633	670	785
Chromium layer thickness (nm)	B	0.5	1.5	2.0
Gold layer thickness (nm)	C	40	45	50
Graphene layer thickness (nm)	D	0.34	0.68	1.02

III. RESULTS AND DISCUSSION

A. Analysis for Surface Plasmon Resonance based Graphene sensor

In this work, all the simulated data have demonstrated a good agreement of analytical data and numerical which is obtained by Lumerical FDTD software [4].

Taguchi method was applied to analyze FWHM by optimizing the control factors in the orthogonal array. All values of FWHM that were simulated are listed in Table II.

TABLE II.   FWHM VALUES AND SNR (η)

Experiment No	Control factor level (L9)				FWHM (degree)	SNR (smaller-the-better)
	A	B	C	D		
1	1	1	1	1	3.6304	-11.20
2	1	2	2	2	3.5868	-11.09
3	1	3	3	3	3.2739	-10.30
4	2	1	2	3	2.1713	-6.73
5	2	2	3	1	1.823	-5.22
6	2	3	1	2	3.646	-11.24
7	3	1	3	2	1.254	-1.97
8	3	2	1	3	1.4746	-3.37
9	3	3	2	1	1.4076	-2.97

Referring to Table II, the rows 7, 8 and 9 give FWHM value of 1.2540, 1.4760 and 1.4076 respectively. Thus, the smaller value of FWHM is preferred due to a narrower and deeper resonance peak for an effective detection. The performance characteristics of the graphene-based biosensor can be judged based on the experiment with the highest SNR values [14]. The SNR values were calculated by applying equation (2) for each experiment. According to Table II, row 7 resulted in the highest SNR value of -1.97 dB corresponding to FWHM value of 1.254° and control factors of A3B1C3D2. The SNR for each level of the

control factor is summarized in Table III. Also, the table shows the total mean of the SNR for the FWHM which is -7.12 dB. The greater the SNR value, the better the quality characteristic of the graphene-based SPR biosensor.

Fig. 3 shows the factor effect of the graph for the SNR (smaller-the-better) of the experiments which the value of the overall mean is -7.12 which represented by a dashed horizontal line. Also, shows the SNR for each level of the parameters and the total mean of the SNR for the FWHM. Referring to the graph, the slopes correspond to the wavelength (Factor A), followed by Chromium layer thickness (Factor B), Gold layer thickness (Factor C) and lastly Graphene layer thickness (Factor D). The factor effect plot shows that the best control factor combination in order to obtain the smallest FWHM is A3B2C3D1.

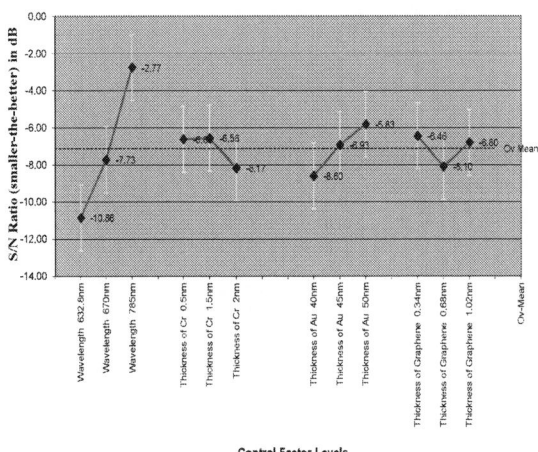

Fig. 3. SNR (smaller-the-better) graph

B. Confirmation of Optimum run

Furthermore, we conducted an analysis of variance (ANOVA) to identify the significant influence of the four control factors on the quality characteristics of SPR biosensor based graphene. The result of ANOVA is presented in Table III. The factor effect percentage on SNR was designated to specify the priority of a control factor to lessen variation. The highest influence on the FWHM is the high percentage value of a factor effect on SNR.

TABLE III.   RESULTS OF ANOVA

Factor	Process parameter	Factor effect on SNR (%)
A	Wavelength	82.54
B	Chromium layer thickness	4.09
C	Gold layer thickness	9.67
D	Graphene layer thickness	3.70

The results clearly show that Factor A which is the incident optical wavelength has the highest percentage of a factor effect on the mean and is the most dominant factor in the performance of FWHM with 82.54 %. The percentage effect for SNR of the other factors which are the Chromium layer thickness, Gold layer thickness, and Graphene layer thickness was much lower, being 4.09 %, 9.67 %, and 3.70 % respectively. This means the FWHM value will change drastically when the optical wavelength is adjusted. Therefore, usage of the most suitable optical wavelength to excite SPR is crucial for designing a sensitive SPR-based biosensor as compared to the other control factors.

In order to finalize the design's control factors in achieving the target, Table of SNR (STB) is a good reference. Based on Table IV, the SNR (STB) values for each level of control factors is selected as the levels were predictable to attain the target which is minimum FWHM. As a result, for the Wavelength (Factor A), level 3 with value of -2.77 dB was selected, for Chromium layer thickness (Factor B) was level 2 by -6.56 dB, while for Factor C which is Gold layer thickness at level 3 with SNR of -5.83 dB and last but not least is Graphene layer thickness at level 1 with SNR of -6.46 dB. The best setting of the control factors for a graphene-based SPR biosensor that affects the FWHM which is recommended by Taguchi method is A3B2C3D1 and this is summarized in Table V. After the optimization approach, the lowest FWHM value for graphene-based SPR biosensor obtained was 1.1261° which is lower than the predicted value of 1.19° by Taguchi's method.

TABLE IV.   BEST SETTING OF PROCESS PARAMETERS

Factor	Control factor	Level	Best value (nm)
A	Wavelength	3	785
B	Chromium layer thickness	2	1.5
C	Gold layer thickness	3	50
D	Graphene layer thickness	1	0.34

Subsequently, the Taguchi optimized graphene-based SPR biosensor was utilized in detecting kidney wastes such as urea and achieved a 36.98% increase in sensitivity for urea detection as compared to previous work [5]. More elaborate results will be published in our subsequent work.

IV. CONCLUSION

As a conclusion, this work emphasized on the usage of numerical analysis using Lumerical's FDTD and Taguchi's optimization analysis as a suitable and reliable method in optimizing the control factors of a graphene –based surface plasmon resonance (SPR) biosensor in order to achieve the optimum FWHM. FWHM was kept as minimum as possible to enhance the performance of the biosensor. The best control factors value that contributes the minimum FWHM

for Wavelength, Chromium layer thickness, Gold layer thickness and Graphene layer thickness are 785 nm, 1.5 nm-thick, 50nm-thick and 0.34 nm-thick respectively. The lowest FWHM value of 1.1261 degree was achieved upon Taguchi optimization with an improved sensitivity percentage of 36.98% for detection of urea.

REFERENCES

[1] M. B. Hossain and M. M. Rana, "Graphene Coated High Sensitive Surface Plasmon Resonance Biosensor for Sensing DNA Hybridization," *Sens. Lett.*, vol. 14, no. 2, pp. 145–152, Feb. 2016.

[2] F.A. Said, P.S. Menon, S. Shaari, B.Y. Majlis, "FDTD analysis on geometrical parameters of bimetallic localized surface plasmon resonance-based sensor and detection of alcohol in water," *International Journal of Simulation: Systems, Science and Technology*, vol. 16, no. 4, pp. 6.1-6.5, 2015.

[3] J. B. Maurya, Y. K. Prajapati, V. Singh, and J. P. Saini, "Sensitivity enhancement of surface plasmon resonance sensor based on graphene-MoS2 hybrid structure with TiO2-SiO2 composite layer," *Appl. Phys. a-Materials Sci. Process.*, vol. 121, no. 2, pp. 525–533, 2015.

[4] F. A. Said, P. S. Menon, V. Rajendran, S. Shaari, and B. Y. Majlis, "Investigation of graphene-on-metal substrates for SPR-based sensor using finite-difference time domain," *IET Nanobiotechnology*, vol. 11, no. 8, pp. 981–986, Dec. 2017.

[5] N. A. Jamil, P. S. Menon, F. A. Said, K. A. Tarumaraja, and G. S. Mei, "Graphene-based Surface Plasmon Resonance Urea Biosensor using Kretschmann configuration," *IEEE Reg. Symp. Micro Nanoelectron.*, pp. 112–115, 2017.

[6] N. A. Jamil, P. S. Menon, S.M. Gan, B. Y. Majlis, "Peningkatan Kepekaan Biosensor Urea Berasakan Resonans Plasmon Permukaan dan tatsusunan Krestschmann dengan struktur hibrid grafin-MoS₂," *Sains Malaysiana*, vol 47, no. 5, pp. 1033-1038, 2018.

[7] M. S. Rahman, M. S. Anower, M. R. Hasan, M. B. Hossain, and M. I. Haque, "Design and numerical analysis of highly sensitive Au-MoS2-graphene based hybrid surface plasmon resonance biosensor," *Opt. Commun.*, vol. 396, pp. 36–43, Aug. 2017.

[8] P. Singh, "SPR Biosensors: Historical Perspectives and Current Challenges," *Sensors and Actuators, B: Chemical*, vol. 229. pp. 110–130, 2016.

[9] Prakash Apte, "TAGUCHI page - Prakash Apte." [Online]. Available: https://www.ee.iitb.ac.in/~apte/CV_PRA_TAGUCHI.htm. [Accessed: 26-Feb-2018].

[10] A. Afifah Maheran, P. S. Menon, I. Ahmad and S. Shaari, "Effect of Halo structure variations on the threshold voltage of a 22nm gate length NMOS transistor," *Materials Science in Semiconductor Processing*, vol. 17, pp. 155-161, 2014.

[11] H. Haroon, S. Shaari, P.S. Menon, H. Abdul Razak, M. Bidin, "Application of statistical method to investigate the effects of design parameters on the performance of microring resonator channel dropping filter," *International Journal of Numerical Modelling: Electronic Networks, Devices and Fields*, vol. 26, pp. 670-679, 2013.

[12] "FDTD Solutions | Lumerical's Nanophotonic FDTD Simulation Software." [Online]. Available: https://www.lumerical.com/tcad-products/fdtd/. [Accessed: 26-Feb-2018].

[13] H. Jussila, H. Yang, N. Granqvist, and Z. Sun, "Surface plasmon resonance for characterization of large-area atomic-layer graphene film," *Optica*, vol. 3, no. 2, pp. 151–158, 2016.

[14] N. A. B. Jamil, P. S. Menon, G. S. Mei, S. Shaari, and B. Y. Majlis, "Urea biosensor utilizing graphene-MoS2 and Kretschmann-based SPR," in *TENCON 2017 - 2017 IEEE Region 10 Conference*, 2017, pp. 1973–1977.

[15] K.-H. Tseng, Y.-F. Shiao, R.-F. Chang, and Y.-T. Yeh, "Optimization of Microwave-Based Heating of Cellulosic Biomass Using Taguchi Method," *Materials (Basel).*, vol. 6, no. 8, pp. 3404–3419, 2013.

[16] P. S. Menon *et al.*, "Urea and creatinine detection on nano-laminated gold thin film using Kretschmann-based surface plasmon resonance biosensor," *PLoS One*, vol. 13, no. 7, p. e0201228, Jul. 2018.

Gap in pagination due to unavailable paper.

Pages 65-68

Separation of Micro Engineered Particle Using Dielectrophoresis Technique

Nur Rabiatul Adawiyah Tajul Othamany, Norazreen Abd Aziz,
Department of Electrical, Electronics and System Engineering,
Universiti Kebangsaan Malaysia (UKM)
Selangor, Malaysia

Muhammad Izzuddin Abd Samad, Muhamad Ramdzan Buyong, Burhanuddin Yeop Majlis,
Institute of Microengineering and Nanoelectronic (IMEN),
Universiti Kebangsaan Malaysia (UKM) Selangor, Malaysia
Email : muhdramdzan@ukm.edu.my

Abstract— The process of manipulation and separation of non-contact particles, which is dielectrophoretic technique, DEP will be described in this paper. This indirect physical contact technique, DEP, focuses on the value of dielectric between its particles and the medium to minimize the problems that occur such as physical and chemical damage to the particles when using the direct physical contact technique. This study was carried out by using polystyrene particles of 1 μm, 4.8 μm and 9.9 μm. The difference of size in this polystyrene particles will determine the change in the pattern the related to the Clausius-Mossotti Factor (CMF) in which the particles will be attracting to the electrodes or repelling from the electrodes. Frequency range selected between 0 to 1.5 MHz The new value of the optimum cross-over frequency will give the more efficient manipulation and separation rates. The force that attracting, PDEP on the y axis and the force that repelling, NDEP on the z axis by the polystyrene particles is observed to see the efficiency and reliability of the process. This separation of polystyrene particles is compared with the separation of the pathogenic bacterial cell particles and blood components in human bodies such as RBC and platelets.

Keywords—Dielectrophoretic, separation of micro particles.

I. INTRODUCTION

The basic technology that is thriving today for the process of manipulating and separating particles can be divided into two techniques, which is i) direct physical contact technique ii) indirect physical contact technique. Direct physical contact technique actually has an adverse impact on the particles and the surrounding medium. For example, physical and chemical damage towards particles, the medium will be contaminated and will give less effective results. Therefore, in order to solve the problem involved in the process direct physical contact technique, the particles manipulation and separation process using indirect physical contact technique is proposed using the application of dielectrophoretic, DEP.

Dielectrophoretic is specially used to describe the movement of particles when exposed to electric field gradients. In the dielectrophoretic technique, the particles will be attracted to the region with high electric field intensities, while the particles will be repelling to the region with low electric field intensities. Dielectrophoretic is very different from electrophoresis because dielectrophoretic using the alternating current and the particles involved do not need to carry electrical charges to induce particle movements. In addition, dielectrophoretic using a non-uniform electric field for each particle, while electrophoresis using a uniform electric field that passes through each particle. In order to separate blood components in human bodies such as platelet, white blood cell, red blood cell (RBC), and others, indirect physical contact technique are suitable for it. The optimum dielectrophoretic force, FDEP based on the dielectric value between the particles and mediums around it will be applied in DEP technique for the process of manipulation and separation of the engineered particles.

Dielectrophoretic can be defined as the movement of suspended particles polarization in the flow of inconsistency electric field [2][3]. Dielectrophoretic force, FDEP can be proved when the particles are exposed to the high intensity of the electric field intensity constructed in the surrounding medium. The incompatibility of the built-in electric field will cause the particle to polarize and the particle mobility occurs based on the dielectric value between the particles and the medium [6]. There are two types of partition manipulation and separation in the DEP. First, positive dielectrophoretic, PDEP and the second is negative dielectrophoretic, NDEP. When particles are more polarized than their medium, they are PDEP. The PDEP particles will move closer to the electrodes and gather in the region with high intensity electric field. Whereas, negative dielectrophoretic NDEP occurs when the particles are less polarized than the surrounding medium. The NDEP particles will be pushed away from the electrode with high intensity electric field and will gather in the region with low electric field intensities.

II. THEORY OF DIELECTRIC

The dielectrophoretic theory is divided into three main factors which is particles properties, suspending medium particles, and the input frequency of the AC electric field generated by microelectrodes [1][5]. The first factor, particles properties, has chemical and physical properties consisting of particle size, external surface, and internal matter. These properties can be defined as value of dielectric where the permittivity and conductivity of the particles are taken into consideration. Next is suspending medium particles. This medium is also known as the nature of the particle environment which means the value of permittivity and the conductivity of the medium is calculated. Last factor, the input frequency of the AC electric field generated by microelectrodes can affects the particle manipulation with the

978-1-5386-5284-8/18 $31.00 © 2018 IEEE

frequency of the alternating current flowing input into the electrode input channel [1].

The dielectrophoretic force, FDEP that related to the value of dielectric between the particles and the medium of the environment can be describe as the motion of a polarized particle in the flow of electric field intensity that is non uniform due to its net force. In the electric field, particle dielectric has an effective dipole with dipole moment (m) proportional to the electric field. The magnitude of the dipole moment for spherical particles is given by [4]:

$$m = 4\pi\varepsilon_0\varepsilon_m R^3 p \tag{1}$$

In equation (1), the factor $\varepsilon_0\varepsilon_m$ is the value of permittivity for the medium. ε_o is the value of permittivity in the vacuum medium of 8.854 x 10^{-12} F/m, while ε_m is the relative permittivity value of the medium known as the dielectric constant. The p parameter is in substitution to Clausius-Mossotti Factor (CMF) where the CMF defines the effective electrical polarizability of the particle relative to that of its surrounding medium [4]. The CMF depends on the intrinsic dielectric properties of the particles, the medium and the input frequency of the AC (electric current) electric field. CMF has a value limit of between -0.5 to +1.0[4]. If the CMF value is greater than 0, CMF> 0, the particle is known as PDEP, where the particles will be attracted to the region with high intensity electric field. On the other hand, if the CMF value is smaller than 0, CMF <0, the particle is known as NDEP, where the particles will be pushed away from the electrode with high intensity electric field. Fig. 1 shows the manipulation particles separation between 1 μm, 4.8 μm and 9.9 μm of engineered particles.

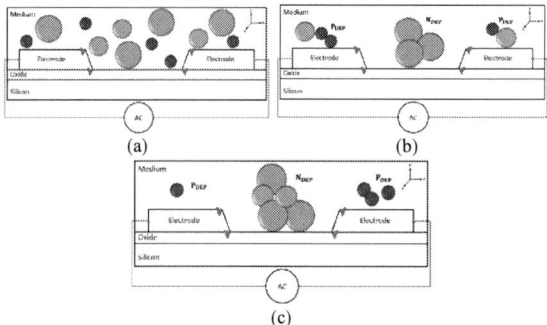

(a) (b)

(c)

Fig. 1 Manipulation particles separation between 1 μm, 4.8 μm and 9.9 μm of engineered particles (a) f=0 Hz (b) f=84 kHz and (c) f=616 kHz

However, the momentum of the dipole moment produces an electric field line and the effects of interaction within the potential energy, W given as [4]:

$$W = -m.E \tag{2}$$

The force acting on the polarized particles in nonuniform electric field known as FDEP. The use of FDEP with a work energy as equation (2), gives [4]:

$$F_{DEP} = -\nabla W = (m.\nabla)E \tag{3}$$

From equation (1) and (3), FDEP given as [2]:

$$F_{DEP} = 4\pi\varepsilon_0\varepsilon_m R^3 p(E.\nabla)E \tag{4}$$

Under normal condition, DEP experiments can be defined by the vector relationship equation:

$(E.\nabla)E = \frac{1}{2}\nabla E^2$; to give equation:

$$F_{DEP} = 2\pi\varepsilon_0\varepsilon_m R^3 p\nabla E^2 \tag{5}$$

The CMF describes the magnitude of particle polarization at a certain frequency value. Particle polarization factor depends on input frequency value [4].

$$CMF^* = \frac{\sigma^*_{particle} - \sigma^*_{medium}}{\sigma^*_{particle} - 2\sigma^*_{medium}} \tag{6}$$

$$\sigma^*_{particle} = \sigma_{particle} + i\omega\varepsilon_{particle}\varepsilon_0 \tag{7}$$

$$\sigma^*_{medium} = \sigma_{medium} + i\omega\varepsilon_{medium}\varepsilon_0 \tag{8}$$

$\sigma_{particle}$ and σ_{medium} is the relative conductivity value of the particles and the medium at low input frequency values.

$$CMF^* = \frac{\varepsilon^*_{particle} - \varepsilon^*_{medium}}{\varepsilon^*_{particle} - 2\varepsilon^*_{medium}} \tag{9}$$

$$\varepsilon^*_{particle} = \varepsilon_{particle} - \frac{i\sigma_{paticle}}{\omega} \tag{10}$$

$$\varepsilon^*_{medium} = \varepsilon_{medium} - \frac{i\sigma_{particle}}{\omega} \tag{11}$$

$\varepsilon_{particle}$ and ε_{medium} is the relative permittivity value of the particles and the medium at high input frequency.

III. METHODOLOGY

In this study, an array of tapered square electrodes with the size of 1 mm x 1 mm with the gap size of 80 um were used. For the this experiment, engineered particles with the size of 1 μm, 4.8 μm and 9.9 μm were chosen because engineered particles suitable in order to learning more detailed about DEP concept.

Based on an understanding of the DEP theory, the research was initially pursued by using the simulation of the analytical software MATLAB. Characterization on cross over frequency was implemented for each particle. The concentration ratio of engineered particles in DI water is 1: 40 for each particle size. Three types of engineered particles with the size of 1 μm, 4.8 μm and 9.9 μm were diluted with DI water with ratio of 1:1:1:1. Rectangle glass slip with thickness of 0.15 mm was placed over the sample to ease the monitoring of the particles. Alternating current (AC) is used in this experiment with 10 Vp-p on the function generator. Input frequency varies in the range 0 Hz to 1.5 MHz was supply to the electrodes. For each test, sample was exposed with non-uniform electric field not more than 1 minutes. Fig. 2 shows the region of interest for the engineered particles separation process.

Fig. 2 Region of interest (ROI) for separation procedure

Olympus STM6 microscope with single light source from top side was used to visualize particles movement. Fig. 3 shows the schematic of experimental setup for particles manipulation observation.

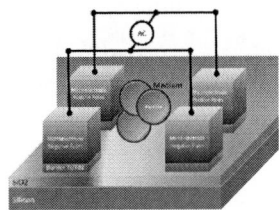

Fig. 3 Schematic experimental setup for particles manipulation observation

IV. RESULT AND DISCUSSION

Before the experiment was conducted, we discovered the expected result by simulation using analytical software MATLAB. We select two cross over frequency as shown in fig. 4 and Fig. 5. Fig.4 shows the graph at 84kHz while Fig. 5 shows the graph at 616kHz.

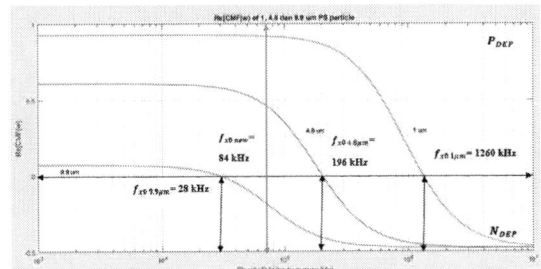

Fig. 4 Frequency dependence of CMF for 1 μm, 4.8 μm, and 9.9 μm at 84kHz

Fig.4 shows frequency dependence CMF factor for 1 μm, 4.8 μm, and 9.9μm particles. Based on the graph, positive CMF at frequencies above cross over frequency, f_{x0} were exhibits by the particles of size 1 μm and 4.8 μm. At this moment, the polarization of the particles was larger than the medium surrounding and indicates PDEP. For the particle of size 9.9 μm, it exhibits the negative CMF at frequencies below its cross over frequency, f_{x0}. It present NDEP as the polarization of the medium surrounding are larger than the particles. The graph also shows that the process of PDEP to NDEP of the particles are form from lower frequency towards higher frequency. The lower frequency was at 0kHz to < 84kHz, while the higher frequency was > 84kHz.

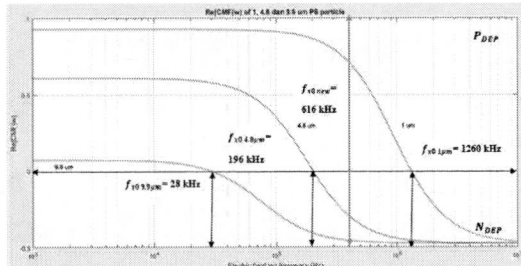

Fig. 5 Frequency dependence of CMF for 1 μm, 4.8 μm, and 9.9 μm at 161kHz

Fig. 5 shows frequency dependence CMF factor for 1 μm, 4.8 μm, and 9.9 μm particles. Based on the graph, positive CMF at frequencies above cross over frequency, f_{x0} were exhibits by the particles of size 1 μm. It shows that the particle was PDEP. The negative CMF at frequencies below its cross over frequency, f_{x0} and exhibits NDEP was particles 4.8 μm and 9.9 μm. Now, the polarization of particle 1μm was larger compare to the medium, meanwhile the polarization of particles 4.8 μm and 9.9 μm was smaller than the medium. The particle was form PDEP at lower frequency to at higher frequency. The lower frequency was at 0kHz to < 616kHz, while the higher frequency was > 616kHz.

Based on the experiments that have been carried out for the frequency of 94kHz, we can observe that the particles on the size of 1μm and 4.8 μm was experienced PDEP and its related to the CMF which is both particles are attracting to the top surface of microelectrodes that having a high intensity electric field. For the particle on the size of 9.9 μm, it was repelling from the electrodes which means it experienced NDEP. It forced the particle to push themselves to the centre of ROI. The results of attraction and repulsion of the particles were influences by the input frequency. Fig. 6 show an image that captured a) before the experiment was conducted b) after the experiment was conducted after 5 second.

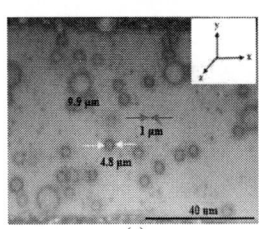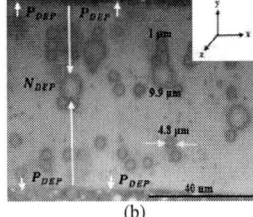

(a)   (b)

Fig. 6 Separation of particles 1 μm, 4.8 μm and 9.9 μm a) f=0 kHz b) f=94 kHz

When we conducted an experiment using the frequency of 616kHz towards the particles, we can see that only the particle 1 μm attracting to the top surface of microelectrodes, meanwhile the particles on the size of 4.8 μm and 9.9 μm were repelling towards inter microelectrodes that a having high intensity electric field. This is mean that PDEP goes to particle 1μm and NDEP was experienced by the particles 4.8 μm and 9.9 μm as it is follow the CMF. Due to induced high electric field at the surface of tapered structure, we noticed that most of the particle of 1 μm were gathered around at that region [6][7]. Fig. 7 shows an image that captured during the experiment.

978-1-5386-5284-8/18 $31.00 © 2018 IEEE

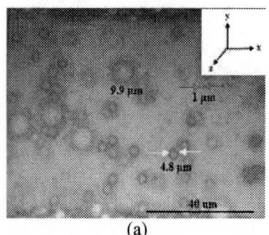 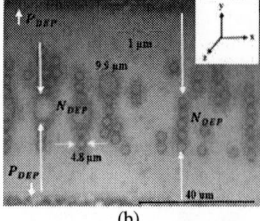

(a)   (b)

Fig. 7 Separation of particles 1 μm, 4.8 μm and 9.9 μm a) f=0 kHz b) f=616 kHz

This experiment was related to the application for separation the physical characteristics of pathogenic bacteria and blood components between red blood cell (RBC) and platelet in the human body in which the physical size of pathogenic bacteria is between the range of 0.5 μm to 3.0 μm. Whereas, the size of the blood component of platelet ranges from 1.5 μm to 3.5 μm, meanwhile the range of size of RBC is between 6μm to 9μm. But, the polarisation factor of the physical characteristics of pathogenic bacteria and blood components between RBC and platelet is different from engineered particle as the PDEP of physical characteristics of pathogenic bacteria and blood components between RBC and platelet are at the higher frequency and NDEP are at the lower frequency while engineered particles are vice versa.

V. CONCLUSION

The experimental work for 1 μm, 4.8 μm and 9.9 μm particles was illustrate the physical characteristics of pathogenic bacteria and blood components between red blood cell (RBC) and platelet in the human body. Although the polarisation factor of the physical characteristics of pathogenic bacteria and blood components between RBC and platelet is different from engineered particle, but the application of the separation still the same. The theory of polarisation factor of CMF are supported by the attraction and repulsion of the particles.

ACKNOWLEDGMENT

Gratefully acknowledge this work was financially supported by Institute of Microengineering and Nanoelectronic (IMEN), Universiti Kebangsaan Malaysia (UKM) and Ministry of Education of Malaysia through the grant project FRGS with the code of FRGS/2017/TK04/UKM/02/14

REFERENCES

[1] M.R Buyong, F Larki,., N. A. Aziz , J. Yunas, A.A. Hamzah, B.Y. Majlis,"Dielectrophoretic force response for lactobacillus casei," in Semiconductor Electronics (ICSE), 2016 IEEE International Conference on, p.p 38-42, 2016.

[2] K.Khashayar, Z.Chen, N.Saeid, T.L. Francisco J., B.Sara, M.Arnan and K. Kourosh, "Size based separation of microparticles using a dielectrophoretic activated system," J. Appl. Phys., vol. 108, no. 3, pp. 1-7, 2010.

[3] M. R. Buyong, N. A. Aziz, A. A. Hamzah, B. Y. Majlis, "Dielectrophoretic characterization of array type microelectrodes," in Semiconductor Electronics (ICSE), 2014 IEEE International Conference on, pp. 240-243, 2014.

[4] R.Pethiq, "Dielectrophoresis: An assessment of its potential to aid the research and particle of drug discover and delivery," Advance Drug Delivery Reviews.,vol. 65,pp.1589-1592, 2013.

[5] M. R. Buyong, N. A. Aziz, A. A. Hamzah, M. F. M. R. Wee, B. Y. Majlis, "Finite Element Modeling of Dielectrophoretic Microelectrodes Based on a Array and Ratchet Type," in Semiconductor Electronics (ICSE), 2014 IEEE International Conference on, 2014, pp. 252-255.

[6] M. R. Buyong, F. Larki, Y. Takanamura, N. A. Aziz, J. Yunas, A. A. Hamzah, B. Y. Majlis, "implementing the concept of dielectrophoresis in glomerular filtration of human kidney," in Semicanductor Electronics (ICSE), 2016 IEEE International Conference on, pp. 33-37, 2016.

[7] M.R Buyong, F. Larki, M.S. Faiz , A.A. Hamzah, J. Yun as, B.Y. Majlis, "A tapered aluminium microelectrode array for improvement of dielectrophoresis-based particle manipulation," Sensors, vol. 15, pp. 10973-10990, 2015.

[8] M.R Buyong, F Larki,., N. A. Aziz , J. Yunas, A.A. Hamzah, B.Y. Majlis, "Design, fabrication and characterization of dielectrophoretic microelectrode array for particle capture," in Microelectronics International, p.p 96-102, 2015

Effect of Different Metal Contact Distance and Light on Electrical Properties of Calcium Carbonate Thin Film

N.H.Sulimai
Faculty of Applied Sciences,
NANO-SciTech Centre,
Universiti Teknologi MARA,
Shah Alam, Malaysia
nurulhidahsulimai@gmail.com

Rozina Abdul Rani,
M.J.Salifairus
NANO-SciTech Centre,
Universiti Teknologi MARA,
Shah Alam, Malaysia
rozina.abdulrani@yahoo.com

Salifairus_mj@yahoo.co.uk

M.H.Mamat, M.F.Malek, A.S.
Zoolfakar

NanoElecTronic Centre, Faculty
of Electrical Engineering,

Universiti Teknologi MARA,
Shah Alam, Malaysia
hafiz_030@yahoo.com

mfmalek07@uitm.edu.my

ahmad074@salam.uitm.edu.my

Z.Khusaimi, S.Abdullah
Faculty of Applied Sciences,
NANO-SciTech Centre,
Universiti Teknologi MARA,
Shah Alam, Malaysia
zurai142@salam.uitm.edu.my

saifollah@salam.uitm.edu

Haseeb Khan, Salman
Alrokayan

Research Chair for Biomedical
Applications of Nanomaterials,
Department of Biochemistry,
College of Science, King Saud
University, Riyadh 11451, Saudi
khan_haseeb@yahoo.com

dr.salman@alrokayan.com

M.Rusop

NANO-SciTech Centre,

NanoElecTronic Centre, Faculty
of Electrical Engineering,

Universiti Teknologi MARA,
Shah Alam, Malaysia
nanouitm@gmail.com

Abstract— **CaCO$_3$ is an inorganic substance with low toxicity. Electrical properties of CaCO$_3$ is rarely reported although it has promising potential as biosensor. Therefore it is our aim to provide information on CaCO$_3$ electrical properties. This work described the effect of different distance of metal contact pad and also the effect of light to the electrical conductivity of CaCO$_3$. CaCO$_3$ thin film was prepared at 1.0M by gas diffusion method. From the results, the value of current is not affected by different distance of metal contact pad. Results showed conductivity of CaCO$_3$ thin films were not affected by the presence of light. Therefore CaCO$_3$ thin film conductivity is stable and suitable for wide application and further study such as for biosensor.**

Keywords—CaCO$_3$ Thin Film; CaCO$_3$ Electrical Properties; Concentration; Metal Pad Distance; Gas Diffusion

I. Introduction

Study of inorganic metal and its physicochemical properties has been extensive and established. Among the most widely applied inorganic metal in the industry is Calcium Carbonate (CaCO$_3$) due to its abundance in nature and its properties. Although it has many strong properties such as high thermal resistance and alkaline pH, its most appealing property is it is non-toxic.

It is an ionic compound hence it posesses electrical properties of a semiconductor There are growing interest on studies of biosensors. Hence there is a promising interest on utilizing CaCO$_3$ as drug carriers or biosensors due to its non-toxic attribute. Calcium (Ca) element exists in the most common form of Calcium Carbonate with the chemical formula (CaCO$_3$). It is widely used in industry due to its availability and cheap price. Naturally CaCO$_3$ exist as solid. Most applications utilized naturally CaCO$_3$ in powder form. There were many studies that investigate physicochemical properties of CaCO$_3$ in powder form. Limited if none, reported on CaCO$_3$ thin film and also its electrical properties. There are other studies that studied on Carbon thin films [1-5] and Copper Iodide thin film [6], ZAO thin film [7] but not much on CaCO$_3$ thin films.

There are other studies that utilized CaCO$_3$ as biosensors. Biosensor used CaCO$_3$ to detect methyl parathion an organophosphate compounds found in pesticide [8] and as protein sensor to detect protein-carbonate ion interaction [9]. More information is needed for the purpose. Stability of CaCO$_3$ thin film is important for the predictability of its suitability as biosensor such as glucose biosensor by crosslinking CaCl$_2$ with alginate under sub zero temperature [10] and electrochemistry biosensors that used CaCO$_3$ as host for

978-1-5386-5284-8/18 $31.00 © 2018 IEEE

CdTe quantum dots upon forming channel-like structure [11].

Other previous work such as Mirwald's was done way too long ago but studied electrical conductivity of calcite at (300-1200)°C in high pressure of 40 bars CO_2 and deduced that electrical conductivity of calcite changes by more than 4 orders of magnitude [12]. They reported electrical value of 7-70.10^{-7} $\Omega^{-1}cm^{-1}$ at 0.1eV at temperature 300-500°C [12]. Thompson and Pownall studied the surface electrical properties of calcite in aqueous solutions in a closed system. The results is to determine the factor that manipulates the electrical properties at calcite solution interface [13].

Other recent works that studies electrical conductivity of $CaCO_3$ are very limited. One was done in high pressure and high temperature to to map the electrical conductivity of the Earth's mantle by Ono and Mibe [14]. They reported electrical conductivity value of aragonite ($CaCO_3$) of 23.9 S/m at high pressure of 4.5GPa. According to their study, the electrical conductivities of carbonates can be described as a function of pressure, temperature, and chemical composition [14]. Wu, Hubbard, Williams and Ajo-Franklin synthesized Calcium Chloride ($CaCl_2$) and Na_2CO_3 in a controlled precipitation rate. The study revealed calcite precipitation has similar electrical response with metallic mineral precipitation [15]. Although $CaCO_3$ is electronically non-conductive, EDL polarization is observed and directly related to volume fraction and grain size [15].

These studies did explained suitability of $CaCO_3$ to be used as biosensor. But none demonstrated steps to prepare $CaCO_3$ thin films and presents its electrical conductivity ranges for the purpose of such applications. This study presented preparation of $CaCO_3$ thin film by gas diffusion method. We also present the effect of distance of metal contact pad and light to $CaCO_3$ electrical conductivity. These informations are crucial since currently no other works reported on that. This information is aimed to contribute towards using $CaCO_3$ for biosensor application. Effect of light and range of current on different voltage is explained

II. METHODOLOGY

A. Materials

Calcium chloride anhydrous granular ≤7.0mm ($CaCl_2$) 93.0% (MW 110.98) and ammonium carbonate ACS reagent ≥30.0% $(NH_4)_2CO_3$ (MW 96.09) and calcium carbonate ($CaCO_3$) are of industrial grade by Sigma Aldrich.

B. Preparation

20mL 1.0M $CaCl_2$ solution is prepared with ethanol used as solvent. Glass substrate is cut and cleaned with piranha solution and dried. Black tape is applied on the glass substrate and is submerged in a petri dish filled with the prepared $CaCl_2$ solution. Another petri dish is filled with 2g of $(NH_4)_2CO_3$ and both petri dish were put in a sealed desiccator. Reaction is left for 24 hours and the glass substrate were later dried in room temperature (RT) for 24 hours. The resulting thin film is coated with 60nm Gold (Au) by sputter coater.

C. Characterization

Samples were characterized with 2-point probe I-V measurement (Advantest) at 1, 5 and 10V. The I-V were

measured in three different distance. A,B and C with A being the furthest and C is the nearest. A,B and C distances is (1, 0.5 and 0.07) mm to study the effect of different metal contact distance. Whereas for the study of effect of light, samples were characterized with 2-point probe I-V measurement (Advantest) at 1, 5 and 10V under the light marked in the chart as "ill". Reference reading was taken in the dark with no light influence marked in the chart as "dark".

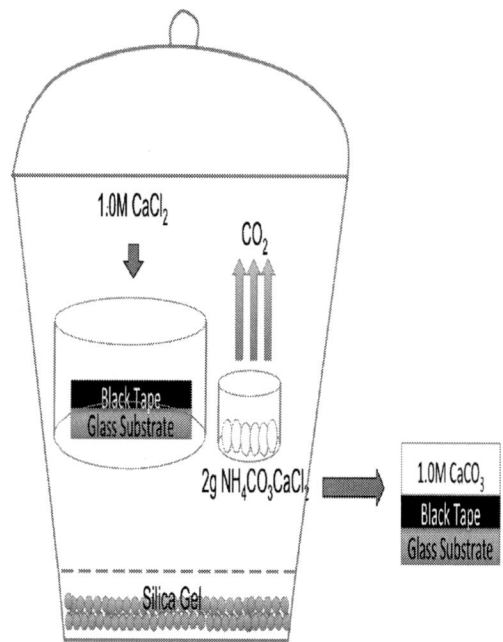

Figure 1. Preparation of 1.0M $CaCO_3$ thin film by gas diffusion method

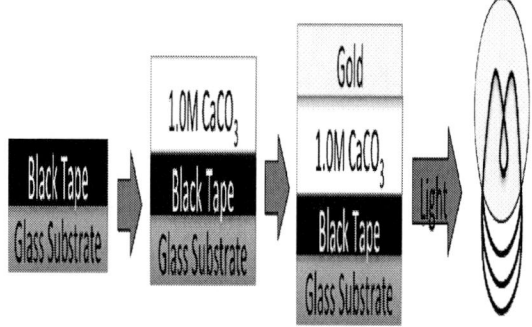

Figure 2. Flow chart of preparation of $CaCO_3$ thin film

III. RESULTS AND DISCUSSION

The value of conductivity of $CaCO_3$ thin film is rare if not never been reported. Therefore, it is crucial for this work to report on electrical properties of $CaCO_3$ thin film. From the results in Figure 3, it is clear the I-V curves of $CaCO_3$ are non-linear. All three different IV-curves are presented in different voltage by all different distances. From the results, it can be seen that $CaCO_3$ portrayed Schottky behavior. The curve surpasses zero intersection and increases gradually forward with extremely small forward current and voltage. Sudden and rapid increase of forward voltage happens when internal barrier voltage is exceeded. Voltage increment is very little resulting in non-linear curve.

Likewise when the semiconductor is reversed biased, it blocks current but only leaks small current and this is at the left quadrant of I-V characteristic curves. The current was blocked continuously by the semiconductor until the reverse voltage becomes greater than breakdown voltage point. Sudden increment in reverse current is observed at the reverse quadrant and produce almost a straight line heading downward.

At low voltage 1V, the electrical properties appeared to have slightly high in resistance for sample A and C compared to sample B. Current measured for 1V applied shown the conductivity ranges from (3 to 10×10^{-7}) A. For $CaCO_3$ thin films measured at 5V and 10V, the measured currents shown no significant differences for all distances and lower resistance is observed compared to in 1V. The magnitude of the current also increased for $CaCO_3$ samples measured at 5V, the currents are all 0.5×10^{-4}A and for 10V, ranges from 2.5 to 3.5×10^{-4}A. It is observed that as the voltage increases, the magnitude of the current also increases. The distance does not shows prominent correlation with measured current. Therefore, distance of the metal contact pad does not affect conductivity of $CaCO_3$ thin film.

To study the effect of light to the conductivity of $CaCO_3$ thin film, they were measured under the influence of light and also in the dark as reference. It is also obvious that presence of light does not affect conductivity at all distances and voltage. This is observed when all I-V curves shows almost identical patterns and magnitude in all forward voltage and also reversed downward voltage in the left quadrant for dark and illuminated samples. Therefore, $CaCO_3$ thin film conductivity is stable in the presence or without light. The electrical properties are stable even under the influence of illumination. This behaviour is indeed helpful upon considering $CaCO_3$ for application as a sensor such as biosensor especially.

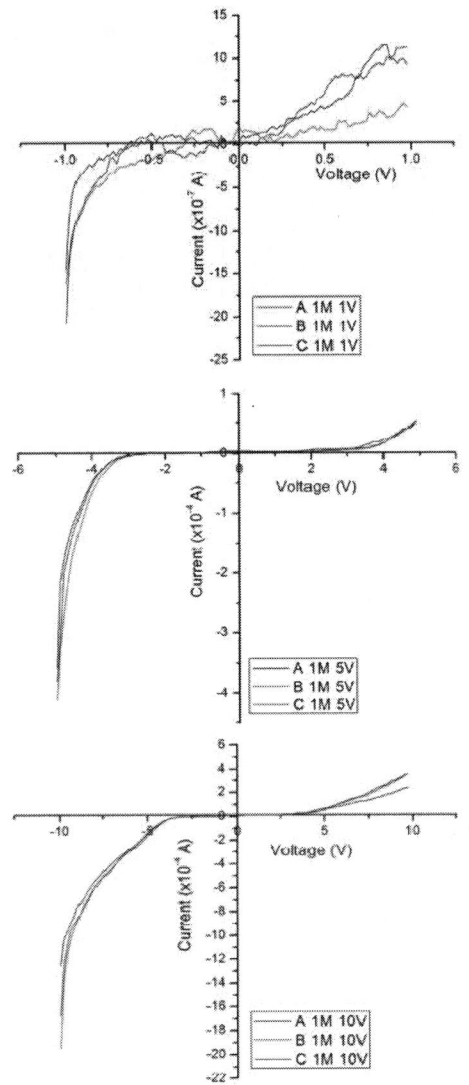

Figure 3. I-V curve of 1.0M $CaCO_3$ thin film at 1,5 and 10V in dark condition at different distances A:1mm, B:0.5mm and C:0.07mm.

I would like to express gratitude to Faculty of Applied Sciences, Universiti Teknologi MARA (UiTM) Shah Alam and Kementerian Pendidikan Malaysia. Thank you to Mr. Azlan, Mdm Nurul Wahida from NANO-SciTech Centre, Institute of Science, UiTM. Cheers to my family and friends. This work has been funded by Geran Bestari Perdana: 600-IRMI/PERDANA5/3 BESTARI (102/2018). Special thanks to Research Chair for Biomedical Applications of Nanomaterials, Department of Biochemistry, College of Science, King Saud University, Riyadh 11451, Saudi Arabia.

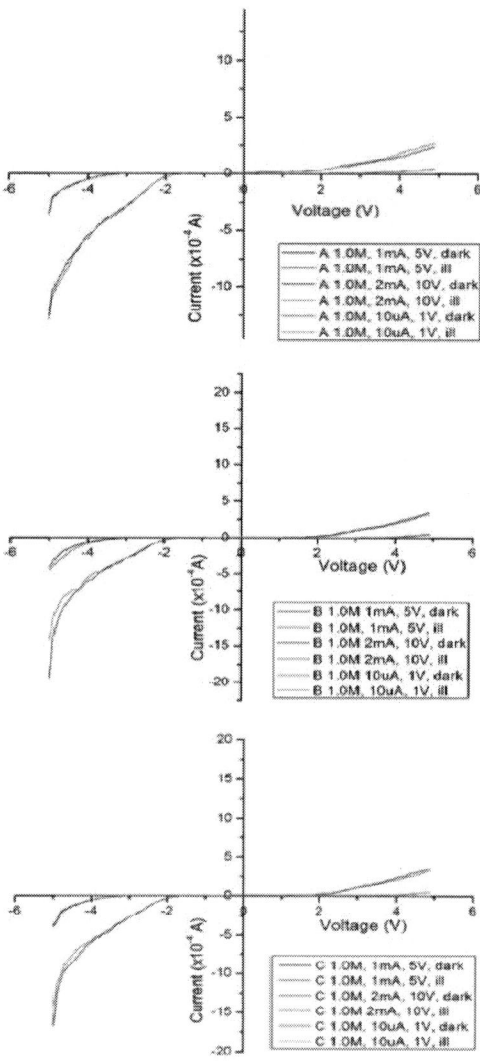

Figure 4. I-V curve of 1.0M CaCO$_3$ thin film at 1,5 and 10V in dark and illuminated condition at different distance of metal contact pad A:1mm, B:0.5mm and C:0.07mm.

IV. CONCLUSION

CaCO$_3$ thin film is prepared by gas diffusion method. From the results, it can be concluded that distance of the metal contact pad does not affect the conductivity of CaCO$_3$. I-V curves shows magnitude of the current increases with increment of voltage. From the I-V curve results, it can be concluded that for all samples, conductivity of CaCO$_3$ thin film was not affected under the presence of light. Therefore CaCO$_3$ thin film is suitable for application as sensor.Therefore CaCO$_3$ is suitable to be considered for further study in application as biosensor. The value of conductivity of CaCO$_3$ is reported in this study.

ACKNOWLEDGMENT

REFERENCES

[1] M. Rusop, Kinugawa, T., Soga, T., Jimbo, T., "Preparation and microstructure properties of tetrahedral amorphous carbon films by pulsed laser deposition using camphoric carbon target", *Diamond and Related Materials,* vol. 13(11-12), pp. 2174-2179, 2004.

[2] M. Rusop,Mominuzzaman, S.M.,Soga, T.,Jimbo, T.,Umeno, M., "Effect of substrate temperature on growth of nitrogen incorporated camphoric carbon films by pulsed laser ablation", *International Journal of Modern Physics B,* vol.16(6-7), 866-870, 2002.

[3] M. Rusop, Adhikari, S., Omer, A.M.M., Jimbo, T., Umeno, M., "Effects of methane gas flow rate on the optoelectrical properties of nitrogenated carbon thin films grown by surface wave microwave plasma chemical vapor deposition", *Diamond and Related Materials,* vol. 15(2-3), pp. 371-377, 2006.

[4] M. Rusop, Tian, X.M., Kinugawa, T., Jimbo, T., Umeno, M., "Preparation and characterization of boron-incorporated amorphous carbon films from a natural source of camphoric carbon as a precursor material", *Applied Surface Science,* vol. 252(5), pp. 1693-1703, 2005.

[5] M. Rusop, Mominuzzaman, S.M., Soga, T., Jimbo, T., Umeno, M., "Nitrogen doped n-type amorphous carbon films obtained by pulsed laser deposition with a natural camphor source target for solar cell applications", *Journal of Physics Condensed Matter,* vol. 17(12), pp. 1929-1946, 2005.

[6] M. Rusop, Soga, T., Jimbo, T., Umeno, M., "Copper iodide thin films as a p-type electrical conductivity in dye-sensitized p-CuI|Dye|n-TiO2 heterojunction solid state solar cells", *Surface Review and Letter,* vol. 11(6), pp. 577-583, 2004.

[7] Malek, M.F.,Mamat, M.H.,Musa, M.Z., Khan, H.A.,Rusop, M., "Metamorphosis of strain/stress on optical band gap energy of ZAO thin films via manipulation of thermal annealing process", *Journal of Luminescence,* vol. 160, pp.165-175, 2015.

[8] K. Yao, J. Li, F. Yao, and Y. Yin, "Chitosan- Based hydrogels functions and applications," CRC Press, pp. 389-390, 2012.

[9] R. A. Rayana, S. R. Hugo Javier, M. G. Maria Liliana, and A. F. Bernardo, "Chemical biosensors based on proteins involved in biomineralization processes biosensors- Emerging materials and applications," 2011.

[10] F. Amin, W. D. Dian, and H. Dadan, "Alginate cryogel based glucose biosensor," *IOP Conference Series: Materials Science and Engineering,* pp. 1-7, 2016.

[11] W. Y. Cai, L. D. Feng, S. H. Liu, and J. J. Zhu, "Hemoglobin-CdTe-CaCO3 @ Polyelectrolytes 3D Architecture: Fabrication, Characterization and Application in Biosensing," *Electrochemistry Biosensor,* vol. 18, pp. 3127-3136, 2008.

[12] P. W. Mirwald, "The Electrical Conductivity of Calcite Between 300 and 1200°C at a CO$_2$ Pressure of 40 Bars," *Phys. Chem. Minerals* vol. 4, pp. 291-297, 1979.

[13] D. Thompson, W., and P. G. Pownall, "Surface Electrical Properties of Calcite," *Journal of Colloid and Interface Science* vol. 131, pp. 74-82, 1989.

[14] S. Ono and K. Mibe, "Electrical conductivity of aragonite in the subducted slab," *European Journal of Mineralogy,* vol. 25, pp. 11-15, 2013.

[15] Y. Wu, S. Hubbard, K. H. Williams, and J. Ajo-Franklin, "On the complex conductivity signatures of calcite precipitation," *Journal of Geophysical Research: Biogeosciences,* vol. 115, pp. n/a-n/a, 2010.

978-1-5386-5284-8/18 $31.00 © 2018 IEEE

Facile Synthesis of N-doped ZnO Nanorod Arrays: Towards Enhancing the UV-sensing Performance

A.S. Ismail
NANO-ElecTronic Centre,
Faculty of Electrical
Engineering, Universiti
Teknologi MARA,
Shah Alam, Selangor, Malaysia
kyrin_sama@yahoo.com

M.H. Mamat
NANO-ElecTronic Centre,
Faculty of Electrical
Engineering, Universiti
Teknologi MARA,
Shah Alam, Selangor, Malaysia
mhmamat@salam.uitm.edu.my

M.F. Malek
NANO-SciTech Centre,
Institute of Science, Universiti
Teknologi MARA
Shah Alam, Selangor, Malaysia
mfmalek07@uitm.edu.my

N.E.A. Azhar
NANO-ElecTronic Centre,
Faculty of Electrical
Engineering, Universiti
Teknologi MARA,
Shah Alam, Selangor, Malaysia
najwaezira@yahoo.com

N.H. Sulimai
NANO-SciTech Centre,
Institute of Science, Universiti
Teknologi MARA
Shah Alam, Selangor, Malaysia
nurulhidahsulimai@gmail.com

R. Abdul Rani
NANO-SciTech Centre,
Institute of Science, Universiti
Teknologi MARA
Shah Alam, Selangor, Malaysia
rozina.abdulrani@yahoo.com

A.S. Zoolfakar
NANO-ElecTronic Centre,
Faculty of Electrical
Engineering, Universiti
Teknologi MARA,
Shah Alam, Selangor, Malaysia
ahmad074@salam.uitm.edu.my

M. Rusop
NANO-SciTech Centre,
Faculty of Electrical
Engineering, Universiti
Teknologi MARA,
Shah Alam, Selangor, Malaysia
rusop@salam.uitm.edu.my

Abstract— **Nitrogen (N) doped zinc oxide (ZnO) nanorod arrays at different doping concentrations from 0 at.% (undoped) to 2 at.% have been prepared using sol-gel immerse method. FESEM images revealed that the average diameter increased when the concentration of N in ZnO increased. The I-V measurement displayed that 1 at.% sample possessed the lowest resistance film and exhibit the highest UV sensing performance with sensitivity of 12.9.**

Keywords—Nitrogen, zinc oxide, nanorod, immersion

I. INTRODUCTION

Zinc oxide (ZnO) is a wide bandgap semiconductor material (3.37 eV) and large exciton binding energy of 60meV [1]. Because of excellent structural, electrical, and optical properties, ZnO nanostructure have been studied for use in many application because of their potential in many applications such as light emitting diodes (LED), sensors, and solar cells. ZnO nanorod arrays are highly recommended because of their high surface area and carrier confinement which is improvement of performance. This promising ZnO nanorod array can be deposited using various methods such as metal-organic chemical vapour deposition (MOCVD), sputtering, pulse-laser deposition (PLD), and the sol-gel method [2]. However, the fabrication method which used high vacuum and high temperature is not a cost effective approach. Thus, a solution-based method is more preferable due to simple process and involves low temperature preparation. In this study, a facile method called sol-gel immersion is used. The preparation involved very low temperature (95 °C). According to previous studies, ZnO nanostructures have been doped with various metals, such as tin (Sn), aluminium (Al), and gallium (Ga) [3-5]. However, the effect of nitrogen (N)-doped ZnO nanorod arrays at different doping concentrations are rarely discussed in literature, especially for ultraviolet (UV)-sensing applications. Therefore, we fabricated N-doped ZnO nanorod arrays at different concentration and analysed their structural, electrical, as well as UV-sensing characteristics.

II. PREPARATION METHOD

A. Synthesis of N-doped Zinc Oxide Nanorods

N-doped ZnO nanorod arrays were synthesized by sol-gel immersion method. The solution for zinc nitrate was composed of a mixture of 0.1M zinc nitrate hexahydrate , 0.1M hexamethylenetetramine, and ammonium acetate. Ammonium acetate acted as a source and the reagents were mixed in 500 ml DI water. The concentration of dopant were varied to achieve 0 (undoped), 0.6, 1, and 2 at.%. After mixed the solution, the solution was sonicated for 30 minute at 50 °C by using an ultrasonic water bath. Then, for 3 hour at room ambient, the solution were stirred. The solution was poured into the Schott bottles which contain ZnO seed layer-coated glass. The sealed bottle was immersed for 50 minutes at 95 °C by using immersion water bath tank. After that the samples were dried by using furnace at 150 °C for 10 minutes. Lastly, by using the same furnace, the sample were anneal for 1 hour at 500 °C.

B. Characterization

N-doped ZnO nanorod arrays were characterized using field emission scanning electron microscopy and x-ray diffractometer for structural properties. Two probing direct current (DC) system was used to characterize the electrical properties. The UV-sensing properties of the thin films were analyzed using ultraviolet (UV) lamp (365 nm, 4 W) combined with DC probing system.

III. RESULT AND DISCUSSION

The surface morphology images of different concentration undoped and N-doped films were displayed in Fig. 1. The films showed that the nanorod arrays are uniformly grown on the substrate and possessed dense arrays. The diameter size

978-1-5386-5284-8/18 $31.00 © 2018 IEEE

of nanorod were observed and estimated to be increased due to dopant increased. The different doping concentration used influenced the growth of N-doped ZnO nanorod diameter, which is 0 at.% is 101 nm, 0.6 at.% is 193 nm, 1 at.% is 181 nm, and 2 at.% is 222nm. In addition, it is observed that the surface of nanorod arrays consist of nanoholes after doped with N. This nanoholes may possibly appear at each of nanorod structure that because effect of nitrogen itself that the incorporation of N into ZnO crystal produced defect to the growth and also may be due to the evaporation of impurities such as HMT from ZnO.

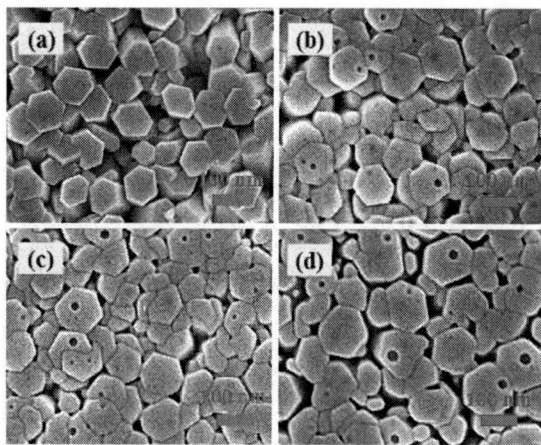

Fig. 1. Surface morphology of (a) undoped and N-doped ZnO nanorod arrays at (b) 0.6, (c) 1, and (d) 2 at.% concentrations.

Fig. 2 shows the crystalline properties of undoped and 1 at.% N-doped ZnO nanorod arrays. Based on the plot, it is observed that the film produced polycrystalline structure with (002) plane orientation possessed the highest intensity. This indicates that the growth of nanorod arrays is along *c*-axis orientation. In addition, it is observed that the intensity of the (002) diffraction peak slightly increased after doping with N. This may be caused by the increment of crystallite size, leading to enhancement of average diameter. Based on the report by Ismail et al., film with large crystallite size produced better crystalline film [6]. Besides, the substitution of larger ionic radius atom (N) into ZnO produced larger crystallite size [7].

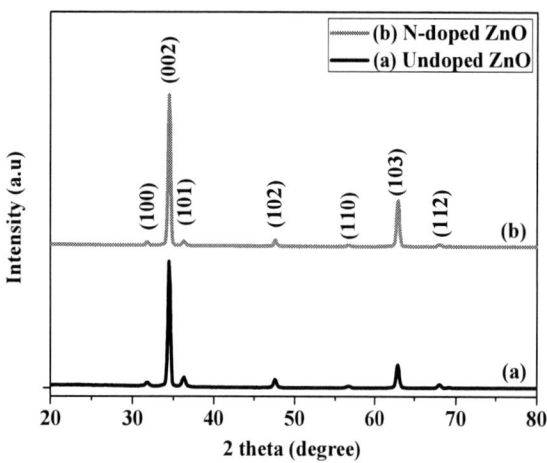

Fig. 2. Surface morphology of (a) undoped and N-doped ZnO nanorod arrays at (b) 0.6, (c) 1, and (d) 2 at.% concentrations.

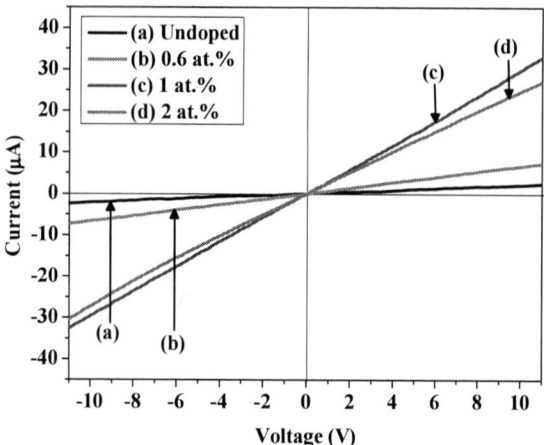

Fig. 3. I-V plot of N doped ZnO nanorod arrays at different concentration.

The I-V characteristics of the samples can be obtained from the I-V plot as shown in Fig. 3. From the plots, it is witnessed that the film produced Ohmic behavior with linear increment of current. It is observed that undoped sample produced the lowest linear current while 1 at.% sample produced the highest linear current values. Using Ohm's law (V=IR), the resistance of films were calculated from the tangent of the linear plots. The calculated resistance of the undoped, 0.6, 1, and 2 at.% N-doped ZnO nanorod arrays are 4.5, 1.5, 0.3, and 0.5 MΩ, respectively, as shown in Fig. 4 plot. From the plot, it is exposed that the resistivity reduced after doping and 1 at.% sample produced the lowest film resistance. This decrement of resistance may be related to the N-related defects which extra free electron in the film and improve the mobility of free carriers.

The photoresponses of the UV sensors are shown in the Fig. 4, which were measured at 12 V bias voltage under 365 nm and UV illumination (750µW/cm²). From the responses, it is found that the films showed good responses to UV. The response plot of the sensors could be described by the excitation of electron from valence to conduction band. Under UV illumination, the film absorbed the illuminated light.

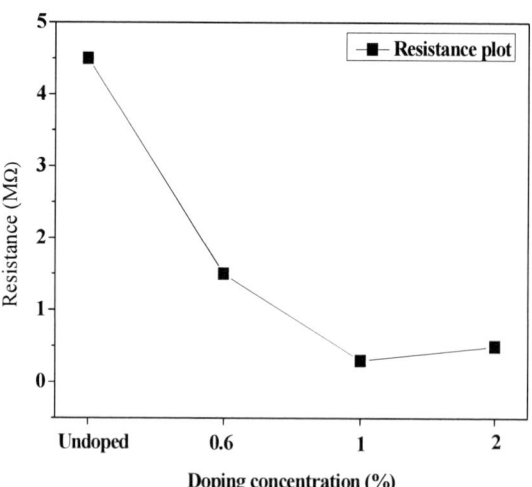

Fig. 4. Resistance plot at different N-doping concentration.

978-1-5386-5284-8/18 $31.00 © 2018 IEEE

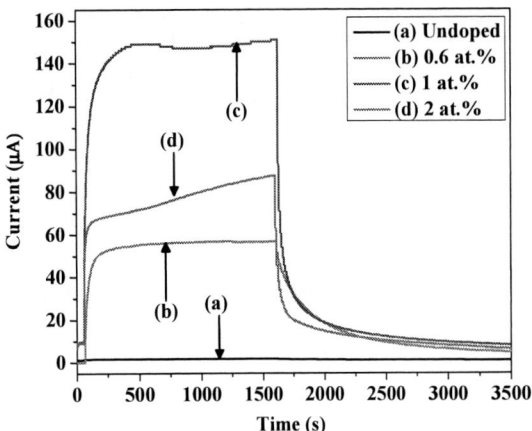

Fig. 5. Photoresponse of undoped and N doped ZnO at different doping concentration.

Due to high photon energy, the electron possessed enough energy to jump from valence band to conduction band, leading to the increment of current values [8]. Once the source of UV light in turned "OFF", the electron dropped from the conduction band and recombined with the hole in valence band. This lead to the reduction of current signal as observed in the responses plot. From the photoresponse result, the responsivity, R of the sample was calculate according to the following relation [9]:

$$R = \frac{I_{ph} - I_{dark}}{P_{op}} \qquad (1)$$

where I_{ph} is the photocurrent, I_{dark} is the dark current and P_{op} is the optical power of the UV source. From the measurement, the responsivity of the sensors are 0.04, 3.51, 9.81, and 4.91 A/µW for undoped, 0.6, 1, and 2.at% N-doped ZnO, respectively, as depicted in Fig. 6. The result has shown that the photoresponse of N-doped ZnO improved after doping. Such improvement may be associated with the increment of surface area produced by the nanoholes and also improvement of crystallinity after doping. Besides, the decrement of resistance also plays an important role to the increment of sensor performance. Further, the sensitivity of the film is estimated from the ratio of dark current to the photocurrent (under illumination). The sensitivity, S is shown by the given equation:

$$S = \frac{I_{ph}}{I_{dark}} \qquad (2)$$

From the calculation, the sensitivity of the sensors to the UV light are 1.39, 8.77, 12.98, and 8.46 for undoped, 0.6, 1, and 2.at% N-doped ZnO-based humidity sensors, respectively, as in Fig. 6 plot. Similar to the responsivity, the 1 at.% doping also displayed the highest sensitivity to UV light.

IV. CONCLUSION

N doped ZnO nanorod arrays at different concentration were prepared on Al doped ZnO coated glass substrates using sol-gel immersion. FESEM images revealed that the average diameter increased when the concentration of N in ZnO increased. The I-V measurement displayed that 1 at.% sample possessed the lowest resistance film of 0.45 MΩ and exhibit the highest UV sensing performance with responsivity and sensitivity of 9.8 A/W and 12.9, respectively. This result indicate that N is suitable dopant to improve the performance of ZnO-based UV sensors.

ACKNOWLEDGMENT

This work was supported by grant no. 600-IRMI/PERDANA 5/3 BESTARI (102/2018). The authors would like to thank the Institute of Research Management and Innovation (IRMI), Universiti Teknologi MARA (UiTM), Malaysia for their support. The authors would also like to acknowledge Mr. Salifairus Mohammad Jafar (UiTM Science Officer), Mrs. Nurul Wahida (UiTM Asst. Science Officer), Mr. Mohd Azlan Jaafar (UiTM assistant engineer) and Mr. Suhaimi Ahmad (UiTM assistant engineer) for their kind support of this research.

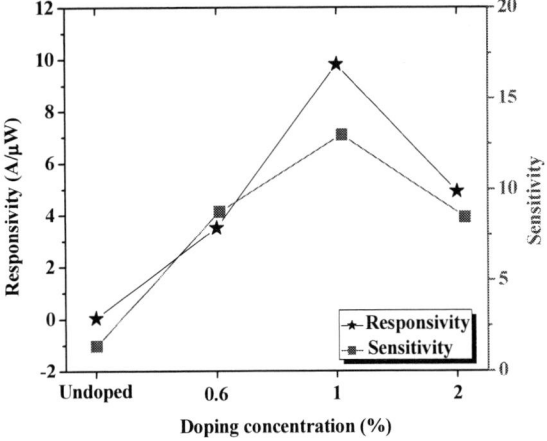

Fig. 6. Responsivity and sensitivity plot at different N-doping concentration.

REFERENCES

[1] A. S. Ismail, M. H. Mamat, M. F. Malek, M. A. R. Abdullah, M. D. Sin, and M. Rusop, "Effect of thermal implying during ageing process of nanorods growth on the properties of zinc oxide nanorod arrays," AIP Conference Proceedings, vol. 1733, p. 020009, 2016.

[2] A. S. Ismail, M. H. Mamat, M. M. Yusoff, M. F. Malek, A. S. Zoolfakar, R. A. Rani, et al., "Enhanced humidity sensing performance using Sn-Doped ZnO nanorod Array/SnO2 nanowire heteronetwork fabricated via two-step solution immersion," Materials Letters, vol. 210, pp. 258-262, 2018.

[3] R. Mohamed, M. H. Mamat, A. S. Ismail, M. F. Malek, A. S. Zoolfakar, Z. Khusaimi, et al., "Hierarchically assembled tin-doped zinc oxide nanorods using low-temperature immersion route for low temperature ethanol sensing," Journal of Materials Science: Materials in Electronics, vol. 28, pp. 16292-16305, 2017.

[4] T. Ivanova, A. Harizanova, T. Koutzarova, and B. Vetruyen, "Optical and structural study of Ga and In co-doped ZnO films," Colloids and Surfaces A: Physicochemical and Engineering Aspects, vol. 532, pp. 357-362, 2017.

[5] N. Krstulović, K. Salamon, O. Budimlija, J. Kovač, J. Dasović, P. Umek, et al., "Parameters optimization for synthesis of Al-doped ZnO nanoparticles by laser ablation in water," Applied Surface Science, vol. 440, pp. 916-925, 2018.

[6] A. S. Ismail, M. H. Mamat, N. D. M. Sin, M. F. Malek, S. A. Saidi, M. M. Yusoff, et al., "Structural and optical properties of N-doped ZnO nanorod arrays prepared using sol-gel immersion method," in

2016 IEEE Student Conference on Research and Development (SCOReD), 2016, pp. 1-6. R.

[7] Perumal and Z. Hassan, "Effect of nitrogen doping on structural, morphological, optical and electrical properties of radio frequency magnetron sputtered zinc oxide thin films," Physica B: Condensed Matter, vol. 490, pp. 16-20, 2016.

[8] M. M. Yusoff, M. H. Mamat, A. S. Ismail, M. F. Malek, Z. Khusaimi, A. B. Suriani, et al., "Enhancing the performance of self-powered ultraviolet photosensor using rapid aqueous chemical-grown aluminum-doped titanium oxide nanorod arrays as electron transport layer," Thin Solid Films, vol. 655, pp. 1-12, 2018.

[9] M. H. Mamat, M. Z. Sahdan, Z. Khusaimi, A. Z. Ahmed, S. Abdullah, and M. Rusop, "Influence of doping concentrations on the aluminum doped zinc oxide thin films properties for ultraviolet photoconductive sensor applications," Optical Materials, vol. 32, pp. 696-699, 2010.

Tip Deflection of a Thermal Bimorph Cantilever Beam with Different Geometrical Structures

Z. H. A. Rahman*, M. H. Md. Khir, M. A. Zakariya
Dept. of Electrical and Electronic Engineering
Universiti Teknologi PETRONAS
Perak, Malaysia
*zatihananirahman@gmail.com

Abstract—The residual stress is often due to a thermal mismatch in the thermal expansion coefficient (CTE) between the two materials. This phenomenon causes device failure such as curling, buckling, and also fracture upon releasing. Hence, strategies are needed to minimize the effect of residual stress on the induced beam bending upon release of the device. In this paper, four different geometrical structures are studied theoretically and simulated. The stress level is deduced by measuring the deflection of the beam. The result has shown that the beam with Al and SiO₂ strip arranged alternately as the top layer and SiO₂ as the bottom layer have the least downward deflection upon applied temperature. For $\Delta T = 50$ K, deflection of 0.8425 μm is measured. Simulation-based of varying thickness also show that having thicker SiO₂ (with a similar total thickness of the beam) contribute to smaller downward deflection and least percentage error between theoretical and simulation result. Surface stress analysis also shows that the stress measured is 107.86 MPa, which is lower than the yield strength of Al and SiO₂ materials.

Keywords—bimetallic, deflections, CTE, residual stress.

I. Introduction

The field of MEMS research has grown rapidly and more applications have emerged based on this technology. MEMS technology offers three generic merits which are miniaturization, monolithic integration and parallel fabrication with high precision. These will allows MEMS devices to achieve greater function, small sizes at lower production costs. To date, the vast majority of MEMS products are either sensors or actuators [1, 2]. Both devices serve as an interface between the system and the physical world. Among these, actuators are of special interest. Actuator works by transforming physical energy into mechanical motion. There are several transduction mechanisms that can generate mechanical movement such as piezoelectricity, electrostatic force, magnetic or thermal expansion. Each actuation methods has its own advantages and disadvantages. The selection of actuation method is usually case dependent.

Recently, actuation through thermal has gain focused attention from the researcher, despite its low-frequency response. In the MEMS world, thermal actuation method is capable of achieving moderately large displacement at moderately fast actuation response and small size [3, 4]. Various actuator based on electrothermal (ET) bimorph beam has been developed [5-7]. For example, in [8], a large range ET bimorph MEMS mirror was studied suitable for various optical imaging and tracking applications. It consists

of three bimorph beams made of copper (Cu) and tungsten (W) materials. The mirror managed to perform both tip-tilt and piston motion. Vertical displacement of 320 μm is achieved when actuated using only 3 V DC with power consumption of 56 W. The maximum optical scan angle measured for this MEMS mirror is ±18°. For biomedical applications, a novel ET actuator based on bimorph beam design was proposed in [9] the optical coherence microscopy (OCM). It is designed to have almost no lateral shift and can generate very large vertical displacement using the three bimorph beam mechanism. Device testing shows that for an actuation voltage of 5.3 V, the vertical displacement is 0.62 mm and both lateral shift and tilting angles are 10 μm and 0.7°, respectively, which can be considered as very small. In [10], a miniature endoscopic based optical coherence tomography (OCT) probe was fabricated using the folded bimorph beam mechanism made of Al and SiO₂. Experimental results recorded the optical scan angle to be ±16° at 3.6 V DC. A 2D MEMS mirror utilizing bimorph beams as actuators have been studied in [11] to be employed in 2D and 3D *in vivo* endoscopic OCT images. The mirror has gimbal-less shape and four actuators on each side. For a device size of 4 mm², it is able to produce large scanning range of up to ±30° for both *x*- and *y*-axes at only 5.5 V DC. Wang *et. al* [12] proposed and fabricated a miniaturized Fourier Transform Spectroscopy (FTS) system. The movable micromirror embedded inside the system is designed with S-shaped inverted series connected (ISC) bimorph beam based electrothermal actuator. The ISC bimorph beams are arranged on each side of the mirror in a folded form as to achieve the S-shaped look. The device is actuated using 0.8 V DC and results show that the micromirror can generate up to 95 μm usable linear optical path difference.

The performance and efficiency of the micromirror usually dependent on the actuator performance. As for cantilever bimorph beam based actuator, the issue that often associated is deformation due to residual stress experience by the thin film materials either in room temperature or zero-loading condition [13-14]. Residual stress results from the mismatch in CTE between the materials consist of a beam. This effect contributes to the undesirable bending upon the release of the cantilever. Efforts have been demonstrated on ways of minimizing this adverse effects [15-16]. This paper investigates the various design and pattern of bimorph beam and its deflection upon the applied temperature in static mode. Four different design is studied based on theoretical and simulation. The effect of each design is validated based on its tip deflection at the free end of the beam.

978-1-5386-5284-8/18 $31.00 © 2018 IEEE

II. THEORY OF BIMORPH EFFECT

The very commonly used the effect in thermal bimetallic or thermal bimorph effect. It consists of two materials joined along their longitudinal axis serving as a single composite beam. Upon actuation, the difference in coefficients of thermal expansion (CTE) between the two layers resulted in transverse beam bending (beam will curve toward the layer made of a material with lower CTE) as shown in Fig. 1,

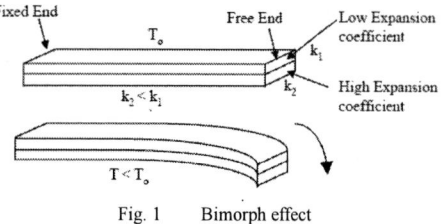

Fig. 1   Bimorph effect

The radius of curvature r can be calculated using (1) as shown below

$$\frac{1}{r} = \frac{6w_1 w_2 E_1 E_2 t_1 t_2 (t_1 + t_2)(\alpha_1 - \alpha_2)\Delta T}{(w_1 E_1 t_1^2)^2 + (w_2 E_2 t_2^2)^2 + 2w_1 w_2 E_1 E_2 t_1 t_2 (2t_1^2 + 3t_1 t_2 + 2t_2^2)} \quad (1)$$

where w, t, E and α are the width, thickness, Young Modulus and CTE for the material for each layer. Once r is evaluated, the deflection d can be determined by (2) as follows

$$d = r - r \cos\theta \quad (2)$$

The next section will discuss the different design and pattern of the bimetallic layer and its deflection upon temperature variation based on results from theoretical and simulation studies.

III. BEAM MODEL

To obtain lowest deflections upon release, several models of bimetallic bimorph cantilever beam which varies in geometric structure and thickness are designed and its mechanical response is simulated and studied with CoventorWare software based on finite element modeling. It consists of two active material of silicon dioxide (SiO_2) and Aluminum (Al) bonded together. However, in fact, due to the physical and mechanical properties of the selected layers, there are some limitations imposed on the deflection. The design constraints may include melting point of the materials and the yield strength of the materials. The models use a thermal application module for the simulation of the mechanical response of the bimetallic cantilever beam. In this models, one end of the cantilever beam is clamped in accordance with the mechanical boundary condition. Hence, the vertical faces of the cantilever at one end are constrained to move (free end). The free end is free to bend along the direction of the applied temperature. Temperature variation, from 10 – 100 K is applied on each surface on the cantilever bimorph beam. Consequently, one layer expands while the other contracts along the length, which results in the deflection of the cantilever beam, along with the z-direction. The material properties of the SiO_2 and Al are taken from the material library of CoventorWare as shown in Table I.

TABLE I.   MATERIALS PROPERTIES FROM COVENTORWARE

Material / Properties	Al	SiO$_2$
CTE (1/K)	2.31e-5	5e-7
Young Modulus (Pa)	7.7e10	7e10
Poisson Ratio	0.3	0.17
Melting point (K)	933.5	1873
Yield strength (MPa)	170	8400

IV. ANALYSIS AND DISCUSSION

To obtain the lowest deflection, the different geometrical structure is tested for the tip deflection as a function of temperature rise. For the first analysis, the thickness for Al and SiO_2 layer is 1.0 and 1.1 μm. The width and the length of the beam are 10 and 100 μm. The first analysis is to compare on the best design which can yield the lowest tip deflection upon temperature rise with each design having similar total thickness. Based on the first analysis, the best design is further analyzed by varying the thickness of each material while maintaining the width and length in order to obtain design parameter that has lowest percentage error between theoretical and simulation result.

A. Analysis of different geometrical design with a similar total thickness

a) Case 1 beam model

Fig. 2 shows bimorph beam in a rectangular pattern. Both active materials are kept at the same length and width but with differences in thickness. The tip deflection of the Case 1 model is shown in Fig. 3. The deflection is -4.118 and -4.031 μm for simulation and theoretical, respectively at $\Delta T = 50$ K. Percentage error between theoretical and simulation is approximately 2.20 %.

Fig. 2   Case 1 beam model

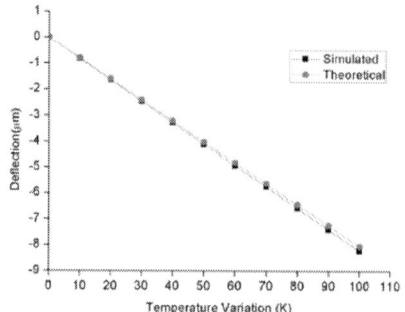

Fig. 3   Deflection vs temperatue variation for case 1

978-1-5386-5284-8/18 $31.00 © 2018 IEEE

b) Case 2 beam model

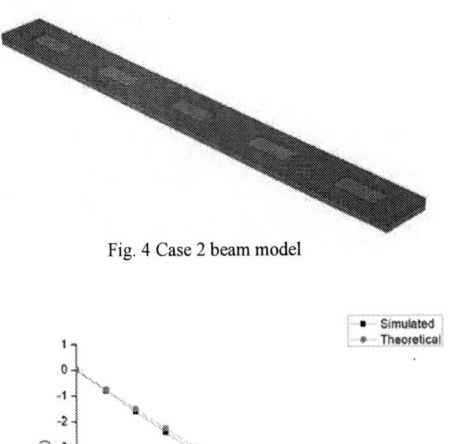

Fig. 4 Case 2 beam model

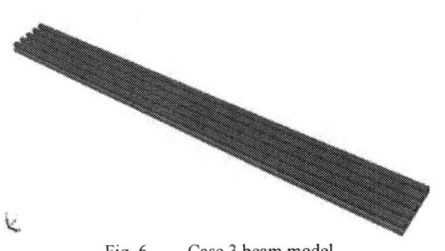

Fig. 5   Deflection vs temperature variation for case 2

Fig. 4 shows cantilever beam in rectangular pattern but the Al layer on top is designed with holes. The holes size is 5 x 10 μm and each holes is 10 μm apart from the consequent holes. The simulation and theoretical results are shown in Fig. 5, depicted that tip deflection for $\Delta T = 50$ K is -4.0221 μm and -3.7792 μm, respectively. Percentage error calculated is 6.421%.

c) Case 3 beam model

Fig. 6   Case 3 beam model

Fig. 7   Deflection vs temperature variation for case 3

Fig. 6 shows the cantilever beam in rectangular beam. However, Al layer is designed as strips. Each strip has the dimension of 100×1 μm. The distance between one strip to the adjacent strip is 1 μm. The tip deflection obtained from both simulation and theoretical results shown in Fig. 7 is -3.9204 μm and -3.8755 μm, respectively. Percentage error is 1.1591%.

d) Case 4 beam model

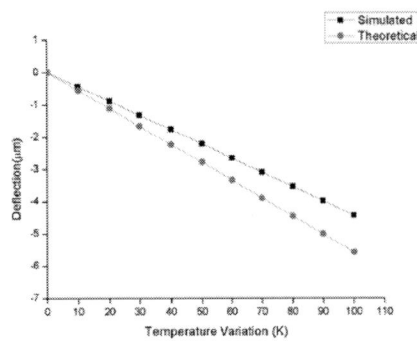

Fig. 8   Case 4 beam model

Fig. 9   Deflection vs temperature variation for case 4

In case 4 beam model, the bimetallic cantilever beam is designed the same as in case 3 except that in this model the trenches between the Al strip is filled with SiO_2 material as shown in Fig. 8. The length and thickness of the two active material layers remain the same. As for case 4 beam model, the simulated and theoretical deflection is -2.213 μm and -2.786 um, respectively at $\Delta T = 50$ K as shown in Fig. 9. Percentage error is 25.86%.

Based on the four case beam models, case 1 has the highest beam deflections. This is due to its moment of inertia. It is well known that the moment of inertia is directly proportional to its width and deflection is inversely proportional to the moment of inertia. Hence, since the width of Al layer in case 3 is smaller than Al layer in case 3, the deflection of beam model in Case 1 is higher compared to others.

As for case 2 beam model, the holes reflect the decrease of length in the Al layer. It is known that the tip deflection is proportional to the length of cantilever beam. Since in case 2, the length is smaller than beam in case 1, the deflection is slightly lower. In case 4 beam model, the combination of both SiO_2 and Al layer at the top layer part, contribute to the low effective CTE value. Hence, with the deflection (negative *z*-direction) upon temperature rise is smaller as the top layer contract and bottom layer expands, hence deflection is in the positive *z*-direction. Thus, this beam model is the best model to describe low tip deflection upon temperature rise. However, in this case, the percentage error between

978-1-5386-5284-8/18 $31.00 © 2018 IEEE   83

theoretical and simulated is high. The next analysis will further analyze the selected beam model as a function of thickness as to choose the best design parameter for the cantilever beam.

B. Analysis of case 4 beam model with Varying Thickness Ratio

Based on the previous analysis, case 4 model is chosen due to its smallest tip deflection upon temperature rise. However, the thickness evaluated lead to higher percentage error between theoretical and simulation. Thus, the thickness ratio between the top and bottom layer with respect to similar total thickness (2.1 µm) is studied to evaluate the best ratio that contributes to the lowest percentage error. The thickness of the top layer, t_T is thinner than the bottom layer, t_B. Table II tabulated the ratio, deflection and percentage error at a temperature rise of 50 K.

TABLE II.   TABULATED DATA FOR $t_T < t_B$

Thickness Ratio	Simulated Tip Deflection (µm)	Theoretical Tip Deflection (µm)	Percentage error (%)
0.1053	-0.8425	-0.8169	3.13
0.2353	-1.4410	-1.5336	6.4
0.4	-1.8654	-2.1064	11.44
0.6154	-2.1252	-2.5209	15.69
0.9091	-2.2133	-2.7858	25.86

Based on the tabulated data in Table II, with $t_T < t_B$, lower tip deflection is obtained upon temperature rise. Furthermore, the percentage error is also low. The best design parameter for this particular cantilever beam is when the thickness ratio between top and bottom is 0.1053. The tip deflection is at its lowest value which is 0.8425 and 0.8169 um in negative z-direction, respectively. Hence, the suitable thickness is 0.2 µm for top layer and 1.9 µm for bottom layer. Thick SiO_2 at the bottom layer will expand less than the thin Al layer, hence contribute to lesser overall deflection. The stress at the surface of the beam also is measured from the simulation studies. It is vital to measure the stress to make sure that the design has appropriate stress upon temperature rise to avoid from break. The stress must be lower than the yield strength of the material. The finding shows that the highest stress occurred at the clamped beam as it is a fixed part. Another location on the cantilever beam has surface stress from 0.70 to 94.72 MPa. It is shown that the surface stress obtained is 107.86 MPa for the case 4 beam model at $\Delta T = 50$ K.

V.   CONCLUSION

The relation of tip deflection for a bimorph cantilever beam with varying geometrical structure, thickness, width, and length has been established based on Euler-Bernoulli beam theory and validated using simulation study. Combination of Al and SiO_2 is a preferable choice for small deflection upon temperature rise. This relation is essential and beneficial for optimizing the design parameters of a bimorph cantilever beam before the actual fabrication of the device. Future works will comprehend the fabrication of the desired design and characterization.

REFERENCES

[1] D. V. Dao, K. Nakamura, T. T. Bui, and S. Sugiyama, "Micro/nano-mechanical sensors and actuators based on SOI-MEMS technology," Advances in Natural Sciences: Nanoscience and Nanotechnology, vol. 1, p. 013001, 2010.

[2] J. Ponmozhi, C. Frias, T. Marques, and O. Frazão, "Smart sensors/actuators for biomedical applications: review," Measurement, vol. 45, pp. 1675-1688, 2012.

[3] K. A. Brown, D. J. Eichelsdoerfer, W. Shim, B. Rasin, B. Radha, X. Liao, et al., "A cantilever-free approach to dot-matrix nanoprinting," Proceedings of the National Academy of Sciences, vol. 110, pp. 12921-12924, 2013.

[4] R. R. A. Syms, H. Zou, J. Yao, D. Uttamchandani, and J. Stagg, "Scalable electrothermal MEMS actuator for optical fibre alignment," Journal of Micromechanics and Microengineering, vol. 14, p. 1633, 2004.

[5] W. Wang, Q. Chen, D. Wang, L. Zhou, and H. Xie, "A bi-directional large-stroke electrothermal MEMS mirror with minimal thermal and temporal drift," in 2017 IEEE 30th International Conference on Micro Electro Mechanical Systems (MEMS), 2017, pp. 331-334.

[6] H. Wang, X. Zhang, D. Zhang, L. Zhou, and H. Xie, "Characterization and reliability study of a MEMS mirror based on electrothermal bimorph actuation," in 2017 International Conference on Optical MEMS and Nanophotonics (OMN), 2017, pp. 1-2.

[7] D. Wang, X. Zhang, L. Zhou, M. Liang, D. Zhang, and H. Xie, "An ultra-fast electrothermal micromirror with bimorph actuators made of copper/tungsten," in 2017 International Conference on Optical MEMS and Nanophotonics (OMN), 2017, pp. 1-2.

[8] X. Zhang, L. Zhou, and H. Xie, "A Fast, Large-Stroke Electrothermal MEMS Mirror Based on Cu/W Bimorph," Micromachines, vol. 6, pp. 1876-1889, 2015.

[9] L. Wu and H. Xie, "A large vertical displacement electrothermal bimorph microactuator with very small lateral shift," Sensors and Actuators A: Physical, vol. 145, pp. 371-379, 2008.

[10] L. Liu, L. Wu, J. Sun, E. Lin, and H. Xie, "Miniature endoscopic optical coherence tomography probe employing a two-axis microelectromechanical scanning mirror with through-silicon vias," Journal of Biomedical Optics, vol. 16, pp. 026006-026006-4, 2011.

[11] J. Sun, S. Guo, L. Wu, L. Liu, S.-W. Choe, B. S. Sorg, et al., "3D In Vivo optical coherence tomography based on a low-voltage, large-scan-range 2D MEMS mirror," Optics Express, vol. 18, pp. 12065-12075, 2010/06/07 2010.

[12] W. Wang, S. R. Samuelson, J. Chen, and H. Xie, "Miniaturizing Fourier Transform Spectrometer With an Electrothermal Micromirror," IEEE Photonics Technology Letters, vol. 27, pp. 1418-1421, 2015.

[13] H. Kupfer, T. Flügel, F. Richter, and P. Schlott, "Intrinsic stress in dielectric thin films for micromechanical components," Surface and Coatings Technology, vol. 116–119, pp. 116-120, 9// 1999.

[14] K. Wu, T.-J. Peters, M. Tichem, F. Postma, A. Prak, K. Wörhoff, et al., "Bimorph actuators in thick SiO2 for photonic alignment," 2016, pp. 975311-975311M. Young, The Technical Writer's Handbook. Mill Valley, CA: University Science, 1989.

[15] H. Kupfer, T. Flügel, F. Richter, and P. Schlott, "Intrinsic stress in dielectric thin films for micromechanical components," Surface and Coatings Technology, vol. 116–119, pp. 116-120, 9// 1999.

[16] A. K. Chinthakindi, D. Bhusari, B. P. Dusch, J. Musolf, B. A. Willemsen, E. Prophet, et al., "Electrostatic Actuators with Intrinsic Stress Gradient: I. Materials and Structures," Journal of The Electrochemical Society, vol. 149, pp. H139-H145, August 1, 2002.

Dielectrophoresis : Characterization of Triple-Negative Breast Cancer using Clausius-Mossotti Factor

Nur Mas Ayu Jamaludin, Muhamad Ramdzan Buyong,
Muhammad Khairulanwar Abdul Rahim, Azrul Azlan
Hamzah, Burhanuddin Yeop Majlis,
Institute of Microengineering and Nanoelectronics (IMEN)
Universiti Kebangsaan Malaysia (UKM)
43600 Bangi, Selangor, Malaysia
Email : muhdramdzan@ukm.edu.my

Badariah Bais,
Centre of Advanced Electronic and Communication
Engineering (PAKET),
Faculty of Engineering and Built Environment,
Universiti Kebangsaan Malaysia (UKM)
43600 Bangi, Selangor, Malaysia

Abstract — **This paper describes a new possible application of dielectrophoresis (DEP) in breast cancer detection. Breast cancer is one of the most common cancer diseases in Malaysia which makes 17.7% of all cases reported between 2007 and 2011. This research aims to determine the dielectric properties of biophysical and biochemical of triple-negative breast cancer (TNBC). The manipulation of TNBC will be on the detection and enumeration based on polarization factor between permittivity and conductivity of TNBC and suspended medium. The results of analytical modelling prove that there are dynamic changes in the dielectric values, which are the permittivity and conductivity values between TNBC cell and medium when the frequency input value is changed. The findings offer a framework for anticipating TNBC dielectric detection response on the basis of structure–function relationship and suggest that dielectrophoresis lab-on-chip (DLOC) application for TNBC detection and enumeration should be widely used as a surface marker-independent method and provide high precision analysis in cancer progression and treatment effectiveness.**

Keywords—dielectrophoresis, triple-negative breast cancer, Clausius-Mossotti factor, dielectric dynamic change

I. INTRODUCTION

Circulating tumor cells (CTCs) are malignant cells that become detached from the primary tumor and travel in circulation to distant sites where they may form new colonies of metastasis. Microscopic residual disease or metastasis can be the cause of fatalities due to breast cancer. Measuring the level of CTCs has been known for predicting clinical outcomes in breast cancer; a high level of CTCs is believed to be an indicator that an intervention is needed to minimize the risk of metastasis [1]. The presence of CTCs was consistently linked to poor prognosis in patients with triple-negative breast cancer (TNBC) [2]. The primary issue that renders difficulty in treating TNBC is that it is non-reliant on the ER, PR, or HER2 pathway for proliferation. However, ER-, PR-, or HER2- expressing CTCs have been isolated from patients with TNBC, contrary to the conventional assumption that CTCs should inherit the characteristics of their parent tumor [3].

A possible explanation is that the primary tumor is composed of a heterogeneous population of cells, in which ER-, PR-, or HER2- expressing members comprise a minor and clinically undetectable fraction. Eliminating ER-, PR-, or

HER2-positive CTCs by using specific agents, such as tamoxifen and trastuzumab, could prevent these cells from establishing metastatic colonies and improve clinical outcomes in patients with TNBC. The challenge for TNBC is to encounter that particular breast cancer subclassification that is more complicated due to the lack of ER, PR and HER2 amplification. For this reason, the use of therapeutic approach will be more challenging. Hence, there are new opportunities for further fundamental and exploratory research in dielectrophoresis lab-on-a-chip (DLOC) application based on dielectrophoresis (DEP).

TNBC has the highest rates of metastatic disease and presents the poorest overall survival rate of all breast cancer subtypes [4]. This means that the development of targeted detection and enumeration for TNBC is urgently needed. Currently, several technologies have been developed to identify CTCs. However, specification in term of contactless methods or biomarker-free for reliable and rapid solution is still in the research and development stage. Fig. 1 shows the graphical differences between localized tumor and advanced tumor, which in this case, is TNBC. Other techniques are used to see whether they can be helpful in developing the cure for TNBC, by using a multi-targeted approach through a blend of agents aiming specifically at particular distinguishable molecules [5].

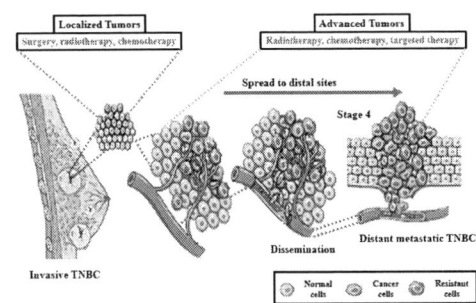

Fig. 1. The flow of multimodality in TNBC [5].

DEP is a phenomenon that produces motion of particles by the action of a non-uniform electric field. It particularly describes the motional behavior of neutral matter as a result of exposure to a non-uniform electric field in a given

978-1-5386-5284-8/18 $31.00 © 2018 IEEE

environment that cause repositioning of particles to a new location [6]. Since 1960 when Pohl performed experiments on the electrical characterization and subsequent separation of biological samples, a vast number of design have been proposed to improve the performance and accuracy of DEP-based devices [7]. With the advances in fabrication technologies, more complicated geometries have been proposed to serve complex applications. Examples in this case are the detection and enumeration for TNBC.

Currently, DEP has been widely used for the determination of cell, bacteria, virus and protein based on the sample dielectric properties characteristics. DEP capabilities in generating DEP forces (F_{DEP}) have caused it to be able to do the detection, manipulation, separation, and isolation of target cell from mixtures of different cells in aqueous suspension medium [7-8]. To ensure the crucial state when detecting the DEP behavior, the consequences of inflicting non-uniform electric field on a TNBC need to be discovered first.

II. DIELECTROPHORESIS AND TRIPLE-NEGATIVE BREAST CANCER

Implementations of DEP for the detection of TNBC can be done by characterizing the dielectric properties for phenotype and physiological state of TNBC and the surrounding medium. The dielectric property of TNBC covers the physical structure (referred to exterior) and chemical composition (referred to interior) [9-10].

Based on the unique identifications of TNBC, physical structure exterior morphologies later will convert those properties into dielectric properties. The dynamic dielectric properties consisting both resistance and conductivity which are TNBC permittivity and conductivity values will change due to the input frequency based on the equivalent circuit as shown in Fig. 2. The parallel combination of capacitance and resistance by the plasma membrane can be expressed by the outer membrane. C_{mem} decreases with the increase of AC resistance which is known as reactance. The resistance element, R_{mem}, presents a large resistance to the passive current short-circuited by the membrane capacitance. Fluctuation of dielectric properties will yield DEP force either in the form of attraction or repulsion form high electric field intensity [11-13].

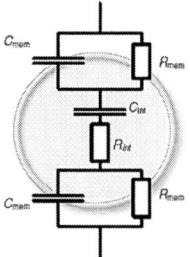

Fig. 2. The electrical equivalent circuit of TNBC cell [14].

Comprehension on the topic of electrical frequency-dependent attributes can be obtained based on the link between the TNBC permittivity and conductivity values. The conductivity is directly correlated with frequency; this means that the increase will subsequently cause the increase of the latter. Higher relative permittivity values shown at smaller frequency values demonstrates exactly the abundance of interfacial polarizations. Polarization charges can only accumulate at the surface of the TNBC as long as it behaves as an electrical insulator. At lower frequencies, this can possibly occur due to the high reactance value constituted by the surface capacitance in order to set a leakage path for the accrued ions. Thereofre, by comparing the aqueous electrolyte, the TNBC is eventually going to have lesser conductivity. The decreasing frequency value will be directly proportional to the effective impedance to charge leakage constituted by the parallel combination of surface capacitance and resistance, in addition to interfacial polarization and effective permittivity. However, the effective conductivity value of the TNBC will be inversely proportional, meaning that the value will increase.

III. CMF MEASUREMENT

In this work, based on the understanding of the frequency-dependent electrical properties of TNBC, the Clausius-Mossotti factor (CMF) will be summarized. DEP is a force that appears in a polarizable particle due to the unique identifications of TNBC in terms of TNBC permittivity and conductivity and suspended medium in a polarizable medium permittivity and conductivity when immersed in a non-uniform electric field. The CMF factor defines the difference in polarizability of the particles and their suspending medium. The CMF not only depends on the physical characteristics of the particle (in this research; TNBC) and the medium but also on the frequency of the AC electric field. The real part of CMF is given by [15]

$$CMF = \frac{(\varepsilon_{particle} - \varepsilon_{medium})}{(\varepsilon_{particle} + 2\varepsilon_{medium})} \quad (1)$$

$\varepsilon_{particle}$ and ε_{medium} are the complex permittivity of the cell particle and the medium, respectively [15]. In Eq. (1), it shows that the CMF is dependent on the electrical properties of the particle and medium. In this work, the CMF will be calculated using MATLAB to find the crossover frequency of the TNBC cell.

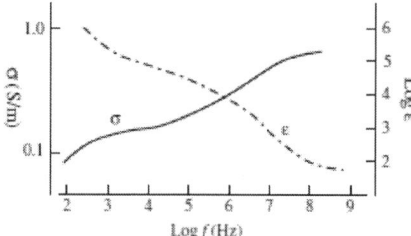

Fig. 3. The effect of conductivity and permittivity towards frequency [14].

Fig. 3 shows the DEP response of a cell to mirror the trends as expected. At low frequencies, a small effective conductivity is expected compared to the medium. From this figure, it also shows that the effective conductivity of TNBC cell will increase with increasing frequency. Therefore, when TNBC is more polarizable than the medium, the dipole induced by the electric field responds to the gradient of this field which attracts the particle to high gradient regions (typically at the borders of the electrodes). This regime is called positive dielectrophoresis (pDEP). On the contrary, when the particle is less polarizable than the medium, the CTCs will be repelled from the high gradient regions. This arrangement is named negative dielectrophoresis (nDEP). This versatility makes DEP a very unique contactless force for attracting and repelling polarizable objects only by tuning the AC frequency.

IV. RESULT AND DISCUSSION

Dielectric dynamic change modeling between TNBC cells and medium was conducted to evaluate the results of dielectric dynamic changes in the medium used for biological cells which is 0.055 S/m. Using MATLAB software, the average size of TNBC cell of 21 um in diameter was analyzed and the results are as shown below [16].

Fig. 4 shows the determination of the polarization factor value by increasing the input frequency value of the TNBC cell size. Based on the polarization factor values plot graph, the cell size and crossover frequency value show the relation between them.

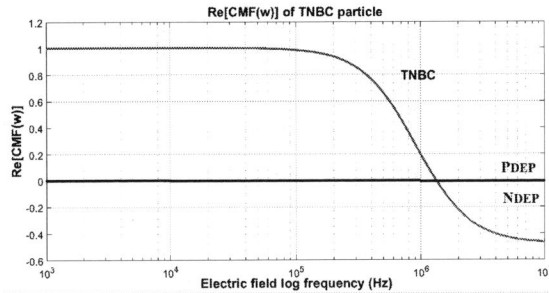

Fig. 4. The CMF plot results for TNBC cell.

At 0 Hz – $1.322e^6$ Hz, P_{DEP} takes place, which is the lateral attraction at y-axis. At $1.32e^6$ Hz – $1.00e^7$ Hz, N_{DEP} takes place, which is the vertical repulsion at z-axis and at y-axis = 0, the point of intersection is the point where $P_{DEP} = N_{DEP}$. This proves that each particle and dielectric value have unique attributes that can be manipulated for the process of particle separation thatis in a singular or two and three different particle sizes mixture. Therefore, based on the input frequency, the process of particle manipulation based on particle size can be done by using the F_{DEP} technique.

Fig. 5 shows the comparison results that illustrate the change in permittivity values between TNBC cell and medium by increasing the input frequency value. The objective of the combination between the two permittivity value changes with the TNBC cell and the medium is to see the reaction of the increase in input frequency value and changes in the permittivity value of TNBC cell and the medium. The permittivity value point of contact of the TNBC cell and the medium is at the frequency intersection point. This means that, the permittivity value of the TNBC cell and medium is the same at that point, where the change in the value of TNBC cell and medium permittivity corresponds to the increase in input frequency.

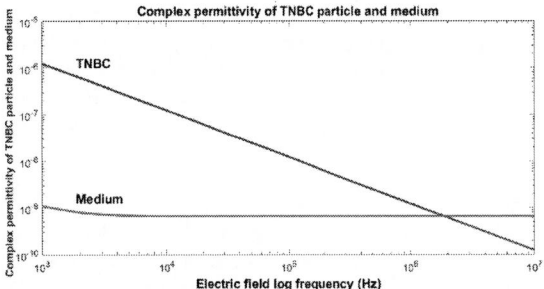

Fig. 5. Comparative results of changes in the value of permittivity between TNBC cell and medium.

Fig. 6 shows the results that illustrate the change in conductivity value between TNBC cell and medium with the increase of input frequency. The objective for the combination of conductivity value of TNBC cell and medium is to look at the effect of increasing the input frequency to the changes in the value of conductivity of TNBC cell and medium. The points of contact between the conductivity value of the TNBC cell and medium are at the intersection of the frequency, parallel to the permittivity value. This means that, the conductivity and medium values at that point are the same, which explains the change in the value of TNBC cell conductivity and medium corresponds to the increase in input frequency.

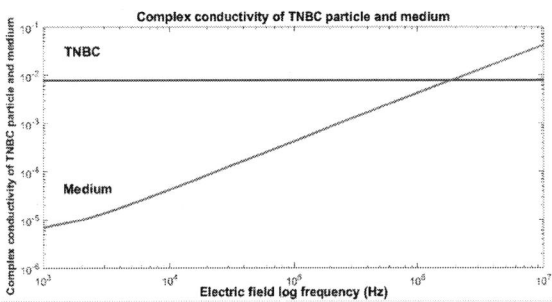

Fig. 6. Comparison result of conductivity values between TNBC cell and medium.

Fig. 7 shows the change through the comparison of permittivity and conductivity values of the TNBC cell and medium with the increase of input frequency. There are similar and different factors between the permittivity and conductivity values of the TNBC cell.

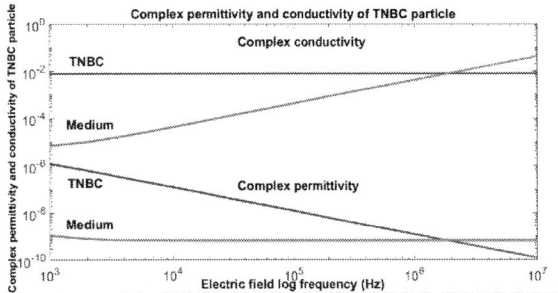

Fig. 7. The results of the comparison of the value of permittivity and conductivity of the TNBC cell and medium.

978-1-5386-5284-8/18 $31.00 © 2018 IEEE

The similar factor between the permittivity and conductivity values is that the change in both elements occurs at the same crossover frequency value. It is evident by looking at the point of intersection between the TNBC cell permittivity and medium values, where both occur at the same frequency value.

The first difference between the permittivity and conductivity value is based on the observations for the increasing frequency inputs. The change in the TNBC cell permittivity value is more prevalent compared to the change in the TNBC cell conductivity value. This is evident by observing the change in the TNBC cell permittivity value that occurs at low frequency input value compared to the change in the conductivity value that occurs at high frequency input value.

The second difference is the TNBC cell permittivity value is higher than the medium permittivity value before the crossover frequency value. Surpassing the crossover frequency value, the permittivity value of the medium gets higher compared to the TNBC cell permittivity value, which is inversely proportional to the frequency input value. It is not the case for the TNBC cell conductivity value, which is higher before the crossover frequency value. When the crossover frequency value is bypassed, the conductivity value of the medium gets higher with a significant increase over the conductivity value, which is directly proportional to the input frequency.

As a conclusion based on the results, there is a dynamic changes of the dielectric values, which are the permittivity and conductivity values between the TNBC cell and medium when changing the frequency input value. This is due to the occurrence of crossover frequency between the TNBC cell and the medium, indicating that polarization occurs between the TNBC cell and medium; either from P_{DEP} to N_{DEP} for medium with low conductivity values or N_{DEP} to P_{DEP} for medium with high conductivity values.

The similarities and differences in values for TNBC cell dielectric dynamic changes of explain the fact that the TNBC cell has different permittivity and electrical conductivity values even from the same type of material, which is the TNBC cell. This supports the statement that each particle with different sizes has its own dielectric value. Therefore, a particle's dielectric value will be uniquely defined based on its dielectric value.

V. CONCLUSION

The results of the analytical modeling of polarization factor CMF plot that was developed using MATLAB show the dynamic changes in dielectric values, namely the permittivity and conductivity values between TNBC cell and the medium with changing frequency input value. The findings provide a framework for anticipating TNBC dielectric detection response on the basis of structure–function relationship and suggest that DEP should be widely applied as a surface marker-independent method for the detection and isolation of TNBC.

ACKNOWLEDGEMENT

The author would like to acknowledge with gratitude the sponsor of Dana Cabaran Perdana (DCP-2017-003/3) and Geran Galakan Penyelidik Muda (GGPM-2017-028) grants funded by Universiti Kebangsaan Malaysia (UKM) as a Research University.

REFERENCES

[1] M. Karhade, C. Hall, P. Mishra, S. Krishnamurthy, and A. Lucci, "Circulating tumor cells in non-metastatic triple-negative breast cancer," pp. 325–333, 2014.

[2] Y. Niu, H. Su, and A. Feng, "Role of Circulating Tumor Cell (CTC) Monitoring in Evaluating Prognosis of Triple-Negative Breast Cancer Patients in China," pp. 3071–3079, 2017.

[3] S. Agelaki, M. Dragolia, H. Markonanolaki, S. Alkahtani, C. Stournaras, V. Georgoulias, and G. Kallergi, "Phenotypic characterization of circulating tumor cells in triple negative breast cancer patients," vol. 8, no. 3, pp. 5309–5322, 2017.

[4] P. Negi, P. A. Kingsley, K. Jain, J. Sachdeva, S. Marcus, and A. Pannu, "Survival of Triple Negative versus Triple Positive Breast Cancers : Comparison and Contrast," vol. 17, pp. 3911–3916, 2016.

[5] M. Kalimutho, K. Parsons, D. Mittal, J. A. López, S. Srihari, and K. K. Khanna, "Targeted Therapies for Triple-Negative Breast Cancer : Combating a Stubborn Disease," *Trends Pharmacol. Sci.*, vol. xx, pp. 1–25, 2015.

[6] N. Marsi, B. Y. Majlis, and A. A. Hamzah, "The Mechanical and Electrical Effects of MEMS Capacitive Pressure Sensor Based 3C-SiC for Extreme Temperature," vol. 2014, 2014.

[7] R. Pethig, "Review Article — Dielectrophoresis : Status of the theory , technology , and applications," pp. 1–35, 2010.

[8] N. Piacentini, G. Mernier, R. Tornay, P. Renaud, and P. Renaud, "Separation of platelets from other blood cells in continuous-flow by dielectrophoresis field-flow- fractionation," vol. 34122, no. 2011, 2014.

[9] S. Shim, K. Stemke-hale, J. Noshari, F. F. Becker, P. R. C. Gascoyne, S. Shim, K. Stemke-hale, and J. Noshari, "Dielectrophoresis has broad applicability to marker-free isolation of tumor cells from blood by microfluidic systems," vol. 11808, 2013.

[10] M. R. Buyong, F. Larki, M. S. Faiz, A. A. Hamzah, J. Yunas, and B. Y. Majlis, "A Tapered Aluminium Microelectrode Array for Improvement of Dielectrophoresis-Based Particle Manipulation," pp. 10973–10990, 2015.

[11] M. R. Buyong, J. Yunas, A. A. Hamzah, B. Y. Majlis, F. Larki, and N. A. Aziz, "Design , fabrication and characterization of dielectrophoretic microelectrode array for particle capture," vol. 2, no. October 2014, pp. 96–102, 2015.

[12] M. R. Buyong, F. Larki, Y. Takamura, and B. Y. Majlis, "Tapered microelectrode array system for dielectrophoretically filtration: fabrication, characterization, and simulation study," vol. 16, pp. 44501–44508, 2017.

[13] M. R. Buyong, F. Larki, C. E. Caille, Y. Takamura, A. A. Hamzah, and B. Y. Majlis, "Determination of lateral and vertical dielectrophoresis forces using tapered microelectrode array," vol. 13, no. 2, pp. 143–148, 2018.

[14] R. Pethig, "Dielectrophoresis: An assessment of its potential to aid the research and practice of drug discovery and delivery," *Adv. Drug Deliv. Rev.*, 2013.

[15] A. Dyda, "Dielectrophoretic separation of bacteria using a conductivity gradient," vol. 51, pp. 175–180, 1996.

[16] A. P. Athreya, A. J. Gaglio, Z. T. Kalbarczyk, R. K. Iyer, J. Cairns, K. R. Kalari, R. M. Weinshilboum, and L. Wang, "Unsupervised Single-Cell Analysis in Triple-Negative Breast Cancer : A Case Study," pp. 556–563, 2016.

Synthesis, Properties and Humidity Detection of Anodized Nb₂O₅ Films

Rozina Abdul Rani
[1]NANO-SciTech Centre,
Institute of Science, [2]NANO-ElecTronic Centre (NET),
Faculty of Electrical
Engineering, Universiti
Teknologi MARA
Shah Alam, Selangor, Malaysia
rozina.abdulrani@yahoo.com

Ahmad Sabirin Zoolfakar
NANO-ElecTronic Centre
(NET), Faculty of Electrical
Engineering, Universiti
Teknologi MARA
Shah Alam, Selangor, Malaysia
ahmad074@salam.uitm.edu.my

Mohamad Fauzee Mohamad
Ryeeshyam
NANO-ElecTronic Centre
(NET), Faculty of Electrical
Engineering, Universiti
Teknologi MARA
Shah Alam, Selangor, Malaysia
mhdfauzee@gmail.com

Najwa Ezira Ahmed Azhar
NANO-ElecTronic Centre
(NET), Faculty of Electrical
Engineering, Universiti
Teknologi MARA
Shah Alam, Selangor, Malaysia
najwaezira@yahoo.com

Mohamad Hafiz Mamat
NANO-ElecTronic Centre
(NET), Faculty of Electrical
Engineering, Universiti
Teknologi MARA
Shah Alam, Selangor, Malaysia
mhmamat@salam.uitm.edu.my

Salman Alrokayan
Research Chair for Biomedical
Applications of Nanomaterials,
Department of Biochemistry,
College of Science, King Saud
University,
Riyadh, Saudi Arabia
salrokayan@ksu.edu.sa

Haseeb A. Khan
Research Chair for Biomedical
Applications of Nanomaterials,
Department of Biochemistry,
College of Science, King Saud
University,
Riyadh, Saudi Arabia
haseeb@ksu.edu.sa

Mohamad Rusop Mahmood
[1]NANO-SciTech Centre,
Institute of Science, [2]NANO-ElecTronic Centre (NET),
Faculty of Electrical
Engineering, Universiti
Teknologi MARA
Shah Alam, Selangor, Malaysia
nanouitm@gmail.com

Abstract—**Synthesis of nanoporous network Nb₂O₅ has been conducted via anodization technique in the flouride-organic based solution. The structure morphology and the properties of the nanoporous films were characterized using field emission scanning electron microscopy (FESEM). In this research, a humidity sensor based on 1.5 micrometer thick nanoporous Nb₂O₅ was developed and their performance was evaluated under 40% to 90% relative humidity at room temperature and different bias voltages of 2, 5 and 8 V. The highest relative sensitivity achieved was 156.2 for humidity sensor operated at bias voltage of 5 V.**

Keywords—**Nb₂O₅, anodization, humidity sensor, nanoporous**

I. INTRODUCTION (*HEADING 1*)

Due to the chemical stability and corrosion resistance, Niobium oxide (Nb₂O₅) is one of the promising materials for humidity sensor application. Those properties are important in humidity sensing as the sensor will be exposed in the air or other gaseous environment to determine the amount of present water content. The humidity sensor are mainly used for humidity control in the semiconductor and textile industry, human breathing monitoring, food industry processes and humidity controlling electronic devices [1]. Some other applications are particularized for humidity monitoring on airplane surfaces and soil moisture monitoring in agriculture [2, 3] . Since each application have different range of operating condition, various types of humidity sensing devices have been developed based on various sensing materials and different working principles [4-7].

In the recent study, humidity sensors based on one-dimensional (1D) nanostructure Nb₂O₅ have been investigated in the form of nanorods structure [8]. This humidity sensor was measured the resistance changes when exposed in the different range of relative humidity (%RH). Furthermore, the characterization of humidity sensor utilizing mixed or doped Nb₂O₅ with other materials (TiO₂, SnO₂, SiO₂, ZnO and PANI) as a sensing layer also have been conducted [8-11]. Most of the current developed Nb₂O₅ based humidity sensors were utilizing grains (powder based Nb₂O₅) and 1D nanostructure as a sensing layer. As there are diverse humidity conditions, a lot of opportunities to discover on sensing layer itself. It is realized that Nb₂O₅ based on three-dimensional (3D) porous structure is still not investigated for humidity sensor. Thus, the true capabilities of the nanoporous structure for humidity application should be explored to provide further insight for other researchers.

Based on the aforementioned motivation, we present for the first time a humidity sensor based on 3D nanoporous Nb₂O₅ films synthesized using anodization technique in fluoride based ethylene glycol solvent. The sensing performance and behaviours of the humidity sensor expose in different %RH and bias at different voltages were evaluated.

II. MATERIALS AND CHARACTERIZATION

A. Synthesis of Nb₂O₅ film and sensor fabrication

Nb₂O₅ was synthesized by anodized niobium (Nb) foil (99.9% purity, Sigma Aldrich) in the electrolyte consists of 50 ml ethylene glycol (98% anhydrous, Sigma Aldrich), 4 vol % deionized water and 0.5% NH₄F (98% purity, Sigma Aldrich). The anodization process was performed at a bias voltage of 10 V in the electrolyte heated at 50°C. Nanoporous Nb₂O₅ with a thickness of ~1.5 μm was

978-1-5386-5284-8/18 $31.00 © 2018 IEEE

obtained after 30 min of anodization. Then, the samples were annealed in air at a temperature of 440°C for 30 min. Orthorhombic crystalline structure of Nb_2O_5 were obtained after undergo the annealing process. Details on the synthesis optimization are presented in the previous work [12, 13]. The humidity sensors were fabricated based on 1 × 1.5 cm foil dimension. Pt electrode with size of 4 × 4 mm and thickness of ~30 nm was deposited on top of the nanoporous Nb_2O_5 films using a thermal evaporator at a deposition pressure of 4×10^{-4} Pa.

B. Structural and sensor characterization

The nanoporous Nb_2O_5 morphological structures were characterized using a JEOL JSM-6700F field-emission scanning electron microscope (FESEM). I-V characteristic of the nanoporous Nb_2O_5 film was conducted using a probe station incorporated with a measurement system (Keithley 2400). The I-V measurement was measured between the Pt electrode and the Nb foil. The humidity characterization measurements of the fabricated sensor were conducted in the humidity chamber (ESPEC-SH261) at different range of relative humidity (40 – 90 %RH) using a measurement system (Keithley 2400). The sensor performance also evaluated using a different bias voltage.

III. RESULTS AND DISCUSSION

The FESEM images in Figure 1 show the cross-sectional and surface morphology of the nanoporous Nb_2O_5 film. From Fig. 1(a), the thickness of the nanoporous Nb_2O_5 film was approximately 1.5 μm thick after anodized for 30 min. It reveals the Nb_2O_5 layer consist of continuous and highly packed criss-cross or vein-like networks. The vein networks with continues lateral interconnection have an internal diameter in the range of 30 to 50 nm. Furthermore, at the bottom part of the nanoporous layer, structure of pseudo semi-spheres was formed. This phenomena was associated to the mechanical stress induced by the effect of the Nb_2O_5 volume expansion which is also happened in Ti and Al anodization [14].

The FESEM image in Fig. 1 (b) shows that the surface morphology of Nb_2O_5 film has a highly organized pore distribution with pores diameter ranging from 30 to 50 nm. The size of the side walls of the pores were approximately 10 to 20 nm thick. The sizes of the pores are almost similar with the diameter of lateral interconnection. The formation of the nano pores and nano veins was mainly due to the field-assisted chemical dissolution during anodization process [13].

The I-V graph in Fig. 2 (a) shows a rectifying behaviour when measured the nanoporous Nb_2O_5 film in configuration of $Pt/Nb_2O_5/Nb$ at room temperature. Under forward bias, the turn-on occurred at approximately 0.8 V. With a +7 V applied bias, the forward current of the device was 2.14×10^{-5} A, and with a −7 V applied bias, the reverse leakage current was 8.05×10^{-6} A. Additionally, the I-V curve showed linear behaviour in the small voltage range, implying that the $Pt/Nb_2O_5/Nb$ junction has ohmic-contact characteristics. In Fig. 2(b), it can be seen that both I-V curves are almost liner between -1.0 to 0.7 V.

The real-time dynamic response of this nanoporous Nb_2O_5 based humidity were evaluated under different RH conditions (40 – 90 %RH) and bias voltages (2, 5 and 8V) by

measuring its current value at room temperature in the humidity chamber. In order to determine the sensitivity of the sensor, the following equation was used;

Relative sensitivity, $\Delta S = (I_x - I_o)/I_o$

where I_x and I_o are the electrical currents at a given and initial humidity level, respectively.

The real-time dynamic response curve of nanoporous Nb_2O_5 based humidity sensor tested using different bias voltages of 2, 5 and 8 V were presented in Fig. 3. Based on the obtained curves, their properties; rise time and recovery time as well as sensitivity, were extracted and presented in Table 1. From the obtained results, the sensitivity of nanoporous Nb_2O_5 based humidity sensor is 58.7, 156.2 and 107.6 when applied bias voltages of 2, 5 and 8 V, respectively. The results indicate that 5 V was the best operating voltage for optimum performance of nanoporous Nb_2O_5 based humidity sensor. The sensitivity of this humidity sensor is much better than those previously reported humidity sensors based on Nb_2O_5 nanomaterials [8, 15].

Beside sensitivity, the rise and recovery times are other significant factors for estimating the humidity sensing performance. In this experiment, the rise time and the recovery time of the humidity sensors is defined as the time required for the current value to increase from 10% to 90% or drop from 90% to 10% from its steady state value.

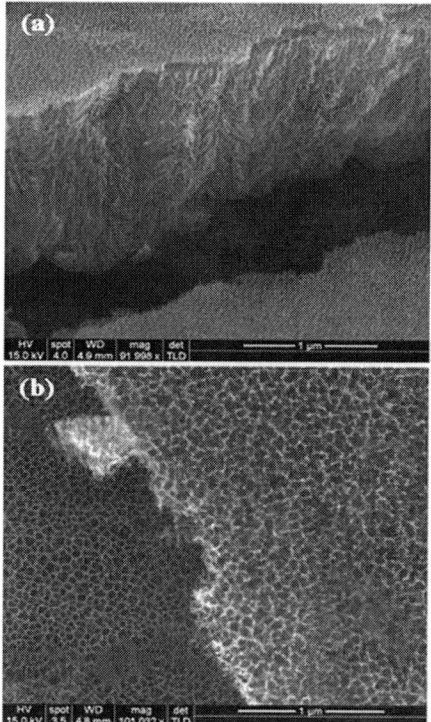

Fig. 1. FESEM images of the nanoporous Nb_2O_5 film: (a) cross-sectional view of the whole nanoporous layer, (b) top view of the nanoporous film.

As listed in Table 1, the rise time of the humidity sensor is in the range of 258 to 309 s. Meanwhile, the recovery time of the humidity sensor is a bit faster than rise time with value in the range of 70 to 123 s. The performance of humidity sensor interm of repeatable and reliabe was evaluated by conducting the repeatability test. Fig. 4 show the recycling response of humidity sensor biased at 5 V under the same experimental conditions. The humidity sensor exhibited an acceptable repeatability with good response-recovery behaviour. The output current of the humidity sensor at 40% RH drifted from ~ 2.85×10^{-10} to 2.80×10^{-10} A after 3 cycles of %RH indicating small changes of steady state signal, which leads to good quality of sensor performance.

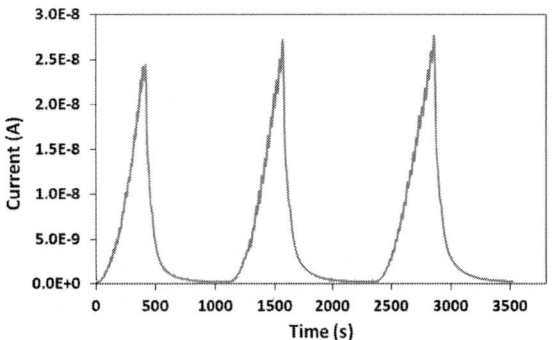

Fig. 4 Repeatability response of 1.5 μm nanoporous Nb_2O_5 humidity sensor for a concentration of 40 – 90 %RH at room temperature and a constant bias voltage of 5 V .

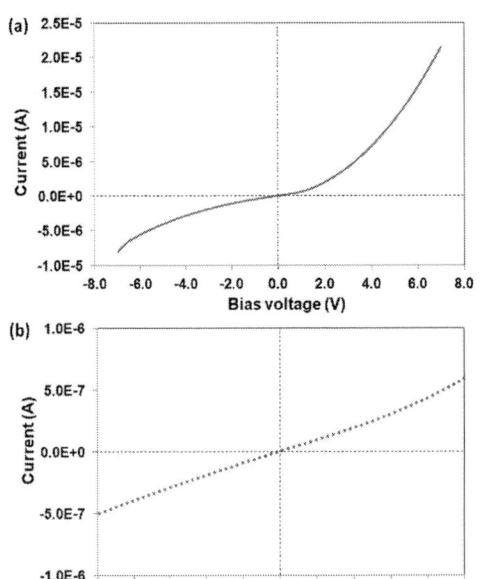

Fig. 2. (a) I-V curve for nanoporous Nb_2O_5 film measured at RT at −7 V to +7 V bias. (b) I-V curve showed a linear behavior in the small voltage range of −1 V to 0.7 V.

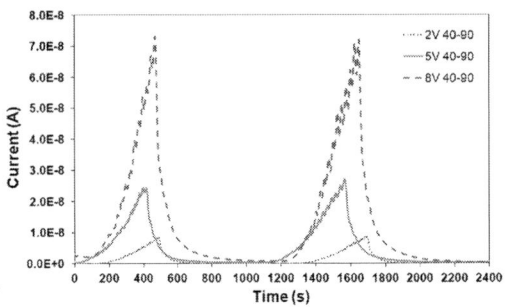

Fig. 3. Dynamic response of 1.5 μm nanoporous Nb_2O_5 based humidity sensors measured in 40 – 90 %RH at room temperature under different bias voltages of 2, 5 and 8 V.

TABLE I. NANOPOROUS Nb_2O_5 BASED HUMIDITY SENSOR PERFORMANCE AT DIFFERENT BIAS VOLTAGES.

Bias voltage (V)	Rise time (s)	Recovery time (s)	Sensitivity
2	258	70	58.7
5	309	120	156.2
8	295	123	107.6

IV. CONCLUSIONS

3D Nanoporous Nb_2O_5 films was successfully synthesized on Nb foil *via* anodization technique in the flouride-ethylene glycol solution. The FESEM results revealed that the Nb_2O_5 films consist of highly-pack organized pores structure with a lateral interconnection vein-like network. Characterizations demonstrate that the humidity sensor based on 1.5 μm thick nanoporous Nb_2O_5 operating at 5 V exhibited highest relative sensitivity of 156.2 than 2 and 8 V of bias voltage. The obtained results also showed that the nanoporous Nb_2O_5 based humidity sensor is highly stable with good repeatability and fast response behaviours. The good performance of this humidity sensor is contributed by high surface to volume ratio of 3D nanoporous Nb_2O_5 which enhanced the interaction of water molecules with the surface of sensing layer.

ACKNOWLEDGMENT

This work is supported by Ministry of Education Malaysia (MOE) under the Bestari Perdana grant, Project code: 600-IRMI/PERDANA 5/3 BESTARI (101/2018). Special thanks to Research Chair for Biomedical Applications of Nanomaterials, Department of Biochemistry, College of Science, King Saud University, Riyadh 11451, Saudi Arabia.

REFERENCES

[1] T. A. Blank, L. P. Eksperiandova, and K. N. Belikov, "Recent trends of ceramic humidity sensors development: A review," *Sensors and Actuators B: Chemical*, vol. 228, pp. 416-442, 2016.

[2] X. Zhang, J. Zhang, L. Li, Y. Zhang, and G. Yang, "Monitoring Citrus Soil Moisture and Nutrients Using an IoT Based System," *Sensors (Basel, Switzerland),* vol. 17, p. 447, 2017.

[3] B. Ingleby, D. Moore, C. Sloan, and R. Dunn, "Evolution and Accuracy of Surface Humidity Reports," *Journal of Atmospheric and Oceanic Technology,* vol. 30, pp. 2025-2043, 2013.

[4] H. Farahani, R. Wagiran, and M. N. Hamidon, "Humidity Sensors Principle, Mechanism, and Fabrication Technologies: A Comprehensive Review," *Sensors (Basel, Switzerland),* vol. 14, pp. 7881-7939, 2014.

[5] M. S. Mansor, M. H. Mamat, Z. Mohamad, and M. Rusop, "Characteristics of Humidity Sensor Fabricated Using Nanostructured ZnO Thin Film by Sol-Gel Method," *Advanced Materials Research* vol. 667, pp. 380-383, 2013.

[6] A. S. Ismail, M. H. Mamat, M. M. Yusoff, M. F. Malek, A. S. Zoolfakar, R. A. Rani, A. B. Suriani, A. Mohamed, M. K. Ahmad, and M. Rusop, "Enhanced humidity sensing performance using Sn-Doped ZnO nanorod Array/SnO$_2$ nanowire heteronetwork fabricated via two-step solution immersion," *Materials Letters,* vol. 210, pp. 258-262, 2018.

[7] N. D. M. Sin, N. Samsudin, S. Ahmad, M. H. Mamat, and M. Rusop, "Zn-Doped SnO2 with 3D Cubic Structure for Humidity Sensor," *Procedia Engineering,* vol. 56, pp. 801-806, 2013.

[8] R. Fiz, F. Hernandez-Ramirez, T. Fischer, L. Lopez-Conesa, S. Estrade, F. Peiro, and S. Mathur, "Synthesis, characterization, and humidity detection properties of Nb$_2$O$_5$ nanorods and SnO$_2$/Nb$_2$O$_5$ heterostructures," *The Journal of Physical Chemistry C,* vol. 117, pp. 10086-10094, 2013.

[9] S. Kotresh, Y. T. Ravikiran, S.C. Vijaya Kumari, H.G. Raj Prakash, and S. Thomas, "Polyaniline Niobium Pentoxide Composite As Humidity Sensor At Room Temperature " *Adv. Mater. Lett.,* vol. 6, pp. 641-645, 2015.

[10] K. H. Katayama, H.; Noda, T.; Akiba, T.; Yanagida, H., "Effect of Alkaline Oxide Addition on the Humidity Sensitivitiy of Nb$_2$O$_5$-Doped TiO$_2$," *Sens. Actuators B Chem.,* vol. 2, pp. 143 - 149, 1990.

[11] X. Hu, X. Hu, Y. Yang, X. Wu, H. M. van Driel, and X. Gu, "Influence of doping Y$_2$O$_3$ and Nb$_2$O$_5$ on ZrO$_2$ characteristics in ZrO$_2$-SiO$_2$-P$_2$O$_5$ semiconductive ceramic humidity sensors," *Proceedings of the SPIE,* vol. 3862, pp. 551-555, 1999.

[12] R. A. Rani, A. S. Zoolfakar, J. Z. Ou, M. R. Field, M. Austin, and K. Kalantar-zadeh, "Nanoporous Nb$_2$O$_5$ hydrogen gas sensor," *Sens. Actuators, B,* vol. 176, pp. 149-156, Jan 2013.

[13] J. Z. Ou, R. A. Rani, M. H. Ham, M. R. Field, Y. Zhang, H. Zheng, P. Reece, S. Zhuiykov, S. Sriram, M. Bhaskaran, R. B. Kaner, and K. Kalantar-Zadeh, "Elevated temperature anodized Nb$_2$O$_5$: A photoanode material with exceptionally large photoconversion efficiencies," *ACS Nano,* vol. 6, pp. 4045-4053, 2012.

[14] A. Ghicov and P. Schmuki, "Self-ordering electrochemistry: a review on growth and functionality of TiO$_2$ nanotubes and other self-aligned MO$_x$ structures," *Chemical Communications,* vol. 0, pp. 2791-2808, 2009.

[15] B. C. Yadav and M. Singh, "Morphological and humidity sensing investigations on Niobium, Neodymium, and Lanthanum Oxides," *Sensors Journal, IEEE,* vol. 10, pp. 1759-1766, 2010.

Impedance based Aluminium Interdigitated Electrode (Al-IDE) Biosensor on Silicon Substrate for Salmonella Detection

R.D.A.A Rajapaksha*
Institute of Nano Electronic Engineering (INEE)
Universiti Malaysia Perlis (UniMAP)
01000 Kangar, Perlis Malaysia
asanka@wyb.ac.lk

M.N. Afnan Uda
School of Microelectronic Engineering
Universiti Malaysia Perlis (UniMAP)
01000 Kangar, Perlis Malaysia
nurafnan92@yahoo.com

U. Hashim*
Institute of Nano Ectronic Engineering
Universiti Malaysia Perlis (UniMAP)
01000 Kangar, Perlis Malaysia
uda@unimap.edu.my

S.C.B. Gopinath
School of Bioprocessc Engineering
Universiti Malaysia Perlis (UniMAP)
02600 Kangar, Perlis Malaysia
gopis11@gmail.com

C.A.N. Fernando
Department of Electronics
Wayamba University of Sri Lanka
60200 Kuliyapitiya, Sri Lanka
canfernando9@gmail.com

Abstract—**Si is one of the most common, and prominent element in electronic field. In this research, Al based interdigitated electrical biosensors was prepared using Si as base material to detect *Salmonella enterica typhi* (*S. typhi*). The usage of the IDE sensors in the biosensor field is tremendous in these days because of the large number of comb structured finger electrodes gain high sensitivity. *S. typhi* is a very harmful food borne pathogen, makes typhoid disease which causes many deaths annually in worldwide. The AutoCAD software was used to design the chrome mask of IDE sensor. The fabrication process was done using conventional photolithography method. The fabricated Al-IDE, morphologically analyzed using SEM and further structurally characterized using EDX. The SEM depicted the well fabricated interdigitated electrode fingers and finger gaps are nearly equal to 50 μm and 1μm, respectively as designed dimensions. Silanization by APTES, immobilization with carboxylic functionalized *S. typhi* ssDNA probes and blocking with tween-20 were the major functionalization steps. The sensitivity measurements were done using different concentration of complementary targets and selectivity measurements were done using complementary, non-complementary and single base mismatch ssDNA samples. Each step was electrically characterized using Nyquist plot by impedance analyzer. Selectivity measurement confirmed, the tween-20 was important for the specificity of biosensor. The sensitivity measurements further verify that the biosensor is appropriate for detection of various concentrations from 1 fM to 1 μM of *S. typhi* targets.**

Keywords—*Al-IDE; S. typhi; DNA immobilization; DNA hybridization; biosensor*

I. INTRODUCTION

The foodborne pathogen is an one of the major public health problems in recent years in the world [1]–[3]. *Salmonella* is one of the major disease pathogen and *Salmonella enterica typhi* (*S. typhi*) is one of the most commonly reported serotypes which causes typhoid fever.

Approximately 17 million people are affected from typhoid fever and nearly 600,000 deaths are recorded annually [4][5]. Typhoid fever also reported as second most reported bacterium among commonly reported laboratory infection. *S. typhi* is a Gram-negative bacterium which causes self-limiting gastroenteritis in human. Fever, Diarrhea, vomiting, abdominal pain, nausea, and loss of appetite are the major symptoms of this bacterium inflection can be seen within 12 to 72h after getting contaminated food or beverage [4], [6].

Colony counting, gel electrophoresis, membrane blots and immunology-based methods are some conventional methods used to detect the disease pathogens. But those methods have many drawbacks because of time consuming, multiple laboratory tests, expensive, qualitative measurements etc. [1], [2], [7]. These conventional methods have many preliminary steps such as pre-enrichment, selective enrichment, selective plating, biochemical screening, and serological confirmation. Polymerize Chain Reaction (PCR) is one of the popular conventional method but it need expensive equipment, skilled labor and labor intensive gel-based detection that shows feeble sensitivity and specificity [8].

Commercial application of biosensors are given significant impact on the medical and food analysis to overcome the drawbacks of conventional methods [9]–[11]. High sensitivity, specificity, repeatability and real-time in-field detection are the main advantages of biosensors [12]–[17]. Even though there are many publications regarding on biosensors under food borne pathogens, only few commercial applications are available in the market [18],[19] due to limited lifetime of biological components, difficulties in mass production and the user-friendliness.

In this research, electrical based Al-IDE on silicon substrate is fabricated. The major steps of functionalization as biosensor are, silanization using APTES, immobilization using carboxylic functionalized ssDNA probes and blocking

978-1-5386-5284-8/18 $31.00 © 2018 IEEE

the non-immobilized area using Tween-20. Finally, the biosensor is tested for specificity and sensitivity detection by hybridizing with synthetic ssDNA samples as complementary, non-complementary and single base mismatch targets.

II. MATERIAL AND METHODS

A. Chemical, Reagents, and Oligonucleotides

(3-Aminopropyl)triethoxysilane (APTES) was purchased from Sigma Aldrich, USA and 30-base synthetic oligonucleotides were purchased from AIT Biotech Pvt. Ltd, Singapore. Other analytical reagent grade chemicals were purchased commercially. Throughout the experiment deionized distilled water was used. target oligonucleotides sequences and carboxylic modified synthetic probes are shown in the table1.

TABLE I. 30-MER OLIGONUCLEOTIDES SEQUENCES FOR S. TYPHI PROBE, COMPLEMENTARY, NON-COMPLEMENTARY AND SINGLE BASE MISMATCH.

Oligonucleotide	Sequences
30-mer probe	5'-(COOH) CGC GCG GCA TCC GCA TCA ATA ATA CCG GCC –3'
30-mer complementary	(5'- GGC CGG TAT TAT TGA TGC GGA TGC CGC GCG –3')
30-mer non-complementary	(5'- CCG GCC ATA ATA ACT ACG CCT ACG GCG CGC –3')
30-mer single base mismatch	(5'- GGC CGG TAT TAT TGT TGC GGA TGC CGC GCG –3')

B. Instrument

Scanning electron microscope (SEM) (JEOL, JSM-610LV) at 20kV with 7000 magnification was used to characterize the surface morphology of the Al-IDE pattern at the room temperature. The biosensor functionalization steps were characterized using EDX (OXFORD instrument). Weifo Electronic dry cabinet was used for incubation process. Electrical properties for different functionalization steps of the biosensor were done using dielectric analyzer (Alpha-A High-Performance Frequency Analyzer, Novocontrol Technologies, Hundsangen, Germany) with sweep frequency of 1 Hz to 1 MHz at an amplitude of 1 Vrms was used to analyze electrical properties of different functionalization steps of the biosensor.

C. IDE Sensor Fabrication

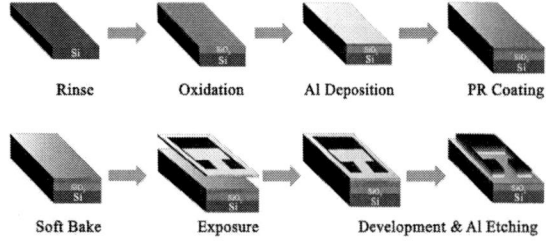

Fig. 1. The fabrication process of Al-IDE on Si substrate using conventional photolithography process.

The mask design of the IDE with 1 μm finger gap was designed using AutoCAD software and transfer the design to chrome mask. Al-IDEs were fabricated on the SiO_2 layer from conventional photolithography method and standard CMOS process. As our previous publications, Si was the base material for this research [20], [21].

Initially, to clean the base material of Si substrate from foreign agent, RCA1 and RCA2 were used. The cleaned wafer was under gone the wet thermal oxidization process inside the wet oxidation furnace at 500 °C for 1 h to prepare 400 nm thickness of the SiO_2 layer. The thermal evaporator was used to deposit the Al on top of the SiO_2 surface. Spin coating method is used to coat the positive photoresist on the well-deposited aluminium surface and it was soft baked to minimize the moisture from the SiO_2 substrate at 60 °C for 90 min. It was affected to minimize the standing wave on the positive PR layer. The fabricated chrome mask with mask aligner was used for patterning process. IDE photomask was aligned with the sample and UV light was exposed for 10 s. In this period IDE pattern was transferred to the positive photoresist. Hard bake process was run for 1 min at 110 °C to remove the moisture and enhance the adhesion force between the aluminium and SiO_2 layer. The development process was conducted for 10 min as the final step of the photolithography. The IDE was immersed in the aqua region for 20 s to remove the unexposed area of the positive photoresist. As the Final step, photoresist was stripped from IDE surface using acetone.

D. Biosensor Functionalization Steps

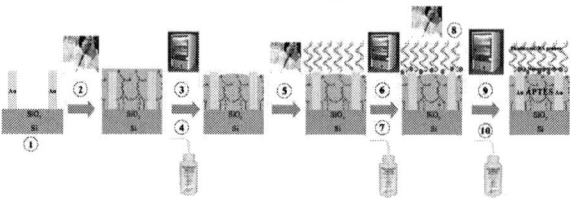

Fig. 2. Functionalization steps of biosensor, (1) Bare Al-IDE (2) Silanization with APTES (3,4) incubation and wash, (5) carboxylic probe S. typhi ssDNA probe immobilization, (6,7) incubation and wash, (8) Tween-20 as blocking agent, (9,10) incubation and wash.

The schematic diagram of the functionalization process of the biosensor is depicted in the Fig.2. The active surface of the bare Al-IDE was immersed in NaOH solution for 2 minutes to enhance the hydroxyl on the SiO_2 layer. The silanization process was done using 0.1 M, 2 μl of APTES. The silanized sensor was incubate 15 min and DI water was used to wash unattached APTES molecules. 10 μM, 2 μl of carboxylic probe ssDNA was used for immobilization process. The immobilized sensor was incubated 1 hour and DI water was used to wash unattached ssDNA probes from APTES layer. Biological grade deionized water (> 18 MΩ) was used to prepare 10 μM of stock solution of complement, non-complement and single base mismatch targets. The prepared samples were kept in freezer at -20 °C. Tween-20 was used to block the non-immobilized area to avoid the attachment of non-specific targets binding. The sensor was incubated 15 min and DI water was used to wash unattached Tween-20 molecules.

The sensitivity test was done using different concentration of S. typhi ssDNA targets and selectivity test was done using complementary, non-complementary and mismatch target ssDNA by dropping 1μl to the active surface area in the biosensor surface. Deionized water was used after each functionalized step to remove un-attached linkers in the active surface.

III. RESULTS AND DISCUSSION

The fabricated Al-IDE was morphologically characterized using SEM and structurally characterized using EDX. The impedance electrical characterization was done using the dielectric analyzer to identify the specific and non-specific *S. typhi* targets of the Al-IDE based biosensor.

A. Morphological and Structural Characterization

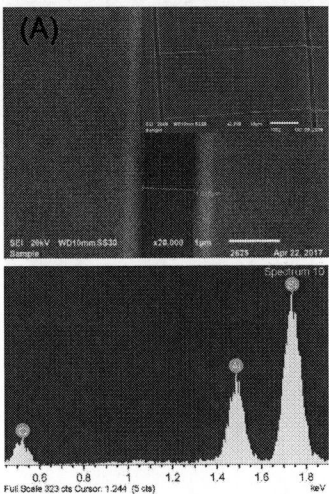

Fig. 3. Morphological and structural analysis using SEM and EDX, (A) SEM image for bare Al-IDE (B) EDX spectrum for bare Al-IDE.

The Fig. 3. (A and B) are depicted the morphological and structural analysis of bare IDE. The SEM image at 20K magnification is clearly shown the well fabricated finger gap near to 1 μm and the inset image shows the well fabricated IDE finger width as 50 μm. Without any other foreign agents, only Al, Si, and O were confirmed the well fabricated Al-IDE. In here, Al for Al electrode, Si and O for SiO_2 layer and Si for Si as a base material.

B. Electrical Characterization

Electrical properties for different functionalization steps of the biosensor were done using dielectric analyzer with sweep frequency of 1 Hz to 1M Hz at an amplitude of 1 Vrms.

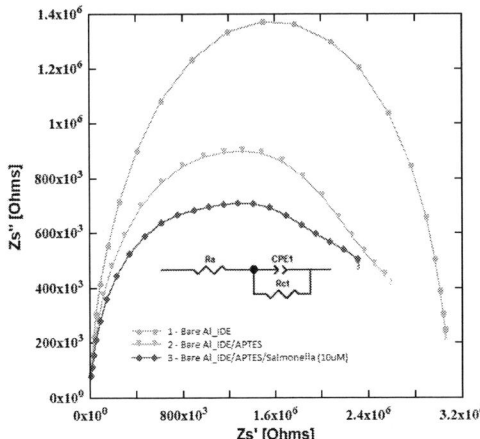

Fig. 4. AC impedance analysis for different functionalization steps of the biosensor.

Fig. 4. shows the Nyquist plot of AC impedance spectra for bare Al-IDE; different biosensor functionalization steps as after salinization and immobilization. Randles equivalent circuit can be used to express the variation of the Nyquist plot according to the changes of Ra and Rct values. Ra and Rct represent the bulk grain resistance and grain boundary charge transfer resistance, respectively. The highest Rct value appeared for bare Al-IDE and the Rct value was decreased respectively to the different functionalization steps as silanization with APTES and immobilization with carboxylic probe ssDNA. Hence, this situation suggests that the reduction of Rct with the functionalization steps. Rct value for bare Al-IDE higher than APTES, and it is further decreased when carboxylic functionalized ssDNA probes immobilized on the APTES layer. In here, APTES layer include amine group which is rich with positive charges. So it is worked as a sensing bridge between two electrodes. Hence, the Rct value tends to decrease after immobilized carboxylic probe ssDNA on top of the APTES layer.

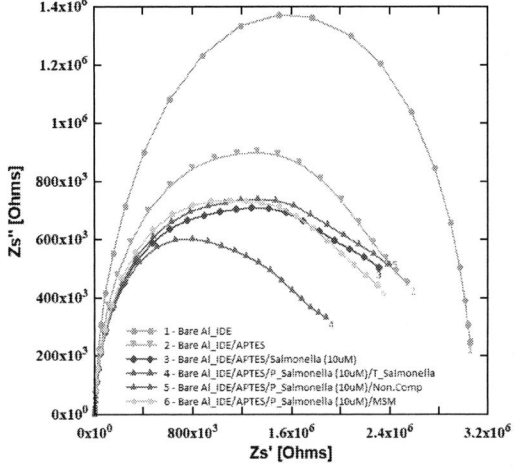

Fig. 5. AC impedance analysis for selectivity analysis as a complement, non-complement, and single base mismatch targets.

The selectivity measurement for complementary, non-complementary and single based mismatches targets of *S. typhi* were shown in Fig. 5. This step is highly important to validate the specficity of biosensor for specific target. The resultant curves of 10 μM concentrated non-complementary, single base mismatch and complementary targets were shown as curve 5, curve 6 and curve 7, respectively. The Rct values of non-complementary and single base mismatch are nearly equal to the Rct value which was generated by ssDNA probes. The tween-20 worked as a blocking agent and it covers unbounded probe area on the APTES layer to avoid the attachment of non-complement and mismatch targets directly. Hence, when wash by deionized waster, non-complement and mismatch targets are removed from active surface and resultant respective curves were fallen to probe level. The resultant Nyquist curve for complementary target was shown significant different Rct value which is compared to non-complement and mismatch targets and it will be proven that the device is suitable to capture specific target of *S. typhi*.

978-1-5386-5284-8/18 $31.00 © 2018 IEEE

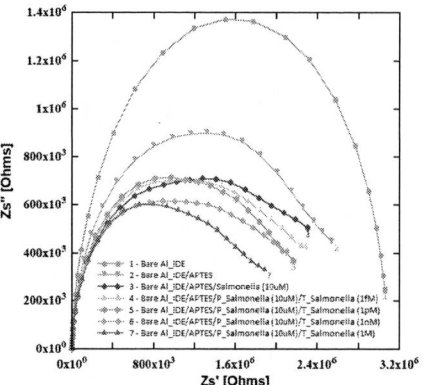

Fig. 6. AC impedance analysis for different concentration of complementary targets as 1fM to 1μM.

Different complementary target concentration of *S. typhi* as 10 fM, 10 pM, 10 nM and 10 μM were used for the sensitivity test. Fig. 6. shows the Nyquist plot for target *S. typhi* with different concentration. The curve number 4 to 7 show a decrement of Rct value while incrementing of target ssDNA concentration. It is proven that the different variation of Rct value because of the electron variation after hybridization of the different ssDNA target concentration with existing ssDNA probes in the biosensor surface. Hence current mobility carriers in the APTES layer increase according to the respective concentration of the complementary target ssDNA. Due to the resultant Nyquist plots, it is proven that presented method is suitable to detect different concentration of *S. typhi* successfully.

CONCLUSION

The Al-IDE on Si substrate was fabricated using conventional photolithography successfully. Selectivity test is ensured that biosensor is specific for *S. typhi* targets. The sensitivity test was done for different concentration of *S. typhi* targets from 1 fM to 1 μM concentration and which was confirmed that complementary targets sensitive for 1fM to 1 μM range. Thus, the implemented mechanism will be helpful for future biosensor applications.

ACKNOWLEDGMENT

The Collaborative Research in Science and Technology Center (CREST), is acknowledged for providing a research grant (CREST project P01C1-16) to successfully complete this research work.

REFERENCES

[1] K. Kant *et al.*, "Microfluidic devices for sample preparation and rapid detection of foodborne pathogens," *Biotechnol. Adv.*, vol. 36, no. 4, pp. 1003–1024, 2018.

[2] V. Velusamy, K. Arshak, O. Korostynska, K. Oliwa, and C. Adley, "An overview of foodborne pathogen detection: In the perspective of biosensors," *Biotechnol. Adv.*, vol. 28, no. 2, pp. 232–254, 2010.

[3] M. Kirk, K. Glass, L. Ford, K. Brown, and G. Hall, *Foodborne illness in Australia: Annual incidence circa 2010*, no. September. 2014.

[4] L. Yang, Y. Li, C. L. Griffis, and M. G. Johnson, "Interdigitated microelectrode (IME) impedance sensor for the detection of viable Salmonella typhimurium," *Biosens. Bioelectron.*, vol. 19, no. 10, pp. 1139–1147, 2004.

[5] G. Kim, S. B. Park, J.-H. Moon, and S. Lee, "Detection of pathogenic Salmonella with nanobiosensors," *Anal. Methods*, vol. 5, no. 20, pp. 5717–5723, 2013.

[6] Y. Ning, Z. J. Li, Y. F. Duan, Z. H. Peng, and L. Deng, "A Novel Biosensor for Detection of Salmonella typhimurium Carrying SSeC Gene Based on the Secondary Quenching Effect of Carbon Nanotubes," *J. Nanomater. Mol. Nanotechnol.*, vol. 02, no. 05, 2013.

[7] K. S. Gracias and J. L. McKillip, "A review of conventional detection and enumeration methods for pathogenic bacteria in food.," *Can. J. Microbiol.*, vol. 50, no. 11, pp. 883–890, 2004.

[8] S. Cagnin *et al.*, "Overview of Electrochemical DNA Biosensors: New Approaches to Detect the Expression of Life," *Sensors*, vol. 9, no. 4, pp. 3122–3148, 2009.

[9] R. D. A. A. Rajapaksha, U. Hashim, M. N. A. Uda, and C. A. N. Fernando, "High-performance Electrical Variable Resistor Sensor for E . coli," *J. Telecommun. Electron. Comput. Eng.*, vol. 10, no. 1, pp. 61–64, 2018.

[10] R. D. A. A. Rajapaksha, U. Hashim, N. Z. Natasha, M. N. A. Uda, V. Thivina, and C. A. N. Fernando, "Gold Nano-particle Based Al Interdigitated Electrode Electrical Biosensor for Specific ssDNA Target Detection," *IEEE Reg. Symp. Micro Nanoelectron.*, pp. 191–194, 2017.

[11] N. Z. Natasha, R. D. A. A. Rajapaksha, M. N. A. Uda, and U. Hashim, "Electrical DNA biosensor using aluminium interdigitated electrode for E.Coli O157:H7 detection," *AIP Conf. Proc.*, vol. 020235, p. 020235, 2017.

[12] R. D. A. A. Rajapaksha, U. Hashim, M. N. Afnan Uda, C. A. N. Fernando, and S. N. T. De Silva, "Target ssDNA detection of E.coli O157:H7 through electrical based DNA biosensor," *Microsyst. Technol.*, vol. 23, no. 12, pp. 5771–5780, 2017.

[13] S. T. Ten *et al.*, "Highly sensitive Escherichia coli shear horizontal surface acoustic wave biosensor with silicon dioxide nanostructures," *Biosens. Bioelectron.*, vol. 93, no. June 2016, pp. 146–154, 2017.

[14] V. Perumal *et al.*, "A new nano-worm structure from gold-nanoparticle mediated random curving of zinc oxide nanorods," *Biosens. Bioelectron.*, vol. 78, pp. 14–22, 2016.

[15] S. R. Balakrishnan *et al.*, "Polysilicon nanogap lab-on-chip facilitates multiplex analyses with single analyte," *Biosens. Bioelectron.*, vol. 84, pp. 44–52, 2016.

[16] S. T. Ten *et al.*, "Highly sensitive Escherichia coli shear horizontal surface acoustic wave biosensor with silicon dioxide nanostructures," *Biosens. Bioelectron.*, no. August, pp. 1–9, 2016.

[17] S. R. Balakrishnan *et al.*, "A point-of-care immunosensor for human chorionic gonadotropin in clinical urine samples using a cuneated polysilicon nanogap lab-on-chip," *PLoS One*, vol. 10, no. 9, pp. 1–17, 2015.

[18] S. P. Mohanty, "Biosensors : A Tutorial Review Biosensors : A Tutorial Review," *IEEE Potentials*, vol. 25, no. APRIL 2006, pp. 35–40, 2015.

[19] A. Hayat, G. Catanante, and J. L. Marty, "Current trends in nanomaterial-based amperometric biosensors," *Sensors (Switzerland)*, vol. 14, no. 12, pp. 23439–23461, 2014.

[20] R. D. A. A. Rajapaksha, U. Hashim, and C. A. N. Fernando, "Design, fabrication and characterization of 1.0 μm Gap Al based interdigitated electrode for biosensors," *Microsyst. Technol.*, vol. 23, no. 10, pp. 4501–4507, 2017.

[21] R. D. A. A. Rajapaksha, U. Hashim, S. C. B. Gopinath, and C. A. N. Fernando, "Sensitive pH detection on gold interdigitated electrodes as an electrochemical sensor," *Microsyst. Technol.*, vol. 24, no. 4, pp. 1965–1974, 2018.

978-1-5386-5284-8/18 $31.00 © 2018 IEEE

Performance Analysis of SAW Gas Sensor Based on ST-Cut Quartz for Breath Analysis

Nur Fatin binti Mohamad
Razali
ECE Department,
Kulliyyah of Engineering,
International Islamic
University Malaysia

Aliza Aini Md Ralib
ECE Department,
Kulliyyah of Engineering,
International Islamic
University Malaysia
alizaaini@iium.edu.my

Rosminazuin Ab Rahim
ECE Department, *Kulliyyah of*
Engineering, International
Islamic University Malaysia
rosmi@iium.edu.my

Abstract— **Breath analysis is the recent technique used to detect diseases. Among the advantages of breath analysis are high sensitivity and fast detection. Surface acoustic wave (SAW) gas sensor is one of the sensors that can be used to analyze breath analysis. This work reports on the design and simulation of SAW gas sensor using a FEM simulator, COMSOL Multiphysics. We analyze the possibilities of increasing the quality factor, coupling coefficient and the wave velocity of the devices by varying the number of interdigitated electrodes (IDTs). We designed and simulated IDTs pf different finger number (IDT= 2, 4, 8, 12, 16, 20, 24, 28 and 32). The measured quality factor, coupling coefficient and wave velocity confirms that varying the number of IDTs fingers enables to reach better electrical performances at resonances frequency around 943 MHz.**

Keywords—SAW gas sensor, number of IDT, COMSOL Multiphysics.

I.   INTRODUCTION

Breath analysis is the recent technique used in detection of diseases such as diabetes, kidney failure, liver disease and lung cancer [1]. Generally, exhaled breath contains volatile and non- volatile organic compound [2]. By collecting the exhaled breath of patients, volatile organic compounds (VOC) are gathered. This VOC was recognized as biomarkers in detecting diseases. The most common chemicals recognized as VOC biomarkers for lung cancer are styrene, 2 decane, isoprene, benzene, 1- hexane and propyl benzene [3]. The main advantage of VOCs is that it have high potential for rapid and non- invasive screening disease detection [4]. Over the past few years, breath analysis has shown many advantages such as non- invasive, high sensitivity, simple and potentially cheaper [5]. Hence, breath analysis has become more popular among researchers as a method for early detection of lung cancer. Compared to current analysis such as urine test and blood test, breath analysis also can provide fast detection. Moreover, breath analysis can be analyzed using various type of sensor such as surface acoustic wave (SAW) sensor [6]–[9], quartz crystal microbalance (QCM) sensor [2], [10], optical sensor and gold nanoparticle (GNP) sensor [11]. However, this work is interested in one of the electroacoustic sensor which is SAW sensor.

The operating principle of SAW sensor is the propagation of acoustic wave along the surface of the piezoelectric material[12]. The basic structure of SAW sensor consists of piezoelectric material, interdigital transducer and sensing layer. The most commonly used piezoelectric materials are lithium niobate and quartz [13]. Various cutting pattern were applied in cutting quartz substrate such as AT- cut quartz, ST-cut quartz and ST-X cut quartz [14]. However, the most common cutting pattern of quartz used in gas sensing application is ST- cut quartz [14]–[18]. Hence, this work has chosen ST-cut quartz for piezoelectric material at resonance frequency of 943 MHz. Another crucial parameter in designing SAW sensor is the number of IDT. According to O. Mortada et al. (2017), using different number of IDT will affect the quality factor (Q) and the electromechanical coupling coefficient (k^2) [19]. The number of IDTs have been varies by 25, 40, 50 and 80. The performance was analyzed by fabricating and measuring the device. However, in this work we are interested in simulating the device and observe the performances accordingly. Section II describes on the design concept and theory of SAW sensor. The finite element simulation is explained in Section III. Finally the conclusion is given in Section IV.

II.   DESIGN CONCEPT AND THEORY

Surface acoustic wave can be generated electrically by using interdigital transducer (IDT). Figure 1 shows the basic structure of SAW gas sensor.

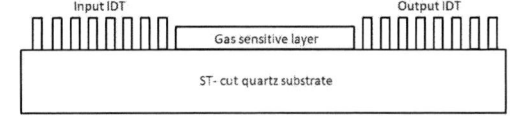

Fig 1. Schematic diagram of SAW gas sensor

Based on Figure 1, it shows there are two set of IDTs. The first set is the input IDT where input signal is applied to generate acoustic wave. Another set of IDT is the output IDT where the received signal can be collected and analyzed. Another important structure in SAW gas sensor is the sensing layer. Sensing layer functioned as a layer to detect the present of the specific gas that act as the biomarker. The dimension of the SAW structure depends on the wavelength of the applied signal. The relationship of the wavelength, frequency and acoustic wave velocity can be derived using the following equation [12]:

978-1-5386-5284-8/18 $31.00 © 2018 IEEE

$$f_0 = \frac{v}{\lambda} \qquad (1),$$

where f_0 is the SAW resonance frequency, v is the acoustic wave velocity of the SAW sensor and λ is the wavelength of the SAW. Table 1 shows the design dimensions used in this work.

TABLE 1. DESIGN DIMENSIONS

Frequency	943 MHz
Wavelength, λ	4 μm
Acoustic wave Velocity, v	3840 m/s
Sensing area, L	2 μm
Piezoelectric material thickness	12 μm
Thickness of IDT	0.2 μm
Number of IDT	2,4,8,12,16,20,24,28,32

The performance of the SAW gas sensor was affected by a few factors which are the resonance frequency, electromechanical coupling coefficient and the quality factor. The resonance frequency is a frequency where the response amplitude is relatively maximum [20]. The electromechanical coupling coefficient of SAW resonators is a dimensionless value that characterizes interaction of two resonators [20]. The coupling coefficient, k^2 can be calculated using equation (2) [19]:

$$k^2 = \frac{\pi^2}{4} \frac{f_a - f_r}{f_a} \qquad (2),$$

where f_a is the anti-resonance frequency and f_r is the resonance frequency. The quality factor is a dimensionless parameter that indicates the energy losses within a resonant element. This losses is links directly into the bandwidth of the resonator with respect to its center frequency [21]. The value of quality factor, Q can be calculated as shown in (3) [22]:

$$Q_{-3\,dB} = \frac{f_0}{\Delta f_{-3\,dB}} \qquad (3),$$

where f_0 is the resonant frequency and $\Delta f_{-3\,dB}$ is the bandwidth at 3 dB. Hence, using all these formulas the performance of this work shall be observed and analyzed to find the optimum design of SAW gas sensor. The next section will be discussing the results obtained from the simulation.

III. RESULTS AND DISCUSSION

The device was modelled and simulated using a finite element (FEM) simulator, COMSOL Multiphysics which has the capabilities to display a two-dimensional model based on the layout and the fabrication process. The aim of the simulation analysis is to simulate the absolute admittance value which gives the value of resonance and anti-resonance frequency to be used in finding the value of coupling coefficient. The simulation also aims to simulate the quality factor of the device to monitors the performance of the device. The applied boundary condition is the bottom of the SAW gas sensor is fixed and the voltage is applied at the first set of the IDTs. Once the model specifications have been completed, it was meshed using a fine physics-controlled mesh, which is preferred for this design.

Two types of studies were applied in this simulation which is eigenfrequency and frequency domain. The eigenfrequency studies the frequencies for surface total displacement. Figure 2 shows the mode shape of the SAW gas sensor after simulation. This work only focuses on modelling and simulating of SAW gas sensor before the presence of gas.

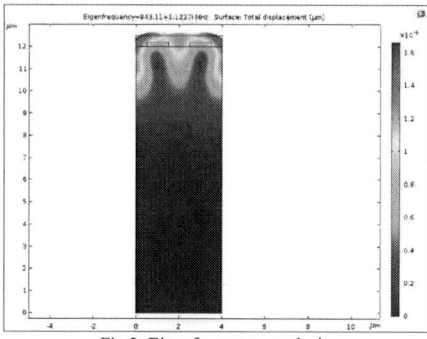

Fig 2. Eigenfrequency analysis

B. Coupling Coefficient and Number of IDT

Figure 4 shows the graph of absolute of admittance. Using this graph, the value of resonance frequency and anti-resonance frequency can be collected. Once the value of f_a and f_r has been collected, the value of the coupling coefficient can be calculated using equation (2). The calculated value is then plotted as shown in the graph of coupling coefficient and number of IDT (Figure 3).

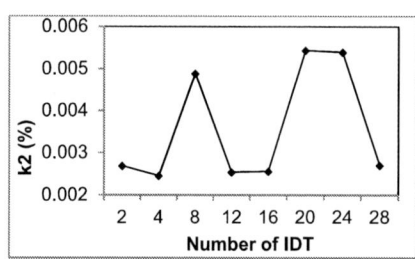

Fig 3. Graph of coupling coefficient and number of IDT

Based on the plotted graph, it can be observed that the highest value of coupling coefficient recorded is at number of IDT 20 with value of $k^2 = 5.435 \times 10^{-3}$ % and the second highest is at IDT 24 with value of $k^2 = 5.393 \times 10^{-3}$ %. However, the uniformity of the k^2 as the number of IDT increased, a fluctuation happened at 12 IDTs and 16 IDTs. This is because small different value of f_a and f_r.

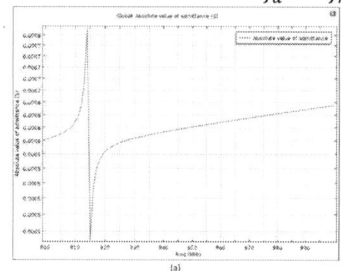

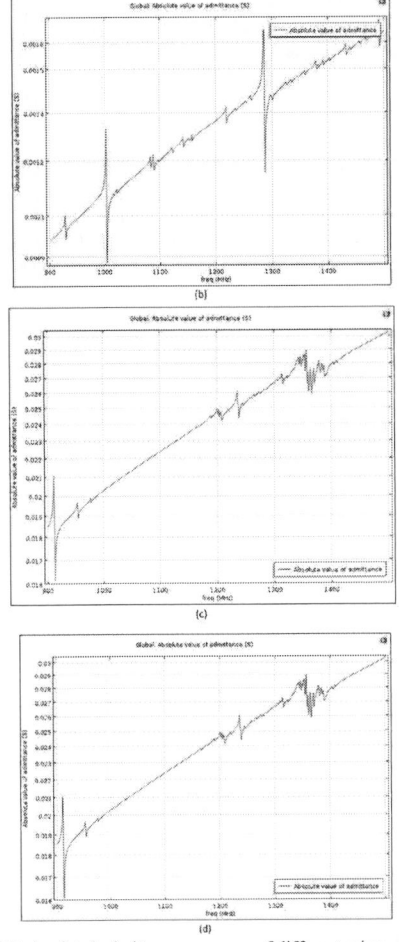

Fig 4. FEM simulated admittance response of different micro-resonators: (a) with 2IDTs, (b) with 4IDTs, (c) with 24IDTs, (d) with 28IDTs.

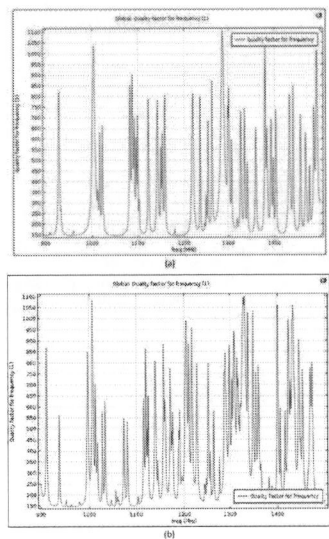

Fig 5. FEM simulated Q- factor of eight different micro-resonators: (a) with 4IDTs, (b) with 8IDTs

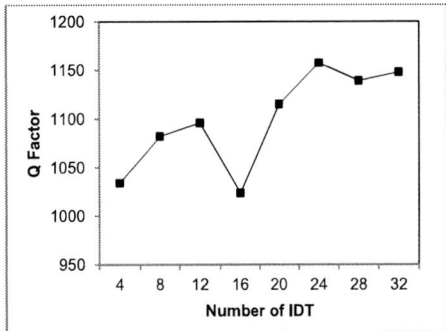

Fig 6. Graph of Q- factor and number of IDT

C. Q- Factor and Number of IDT

Theoretically the value of quality factor can be calculated using equation (3) as discussed previously. However, the FEM simulator has an advanced function where it can compute the value of quality factor using its own program. Figure 5 shows the value of quality factor collected as the number of IDT varies. The results displayed are the multiple values recorded for different mode. However, we choose the value of quality factor that shows at the resonant frequency. Hence, the value of quality factor as shown in Figure 6 is collected. Based on data collected, the highest value of quality factor recorded is at 24 IDT which is $Q = 1157.5$. The uniformity of the graph of quality factor as the number of IDT increased, a fluctuation happened at IDT= 16. This is because at IDT= 16 the value of resonance frequency decreases. In previous section, equation (2) has discussed that Q- factor is related to resonant frequency and bandwidth. However, the resonant frequency of IDT= 16 has decreased. Hence, the value of Q- factor also decreases due to the decreases of resonant frequency. Based on observation, it can be concluded that the increasing number of IDT does affect the quality factor of SAW gas sensor.

C. Velocity and Number of IDT

The effect of the variation number of IDT on the performances of SAW gas sensor can be analyzed further using the wave velocity plots as shown in Figure 7. Based on Figure 7, it can be observed that wave velocity is highest at IDT 4 and lowest at IDT 20. However, the wave velocity becomes constant at IDT 20 onwards. This shows that as the number of IDT increases the wave velocity increases, until at one point it started becoming almost constant. However, by comparison, the most common wave velocity used to operate SAW gas sensor is within 3100 m/s and 3700 m/s. Therefore, the preferred number of IDT in this simulation is from IDT 20 onwards.

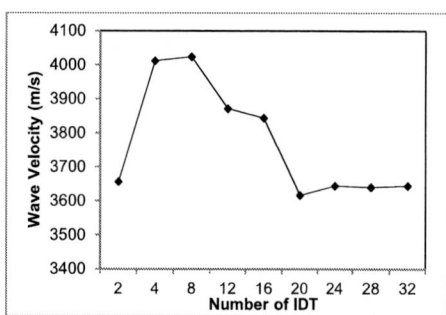

Fig 7. Graph of velocity and number of IDT

TABLE 2. SUMMARY OF FACTORS AFFECTED THE PERFORMANCE OF SAW GAS SENSOR

No of IDT	f_a (MHz)	f_r (MHz)	k^2 (%)	Q factor	Velocity (m/s)
2	914	915	2.694×10^{-3}	1170	3656
4	1003	1004	2.455×10^{-3}	1034	4012
8	1006	1008	4.886×10^{-3}	1082	4024
12	968	969	2.544×10^{-3}	1095.7	3872
16	961	962	2.562×10^{-3}	1023.8	3844
20	904	906	5.435×10^{-3}	1114.95	3616
24	911	913	5.393×10^{-3}	1157.5	3644
28	910	911	2.705×10^{-3}	1138.9	3640
32	911	913	5.393×10^{-3}	1147.8	3644

Table 2 shows the summarization of the collected values of different factors that affect the performance of SAW gas sensor. As discussed previously, it can be concluded that the performance of the SAW gas sensor is at its optimum when the number of IDT is 24. Hence, the optimum modelled and simulated SAW gas sensor can later be used during fabrication process.

IV. CONCLUSION

In this work, SAW gas sensor was modelled and simulated using a FEM simulator, COMSOL Multiphysics in order to assess the effect of number of IDT on the electromechanical coupling coefficient, quality factor and wave velocity. ST- cut quartz was chosen as piezoelectric material because of its excellence bonding to substrate material. Simulation analysis was done to simulate the resonance frequency to observe the performance of SAW gas sensor. The results show that number IDT 24 is the optimum design for SAW gas sensor. This is because it gives the highest value of coupling coefficient, quality factor and the reasonable wave speed.

ACKNOWLEDGMENT

This research was supported by Fundamental Research Grant Scheme (FRGS 17-032-0598) and Research Initiative Grant Scheme (RIGS 16-083-0247) under International Islamic University Malaysia and Ministry of Higher Education Malaysia

REFERENCES

[1] K. H. Kim, S. A. Jahan, and E. Kabir, "A review of breath analysis for diagnosis of human health," *TrAC - Trends Anal. Chem.*, vol.

33, no. September 2017, pp. 1–8, 2012.

[2] A. G. Dent, T. G. Sutedja, and P. V. Zimmerman, "Exhaled breath analysis for lung cancer," *J. Thorac. Dis.*, vol. 5, no. SUPPL.5, 2013.

[3] H. Yu *et al.*, "Detection volatile organic compounds in breath as markers of lung cancer using a novel electronic nose," *Proc. Ieee Sensors 2003, Vols 1 2*, pp. 1333–1337, 2003.

[4] G. Konvalina and H. Haick, "Sensors for breath testing: From nanomaterials to comprehensive disease detection," *Acc. Chem. Res.*, vol. 47, no. 1, pp. 66–76, 2014.

[5] W. Li *et al.*, "Advances in the early detection of lung cancer using analysis of volatile organic compounds: from imaging to sensors.," *Asian Pac. J. Cancer Prev.*, vol. 15, no. 11, pp. 4377–4384, 2014.

[6] N. Barié, A. Voigt, M. Rapp, and J. Marcoll, "Fast SAW based sensor system for real-time analysis of volatile anesthetic agents," *Proc. IEEE Sensors*, pp. 958–961, 2007.

[7] D. Wang, Y. Wang, K. Yu, and P. Wang, "Systems based on surface acoustic wave sensors for detection of gaseous and liquid exhaled breath," pp. 317–320, 2012.

[8] He Shi-tang, Y. Gao, J. Shao, and Y. Lu, "Application of SAW gas chromatography in the early screening of lung cancer," *2015 Symp. Piezoelectricity, Acoust. Waves, Device Appl.*, pp. 22–25, 2015.

[9] S. Li, Y. Wan, Y. Su, C. Fan, and V. R. Bhethanabotla, "Gold nanoparticles amplified surface acoustic wave biosensors for immunodetection," *Proc. IEEE Sensors*, pp. 3–5, 2017.

[10] N. Queralto, A. N. Berliner, B. Goldsmith, R. Martino, P. Rhodes, and S. H. Lim, "Detecting cancer by breath volatile organic compound analysis: a review of array-based sensors," *J. Breath Res.*, vol. 8, no. 2, p. 027112, 2014.

[11] C. Di Natale *et al.*, "Lung cancer identification by the analysis of breath by means of an array of non-selective gas sensors," *Biosens. Bioelectron.*, vol. 18, no. 10, pp. 1209–1218, 2003.

[12] A. Afzal, N. Iqbal, A. Mujahid, and R. Schirhagl, "Advanced vapor recognition materials for selective and fast responsive surface acoustic wave sensors: A review," *Anal. Chim. Acta*, vol. 787, no. July, pp. 36–49, 2013.

[13] M. N. Hasan, S. Maity, A. Sarkar, C. T. Bhunia, D. Acharjee, and A. M. Joseph, "Simulation and Fabrication of SAW-Based Gas Sensor with Modified Surface State of Active Layer and Electrode Orientation for Enhanced H2 Gas Sensing," *J. Electron. Mater.*, vol. 46, no. 2, pp. 679–686, 2017.

[14] J. Devkota, P. R. Ohodnicki, and D. W. Greve, "SAW sensors for chemical vapors and gases," *Sensors (Switzerland)*, vol. 17, no. 4, pp. 13–15, 2017.

[15] K. M. M. Kabir *et al.*, "Investigating the cross-interference effects of alumina refinery process gas species on a SAW based mercury vapor sensor," *Hydrometallurgy*, vol. 170, pp. 51–57, 2017.

[16] X. Zhang, Y. Zou, C. An, K. Ying, X. Chen, and P. Wang, "Sensitive detection of carcinoembryonic antigen in exhaled breath condensate using surface acoustic wave immunosensor," *Sensors Actuators, B Chem.*, vol. 217, pp. 100–106, 2015.

[17] C. Viespe and C. Grigoriu, "SAW sensor based on highly sensitive nanoporous palladium thin film for hydrogen detection," *Microelectron. Eng.*, vol. 108, pp. 218–221, 2013.

[18] W. Li *et al.*, "Room-temperature ammonia sensor based on ZnO nanorods deposited on ST-cut quartz surface acoustic wave devices," *Sensors (Switzerland)*, vol. 17, no. 5, 2017.

[19] O. Mortada, A. H. Zahr, J. C. Orlianges, A. Crunteanu, M. Chatras, and P. Blondy, "Analysis and optimization of acoustic wave micro-resonators integrating piezoelectric zinc oxide layers," *J. Appl. Phys.*, vol. 121, no. 7, 2017.

[20] A. Oberoi and R. Sinha, "A Novel MEMS based Surface Acoustic Wave Gas Sensor for Carbon Dioxide Detection in Hot-Process Areas," *Proc. Int. Electron. Conf. Sensors Appl.*, p. e001, 2014.

[21] K. Lee, W. Wang, T. Kim, and S. Yang, "A novel 440 MHz wireless SAW microsensor integrated with pressure-temperature sensors and ID tag," *J. Micromechanics Microengineering*, vol. 17, no. 3, pp. 515–523, 2007.

[22] X. Zhao *et al.*, "Protein functionalized ZnO thin film bulk acoustic resonator as an odorant biosensor," *Sensors Actuators, B Chem.*, vol. 163, no. 1, pp. 242–246, 2012.

2018 IEEE International Conference on Semiconductor Electronics (ICSE)

Comparative Study of the Calcium Ferrite Nanoparticles (CaFe2O4-NPs) Synthesis Process

Jumril Yunas[1], Noor Humam Sulaiman[2] and Mariyam Jameelah Ghazali[2]
[1]Institute of Microengineering and Nanoelectronics
Universiti Kebangsaan Malaysia
[2]Department of Mechanical and Materials Engineering, Faculty of Engineering and Built Environment,
Universiti Kebangsaan Malaysia,
43600 Bangi, Malaysia
Email: jumrilyunas@ukm.edu.my

Abstract—A comparative study of the synthesis process of calcium ferrite nanoparticles (CaFe2O4 NPs) using autocombustion method and solgel method is reported. The study is aimed to find an appropriate process with optimum nanoparticles specification. The process was started by mixing the Ca(NO2)3 solution with Fe(NO3)3 in 150mL of distilled water. A chelating agent of citric acid was then added to the mixture. In addition, the mixture was synthesized through sol-gel and auto-combustion method and then calcined at 550°C to obtain powders. The crystalline structure and surface morphology of CaFe2O4 NPs were observed and compared by using XRD and FESEM measurement. The magnetic property of the synthesized particles was analyzed by using VSM. XRF analysis results showed that both methods show an orthorhombic structure of calcium ferrite NPs. The analysis using the transmission electron microscopy (FESEM) and VSM measurement showed that the particle size of CaFe2O4 nanoparticles ranging from 5 to 20 nm was obtained. While the magnetic intensity at saturation condition (Ms) of 31.1emu/g for auto-combustion method was observed with good magnetic properties. This work will give important information about an alternative technique to synthesize the Calcium Ferrite magnetic nanoparticles used for biomedical application.

Keywords—Synthesis; magnetic nanoparticles; CaFe2O4; auto-combustion; sol-gel method.

I. INTRODUCTION (*HEADING 1*)

Calcium Ferrites magnetic nanoparticles ($CaFe_2O_4$-NPs) have attracted great intention in the biomedical field, especially as drug carriers in drug delivery, attributed to their small size, unique physical properties and excellent magnetic property. In the previous reports, the ferrite based magnetic nanoparticles exhibited significant properties including the high magnetic saturation and low coercivity, magnetization [1]. These particle properties change its magnetic property as particle sizes decrease to the nanometer range. The very low particles size enables the particles to become a superparamagnetic material [2,3]. Furthermore, the superparamagnetic ferrites based magnetic nanoparticles can respond sensitively to external magnetic fields [4].

The magnetic nanoparticles found their potential applications in various fields of biomedical application, such as in cell separation and detection [5, 6, 7], to support gene delivery [8] also for cell diagnostic using magnetic resonance imaging [9]. This potential applications shows that the superior property of the magnetic nanoparticles would play important role for the development of the future biomedical

technology especially for the diagnostics, therapy as well as for targeting drugs purposes. Furthermore, the magnetic nanoparticles are also utilized to create localize heating effect for selective cancer cells treatment [10] and to transport drug in blood vessel as magnetic drug carrier in drug delivery application [11] as well.

Various methods to synthesize the magnetic NPs, such as solution combustion [12], sol-gel [13], coprecipitation [14], autocombustion [15], and aerosolization processes [16] have been developed. Among all the methods, sol-gel auto-combustion is found as the simplest method and the most attractive method due to the common process used for the nanoparticles synthesis process. On the other hand the process used a low temperature that could prevent the oxidation of the particles during the process.

Therefore, we report here the auto-combustion process for the synthesis of $CaFe_2O_4$-NPs. In this study, the Calcium Ferrite-NPs were initially synthesized using sol-gel method followed with auto-combustion at 250°C and calcination at 550°C to produce the nano-powder. The physical properties as well as the magnetical properties of the synthesized materials were also studied in detail.

II. MATERIALS AND METHODS

The Calcium Nitride solution (99%), Ferrite Nitride (99.5%), Citric Acid (99.5%) and Ethylene Glycol (99.5%) were used in the process. The reagents were utilized without any additional purification. In addition, there are some equipment's utilized for synthesized $CaFe_2O_4$ NPs we used a hot plate device for proceeding the experiment by place a small magnetic stirring bar in the beaker, then the solution heated at certain temperature, and continuously stirred until it is boiling completely. Moreover, the next step drying the sample, we utilized the oven and set at the required temperature on it. Last step, the heat treatment using a furnace for the calcination process.

The detailed process flow of the NP's synthesis using solgel and autocombustion methods are revealed in Figure 1 and Figure 2, respectively. The Calcium-Ferrites nanoparticles were prepared in two ways namely solgel, and autocombustion methods. The process started by mixing the Calcium Nitrate $Ca(NO_2)_3$ with Ferrite Nitrate $Fe(NO_3)_3$. In the case of sol-gel method, the solution was stirred constantly using magnetic stirrer at 80°C followed with added 6mL of ethylene-glycol. During this process, the solution should convert to the gel, then the dried gel at 80°C.

978-1-5386-5284-8/18 $31.00 © 2018 IEEE

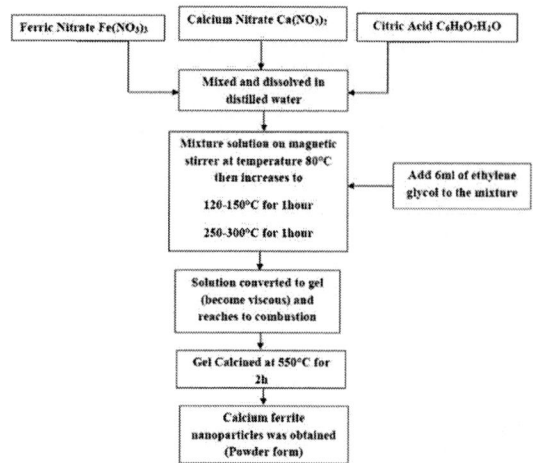

Fig. 1. The schematic of the process flow using auto combustion method for the synthesis of CaFe$_2$O$_4$ NPs.

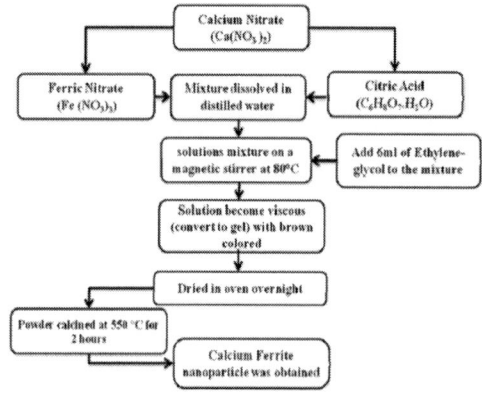

Fig. 2. The Schematic of the process flow using the sol-gel method for the synthesis of the CaFe$_2$O$_4$ NPs.

Moreover, the same steps was used to prepare CaFe$_2$O$_4$ NPs by auto-combustion method, but in this case the gel will be combusted at 250°C, The gel will be burnt completely and finally become powder.

At the final step of the process (for both methods), the powder was dried (in a calcination process at 550°C for 2hr) to remove the water contents and to obtained the nanoparticles product.

III. RESULTS AND DISCUSSION

The physical properties including the structure of the crystallite and the size of the CaFe$_2$O$_4$ NPs samples were determined by an X-Ray Diffractometer (XRD). While the magnetic properties of the powders was examined using a vibrating sample magnetometer (VSM).

The XRD patterns of the CaFe$_2$O$_4$ NPs after calcination at 550 °C is shown in Figure 3(a). The diffraction peaks were attributed to the orthorhombic structure of CaFe$_2$O$_4$ with crystal sizes of approximately 16.8 nm. During the heat treatment, the crystal sizes of the samples increases as the calcination temperature increases, which confirm these results and reported by [17]. This analysis shows that the calcination process play also significant effect on the crystal size of the particles.

The FESEM image shows that the majority of particles synthesized using auto-combustion methods have the size ranging between 5 to 20 nm which is relatively smaller than that synthesized using sol-gel method as reported previously [18-20] (Fig. 3b).

Figure 4 shows the magnetic property of the Calcium-Ferrites NPS synthesized using sol-gel method and its comparison with that using auto-combustion method. The comparison is aimed to see the influence of the particle size to the magnetic property. The thin hysteresis curve line indicates to a superparamagnetic property of the nanoparticles. It can be seen that the coercivity value (H$_c$) of the magnetic field is about 100 Gauss.

It is also shown that the magnetic saturation of auto-combustion sample changes in the magnetic properties attributing to crystallite sizes effect.

Compared to the sample prepared using solgel method it is clearly seen that the synthesis process using autocombustion method produced the nanoparticles having the magnetic saturation M$_s$ value of 31.1emu/g which is about 50% lower than that synthesized using sol-gel method. It can be analyzed that the particle size affect to the response to the external magnetic field, which means that the energy of a magnetic particle in the external field is proportional to its particle sizes. Therefore, the M$_s$ values decrease as particle sizes decreases because of smaller surface effects that is resulted from the finite-size scaling of nanocrystallites [19]. This observation shows that there is strong correlation between the physical property and the magnetic property of the magnetic nanoparticles

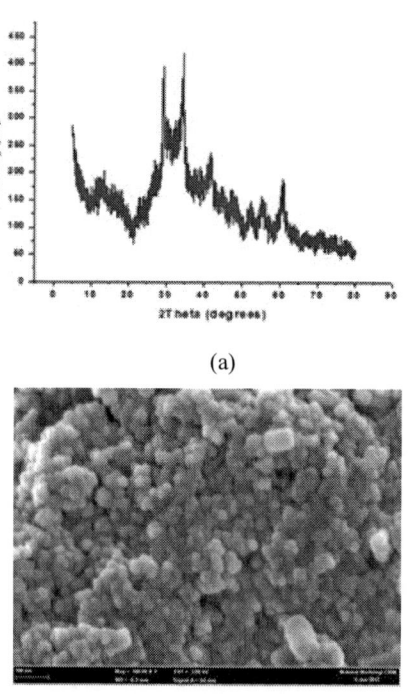

(a)

(b)

Fig. 3. (a) XRD Pattern for CaFe2O4 nanoparticles calcined at 550 °C, (b) FESEM image of the synthesized magnetic NPs

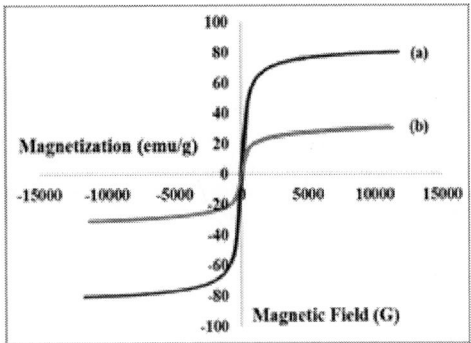

Fig. 4. Hysteresis curve of calcium ferrite nanoparticles, (a) synthesized using standard sol-gel method, (b) synthesized using auto-combustion.

IV. CONCLUSION

The calcium ferrite magnetic nanoparticles have been synthesized using autocombustion method and compared with solgel method. The structure, the physical and magnetic properties of the synthesized nanoparticles have been analyzed using XRD, FESEM and VSM measuremet. The process comparison is aimed to see the strength and weakness of the process and to find appropriate method to synthesized magnetic nanoparticles. The characterization results shows that the orthorhombic structure with crystal size of approximately 16.8 nm was observed. The FESEM results showed that the synthesized NPs have the particle size ranging from 5 to 20 nm. While the VSM results revealed that the calcium ferrite nanoparticles synthesized using autocombustion method exhibit the magnetic saturation of 31.1emu/g, which are 50% lower than other method. Both method show a nanoparticles product having superparamagnetic property. It can be concludd that the autocombustin method is then preferred method due to the simplicity of the process. The study on the process method for the synthesis of CaFe2O4 NPs would beneficial for the development of magnetic nanoparticles for various biomedical applications.

ACKNOWLEDGMENT

This project is financially supported under FRGS Project, No. FRGS/1/2016/TK05/UKM/02/2.

REFERENCES

[1] L. Dan, W.Y. Teoh, C. Selomulya, C. Woodward, P. Munroe, and R. Amal," Insight into Microstructural and Magnetic Properties of Flame-Made Γ-Fe$_2$O$_3$ Nanoparticles," Journal of Materials Chemistry, vol. 17, 2007, pp. 4876- 4884.

[2] V. Kumara, A. Ranaa, M.S. Yadavb, and R.P. Pant," Size-induced effect on nano-crystalline CoFe$_2$O$_4$," Journal of Magnetism and Magnetic Materials, vol. 320, 2008, pp. 1729-1734.

[3] M.M. Said, J. Yunas, R.E. Pawinanto, "B.Y. Majlis, Physical PDMS based electromagnetic actuator membrane with embedded magnetic

particles in polymer composite," Sensors and Actuators A., vol. 245, 2016, pp. 85–96.

[4] C.P. Robert, J.M. Idée and M. Port," Recent advances in iron oxide nanocrystal technology for medical imaging," Advanced Drug Delivery Reviews, vol. 58, 2006, pp. 1471-1504.

[5] J.W. Bulte, T. Douglas, B. Witwer, S.C. Zhang, E. Strable, B.K. Lewis," Magnetodendrimers allow endosomal magnetic labeling and in vivo tracking of stemcells," Nat. Biotechnol., vol. 19, 2001, pp. 1141–1147.

[6] O. Olsvik, T. Popovic, E. Skjerve, K.S. Cudjoe, E. Hornes, J. Ugelstad," Magnetic separation techniques in diagnostic microbiology," Clin Microbiol Rev., vol. 7, 1994, pp. 43–54.

[7] M.R. Buyong, F. Larki, M.S. Faiz, A.A Hamzah, J. Yunas and B.Y Majlis," A Tapered Aluminium Microelectrode Array for Improvement of Dielectrophoresis-Based Particle Manipulation," Sensors vol. 15, 2015, pp. 10973-10990

[8] F. Scherer, M. Anton, U. Schillinger, J. Henke, C. Bergemann and A. Kruge," Magnetofection: enhancing and targeting gene delivery by magnetic force in vitro and in vivo," Gene Ther,vol. 9, 2002, pp. 102–9.

[9] Y.X.Wang, S.M. Hussain, G.P. Krestin," Superparamagnetic iron oxide contrast agents:physicochemical characteristics and applications in MR imaging," Eur Radiol., vol. 11, 2001, pp. 2319–31.

[10] M. Johannsen, U. Gneveckow, L. Eckelt, A. Feussner, N. Waldofner, R. Scholz," Clinical hyperthermia of prostate cancer using magnetic nanoparticles: presentation of a new interstitial technique," Int J Hyperthermia, vol. 21, 2005, pp. 637–47.

[11] C. Alexiou, W. Arnold, R.J. Klein, F.G. Parak, P. Hulin, C. Bergemann," Locoregional cancer treatment with magnetic drug targeting," Cancer Res., vol. 60, 2000, pp. 6641–6648.

[12] Z. Zhang, and W. Wang," Solution combustion synthesis of CaFe$_2$O$_4$ nanocrystal as a magnetically separable photocatalyst." Materials Letters, vol. 133, 2014, pp. 212-215.

[13] J. Wan, X. Chen, Z. Wang, X. Yang, Y. Qian," A soft-template-assisted hydrothermal approach to single-crystal Fe3O4 nanorods," J. Cryst. Growth, vol. 276, 2005, pp. 571–576.

[14] Z. Yuanbi, Q. Zumin, J. Huang," Preparation and analysis of Fe3O4 magnetic nanoparticles used as targeted-drug carriers," Chin. J. Chem. Eng., vol. 16, 2008, pp. 451–455.

[15] M. Faraji, Y. Yamini, M. Rezaee," Magnetic nanoparticles: synthesis, stabilization, functionalization, characterization, and applications," J. Iran. Chem. Soc., vol. 7, 2010, pp. 1–37.

[16] E. Ruiz-Hernández, A. Lopez-Noriega, D. Arcos, I. Izquierdo-Barba, O. Terasaki, M. Vallet-Regí," Aerosol-assisted synthesis of magnetic mesoporous silica spheres for drug targeting," Chem. Mater., vol. 19, 2007, pp. 3455–3463.

[17] M. Gharagozlou, "Influence of calcination temperature on structural and magnetic properties of nanocomposites formed by Co-ferrite dispersed in sol-gel silica matrix using tetrakis (2-hydroxyethyl) orthosilicate as precursor," Chemistry Central Journal, vol. 5, 2011, pp. 19.

[18] N.H. Sulaiman, M.J. Ghazali, J. Yunas, "Superparamagnetic calcium ferrite nanoparticles synthesized using a simple sol-gel method for targeted drug delivery," Bio-Med. Mater. Eng., Vol. 26, 2015, pp. S103–S110.

[19] N.H. Sulaiman, M.J. Ghazali, J. Yunas, A. Rajabi, B.Y. Majlis, M. Razali, "Synthesis and characterization of CaFe$_2$O$_4$ nanoparticles via co-precipitation and auto-combustion methods," Ceramics International, vol. 44, 2018, pp. 46–50

[20] N.D.S. Mohallem, L.M. Seara, "Magnetic nanocomposite thin films of NiFe$_2$O$_4$/SiO$_2$ prepared by sol-gel process," Applied Surface Science, vol. 214, 2003, pp. 143-150.

Dynamic Behavior of Condenser Microphone Under the Influence of Squeeze Film Damping

Siti Aisyah Zawawi
Institute of Microengineering
and Nanoelectronics (IMEN)
Universiti Kebangsaan Malaysia
Bangi, Malaysia
aisyahzawawi@gmail.com

Azrul Azlan Hamzah
Institute of Microengineering
and Nanoelectronics (IMEN)
Universiti Kebangsaan Malaysia
Bangi, Malaysia
azlanhamzah@ukm.edu.my

Reena Sri Selvarajan
Institute of Microengineering
and Nanoelectronics (IMEN)
Universiti Kebangsaan Malaysia
Bangi, Malaysia
sreenat90@gmail.com

Burhanuddin Yeop Majlis
Institute of Microengineering
and Nanoelectronics (IMEN)
Universiti Kebangsaan Malaysia
Bangi, Malaysia
burhan@ukm.edu.my

Faisal Mohd-Yasin
Queensland Micro-and
Nanotechnology Centre
Griffith University
Brisbane, QLD, Australia
f.mohd-yasin@griffith.edu.au

Abstract— This paper investigates the effect of squeeze film damping on a micromechanical structure that oscillates due to the pressure distribution on moveable top plate that pair with perforated backplate. The damping characteristics are analytically evaluated to establish the optimum design that increase dynamics performance. The squeezed film damping governing equation, the Reynolds equation, is the fundamental equation of this work. The damping coefficient, air gap damping and viscous damping are calculated by using Skvor's formula. The size of perforation holes and air gap thickness that influencing the squeeze film damping were studied. Analytical solution with various holes structure are summarized. Overall, the 3µm of hole radius, 50µm of holes center-to-center distance and 4µm of air gap thickness were found to be the best in order to reduce squeeze film damping in capacitive microphone.

Keywords—squeeze film damping, viscous damping, perforated structure.

I. INTRODUCTION

Micro-electromechanical (MEMS) oscillating structure operates in surrounding air between the two plates that can change the dynamic characteristics. The basic operation of parallel-plates microstructure is the pressure flow normally against the top plate (moveable plate) increased the air damping between the plates. The bottom plate designed as perforated plate to reduce the acoustic damping [1]. The schematic view of squeeze film damping for parallel-plates microstructure is shown in Fig. 1.

The pressure distribution force flow normally on the surface of moveable diaphragm, the air trapped between the two plates caused the air sucked into the gap, and a significant pressure fields arises into the channel. This phenomenon creates a resistive force on the moveable plate that affect the dynamics characteristics of the condenser microphone. Therefore, the evaluation of squeeze film damping has important roles in determining the device performance. Some methods have been proposed to reduce the squeeze film

damping such as vacuum packaging and perforating the plate. However, squeeze film damping effect remain unchanged in vacuum packaging because of air leakage [2]. Employing holes on the one plate is the best way to reduce the damping effect hence increase the dynamics behavior of microphone.

Some research has been reported on perforated model to address the damping effects. The basic Reynolds equation applied to evaluate the analytical model for perforated plate. Bao et al.[3] proposed a modified Reynolds equation to introduced 'effective damping width' that determine the damping force for circular plates. In addition, an extended of compressibility effect and rarefaction effect were introduced to the perforation model by Pandey et al. [4]. The same researcher presents an improved model by integrating more accurate losses through holes and treating boundary cells and interior cells differently in order to analyze squeeze film damping for rectangular plate with transverse motion [5]. The first squeeze film damping analysis employing perforated circular plates with transverse motion was reported by Li et al. [6]. Another approach to evaluate damping effect was proposed by Mohite et al. and Homentcovschi et al. They separated the whole plate into a few uniform cells which each cell contains one hole. The damping results analyzed by calculating the model for one cell and multiplied with the remaining cells [7]–[9]. Currently, the analysis of squeeze film damping for microphone was studied by Ishfaque et al. [10]. The damping effect was performed by vary the hole radius and air gap thickness.

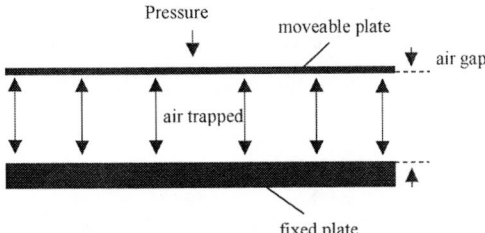

Fig. 1. 2D cross-section of condenser microphone.

In this paper, the square design of MEMS condenser microphone was proposed. This novelty condenser microphone utilizing silicon carbide (SiC) as a moveable diaphragm and silicon (Si) as a substrate (fixed back plate). Silicon carbide (SiC) material has several advantages over other material; it has superior mechanical properties in terms of density, thermal conductivity and Young's modulus [1]. Most of MEMS devices application focused on harsh extrinsic vibration environment that lead to poor performance, failure and crush of the structure. Hence the emphasized of high vibration damping material must be considered in order to protect MEMS devices from mechanical vibrations and shocks [11]. SiC has a high vibration damping due to the higher Young's modulus compare to other common metals and ceramics as summarized in Table 4 of [12]. Moreover, damping effect can be reduced by employing material with large tensile stress [12]. Furthermore, one of the strategy to lower the damping effect is through controlling dissipation due to internal friction by reduce the mobility of defects in material selection [12]. Thus, SiC in a group of brittle ceramic is less defects mobile compared to common metals and alloys [13].

The feature size of the microstructure should be reduced to increase the surface area to volume ratio, hence the squeeze film damping significantly affected [14][15]. At present, numerous microstructure design have been developed to understand the relative importance of damping effect as long as the air gap thickness is smaller than one-third of the plate width [16]. Formation of perforation holes in one of the plate is the best way to reduce the excessive air damping problem. Based on the changes in size of perforation holes on the backplate, the damping effect can detect the dynamics behavior development and important transition in microstructure design.

In this study, the holes size and damping coefficient considered as the fundamental parameter to assess dynamics performance of condenser microphone. The analytical study of air gap damping and viscous damping are analyzed and summarized. Furthermore, the tensile stress of the damping material can be reduced the damping effect of the device. Therefore, the consideration of squeeze film damping plays an important role in order to get optimum dynamic behavior of the design [17].

II. THEORY

A. Squeeze film air damping of parallel plates.

The squeeze film air damping can be reduced by introducing the perforated holes on the back plate. The air flow through the holes on the back plate should be considered as damping resistance which adds to the squeeze film damping. There are two main factors that affect the squeeze film air damping; the air gap damping, R_{ag} that caused the compression of the air film and viscous flow damping, R_{vis} that cause the viscous air flow resistance along holes on the back plate as indicated in Fig. 2.

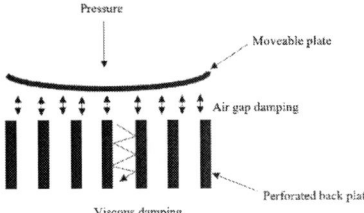

Fig. 2. Schematic view of squeeze film damping between moveable and perforated plates.

An analytical formula in explaining the squeeze film damping for acoustical devices have been simplified by Skvor [18] and relevant to apply for incompressible fluids and neglects any added damping due to the flow through the holes [8]. At present, numerous models have been developed to investigate the film damping incorporating the holes resistance in the linearized Reynolds equation [16].

B. Linearized Reynolds equation for squeeze film air damping of parallel plates.

Theory for the film between two plates in relative motion was first formulated by Osborne Reynolds in more than a century ago [19]. Reynolds equation is a two-dimensional partial differential equation which is combination of the Navier-Stokes equation [4]. This linearized Reynolds equation is only applied for small air gap thickness compared to the plate thickness for microstructure design.

The most nonlinear Reynolds equation for normal motion of parallel plate can be simplified as [16]

$$\frac{\delta}{\delta x}\left(P\frac{\delta P}{\delta x}\right) + \frac{\delta}{\delta y}\left(P\frac{\delta P}{\delta y}\right) = \frac{12\mu}{h^3}\frac{\delta(hP)}{\delta t} \tag{1}$$

or

$$\frac{\delta^2}{\delta x^2}P^2 + \frac{\delta^2}{\delta y^2}P^2 = \frac{24\mu}{h^3}\frac{\delta(hP)}{\delta t} \tag{2}$$

where P is a pressure in the plate, h and μ are the thickness of the plate and viscosity of the fluid respectively.

Equation (1) can be linearized in order to apply for small plate deflection at its equilibrium position ($\Delta h << h_o$ and $P << P_a$) as [20]

$$\tag{3}$$

$$P_a \left(\frac{\delta^2 P}{\delta x^2} + \frac{\delta^2 P}{\delta y^2} \right) - \frac{12\mu}{h_o^2} \frac{\delta P}{\delta t} = \frac{12\mu P_a}{h_o^3} \frac{dh}{dt}$$

where h_o and P_a are the initial thickness of the plate and ambient pressure respectively. However, the deflection of the diaphragm greatly depends on the material's properties such as Young's Modulus, Poisson's ratio and prestress [21].

The deflection of the moveable diaphragm due to pressure distribution caused the change in capacitance which are converted to the electrical performance of the microphone. The open circuit sensitivity is generally defined as [22]

$$S = S_e . S_m \tag{4}$$

$$S = \frac{V_b}{g} . \frac{\Delta g}{\Delta P} \tag{5}$$

where S_e and S_m are electrical and mechanical sensitivity respectively, while V_b, g, Δg and ΔP are bias voltage, air gap thickness, air gap changes and pressure changes respectively. The mechanical sensitivity determined from the deflection of the diaphragm by the various given pressure. Recently, the standard requirement of most commercial MEMS microphone these days is about -38 dBV/Pa (0.0126 volts/Pa) [22].

C. Squeeze film air damping of perforated plate.

Perforated holes introduced in the back plate to reduce the excessive air damping that trapped between the plates. Moreover, the most important factors that affect the perforated back plate are to reduce the squeeze-film damping effect and enhance the etching of underlying sacrificial layers in the microfabrication process [8].

The total flow resistance between plates determined by air gap resistance, R_{ag} and viscous flow resistance of the perforation back plate, R_{vis}. These damping resistances can be expressed using Skvor's formula as [23], [24]

$$R_{ag} = \frac{12\mu}{n\pi g^3} B(Ar) \tag{6}$$

$$R_{vis} = \frac{8\pi \mu L}{n A^2} \tag{7}$$

where μ = viscosity of air ($18.6 \times 10^{-6}\ Ns/m^2$), L = hole length (back plate thickness), n = number of holes, g = air gap thickness and A = individual hole area. The coefficient of the effective back plate area is given by

$$B(Ar) = \frac{1}{4}\ln\left(\frac{1}{Ar}\right) - \frac{3}{8} + \frac{1}{2}(Ar) - \frac{1}{8}(Ar)^2 \tag{8}$$

where (Ar) is the ratio of hole area to closed area. The air gap damping as in (6) was derived with an assumption that the air pressure at the inner edge of the hole is zero. The viscous air resistance should be taken into consideration due to the pressure change in and out through the definite length of back plate thickness. The air resistance through the perforation holes can be simply modeled with Hagen Poiseuille viscous flow as in (7) [16].

The overall damping coefficient can be derived by modifying (6) and (7) and can be rewritten as:

$$B_k = \left[\frac{1}{4}\ln\left(\frac{0.319b^2}{r_h^2}\right) - \frac{3}{8} + \frac{r_h^2}{0.638b^2} - \frac{r_h^4}{0.815b^4} \right] \tag{9}$$

where b is holes center-to center distance and r_h is hole radius. The damping coefficient greatly depends on the dimensions and distance between the perforation holes in the back plate.

D. Controlling temperature and pressure for squeeze film air damping.

Other than characteristics mentioned above, the magnitude of the damping is controlled by temperature and pressure as well, as listed in Table 1 of [12]. Sun et al. revealed that the squeeze air damping effect is very sensitive to ambient pressure between 1 torr to 10 torr for MEMS resonator due to the natural frequency of damping model change when varying the pressure [25]. Meanwhile, the ambient temperature is not as influence as the effects introduced from the gap thickness and ambient pressure [25]. However, the temperature is an important parameter as the condenser microphone are applied in a zone of high temperature gradients that effect the stiffness of moveable plate.

III. RESULT AND ANALYSIS

The damping coefficient as a function of b and r_h is plotted in Fig. 3. The hole radius are varied from 3μm to 8μm and holes center-to-center distance are varied from 20μm to 50μm. It is observed that the air gap damping coeffieient increases with the increase of hole center-to-center distance. However, the air gap damping coefficient reduced with the increase of hole radius due to the decreases of effective plate area. In this case, the pressure distribution affected the air flow between the plates hence the system undergo underdamping [26].

As can be seen, the air damping coefficient consistantly increased with decreasing of hole radius. The smallest hole radius (3μm) gives 0.32 and 0.75 of damping coefficient when holes center-to-center distance are 20μm and 50μm respectively. The largest hole radius (8μm) gives lower damping coefficient as 0.02 and 0.29 when holes center-to-center distance are 20μm and 50μm respectively. As discussed earlier, the higher value of the air damping coefficient, the better the flow resistance between plates hence achieved optimum capacitance performance of the device [27]. From the plotted graph in Figure 3, the maximum air damping coefficient is at 0.72 when hole radius at 3μm and holes center-to-center at 50μm.

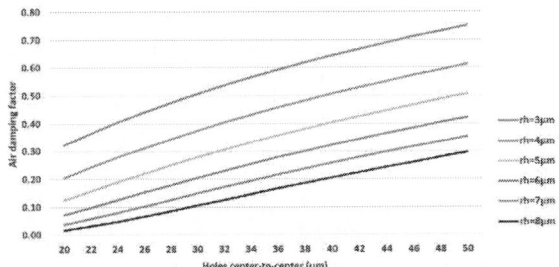

Fig. 3. Air damping coefficient versus holes center-to-center distance at various hole radius.

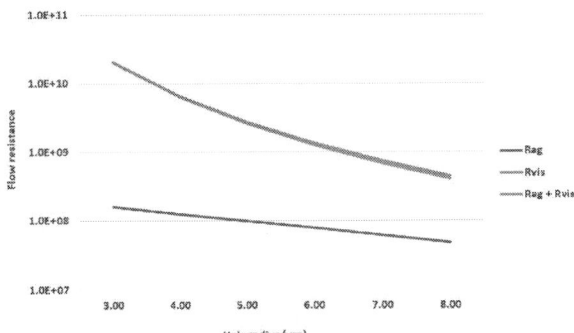

Fig. 4. Total backplate flow resistance as a function of hole radius.

Based on (6) and (7), the consideration of hole radius and perforation hole area on a backplate play an important role in order to monitor the air damping resistance between two parallel plates. However, the number of holes, n was fixed due to less significant compared to location of acoustics holes on the microphone performance [28]. Total backplate flow resistance can be describe as a sum of air gap damping, R_{ag} and viscous flow resistance, R_{vis} along perforation holes on the backplate [23], [24]. The result for backplate resistance as a function of hole radius for squeeze film damping are depicted in Fig. 4.

The decrease of backplate flow resistance as the hole radius increases is mainly due to decreasing of air gap damping and viscous flow damping. The flow resistance by R_{ag} is obviously less compared to R_{vis}. Therefore, it reveals the total backplate flow resistance value is closed to the value of R_{vis}. Hence, the viscous flow resistance had the greatest influence on backplate resistance followed by air gap damping.

As in (5) calculated, when the mechanical sensitivity gets fixed, the open circuit sensitivity is greatly depending on the electrical sensitivity of the microphone. This represents the increase of open circuit sensitivity as the result of decreasing the air gap thickness. However, decreasing of air gap thickness tend to increase the air gap damping, R_{ag} as plotted in Fig. 6. As previously discussed in Figure 4, the total damping resistance highly depends on the viscous damping compared to air gap damping. Therefore, it can be concluded that the lower air gap thickness is selected to achieve the higher open circuit sensitivity of the microphone.

Fig. 6. Air gap flow resistance at various air gap thickness.

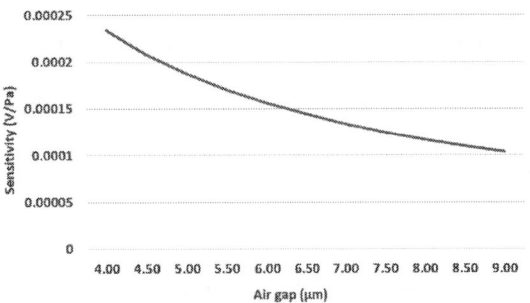

IV. CONCLUSION

The dynamics behavior of condenser microphone has been presented. The analytical results of air gap thickness, center hole-to-hole distance and hole radius are strongly affect the squeeze film damping for two parallel plates. Results obtained that 3µm of hole radius at 50µm of holes center-to-center give the higher damping coefficient and higher air gap resistance (R_{ag}). However, the air gap resistance gives some effect on squeeze film damping compared to viscous resistance. The air gap resistance showed the significant influence on air gap thickness since lower air gap thickness give the higher sensitivity of the microphone. It was found that 4µm of air gap thickness give $1.59 x 10^8 \, Ns/m$ of R_{ag} and $0.00023 \, V/Pa$ of sensitivity.

ACKNOWLEDGMENT

The authors would like to acknowledge the research grant provided by the IRU-MRUN Collaborative Research Programme for this project (UKM's Grant Number MRUN-2015-004, Griffith's Project Number 218997). This work was performed in part at the Queensland node of the Australian National Fabrication Facility, a company established under the National Collaborative Research Infrastructure Strategy to provide nano- and micro-fabrication facilities for Australia's researchers.

DPP-2018-006. Dana Pembangunan Penyelidikan PTJ. Peranti bioperubatan dengan IOT.

PRGS/1/2017/TK05/UKM/02/1. Skim Geran Penyelidikan Pembangunan Prototaip (PRGS). Design and Fabrication of Silicon Membrane Filtration System for Artificial Kidney.

REFERENCES

[1] S. A. Zawawi, A. A. Hamzah, F. Mohd-Yasin, and B. Y. Majlis, "Mechanical performance of SiC based MEMS capacitive microphone for ultrasonic detection in harsh environment," in *Proceedings of SPIE - The International Society for Optical Engineering*, 2017, vol. 10354.

[2] Y. Mo, H. Zhou, G. Xie, and B. Tang, "Investigation of air damping effect in two kinds of capacitive MEMS accelerometers," *Microsyst. Technol.*, 2017.

[3] M. Bao, H. Yang, and Y. Sun, "Modified Reynolds' equation and analytical analysis of squeeze-film air," vol. 795, 2003.

[4] A. K. Pandey, R. Pratap, and F. S. Chau, "Analytical solution of the modified Reynolds equation for

978-1-5386-5284-8/18 $31.00 © 2018 IEEE

squeeze film damping in perforated MEMS structures," *Sensors Actuators, A Phys.*, vol. 135, no. 2, pp. 839–848, 2007.

[5] A. Kumar and P. Æ. Rudra, "RESEARCH PAPER A comparative study of analytical squeeze film damping models in rigid rectangular perforated MEMS structures with experimental results," pp. 205–218, 2008.

[6] P. Li, Y. Fang, and F. Xu, "Analytical modeling of squeeze-film damping for perforated circular microplates," *J. Sound Vib.*, vol. 333, no. 9, pp. 2688–2700, 2014.

[7] S. S. Mohite, H. Kesari, V. R. Sonti, and R. Pratap, "Analytical solutions for the stiffness and damping coefficients of squeeze films in MEMS devices with perforated back plates," *J. Micromechanics Microengineering*, vol. 15, no. 11, pp. 2083–2092, 2005.

[8] D. Homentcovschi and R. N. Miles, "Viscous damping of perforated planar micromechanical structures," vol. 119, no. October 2004, pp. 544–552, 2005.

[9] D. Homentcovschi and R. N. Miles, "Viscous microstructural dampers with aligned holes: design procedure including the edge correction.," *J. Acoust. Soc. Am.*, vol. 122, no. 3, p. 1556, 2007.

[10] A. Ishfaque and B. Kim, "Analytical modeling of squeeze air film damping of biomimetic MEMS directional microphone," *J. Sound Vib.*, vol. 375, pp. 422–435, 2016.

[11] N. Choudhary and D. Kaur, "Vibration Damping Materials and Their Applications in Nano/Micro-Electro-Mechanical Systems: A Review," *J. Nanosci. Nanotechnol.*, vol. 15, no. 3, pp. 1907–1924, 2015.

[12] S. Joshi, S. Hung, and S. Vengallatore, "Design strategies for controlling damping in micromechanical and nanomechanical resonators," *EPJ Tech. Instrum.*, vol. 1, no. 1, p. 5, 2014.

[13] V. B. (Vladimir B. Braginskiĭ, V. P. Mitrofanov, V. I. (Vladimir I. Panov, K. S. Thorne, and C. Eller, *Systems with small dissipation.* University of Chicago Press, 1985.

[14] J. B. Starr, "Squeeze-Film Damping in Solid State Accelerometers.pdf," *Solid-State Sensor and Actuator Workshop.* pp. 44–47, 1990.

[15] T. Veijola, "Equivalent Circuit Models for Micromechanical Inertial Sensors," no. January, 1999.

[16] M. Bao and H. Yang, "Squeeze film air damping in MEMS," *Sensors Actuators, A Phys.*, vol. 136, no. 1, pp. 3–27, 2007.

[17] C. Bai and J. Huang, "Improving the validity of squeeze film air-damping model of MEMS devices with border effect," *J. Appl. Math.*, vol. 2014, 2014.

[18] J. S. Krause, R. D. White, M. J. Moeller, J. M. Gallman, and R. De Jong, "MEMS Pressure Sensor Array for Aeroacoustic Analysis of the Turbulent Boundary Layer on an Airplane Fuselage," *Une*, vol. 13, p. 15, 2016.

[19] W. E. Langlois, "Isothermal squeeze films," p. 50, 1961.

[20] J. J. Blech, "On Isothermal Squeeze Films," *J. Lubr. Technol.*, vol. 105, no. 4, p. 615, 1983.

[21] N. Marsi, B. Y. Majlis, F. Mohd-Yasin, and A. A. Hamzah, "The fabrication of back etching 3C-SiC-on-Si diaphragm employing KOH + IPA in MEMS capacitive pressure sensor," *Microsyst. Technol.*, vol. 21, no. 8, pp. 1651–1661, 2014.

[22] B. Kim and H. Lee, "Acoustical-Thermal Noise in a Capacitive MEMS Microphone," vol. 15, no. 12, pp. 6853–6860, 2015.

[23] H. Gharaei and J. Koohsorkhi, "Design and characterization of high sensitive MEMS capacitive microphone with fungous coupled diaphragm structure," *Microsyst. Technol.*, vol. 22, no. 2, pp. 401–411, 2016.

[24] D. T. Martin, J. Liu, K. Kadirvel, R. M. Fox, M. Sheplak, and T. Nishida, "A micromachined dual-backplate capacitive microphone for aeroacoustic measurements," *J. Microelectromechanical Syst.*, vol. 16, no. 6, pp. 1289–1302, 2007.

[25] S. Sun, C. Chung, C. Hsu, and J. Kuang, "The Squeeze Film Damping Effect on Dynamic Responses of Micro-electromechanical Resonators," vol. 287, pp. 1961–1965, 2013.

[26] A. Ishfaque and B. Kim, "Squeeze film damping analysis of biomimetic micromachined microphone for sound source localization," *Sensors Actuators, A Phys.*, vol. 250, pp. 60–70, 2016.

[27] A. Ranjbar, M. T. Mehrabani, and F. T. Pary, "A numerical study on the viscous damping effect for a condenser microphone," *IEEE Sens. J.*, vol. 11, no. 6, pp. 1307–1316, 2011.

[28] C. W. Tan, Z. Wang, J. Miao, and X. Chen, "A study on the viscous damping effect for diaphragm-based acoustic MEMS applications," *J. Micromechanics Microengineering*, vol. 17, no. 11, pp. 2253–2263, 2007.

Optimization of p-type Emitter Thickness for GaSb-Based Thermophotovoltaic Cells

W. Emilin Rashid, Pin Jern Ker*, M. Z. Jamaludin,
N. A. Rahman and Mahdi All Khamis
Institute of Power Engineering,
Universiti Tenaga Nasional,
Jalan IKRAM-UNITEN,
43000 Kajang, Selangor, Malaysia
*pinjern@uniten.edu.my

Abstract—**Thermophotovoltaic (TPV) cells that convert thermal heat directly into electricity are attracting attention as they potentially produce high output power densities. Owing to its capability to convert with a Carnot efficiency, an optimization of these cells is essential to further enhance their performance and efficiency. This paper focuses on the optimization of p-type emitter thickness of Gallium Antimonide (GaSb) based TPV cell using Silvaco TCAD simulation software. The simulation works in this paper were validated by having a good agreement with those from the experimental work in terms of the electrical characteristics and efficiency of the GaSb TPV cell. Further simulation was done with different p-type emitter thicknesses ranging from 0.15 μm to 1.20 μm. It was demonstrated that the open circuit voltage (V_{oc}) of the cell increases while the short-circuit current density (J_{sc}) decreases with increasing p-type emitter thickness. Since the rate of increasing V_{oc} is faster than that of decreasing J_{sc}, higher maximum power efficiency was obtained at an optimum thickness of 0.85 μm. It was found that, under AM1.5 illumination condition, an increment of power efficiency from 5.91 % to 6.63 % was achieved when increasing p-type emitter thickness from 0.15 μm to 0.85 μm.**

Keywords— *emitter thickness; Gallium Antimonide; thermophotovoltaic*

I. INTRODUCTION

Thermophotovoltaic (TPV) device can be utilized for clean energy conversion from thermal heat into electricity. Recently, the advances in semiconductor materials eventually broaden up the opportunity for the researchers to explore on the use of TPV cells in a wide range of applications. In principle, TPV cells could operate at their optimum efficiency in the condition where the TPV semiconductor energy bandgap spectrally matched to the blackbody spectrum generated by the heat source [1].

Since TPV cells are ideally designed to convert radiation heat from the infrared region, a semiconductor material with a higher cut-off wavelength which corresponds to a lower bandgap energy (typically <0.75 eV) is more desirable to be used as the material for the fabrication of TPV cells [2]. Therefore, research efforts have been focused on group III-V semiconductor materials as they have the possibility of growing lattice-matched compound with bandgap energies ranging from 0.53 to 0.73 eV [3].

The authors gratefully acknowledge the Tenaga Nasional Berhad (TNB) Seeding fund (Project code: U-TG-RD-18-04) and the UNITEN Internal Grant (J510050710) for the access to the simulation software.

At present, Gallium Antimonide (GaSb) material is often regarded as one of the most ideal choices for fabricating an infrared device. This is because of its low bandgap energy (~0.72 eV) which allows the material to absorb photons at longer wavelengths (up to 1720 nm) from the blackbody radiation spectrum. The conventional ways of fabricating GaSb cells are either by Zinc (Zn) diffusion [4] or epitaxial growth [5] method. Numerous researches [6, 7, 8] have been focusing on improving the GaSb cell performance by optimizing various parameters such as doping concentration, layer thickness, and operating temperature. This paper will only focus on the effect of p-type emitter layer thickness of GaSb TPV cells.

Most of the literatures reported that a thin layer of p-type emitter for their GaSb TPV cells was found to be optimum in producing good quantum efficiency. In these studies, the optimum emitter thickness was investigated based on the factors of internal quantum efficiency (IQE), external quantum efficiency (EQE), short-circuit current density (J_{sc}) and open-circuit voltage (V_{oc}). Table I shows the range of p-type emitter thickness studied in previous literatures, their optimum thicknesses as well as the factors they considered for the chosen optimum thicknesses.

TABLE I.   TYPICAL FACTORS CONSIDERED IN CHOOSING THE OPTIMUM P-TYPE EMITTER THICKNESS

Range of emitter thickness studied (μm)	Optimum thickness (μm)	Factors considered	References
0.2 to 0.60	0.40	EQE and V_{oc}	[9]
0.0 to 0.40	0.23	J_{sc} and V_{oc}	[5]
0.05 to 0.30	0.18	IQE	[7]
0.05 to 0.10	0.05	IQE	[10]

However, to date, the relationship between the p-type emitter thicknesses of GaSb TPV cell and the maximum power efficiency was not fully elucidated. In this paper, a comprehensive study and detailed analysis were carried out for optimizing p-emitter thickness of GaSb cell. The objective of this work is to investigate the effect of p-emitter layer thickness of GaSb TPV cell on its maximum power efficiency using Silvaco TCAD software as a simulation tool.

II. METHOD OF SIMULATION

Silvaco TCAD software was used as a simulation tool due to its ability to predict the electrical behavior of specified semiconductor material. The software consists of a Virtual Wafer Fabrication package which includes ATLAS, BLAZE and Dev Edit modules that allow the user to easily optimize various of parameters on their semiconductor device prior to the fabrication process [11]. In this study, the GaSb TPV cell structure was modeled using Dev Edit interface and further analysis on its electrical characteristic with different p-type emitter layer thicknesses was performed using the ATLAS module.

A. The GaSb-based TPV Cell Structure

Fig. 1 shows the schematic diagram of the GaSb TPV cell constructed using Dev Edit interface. The structure modeled is mostly similar to that of the experimental work done in [4]. In their experiment, a Zn diffusion method was used to form the p-type emitter on the GaSb n-type substrate. Since a Zn diffusion depth of ~0.25 μm was achieved, a thickness of 0.25 μm p-type emitter was used in this simulation.

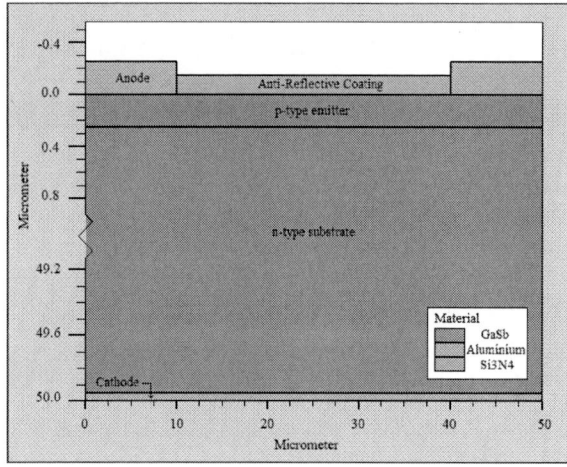

Fig. 1. Schematic diagram of GaSb TPV cell.

The doping concentrations of acceptors (p-type) and donors (n-type) used for our model structure were 2×10^{19} cm^{-3} and 3×10^{17} cm^{-3} respectively. Since the n-type layer was highly doped, the electric field was formed near the p-n junction. In this case, a thick n-type substrate would not be necessary as it doesn't cause significant difference in the simulation results. In this work, 50 μm of n-type substrate was used. Besides, a 0.15 μm thick Si$_3$N$_4$ anti-reflective coating was included in the structure which was similarly deposited to the experimental cell structure done in [4].

B. Material Parameters of GaSb for Simulation

Precise and reliable GaSb material parameters are needed to ensure the validity of the simulation results. Previous studies in [12] and [13] have successfully developed a reliable set of GaSb material parameters as an input for the semiconductor simulation software. In this work, a set of the GaSb material parameters were determined and the values were within the reported range by several literatures. Table II summarizes the material parameters of GaSb cell used in this simulation at an operating temperature of 300 K.

TABLE II. LIST OF GaSb PHYSICAL PARAMETERS USED IN THIS WORK AND THEIR VALUES REPORTED IN OTHER LITERATURES

GaSb parameter	Numerical value	Reference	Value used in this work
Intrinsic carrier concentration	1.7×10^{12} cm^{-3}	[12]	1.4×10^{12} cm^{-3}
	1.405×10^{12} cm^{-3}	[13]	
	1.4×10^{12} cm^{-3}	[7], [14]	
Electrons density of states	5.68×10^{18} cm^{-3}	[15]	2.1×10^{17} cm^{-3}
	2.1×10^{17} cm^{-3}	[16]	
Holes density of states	2.95×10^{18} cm^{-3}	[15]	1.8×10^{19} cm^{-3}
	1.8×10^{19} cm^{-3}	[16]	
Shockley-Read-Hall lifetime electrons	10×10^{-9} s	[12], [7], [13], [14]	10×10^{-9} s
	1×10^{-9} s	[15], [16]	
Shockley-Read-Hall lifetime holes	600×10^{-9} s	[12], [7], [13], [14]	600×10^{-9} s
	1×10^{-9} s	[15], [16]	
Electrons mobility	6600 cm^2/Vs	[12], [7], [14]	6600 cm^2/Vs
	5650 cm^2/Vs	[13]	
Holes mobility	1250 cm^2/Vs	[12], [7], [14]	1250 cm^2/Vs
	875 cm^2/Vs	[13]	
Auger coefficient	5×10^{-30} cm^6/s	[12], [7], [13], [14]	5×10^{-30} cm^6/s
Permittivity	14.4	[15], [16]	14.4
Affinity	4.06 eV	[15], [16]	4.06 eV

Another important parameter that shall be considered is the absorption coefficient ($\propto$) of the GaSb material. The $\propto$ determines the percentage of absorbance and the penetration depth of the photons at a particular wavelength in the material. In Silvaco, the absorption coefficient of a semiconductor material is given by equation (1) where λ is the wavelength of the incoming photon and k is the imaginary part of the optical index of refraction for GaSb material.

$$\propto = \frac{4\pi}{\lambda} k \qquad (1)$$

From this equation, it can be clearly seen that the k values are needed as an input parameter to Silvaco simulation. Therefore, the k values used in this work was obtained from a study reported in [17].

C. Model Validation

In order to increase the accuracy of the simulation results, several physical models were used and the Newton numerical solution method was utilized. The physical models used were the Shockley-Red-Hall (SRH) recombination, Auger (AUGER) recombination as well as optical recombination (OPTR) models. More detailed information about these physical models can be found in [11]. The performance analysis was carried out under a standard irradiance spectrum with an air mass of 1.5 (AM1.5) illumination condition. The simulation results obtained in this work were compared to the experimental work done in [4] for validation purposes.

TABLE III. COMPARISON OF PERFORMANCE PARAMETERS

Performance Parameter	Experiment [4]	JX Crystal Inc (JXC) [4]	Present work
J$_{sc}$ (mA/cm^2)	29.0	32.3	26.76
V$_{oc}$ (V)	0.281	0.326	0.316
Efficiency, η (%)	3.90	5.50	6.19

978-1-5386-5284-8/18 $31.00 © 2018 IEEE   110

Table III shows the performance comparison between the experimental work done in [4] and the present work. The performance parameters were extracted when the p-type emitter thickness of GaSb TPV cell was at 0.25 μm. It can be seen that the J_{sc}, V_{oc} and η of the proposed model are comparable to those of both experimental work and JXC's GaSb cell, thus confirms the validity of the present work. Since the GaSb model structure was created in Dev Edit interface, the model developed follows an epitaxially grown fabrication's steps. Hence, the simulated results in the present work indicates a smaller J_{sc} value compared to others and a slight increment in η can be observed. As claimed in [5], epitaxially-grown GaSb cells produced better performance than the Zn-diffused GaSb cells.

D. Optimization of p-type Emitter Thickness of GaSb TPV Cell

The validation result shows a close agreement between experimental work and present work, thus validates the set of parameters used in the simulation. Further optimization of the p-type emitter thickness was carried out using thicknesses ranging from 0.15 μm to 1.20 μm. The electrical characteristics and performance of the modeled structure were analyzed for each thickness. The doping concentrations, n-type substrate thickness as well as the physical parameters remained constant throughout the optimization.

III. RESULTS AND DISCUSSIONS

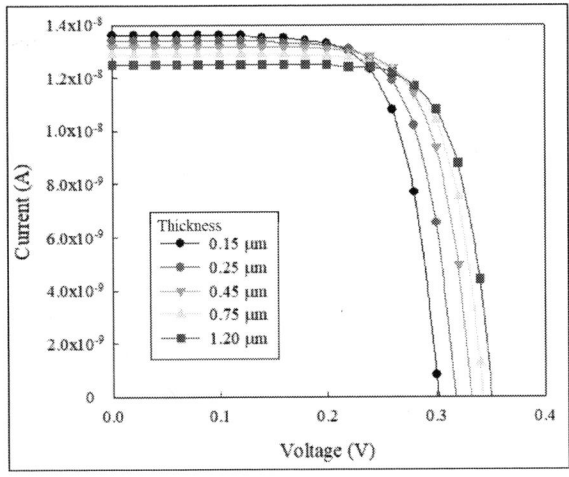

Fig. 2. Current-Voltage characteristic for different p-type emitter GaSb thickness.

Fig. 2 shows the current-voltage characteristic at different p-type emitter GaSb layer thicknesses. The graph clearly shows that increasing the p-type emitter thickness would greatly increase the V_{oc} but slightly reduce the short circuit current of the GaSb TPV cell. These results are similar to those reported by [12] and [14]. Both claimed that the increase in V_{oc} correlates to the decreasing dark current in the GaSb cell. The dark current is commonly defined as the sum of diffusion current defects and generation-recombination current occurs in the depletion region of the p-n junction. As stated in [9], a thick emitters are advantageous as they decrease the dark current in the device due to the influence of the infinite surface recombination at the interface.

In terms of the photocurrent, increasing the thickness of cell layer could intentionally cause an increase in series resistance at the p-n junction interface thus decreasing the short circuit current [18]. Another reason is, thicker p-type emitter layer tends to reduce the spectral response due to higher free-carrier absorption. Additionally, the increase in sensitivity of the front surface minority carrier recombination by thickening the layer would also contribute to the lower I_{sc} generated [19]. Therefore, it is clear that the variation of p-type emitter thickness could affect many crucial parameters which determine the device performance.

To date, most of the optimization of p-type emitter thickness based upon the factors of IQE and EQE. These research studies claimed that thinning of the p-type emitter thickness tends to enhance the quantum efficiency of the device [14, 15, 16]. However, the quantum efficiency only relates to the photocurrent generation by the TPV cell. As per definition, it is the ratio of the number of carriers collected by the TPV cell to the number of photons of a given energy incident on the cell. While the quantum efficiency of the TPV cell is important, thinning the p-type emitter layer would eventually decrease the V_{oc} of the device which contributes to lower power output. Hence, the optimization of p-type emitter of GaSb layer thickness should not be done solely based on either the quantum efficiency or V_{oc}. The optimization should instead consider the overall maximum power output or the overall efficiency.

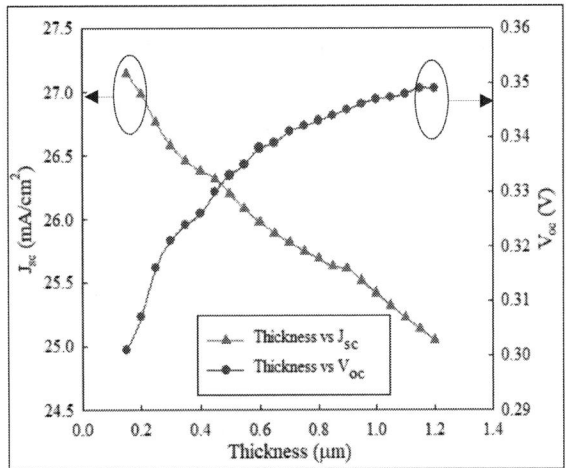

Fig. 3. Effect of p-type emitter thickness on V_{oc} and J_{sc}.

Fig. 3 illustrates the rate of increasing V_{oc} and the rate of decreasing J_{sc} while increasing the p-type emitter thicknesses ranging from 0.15 μm to 1.20 μm. It can be observed that the decreasing rate of J_{sc} was almost constant throughout the range of thicknesses. Whereas, the increasing rate of V_{oc} was higher at the beginning and start to slow down for thicknesses ≥ 0.80 μm. Since the rate of increasing V_{oc} is faster than that of decreasing J_{sc}, the maximum power efficiency would also be increased.

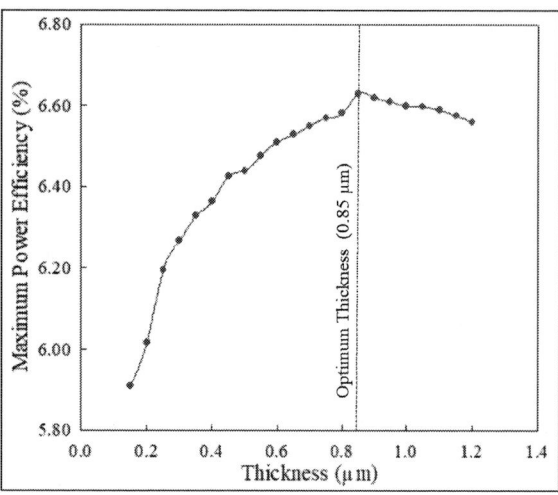

Fig. 4. Effect of p-type emitter thickness on the maximum power efficiency.

Fig. 4 shows the impact of varying the p-type emitter thickness on the maximum power efficiency of the GaSb TPV cell. As can be seen, an increment in power efficiency from 5.91 % to 6.63 % was achieved when the p-type emitter thicknesses were increases from 0.15 μm to 0.85 μm. In this work, the optimal thickness of the p-type emitter was found to be at 0.85 μm, with a maximum power efficiency of 6.63%. This is the optimum thickness where the highest maximum power output can be obtained.

IV. CONCLUSION

The effect of p-type emitter thickness on maximum power efficiency was investigated and presented in this paper. The simulated J_{sc} and V_{oc} values were in a good agreement with those obtained from experimental work and JX Crystal Inc, hence validates the set of parameters used for the GaSb model. A range of p-type emitter thicknesses from 0.15 μm to 1.20 μm were used to analyze the electrical characteristic and performance of the modeled structure. The result shows an increment in maximum power efficiency from 5.91% to 6.63% that is 1% increase in efficiency when the p-type emitter layer thickness increases to 0.85um. It was found that increasing the p-type emitter thickness would eventually increase the V_{oc} of the cell, hence improving the overall power efficiency of the GaSb TPV cell. Instead of considering quantum efficiency as a factor for optimizing the p-type emitter thickness, the results of this work clearly demonstrated that the overall power efficiency is the most suitable performance parameter for the optimization study.

REFERENCES

[1] T. Bauer, *Thermophotovoltaics*. Berlin, Heidelberg: Springer Berlin Heidelberg, 2011.

[2] A. S. Licht, D. F. Demeo, J. B. Rodriguez, and T. E. Vandervelde, "Decreasing Dark Current in Long Wavelength InAs / GaSb Thermophotovoltaics via Bandgap Engineering," *40th IEEE Photovoltaics Spec. Conf.*, pp. 482–486, 2014.

[3] N. Rahimi *et al.*, "Epitaxial and non-epitaxial large area GaSb-based thermophotovoltaic (TPV) cells," *2015 IEEE 42nd Photovolt. Spec. Conf. PVSC 2015*, pp. 2–4, 2015.

[4] L. Tang, H. Ye, and J. Xu, "A novel zinc diffusion process for the fabrication of high-performance GaSb thermophotovoltaic cells," *Sol. Energy Mater. Sol. Cells*, vol. 122, pp. 94–98, 2014.

[5] D. Martín, "Theoretical Comparison between Diffused and Epitaxial GaSb TPV Cells," *AIP Conf. Proc. 738; TPV6 Sixth World Conf. Thermophotovoltaic Gener. Electr.*, vol. 738, no. 1, pp. 311–319, 2004.

[6] L. Tang, L. M. Fraas, Z. Liu, H. Duan, and C. Xu, "Doping Optimization in Zn-Diffused GaSb Thermophotovoltaic Cells to Increase the Quantum Efficiency in the Long Wave Range," *IEEE Trans. Electron Devices*, vol. 64, no. 12, pp. 5012–5018, 2017.

[7] L. Tang, L. M. Fraas, Z. Liu, C. Xu, and X. Chen, "Performance Improvement of the GaSb Thermophotovoltaic Cells with n-Type Emitters," *IEEE Trans. Electron Devices*, vol. 62, no. 9, pp. 2809–2815, 2015.

[8] L. G. Ferguson and L. M. Fraas, "Theoretical study of GaSb PV cells efficiency as a function of temperature," *Sol. Energy Mater. Sol. Cells*, vol. 39, no. 1, pp. 11–18, 1995.

[9] G. Rajagopalan *et al.*, "A Simple Single-Step Diffusion and Emitter Etching Process for High-Efficiency Gallium-Antimonide Thermophotovoltaic Devices," vol. 32, no. 11, pp. 1317–1321, 2003.

[10] H. Ye, Y. Shu, and L. Tang, "The simulation and optimization of the internal quantum efficiency of GaSb thermophotovoltaic cells with a box-shaped Zn diffusion profile," *Sol. Energy Mater. Sol. Cells*, vol. 125, pp. 268–275, 2014.

[11] "Atlas User ' s Manual," no. 408, pp. 567–1000, 2017.

[12] G. Stollwerck, "Characterization and simulation of GaSb device-related properties," *IEEE Trans. Electron Devices*, vol. 47, no. 2, pp. 448–457, 2000.

[13] D. Martin and C. Algora, "Temperature-dependent GaSb material parameters for reliable thermophotovoltaic cell modelling," *Semicond. Sci. Technol.*, vol. 19, no. 8, pp. 1040–1052, 2004.

[14] O. V. Sulima and A. W. Bett, "Fabrication and simulation of GaSb thermophotovoltaic cells," *Sol. Energy Mater. Sol. Cells*, vol. 66, no. 1–4, pp. 533–540, 2001.

[15] K. J. Singh, "Modeling of an efficient Thermo-Photovoltaic (TPV) cell as a power source for space application," no. December, 1970.

[16] B. P. Davenport, "Advanced Thermophotovoltaic Cells Modeling, Optimized for use in Radioisotope Thermoelectric Generators (RTGS) for Mars and Deep Space Missions," *Comput. Eng.*, no. June, pp. 2012–2013, 2004.

[17] R. Ferrini, M. Patrini, and S. Franchi, "Optical functions from 0.02 to 6 eV of AlxGa1-xSb/GaSb epitaxial layers," *J. Appl. Phys.*, vol. 84, no. 8, pp. 4517–4524, 1998.

[18] C. B. Singh, V. Singh, S. Bhattacharya, P. B. Bhargav, and N. Ahmed, "Effect of ZnO:Al Thickness on the Open Circuit Voltage of Organic/a-Si:H Based Hybrid Solar Cells," *Conf. Pap. Med.*, vol. 2013, pp. 1–4, 2013.

[19] L. M. Fraas, J. E. Avery, P. E. Gruenbaum, V. S. Sundaram, K. Emery, and R. Matson, "Fundamental characterization studies of GaSb solar cells," *Conf. Rec. Twenty-Second IEEE Photovolt. Spec. Conf. - 1991*, vol. 1, no. 5, pp. 80–84, 1991.

978-1-5386-5284-8/18 $31.00 © 2018 IEEE

2018 IEEE International Conference on Semiconductor Electronics (ICSE)

Capacitance Effects of Ring Oscillator's Waveform Quality in Designing Physically Unclonable Functions

Zulfikar Zulfikar[1], Norhayati Soin[2], Sharifah Wan Muhamad Hatta[3]
[1]Department of Electrical and Computer Engineering, Syiah Kuala University
Banda Aceh, Indonesia
[1,2,3]Department of Electrical Engineering, Faculty of Engineering, University of Malaya
[2,3]Center of Printable Electronics, University of Malaya
Kuala Lumpur, Malaysia
[1]zulfikarsafrina@unsyiah.ac.id, [2]norhayatisoin@um.edu.my, [3]sh_fatmadiana@um.edu.my

Abstract— **The paper discusses the effect of capacitance changes on the ring oscillator to its output wave for a physically unclonable function (PUF) application. The wave quality generated by the ring oscillator (RO) will largely determine the response of PUF. We have analyzed the resulting RO waveforms to determine the capacitor values and the ideal number of RO stages. The peak voltage analysis of the capacitance change has been done with the help of the Mentor Graphics software package. High performance PTM model transistors (16nm, 22nm, 32nm, and 45nm technologies) are selected to form the RO circuit. Based on the results of the study, we recommend that the number of RO stages is at least 5. Then, we also recommend the ideal capacitance value as a compromise between quality and delay. The capacitances should be 0.7fF, 0.8fF, 0.9fF, and 1fF for 16nm, 22nm, 32nm, and 45nm technologies, respectively.**

Keywords—ring oscillator, challenge time, peak voltage, square wave, PUF

I. INTRODUCTION

The process of identification and authentication is an essential requirement of an electronic device today. The identification process requires an encrypted key, i.e., a private key corresponding to the device identity [1]. In fact, if a cryptographic function can guarantee mathematically calculated security, secure storage is required to store the key itself. Unfortunately, the storage memory is vulnerable to attack, which successfully extracts the privacy data stored [2]. So this is a danger to the security model that is based on a secret. Currently, some security mechanisms have used the concept of Physically Unclonable Functions (PUF) on silicon materials, because internally they have a safety model that attracts attention to apply [3]. The main principle of PUF is to take advantage of self-existing variations in integrated circuit production processes. The difference is random and unpredictable. Therefore, there are no two chips to be the same on the nanoscale. Thus, PUF is suitable to be the digital identity of an integrated circuit or device [3].

The concept of PUF was first proposed in 16 years ago [4]. Since then, the idea of PUF has begun to attract the attention of researchers. Various types of PUF have been developed, and until now the concept has not been applied. Broadly speaking, PUF can be classified as follows: PUF based on memory, PUF based on delay as discussed on [5], physically PUF, and other PUF. But the grouping of the PUF is not final yet, as the scientists are continually looking for new, better PUF models to apply. Here are some models of PUF whose operations are based on delay: Ring Oscillator (ROPUF) [6, 7], and PUF Arbiter.

In its development, researchers have proposed various types of PUF variants. However, there is no single type of PUF as mentioned has stability or reliability 100%. Stability value is one of the absolute parameters to be achieved; otherwise, the PUF cannot be applied. Lately, researchers are increasingly active in researching RO type PUF and Arbiter. This is evidenced by the number of publications published on research on the concept of PUF.

This research will explore the circuit of the ring oscillator (RO) as it is the most promising PUF type today. The wave quality generated by RO will be assessed in various process technologies to find the optimal load capacitance and number of stage.

II. PUF LITERATURE REVIEW

Researchers have developed RO type PUF. The PUF types based on these frequency differences are CRU PUF [8], C-CRO PUF [9], aging resistant RO (AROPUF) [10] and frequency signature base (FSPUF) [11]. The fundamental principle of PUF based on RO has been discovered since 2002 [4] by differentiating the specific characteristics of each device based on its unique frequency. Since then, many researchers have developed the concept of ROPUF. Rahman et. al. Propose the first aging-resistant RO-PUF design which they refer to as ARO-PUF. The PUF over time to produce a more reliable output [10]. Next, an improvement of ROPUF's uniqueness and reliability parameters by manipulating the structure of Ros have been developed. The authors provide an algorithm that can dismiss the systematic variation effect on ROPUF and also develop an algorithm which will enhance the security and reliability of the ROPUF [6].

Other research in RO PUF includes a novel Current controlled CRO (C-CRO) PUF in which inverters of RO uses different logic styles for security improvement. The idea is using static CMOS and Feedthrough logic (FTL). Simulations are carried out in Cadence Virtuoso environment using SPECTRE SPICE by using a library from UMC foundry of CMOS 90 nm process [9]. Then, a crossover RO PUF to improve flexibility and reliability and reduce hardware overheads has also been published. The basic idea is to implement one-to-one input-output mapping with Lookup Table (LUT)-based inter stage crossing structure in each level of inverters [8]. Later than, Barbareschi et al. proposed the Frequency Signature-based PUF (FSPUF), an essential mechanism for authentication and identification of

978-1-5386-5284-8/18 $31.00 © 2018 IEEE

digital devices based on ROs, which is immune to working conditions and aging [11].

III. RING OSCILLATOR

In PUF, the output of RO is connected to a counter via the multiplexer and considered as a clock. The quality of waveform generated by RO should be good enough and able to be detected properly by a counter under any circumstances. This study aims to analyze the factors that affect the quality of the RO output. Analysis of the period change and the minimum voltage at high state will be taking into account as it is determined by the input capacitance and the load capacitance of the next circuit.

Therefore, we will test the effect of variating load capacitance against waveform quality. It might also be directly related to the ideal number of RO stages in various processing technologies. Delay in an RO depends on the size of the transistor, the input capacitance, and the load capacitance [12]. Therefore, the quality of the RO output wave is determined by the load capacitance if it is assumed that the input capacitance was fixed, or vice versa.

Several studies have been conducted regarding the effect of capacitance on the ring oscillator has been recorded. For example, in 2008, an analysis of the ring oscillator on 0.5μm technology was performed. The effect of both input and load capacitance has been simulated [13]. Many similar studies have looked at how far the impact of capacitance on the ring oscillator has been published. A novel RO test structure has been designed to mimic aging effects over a wide range of frequencies [14]. This research concludes that degradation slope cannot reflect the dominant degradation mechanism of RO. Furthermore, the influence of capacitance has also been investigated on PUF, Barbareschi et al. proposed a new mechanism for identifying RO. The author expected that their idea is immune to aging and working condition [11].

RO is often used to measure IC performance on various technologies. In this case, RO is composed of inverters in odd numbers ranging from 3, 5, ... and so on. MOSIS uses 31 stages to compare the speed of the ICs produced. The output of RO will oscillate at the frequency corresponding to equation (1)

$$f = 1 / (2.n.d) \tag{1}$$

where n is the number of inverters and d is the delay of the inverter.

ROs that are used for PUF is slightly different from general usage. It is because the RO needs to be activated within a specific time range. This time is known as *challenge time*. The RO circuit for the PUF application is shown in Fig. 1. The inverter in the first stage is replaced to NAND 2 input gate. One of its inputs is used to enable RO. The generated wave oscillates as long as the signal (*Vin*) exists. The RO delay will be slightly higher due to inverter gate replacement with NAND, but the increase is not significant, especially in circuits with a large number of stages.

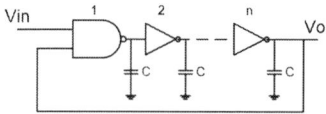

Fig. 1.   RO circuit for PUF application

Fig. 2 shows the output wave of the RO21 circuit (21 stages) when a response is given to the input. The RO, designing under 16nm technology using 5fF capacitance, would oscillate during *Vin* at high conditions and forming 13 pulses during the 25ns challenge time. The wave is approaching the square wave. However, it is complicated to achieve, especially in nanoscale technology and RO circuit with a small number of stages. With the addition of a capacitor will produce an output wave that is closer to the square wave. Therefore, in the following sections, we will examine the effect of added capacitance and the stage change on the delay and waveform generated by RO.

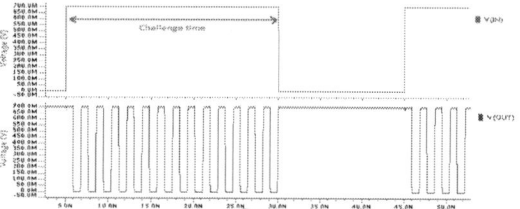

Fig. 2.   RO21 output waveform (16nm technology, and 5fF capacitance) for 25ns challenge time

IV. CAPACITANCE EFFECT ON RO WAVEFORM

In nanoscale technology, capacitors exist by themselves on each transistor [12]. Therefore, in the IC design process, the capacitance is always taken into account. The smaller the capacitance of a circuit, the delay will be lower as well. Therefore, to get a faster circuit, it is necessary to reduce capacitance in addition to resistance. Thus, transistors that are placed in ICs are designed to become smaller in size. However, it does not necessarily lead to reduced delay. In technology smaller than 45nm, it is challenging to realize the transistor (and wire) on the scale.

We tried to analyze the output waveform of RO to the change of capacitance value on 16nm, 22nm, 32nm, and 45nm process technology. The observations are done with the Mentor Graphics simulator package. Transistor model used is PTM high performance. The size of the transistor on each technology has been adapted to that technology. The size of the transistor on the inverter is for nMOS W/L = 4/2, and for pMOS is set to W/L = 8/2. While the size of nMOS and pMOS on NAND is the same, ie, W/L = 8/2. Supply voltage (*Vdd*) is selected according to the recommendation by the PTM library, ie, 0.7V, 0.8V, 0.9V, and 1V for 16nm, 22nm, 32nm, and 45nm technologies, respectively. SPICE program is executed using Eldo, then the output wave is observed using Ezwave program.

Fig. 3 shows the 3-stage RO (RO3) output waveform using 1fF capacitance in 16nm process technology. Visually, the four waves are different and not similar to the square wave. The stream is the result of capture when the output condition has stabilized. The peak voltage (*Vp*) used as the reference in this measurement is the peak voltage of the lowest peak wave. It aims to ensure that the resulting wave is detected by the counter (assumed high) or by another sequence after that. As shown in Fig. 3, the value of *Vp* does not reach *Vdd* = 0.7V voltage under PTM reference.

Further observations were made on the output responses of all RO (RO3, RO5, RO7, RO9, and RO21). Inspection is done to see how far the effect of capacitance change to RO output waveform. Fig. 4 shows the variation output waveform of RO21 in 16nm process technology. When the value of the capacitance is increased, the wave widens (delay increases) and the smaller the defect or spike. In other words, the larger the capacitance, the output wave approaches the square wave.

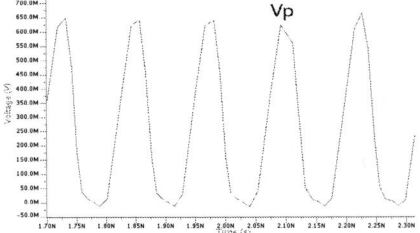

Fig. 3.   The response form is RO3 (16nm technology, and 1fF capacitance) with $Vdd = 0.7V$

Fig. 4(a) shows the output waveform of RO21 with no capacitance value ($C = 0$). Fig. 4(b) shows the RO wave when using 0.5fF capacitance at each stage. While Figs. 4(c) and 4(d) are RO outputs are the results of using 2fF and 5fF capacitances, respectively. Among all the images, the waves in Fig. 4(d) are closest to the square wave. Please note that the capacitance changes are made at all stages. Thus, if we increase the capacitance suppose from 1fF to 2 fF, the total capacitance in the RO21 circuit has changed from 21fF to 42fF.

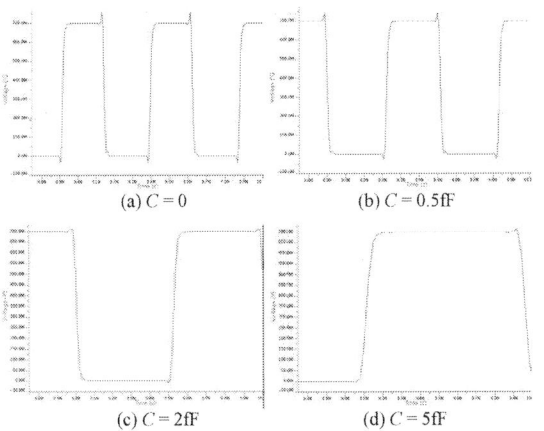

(a) $C = 0$ (b) $C = 0.5fF$

(c) $C = 2fF$ (d) $C = 5fF$

Fig. 4.   The output waveform changes in RO21 versus capacitance on 16nm technology

Next, we analyzed the trend of RO voltage change (Vp) at all stages (3, 5, 7, 9, and 21) and on all technological processes. The observation result shows that the output voltage response of RO3 on all the technology is not close to the Vdd voltage although the capacitance has been raised. The trend is shown in Fig. 5. Note that the Vdd voltage applied here is 0.7V, 0.8V, 0.9V, and 1V for 16nm, 22nm, 32nm, and 45nm technologies, respectively.

As shown in Fig. 5, the voltage value (Vp) of RO3 on the 16nm technology is well below 0.7V. The same thing is also seen in other process technologies, that all Vp is smaller than Vdd. As the capacitance is increased slowly, from $C = 0$ to $C = 8fF$, then Vp becomes even lower. In such conditions, the output waveform of RO3 will not be detected by the counter because it is not considered at high (logic 1). Therefore, we conclude that RO3 cannot be used for PUF applications.

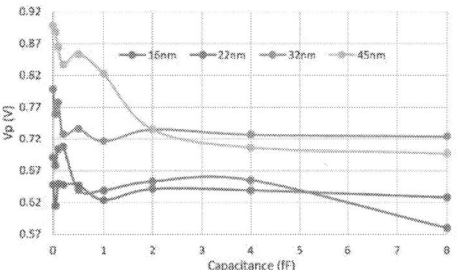

Fig. 5.   The trend of peak voltage change (Vp) of RO3

Furthermore, we analyze the change in Vp along with the increase of capacitance in the circuit other than RO3. Fig. 6 show the pattern of voltage change (Vp) to the increase in capacitance across of RO5, RO7, RO9, and RO21 on the 16nm process technology. When there is no additional capacitor ($C = 0$), Vp is higher than Vdd. Then, the Vp voltage will decrease as the capacitance is increased. The voltage drop is toward (approaching) to the Vdd voltage. The trend of reducing the output voltage of RO21 is better than the other RO. But keep in mind that the total capacitance at RO21 is 21C, undoubtedly more significant than the total capacitance in other RO circuit. The voltage drop pattern of Vp in the circuit other than RO21 is almost the same.

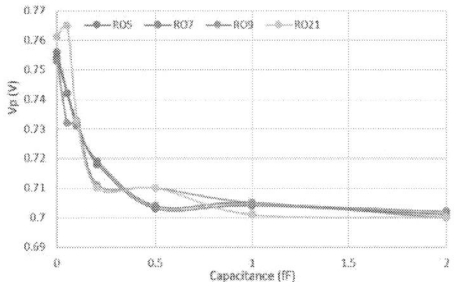

Fig. 6.   Change in peak voltage (Vp) on 16nm technology

As previously explained that measurement of peak voltage (Vp) is done when the output wave is stable. In the initial conditions, when the stream starts to produce, the RO response has not yet formed a perfect wave. The peak voltage (Vp) varies, even on some RO, the output voltage is below 50% of Vdd.

V. CAPACITANCE EFFECT ON RO DELAY

A change in capacitance value will affect the delay of the output waveform RO. In general, with increasing stages on RO, then the delay will also increase. Therefore, particular observations are needed to monitor the increase of delay whether the changes affect the output waveform RO or not. The delay measurements of RO generated waves are done manually using the cursor available in the Ezwave program. The measurement results may also be less accurate because of the uniformity of the waveform especially on RO3 and at the minimum capacitance.

Fig. 7 shows the increasing pattern of the output delay of RO21 as the capacitance value increases in various process technologies. The trend of delay change to the capacitance is linear. The results showed that the highest delay occurred on 16nm technology. However, under conditions where there is no additional capacitance ($C = 0$), the delay of all technologies is almost the same.

978-1-5386-5284-8/18 $31.00 © 2018 IEEE

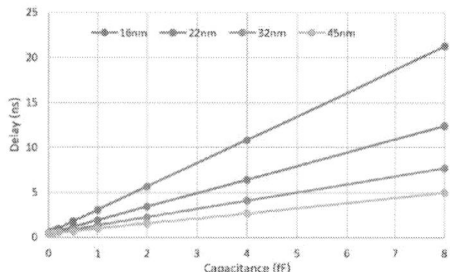

Fig. 7. Delay of the RO21 wave on various technologies

Since the desired waveform is the closest to the square wave shape, larger capacitors are required to produce these waves. But along with the addition of capacitance, then by itself will increase the delay. If the delay increases, then it takes a more prolonged time challenge so that the small frequency difference between RO can be detected adequately. This will slow down the process of identifying devices using PUF as its security system.

Therefore, we try to find the intersection so that the waves generated close to the square wave but with a delay that is not too large. Fig. 8 shows the change of waveforms against capacitance and delay on RO21 in 16nm technology. From the picture, it can be seen that there is a defect (spike) of the resulting waveform so that Vp is much larger than Vdd = 0.7V. Vp value of wave without additional capacitance is highest reaching 760 mV. Then the value of Vp decreases with the addition of capacitance. To get a wave that is very close to square, required a large enough capacitor. However, the delay of RO would also increase. This phenomenon is as seen from the waveform with a capacitance value of 2 fF in Fig. 8. The delay of the wave is twice higher than the delay of those without additional capacitance.

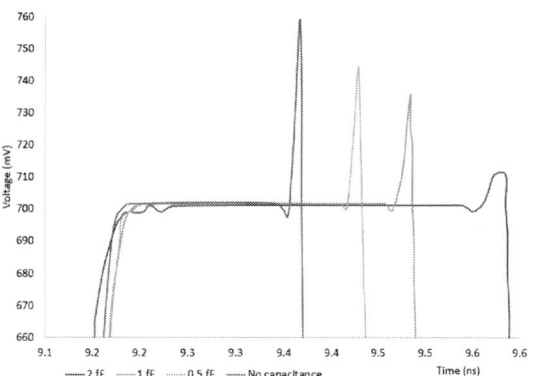

Fig. 8. The curve of peak voltage (Vp) and delay in various capacitance on RO21 (16nm)

Based on the simulation and observation of the intersection between the delay and peak voltage (Vp), we recommend that the minimum stage number of the RO circuit is 5, and a right capacitor size is at least 0.7fF for RO5 on 16nm technology (see Table I for other RO and other technologies). The interesting thing seen from Table I is that the recommended capacitor size is the same for all RO on a particular technology. But the measure does not necessarily indicate that the total capacitance for all RO is the same because the capacitance also depends on the number of stages.

TABLE I.   RECOMMENDED CAPACITOR SIZE AT RO

RO Stage	5	7	9	21
16nm	0.7fF	0.7fF	0.7fF	0.7fF
22nm	0.8fF	0.8fF	0.8fF	0.8fF
32nm	0.9fF	0.9fF	0.9fF	0.9fF
45nm	1fF	1fF	1fF	1fF

VI. CONCLUSIONS

An analysis of the output quality of the ring oscillator due to changes in capacitance at various stages using various process technologies has been performed. Mentor Graphics software packages are used to run SPICE programs regarding high performance PTM model transistors. Observation of peak voltage change and RO output wave delay is done as a result of capacitance change. Based on these observations, the 3-stage ring oscillator is not suitable for PUF applications. Then, the ideal capacitance value should be 0.7fF, 0.8fF, 0.9fF, and 1fF for 16nm, 22nm, 32nm, and 45nm technologies, respectively. These values are chosen for optimal wave delay. The result of this research is the initial study of RO wave quality. Therefore, further research should be conducted to improve wave quality. The challenge time should although be determined further by the quality of wave and security level of PUF.

REFERENCES

[1] European Union. (2013). *Feasibility study on an electronic identification, authentication and signature policy (IAS)*.

[2] L. Guan, J. Lin, B. Luo, J. Jing, and J. Wang, "Protecting private keys against memory disclosure attacks using hardware transactional memory," in *Security and Privacy (SP), 2015 IEEE Symposium on*, 2015, pp. 3-19.

[3] R. Maes, *Physically unclonable functions: Constructions, properties and applications*: Springer Science & Business Media, 2013.

[4] B. Gassend, D. Clarke, M. Van Dijk, and S. Devadas, "Silicon physical random functions," in *Proceedings of the 9th ACM conference on Computer and communications security*, 2002, pp. 148-160.

[5] W. Wang, A. Cui, G. Qu, and H. Li, "A low-overhead PUF based on parallel scan design," in *Proceedings of the 23rd Asia and South Pacific Design Automation Conference*, 2018, pp. 715-720.

[6] F. Kodýtek, R. Lórencz, and J. Buček, "Improved ring oscillator PUF on FPGA and its properties," *Microprocessors and Microsystems*, vol. 47, pp. 55-63, 2016.

[7] F. Kodýtek and R. Lórencz, "A design of ring oscillator based PUF on FPGA," in *Design and Diagnostics of Electronic Circuits & Systems (DDECS), 2015 IEEE 18th International Symposium on*, 2015, pp. 37-42.

[8] Z. Pang, J. Zhang, Q. Zhou, S. Gong, X. Qian, and B. Tang, "Crossover Ring Oscillator PUF," in *Quality Electronic Design (ISQED), 2017 18th International Symposium on*, 2017, pp. 237-243.

[9] S. R. Sahoo, K. S. Kumar, and K. Mahapatra, "A novel current controlled configurable RO PUF with improved security metrics," *Integration, the VLSI Journal*, vol. 58, pp. 401-410, 2017.

[10] M. T. Rahman, F. Rahman, D. Forte, and M. Tehranipoor, "An aging-resistant RO-PUF for reliable key generation," *IEEE Transactions on Emerging Topics in Computing*, vol. 4, pp. 335-348, 2016.

[11] M. Barbareschi, G. Di Natale, L. Torres, and A. Mazzeo, "A Ring Oscillator-Based Identification Mechanism Immune to Aging and External Working Conditions," *IEEE Transactions on Circuits and Systems I: Regular Papers*, vol. 65, pp. 700-711, 2018.

[12] N. Weste and D. Harris, "CMOS VLSI Design: A Circuits and Systems Perspective," ed: Pearson, 2010.

[13] F. Sandoval-Ibarra and E. Hernández-Bernal, "Ring CMOS NOT-based Oscillators: Analysis and Design," *Journal of applied research and technology*, vol. 6, pp. 54-64, 2008.

[14] M.-H. Hsieh, Y.-C. Huang, T.-Y. Yew, W. Wang, and Y.-H. Lee, "The impact and implication of BTI/HCI decoupling on ring oscillator," in *Reliability Physics Symposium (IRPS), 2015 IEEE International*, 2015, pp. 6A. 4.1-6A. 4.5.

Numerical study of Zigzag Micro Mixer with 3D Channel Dimension

Wan Ammar Fikri Wan Ali[1], Azrul Azlan Hamzah[1], Kamarul `Asyikin Mustafa[1,2], Burhanuddin Yeop Majlis[1], Jumril Yunas[1]
[1]Institute of Microengineering and Nanoelectronics, Universiti Kebangsaan Malaysia, Bangi, Malaysia
[2]Department of Electrical and Electronics Engineering, Faculty of Engineering,
Universiti Pertahanan Nasional Malaysia, Sungai Besi, Malaysia
jumrilyunas@ukm.edu.my

Abstract – **In this paper a micro mixer device with 3D channel dimension is simulated and analysed. Typical planar zigzag micro mixers (ZZM) with the angle of intersection 30°, 60°, and 90° are first simulated using FEM analysis to study its mixing performance. Then, the zigzag micro mixers are modified into 3-dimensional zigzag micro mixer (3DZZM) by moving up 0.1mm several channels. The results show that 3DZZM with 60° angle of intersection revealed the highest mixing index performance which is above 98% in all ranges of Reynolds numbers. This study can be used as a guidance in designing a passive micro mixer for various biomedical device applications.**

Keywords—micro mixing, passive micro mixer, Reynolds number, mixing performance

I. INTRODUCTION

The development of MEMS and micro technologies have offer significant improvements in the miniaturization of the analysis system in biochemical field. One of the fields that has been consistently study during the past 5 years is the microfluidic system, which could manipulate the behaviour of the fluids in micro scale [1]. The study has attracted many researchers leading to the improvement in polymerase-chain-reaction to multiply DNA sequence with the limited samples [2] as well as bioparticle detection [3].

One of the components that have been playing important role in the microfluidic system is the micro mixer. Mixing mechanism is not such an issue in the macro scale as the fluid flow is dominated by the inertia force. Passive micro mixers have been simulated and investigated intensively in several experiments as it does not use external force in enhancing mixing mechanism as per active micro mixer [2]. This method of mixing could lead to the excessive heat received by the samples in the micro mixer channel that could damage the property of the biological samples during the mixing process. On the other hand, passive micro mixer uses its channel geometrical complexity to create a mixing mechanism [4].

In the micro scale, fluid flow is dominated by the viscous force which controlling the direction and behaviour of the fluid flow [5]. With the absent of the water turbulent fluctuation, the stream flow become straight and smooth. This phenomenon is called laminar flow. Many researchers have conducted experiment-in-depth in eradicating the problem associated with laminar flow in micro mixing.

One of the major parameter that plays important role in mixing performance is the channel geometry modification. [5] has studied the performance of a planar serpentine micro mixers to investigate the serpentine modification effect on mixing index. Transverse flow has been induced in numerical experimental by using zigzag trapezoidal channel [6]. However, simple planar micro mixer doesn't offer a complete range of interaction between fluids in a short channel. A 3D modification of the planar micro mixer is expected to be able to effectively manipulate the fluids position by folding mechanism in order to increase interfacial interaction of the fluids. One of the common techniques in fabricating a micro mixer is the MEMS based microfabrication technique that provides simple and cost effective process [7][8].

In this paper, a computational fluid dynamics (CFD) was used as a tool to study the effects of channel modification for a single phase fully developed laminar flow. The zigzag micro mixer has been design to agitate the stream line in order to create transverse and chaotic flow. By creating the transverse and chaotic flow, we can increase the interfacial area between samples which lead to the increment in mixing performance. The effect of the various ranges of low Reynolds number (Re), channel geometry and intersection angle have been studied.

II. MICRO MIXER MODEL

The planar geometric designs used are zigzag micro mixers (ZZM) as shown in Figure 1. The angle studied between the micro channels to induced mixing mechanism in this experiment were 30° (ZZM30), 60° (ZZM60), and 90° (ZZM90). The micro mixers then modified into 3-dimensional micro mixer (3DZZM) which are 3DZZM30, 3DZZM60 and 3DZZM90 to study the effect of the enhancement through the modification of geometrical channel. Figure 2 shows the modified zigzag micro mixer. The ZZM and 3DZZM micro mixers simulated in this numerical investigation have Y-shaped inlet and approximately 9 millimetres in total length.

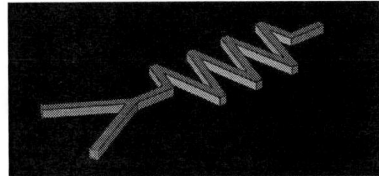

Figure 1. Zigzag Micro Mixer (ZZM)

---

1) DPP-2018-006 Dana Pembangunan Penyelidikan PTJ. Peranti bioperubatan dengan IOT
2) PRGS/1/2017/TK05/ukm/02/1 SKIM GERAN PENYELIDIKAN PEMBANGUNAN PROTOTAIP (PRGS). Design and Fabrication of Silicon Membrane Filtration System for Artificial Kidney

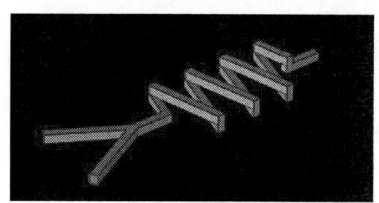

Figure 2. 3D Zigzag Micro Mixer (3DZZM)

Figure 3 shows the angle,a has been manipulated during this experiment. At the inlet 1, the concentration used was 0 mol/m³ whereas at the inlet 2, the concentration used was 10 mol/m³. All ZZMs channels (modified channel) were moved up to 0.1mm to produce 3D effect. First corner of the micro mixer were analysed the effect of the 3D modification that has been made.

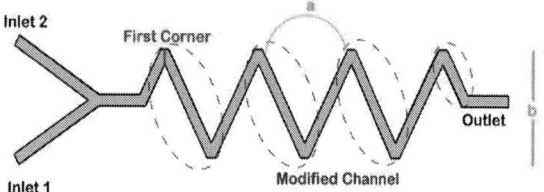

Figure 3. Research parameter

III. METHODOLOGY

FEM analysis was used in the simulation using two physics modules which are laminar flow (spf) and transport of diluted species (chds). The spf model used to simulate the fluid flow profile in the microchannel described by Navier-Stokes equation (1) and continuity equation (2) as follow:

$$\rho \left(\frac{\partial u}{\partial t} + (u \cdot \nabla)u \right) = f - \nabla p + v\nabla^2 u \qquad (1)$$

$$\nabla u = 0 \qquad (2)$$

where u is the velocity, f the body force, p is the pressure, ρ is the density of the fluid, t is the time, and v is the dynamic viscosity of the fluid.

For mixing mechanism, chds model was used to study the convection-diffusion equation as in equation (3).

$$\frac{\partial c}{\partial t} + (u \cdot \nabla)c = D\nabla^2 c \qquad (3)$$

where c is the concentration and D it the coefficient diffusion of the species.

In order to investigate mixing performance, Reynolds Number (Re) is used as a manipulative variable by using the following equation:

$$Re = \frac{\rho u D}{v} \qquad (4)$$

where D is the characteristic linear dimension. For non-circular channel, D indicate hydraulic diameter.

Mixing index performance by using the following equation;

$$M = 1 - \sqrt{\frac{1}{N}\sum_{i=1}^{N}(c_i - c_m)^2} \qquad (5)$$

where N is the sampling points inside cross section; c_i is the mass fraction at the sampling point i, and c_m is the mean of the mass fraction. Value of M ranges from 0 (indicating the fluids not mixed at all, 0%) to 1 (indicating fully mixed, 100%) [9]. Mixing performance is proportional to the value of the Re. However, in the case of very low Re (<Re10), the mixing performance is high due to the time consume in the micro mixer [6].

IV. RESULT AND DISCUSSION

Experimental models dimensions and manipulated variables are shown in the Table 1.

Model	Height (μm)	b (mm)	a (°)	Length (mm)
ZZM30	100	0.966	30	9.0
ZZM60	100	0.866	60	9.0
ZZM90	100	0.707	90	9.0
3DZZM30	200	0.966	30	9.8
3DZZM60	200	0.866	60	9.8
3DZZM90	200	0.707	90	9.8

The physics modules were simulated with normal mesh. The flow rate at the inlet was based on the Re. The Re used ranging from 10 to 150 as the fluid flow in laminar within this range. The value of Re was based on equation 4, where all the parameters were kept as a constant except for the velocity, u. Cut plane at 1st corner was observed for both types of micro mixers to study the 3D modification's effect on the fluid flow.

From the figure 4, it is shown that at Re 10, the fluids interaction area in zzm30, zzm60 and zzm90 is not convective and chaotic as the viscosity of the fluid is dominance. However, at the Re 10, the fluid flow in 3DZZM was convective because of the disturbance due to the channel geometry that changes the direction of the fluid flow [9]. The disturbed stream caused the transverse flow mechanism which increases the interfacial area, thus, leading to higher mixing performance. As the Re increased, the fluids behaviour at the first corner became more chaotic especially for 3D micro mixer. The chaotic flows induced by the geometrical effect enhance the mixing mechanism of the micro mixer [5].

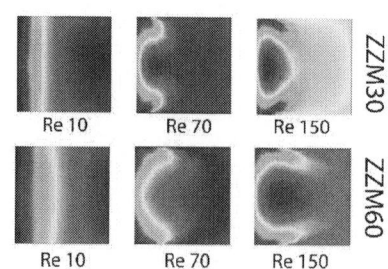

2018 IEEE International Conference on Semiconductor Electronics (ICSE)

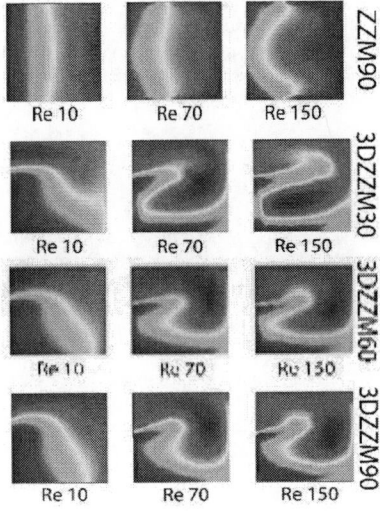

Figure 4. Interaction of the fluid at the first corner

Figure 5 shows the fluids interaction throughout the micro mixer (ZZM60 and 3DZZM60). The modified micro mixer was able to shift the position of the fluids. The interaction pattern showed in the Figure 4 justified that area of interaction of the fluids is wider and covers both side of the stream thus encouraging the mixing mechanism to take place.

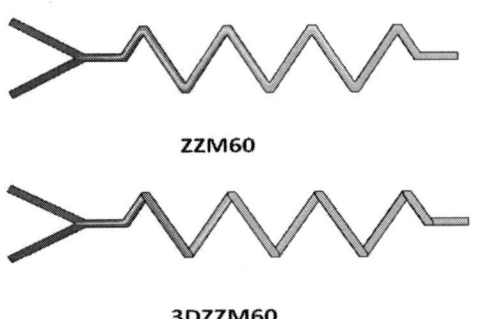

Figure 5. Interaction of the fluids in ZZM60 and 3DZZM60

Figure 6 shows the graph of mixing performance of the ZZM60 and 3DZZM60 against their respective length. It is clearly sen that at the channel distance of 3.2mm, the 3DZZM60 has already achieved mixing performance above 90% whereas the channel design of ZZM60 need at least 5mm to get uniformity above 90%. The channel modification of 3DZZM shows that the 3D design is able to reduce time needed as well as the channel length to achieve high mixing performance compare to ZZM. The chaotic flow induced at the corner of the 3DZZM has speed up the mixing process via folding mechanism that increases the interaction area between fluids.

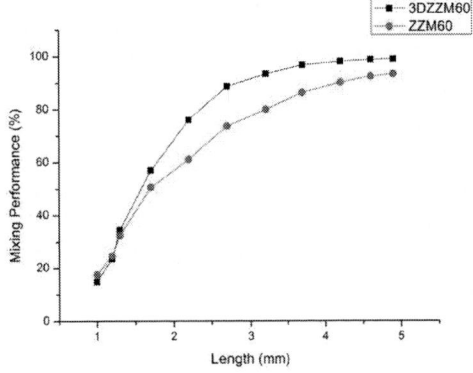

Figure 6. Mixing performance throughout the micro mixers.

Figure 7 shows mixing index performance of all the zigzag micro mixers. All micro mixer showed mixing performance above 90 %. From the result, value of Re play important role in mixing mechanism as the higher of Re, the better the mixing index performance. At low Re, the fluid flow only rely on molecular diffusion as it cannot induced chaotic and transverse flow even though the channel is geometrically modified. As the Re increases, the flows become more chaotic especially at the corner of the micro mixer thus increase mixing performance. However, extremely high values of Re are not preferable in micro systems as fluid body pressure will damage the device to function properly [10][11]. In a photolithography technique, a few layers of polymer are needed to fabricate a 3D micro mixer. High input pressure will lead to the high probability of leakage [12]. At the low Re, the kinematic force of the fluids could not induced effective chaotic and transverse flow at the turns of the channel compare to high Re where the force has driven the fluids to create enough momentum in inducing chaotic convection at the turns especially for 3DZZMs that folding mechanism is also introduced [9]. The 3D modification at the turns of the micro channel has added more effective mechanism in mixing compare to 3D modification at the straight micro channel.

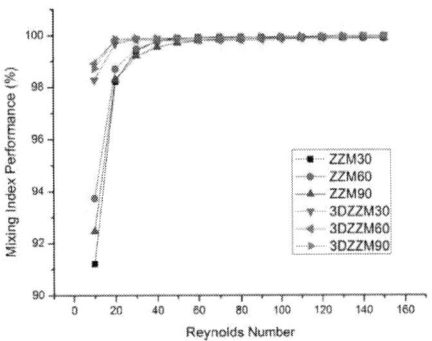

Figure 7. Mixing index performance of all micro mixers.

V. CONCLUSION

Micromixers with 3D channel dimension have been numerically studied using finite element analysis (FEA). The micro channel has Zigzag configurations (ZZM) with channel angle ranging from 30º to 90º. The mixers were then modified to be 3D structure. It is shown that the 3DZZM has channel design contributing to less channel length needed to achieve high mixing performance. The channel geometry of 3DZZMs is capable inducing folding mechanism from changing the fluid flow direction thus enhancing mixing performance. From the Figure 7, the 3DZZMs mixing index performances are considered as higher than ZZMs in all Re range. The study can be used in considering the parameter of geometrical design of passive micro mixer for the future lab-on-chip and biomedical application.

REFERENCES

[1] Bahadorimehr A., Y. Jumril Y:, Majlis B.Y., "Low cost fabrication of microfluidic microchannels for lab-on-a-chip applications, 2010 Int. Conf. Electronic Devices, Systems and Applications (ICEDSA2010), pp. 242-244.

[2] V.E. Papadopoulos, I.N. Kefala, G. Kaprou, G. Kokkoris, D. Moschou, G. Papadakis, E. Gizeli and A. Tserepi "A passive micromixer for enzymatic digestion of DNA," Microelectronic Engineering. 2014. 124: 42-46

[3] Masrie, M.; Majlis, B.Y.; Yunas, J. Fabrication of multilayer-PDMS based microfluidic device for bio-particles concentration detection. Bio-Med. Mater. Eng. 2014, 24, 1951–1958.

[4] M.N. Sabry, S.H. El-Emam, M.H. Mansour and M.A. Shouman, "Development of an efficient uniflow comb micromixer for biodiesel

production at low Reynolds number;" Chemical Engineering and Processing – Process Intensification Vol 128. 2018. 162-172

[5] X. Chen, T. Li, H. Zheng, Z. hu and B. Fu. "Numerical and experimental investigation on micromixers with serpentine microchannel," International Journal of Heat and Mass Transfer. 2016. 98:131-140

[6] H.L. The, B.Q. Ta, H.L. Thanh, T. Dong, T.N. Thoi and F. Karlsen "Geometric effects on mixing performance in a noverl passive micromixer with trapezoidal-zigzag channels," J. Micromech. Microeng. 25. 2015

[7] J. Yunas, A.A. Hamzah and B.Y. Majlis, "Fabrication and characterization of surface micromachined stacked transformer on glass substrate," Microelectronic Engineering 86. 2009. 2020-2025

[8] J. Yunas, A.A. Hamzah and B.Y. Majlis, "Surface micromachined on-chip transformer fabricated on glass substrate," Microsyst Technol 15. 2009. 547-552

[9] X. Chen and Z. Zhao, "Numerical investigation on layout optimization of obstacles in a three-dimensional passive micromixer," Analytica Chimica Acta. 2017

[10] N. Marsi, B.Y. Majlis, F. Mohd-Yasin and A.A. Hamzah, "The fabrication of back etching 3C-SiC-on-Si diaphragm employing KOH+IPA in MEMS capacitive pressure sensor," Microsystem Technologies 21. 2015. 8: 1651-1661

[11] N. Marsi, B.Y. Majlis, A.A. Hamzah and F. Mohd-Yasin, "High reliability of MEMS packaged capacitive pressure sensor employing 3C-SiC for high temperature," Energy Procedia 68. 2015. 471-479

[12] N. Marsi, B.Y. Majlis, A.A. Hamzah and F. Mohd-Yasin, "The mechanical and electrical effects of MEMS capacitive pressure sensor based 3C-SiC for extreme temperature," Journal of engineering 2014. 2014

978-1-5386-5284-8/18 $31.00 © 2018 IEEE

A Novel Digital Etch Technique for p-GaN Gate HEMT

Yuan Lin[1], Yueh Chin Lin[1], Franky Lumbantoruan[1], Chang Fu Dee[4], Burhanuddin Yeop Majilis[4], and Edward Yi Chang[1,2,3]

[1]Department of Materials Science and Engineering, National Chiao-Tung University (NCTU)
R407, MIRC Building, 1001, Ta Hsueh Rd., Hsin-Chu, Taiwan 30010
Phone: +886-3-571-2121#52999 E-mail:nctuharuka@gmail.com
[2]Department of Electronics Engineering, National Chiao-Tung University (NCTU)
[3]International college of Semiconductor Technology, National Chiao-Tung University (NCTU).
[4]Institute of Microengineering & Nanoelectronics, Universiti Kebangsaan Malaysia (UKM)

Abstract

We demonstrate the digital etching (DE) process to fabricated E-mode p-GaN/AlGaN/GaN HEMT. DE process comprising low power oxygen (O_2) plasma oxidizing and low power boron trichloride (BCl_3) plasma etching to selectively remove p-GaN layer. The atomic layer etching (ALE) has an etching rate of 1.62 nm/cycle to achieved depth of 70nm. The 5-μm source-drain offset length (L_{SD}) device with Ni/Au gate metal demonstrated 365 mA/mm drain current density with threshold voltage (V_{TH}) of +1.8V, on/off current ratio of $1.6x10^6$, breakdown voltage (BV) of 154V, and static on-resistance (R_{ON}) of 8.47 Ω•mm. The 20-μm L_{SD} device with Ni/Au gate metal demonstrated 211 mA/mm drain current density with V_{TH} of +2V, and on/off current ratio of $1.2x10^6$, BV of 426V, and static R_{ON} of 17.3 Ω•mm.

Keyword: p-GaN, AlGaN/GaN, HEMT, Digital Etching. Normally-off, E-mode, Atomic layer etching.

1. Introduction

Gallium nitride (GaN) high-electron-mobility-transistors (HEMTs) have attracted many interests for power applications due to its unique physical properties, such as high band-gap, high electron mobility, and high breakdown field. A two dimensional electron gas (2DEG) is naturally formed at the AlGaN/GaN hetero-interface because of spontaneous and piezoelectric polarization effect. For the safe operations and the low power consumption, the normally-off characteristic is strongly required. Therefore, several techniques for enhancement mode (E-mode) or normally-off characteristic have been achieve, such as gate recess [1], fluorinated treatment [2], and p-GaN cap layer [3]. Among the E-mode structures, the p-GaN gate HEMT is a promising candidate for normally-off device. The benefits of digital etching (DE) have been investigated by others. Gate recess with DE process shows precise thickness control and low surface damage [4]. For vertical nanowire by DE process, smooth, vertical sidewall and high aspect ratio has been developed [5]. Nevertheless, there is less literature study such as directional type of atomic layer etching (ALE) [6], anodic oxidation and dissolution oxide [7] to discuss p-GaN gate HEMT fabricated by DE process.

In this study, we fabricate the p-GaN gate HEMT by two-step DE to remove thin oxidation layer (1~2nm) per cycle with minimum damage to the remaining material. Finally the DC characteristic of p-GaN Gate HEMT such as gate breakdown voltage and off-state breakdown voltage (BV) threshold voltage (V_{TH}), static on-resistance (R_{ON}), will be demonstrated

2. Device Fabrication

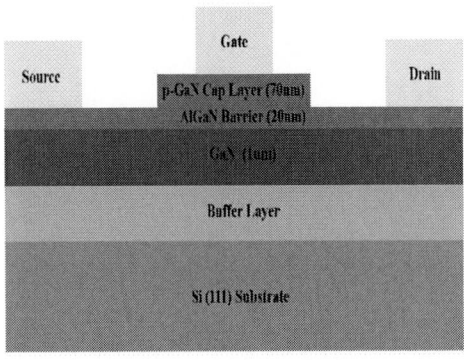

Figure 1. Schematics cross section of the p-GaN gate HEMT structure

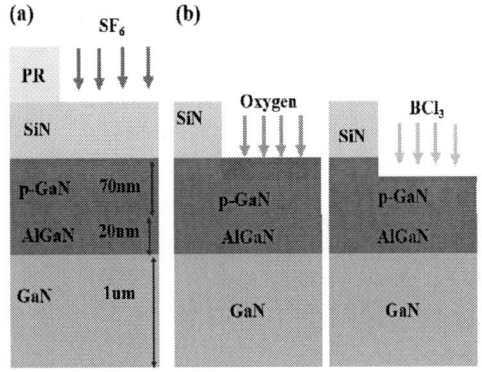

Figure 2. The process of digital etching (DE). (a) SiN hard mask removal by SF_6 in the access region followed by (b) oxidation and etching cycle by oxygen and boron trichloride, respectively.

Fig. 1 shows the basic transistor of the epitaxial grown. The devices were grown on a Si (111) substrate by metal organic chemical vapor deposition (MOCVD), which is composed of a buffer layer, a 1 um undoped GaN layer, a 20 nm AlGaN barrier layer, and a 70 nm p-type GaN layer with Mg concentration of 3 x 10^{19} cm^{-3}.

Our p-GaN gate HEMT process started from etching the mesa area by inductively coupled plasma reactive ion etching (ICP-RIE) with etching gas Cl_2/BCl_3. Fig. 2 displays the DE process consists of two step. First step was defined the DE area by deposited 50 nm SiN as a hard mask and removed the access region SiN by SF_6 plasma. The DE process of second step is oxidized and removed oxidation cycles, forming a thin oxide layer on top p-GaN layer surface by low power O_2 plasma 150 second and etching oxidation by low power BCl_3 plasma 20 second. The DE etching process recipes of the an-

978-1-5386-5284-8/18 $31.00 © 2018 IEEE

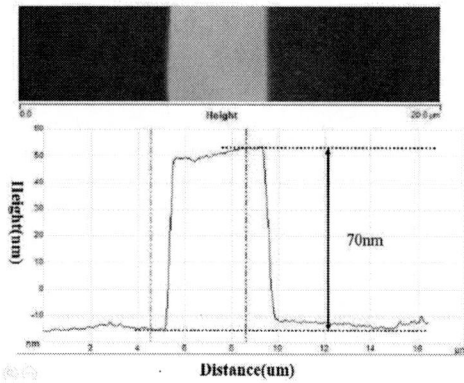

Figure 3. Atomic force microscope (AFM) image of the p-GaN gate region and depth profile of the p-GaN gate (W_G=20um).

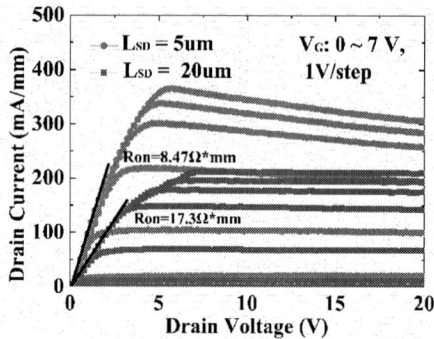

Figure 4. DC output characteristics of the p-GaN gate HEMT with V_{DS} sweeping from 0 to 20 V and stepping from 0 to 7 V. Static R_{ON} of L_{SD} of 5um and 20um with 8.47Ω•mm and 17.3 Ω•mm.

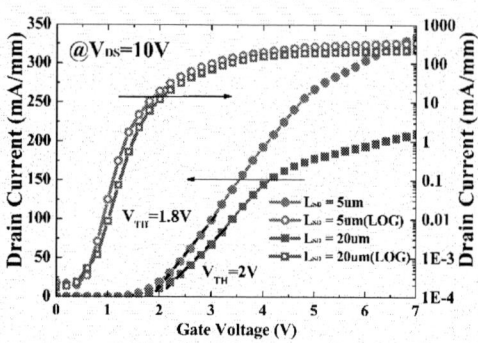

Figure 5. Transfer characteristics under V_{DS}=10 V. The drain current is shown in logarithmic (right axis) and linear scales (left axis). The V_{TH} of L_{SD} of 5um and 20um is +1.8V and +2.0V.

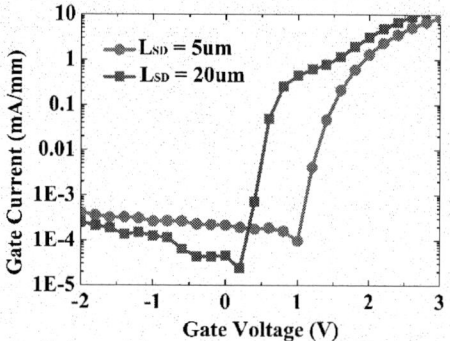

Figure 6. Gate voltage vs. gate current and gate breakdown characteristics at positive-bias conditions of the p-GaN gate HEMTs at V_D=V_S=0V.

tenna RF/bias RF was 200W/20W, the chamber pressure was 1.3pa, chamber temperature was 0˚C, and using O_2 and BCl_3 gas with flow rate is 60 and 40 sccm. The p-GaN layer over AlGaN barrier was etched layer-by-layer by repeating the oxidation and oxide removed. Fig.3 indicates the depth of selective removed the top p-GaN layer by two-step DE process after 43 cycle of DE is 70nm. For source and drain contacts, Ti (20 nm)/Al (120 nm)/Ni (25 nm)/Au (100nm) metal stacks was deposited by electron beam evaporation, and alloyed by rapid temperature annealing (RTA) at 800˚C for 30 second in a nitrogen (N_2) ambient. The Ni (50 nm)/Au (300 nm) were used as gate metal. Finally, deposited 25 nm thickness SiN as the passivation layer. We design device with two different source and drain offset length (L_{SD}) were 5 um and 20 um. The gate-source offset length (L_{GS}), gate length (L_G), and gate width (W_G) is 1.5, 2.2, 20 um, respectively.

3. Device results and discussion

Fig. 4 shows I_D-V_D characteristics of the p-GaN gate HEMTs with L_{SD} of 5um and L_{SD} of 20um with static R_{ON} of 8.47 Ω.mm and 17.3 Ω.mm at V_{DS}=1V. The maximum saturation drain current I_{DS} is 365 mA/mm and 211mA/mm with L_{SD} of 5um and 20um at V_G=7V. As L_{SD} increased, saturation drain current reduced and static R_{ON} increased.

Fig.5 displays V_{TH} of +1.8 and +2V with L_{SD} of 5um and 20um. As seen in log-scale, the on/off current ratio were 1.6x10^6 and 1.2x10^6 with L_{SD} of 5um and 20um. The subthreshold swing (SS) were 187 and 164 mV/dec with L_{SD} of 5um and 20um.

Fig.6 indicates the gate breakdown voltage, defined as the gate voltage when the gate current is 1 mA/mm at V_{DS}=0V [8]-[9], were 1.9 and 1.6 V with L_{SD} of 5um and 20um.

Fig.7 shows the off-state leakage current characteristic for BV of 154 and 426V with L_{SD} of 5um and L_{SD} of 20um at V_{GS}=0V. For power application, our best performance of L_{SD} of 20um (Static R_{ON} of 17.3Ω•mm and W_G of 20um) is demonstrated with lower specific $R_{ON}A$ of 3.46mΩ•cm^2 and BV of 426V.

4. Conclusions

In summary, we demonstrated the two-step digital etching (DE) can success fabricated p-GaN Gate HEMT. Our p-GaN gate HEMT device performance which L_{SD} of 5um of V_{TH} was 1.8V, max saturation drain current was 365 mA/mm, and R_{ON} was 8.47Ω*mm. To our knowledge, this L_{SD} of 20um result displays a best performance for E-mode power application with lower specific $R_{ON}A$ of 3.46 mΩ*cm^2 and BV of 426V.

978-1-5386-5284-8/18 $31.00 © 2018 IEEE

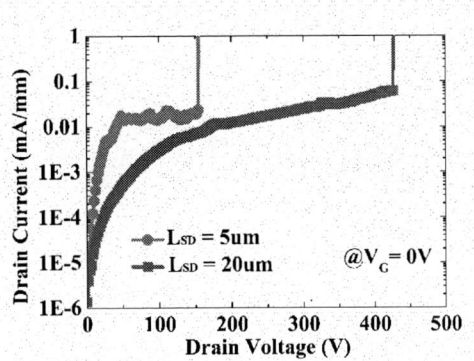

Figure.7 Off-state drain leakage under $V_G = 0$ V.

5. Acknowledgements

This work was financially supported by the "Center for Semiconductor Technology Research" from The Featured Areas Research Center Program within the framework of the Higher Education Sprout Project by the Ministry of Education (MOE) in Taiwan. Also supported in part by the Ministry of Science and Technology, Taiwan, under Grant MOST-107-3017-F-009-002.

5. References

[1] T. Oka, "AlGaN/GaN Recessed MIS-Gate HFET With High-Threshold-Voltage Normally-Off Operation for Power Electronics Applications", IEEE Electron Device Lett. Vol. 29, No 7, July 2008.

[2] Y-H. Wang et al, "6.5 V High Threshold Voltage AlGaN/GaN Power Metal-Insulator-Semiconductor High Electron Mobility Transistor Using Multilayer Fluorinated Gate Stack" ,IEEE Electron Device Lett., Vol. 36, No 4, April 2015.

[3] K-J. Chen et al., "GaN-on-Si Power Technology: Devices and Applications", IEEE Trans. Electron Dvices, Vol. 64, No 3, March 2017.

[4] S-D. Burnham et al., "Gate-recessed normally-off GaN-on-Si HEMT using a new O2-BCl3 digital etching technique", Phys. Status Solidi, Vol. 7, No 7-8, July 2010.

[5] X. Zhao et al., "Nanometer-Scale Vertical-Sidewall Reactive Ion Etching of InGaAs for 3-D III-V MOSFETs", IEEE EDL, VOL. 35, NO. 5, MAY 2014.

[6] T. Ohba et al., "Atomic layer etching of GaN and AlGaN using directional plasma-enhanced approach",JJAP 56, 06HB06 (2017).

[7] T. Sato et al., "Interface control technologies for high-power GaN transistors Self-stopping etching of p-GaN layers utilizing electrochemical reactions ", Proc. SPIE 9748, Gallium Nitride Materials and Devices XI, 97480Y, 26 February 2016.

[8] I. Hwang et al., "p-GaN Gate HEMTs With Tungsten Gate Metal for High Threshold Voltage and Low Gate Current", IEEE Electron Device Lett., Vol. 34, No2, Feb. 2013.

[9] T-F. Chang et al, "Phenomenon of Drain Current Instability on p-GaN Gate AlGaN/GaN HEMTs", IEEE Trans. Electron Devices, Vol. 62, No 2, Feb. 2015.

pH Sensing Characteristics of Silicon Nitride As Sensing Membrane based ISFET Sensor For Artificial Kidney

Mas Syarafina Norzin, Azrul Azlan Hamzah, Farahdiana Wan Yunus, Jumril Yunas, Burhanuddin Yeop Majlis

Institute of Microengineering and Nanoelectronic (IMEN),
Universiti Kebangsaan Malaysia (UKM),
43000 Bangi, Selangor, MALAYSIA.
Email: P90452@siswa.ukm.edu.my

Abstract— **Field Effect sensor are widely used for detecting various target of analytes in chemical and biological solutions. The electrochemical sensor like Ion Sensitive Filed Effect Transistor (ISFETs) is the primitive structure for biosensor. This paper presents ISFET based on pH sensor modelled using COMSOL Multiphysiscs for artificial kidney. In this paper, pH characterizations of ISFET sensor with silicon nitride (Si₃N₄) as sensing membrane were discussed. The measurement were includes various pH value from 3 to 11. The pH sensitivity of ISFET by using Si₃N₄ as sensing membrane give 44.4 mV/pH. Besides, this ISFET sensor is made to relate with microelectrode arrays with a tapered profile which the flow of red blood cell from the channel will flow into the electrode and ISFET biosensor then take place as ions determination.**

Keywords—ISFET sensor, pH characterization, microelectrode arrays, artificial kidney.

I. INTRODUCTION

Artificial kidney is being introduced by using biomedical micro electromechanical systems (Bio-MEMS). Bio-MEMS technology can improve human health in application of medicine as it can integrate microscale sensor and actuator, micro optics and microfluidic [1]. As in artificial kidney, artificial process of filtering, eliminating waste and unwanted water from the blood will occur [2]. There are several research on sensor have been made. Using these sensors and controls, the dialysis machine can be designed to automatically make adjustments during the dialysis treatment, or a health technician monitoring the sensors and controls can make the necessary adjustments [3].

According to several potential for miniaturization, parallel sensing and fast response time, field effect transistors (FETs) has the most attention to be considered as a sensing based. [4]. The most well-known field effect transistors (FETs) among various potentiometric techniques is Ion-sensitive field-effect transistors (ISFETs) sensor. [5]. As in industries, ISFETs were used in food and beverage, chemical processing, biotechnology and others, that used in pH measurement which ISFETs itself are miniature and near-instant robust [6]. ISFETs can be used for chemical as well as biochemical sensing [7]. In addition, ion concentrations in the solution can also be measured by ISFETs, where the current through transistor change as the ion concentration is measured [8]. Due to the ion sheath, the increasing of voltage between substrate and oxide surfaces will occur [8]. Based

on the pH value, the surface concentration of the -OH groups of the gate materials varies in the aqueous solutions [8].

According to research by R. Rani,, the potential developed across the insulator layer directly depends on the number of H+ ions in contact with it [9]. Thus, the ISFET channel would be affected by the potential at the gate, which would modulate the current flow across the source and drain when the device is turned on [9]. Besides, calibrating the amount of current flow give the measurement of the concentration of the H+ ions [9]. As the current value measured, the relationship between the current and H+ ion concentration will allow the pH value [9].

In this paper, the pH characteristics of Si₃N₄ as sensing membrane in ISFET sensor will be calculated by using simulation of COMSOL Multiphysics. The electrical measurement that are electrolyte potential with different values of pH, drain current-gate voltage (Id-Vg), drain current-drain voltage (Id-Vd) and the pH sensitivity were measured. A potential at the oxide-silicon interface is generated, by using silicon nitride, Si₃N₄ as sensing membrane to detect specific ion concentration, that correspond a drain-source current change in semiconductor channel [10].

II. THEORY OF ISFET

An ISFET, is a microelectronic device, that have feature in ability of chemically modifying the threshold voltage via the interfacial potential at the electrolyte/oxide interface [11]. The threshold voltage of ISFET as below :

$$V_{TH} = E_{ref} - \psi + \chi^{sol} - \varphi_{Si}/q - \{(Q_{ox} + Q_{ss} + Q_B)/C_{ox}\} + 2\varphi_f \quad (1)$$

where E_{ref} is the constant potential of the reference electrode, $-\psi + \chi^{sol}$ is the interfacial potential at the solution/oxide interface of which ψ is the chemical input parameter, a function of the solution pH and χ_{sol} is the surface dipole potential of the solvent and thus having a constant value [11]. φ_{Si} is the work function of silicon and q is the elementary electronic charge [11]. The fifth term on right-hand side is due to accumulated charge in the oxide (Q_{ox}), at the oxide-silicon interface (Q_{ss}), and the depletion charge in the silicon (Q_B), with Cox representing the oxide capacitance per unit area, whereas the last term determines the onset of inversion depending on the doping level of silicon ; it is given by:

$$\varphi_f = (kT/q) \ln (N_A/n_i) \qquad (2)$$

k is the Boltzmann constant, T is the temperature in kelvin scale, N_A is the acceptor concentration of p-type wafer and n_i is the intrinsic carrier concentration of silicon [11]. The ion adsorption processes at the electrolyte-SiO_2 interface are described by the well-known site-binding model [11]. For silicon dioxide dielectric, the oxide surface (i.e. pH sensitive surface) contains three types of sites: Si−O⁻, Si−OH, and Si−OH⁺ [11]. Therefore, the oxide surfaces are amphoteric, meaning that the surface hydroxyl groups can be neutral, protonized (thus positively charged) or deprotonized (thus negatively charged) depending on the pH of the bulk solution; and the surface potential ψ is:

$$\psi = 2.3(kT/q)\{\beta/(\beta+1)\}(pH_{pzc} - H) \qquad (3)$$

where pH_{pzc}, known as the pH at the point of zero charge, is the value of the pH for which the oxide surface is electrically neutral and β, the sensitivity parameter, determines the final sensitivity of the device [11]. The value of β is expressed in terms of the acidic and basic equilibrium constants of the surface reactions, for which a parameter $[H^+]_s$ is introduced, which represents the surface concentration of H^+ ions, being related to the $[H^+]$ bulk value by Boltzmann statistics [11].

III. EXPERIMENTAL SETUP

The illustration of the fabrication process flow of ISFET as shown in Fig. 1 [12]. Usually, silicon substrate is used in fabrication of sensor or actuator in MEMS devices [13]. In this paper, it is possible to relate with dielectrophoresis concept that will be implemented on artificial kidneys [14].

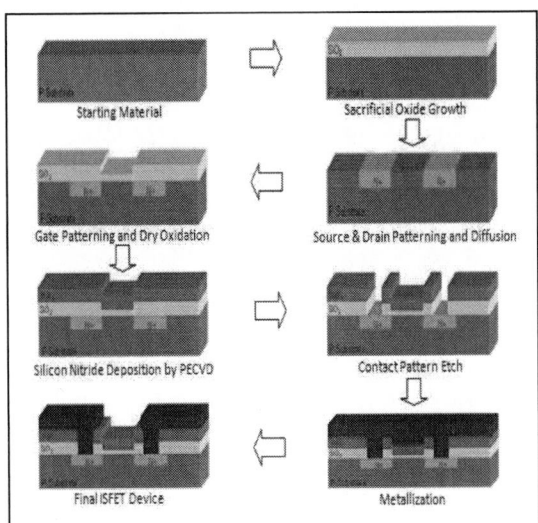

Fig. 1. Illustration of the fabrication process flow of ISFET.

From previous research, non-uniform electric field in the medium was introduced to improve the sensitivity also selectivity of F_{DEP} technique in microelectrode profile [15]. The device named as Tapered Aluminium microelectrode arrays (TAMA) which was designed based on microelectrode arrays with a tapered profile [15] as shown in

Fig. 2(a). The flow of the red blood cell from channel as shown in Fig. 2(b), which is the length is $80\mu m$, into the microelectrode will left the filtered red blood cell which ISFET can take the next step where the ISFET will detect the ions from the filtered blood. Moreover, parameter of sensitivity, selectivity and linearity is most important to be include in ions detection [16]. Further research will be done to determine selectivity of specific ions in blood. This relation in future will develop in artificial kidney.

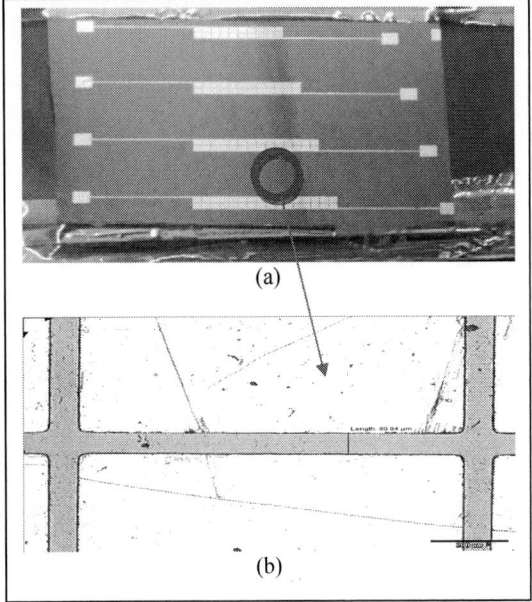

Fig. 2. (a) Tapered aluminium microelectrode arrays (b) Figure of microelectrode from microscope view.

IV. COMSOL MODELLING DESIGN

The simulation model as shown in Fig. 3 had been modified by using COMSOL Multiphysics software. The model is composed into two main domain which are semiconductor domain and an electrolyte domain. The physics involved in this ISFET model very similar to those in MOSFET model. This ISFET is constructed by replacing the gate contact of MOSFET with an electrolyte.

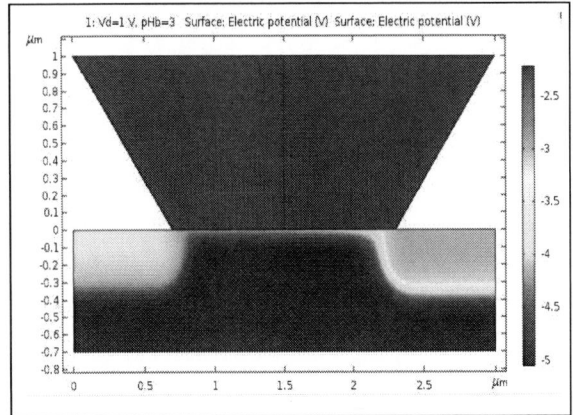

Fig. 3. Diagram of electric potential in an ISFET for Vd=1, pHb=3.

V. RESULTS AND DISCUSSIONS

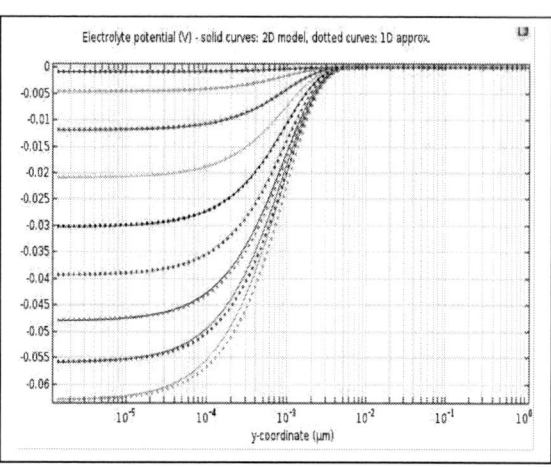

Fig. 4. Electrolyte potential along the center line of the electrolyte domain.

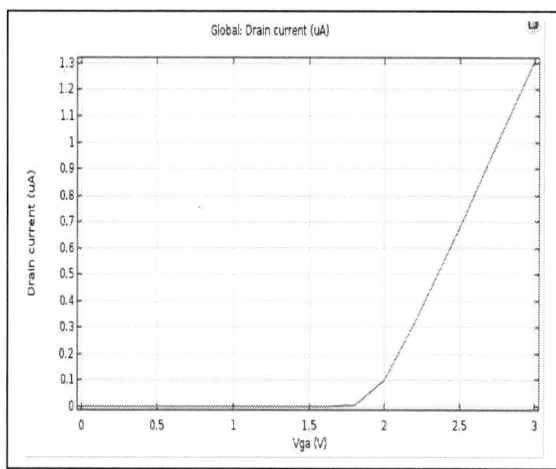

Fig. 5. Drain current-gate voltage (Id-Vg) curve.

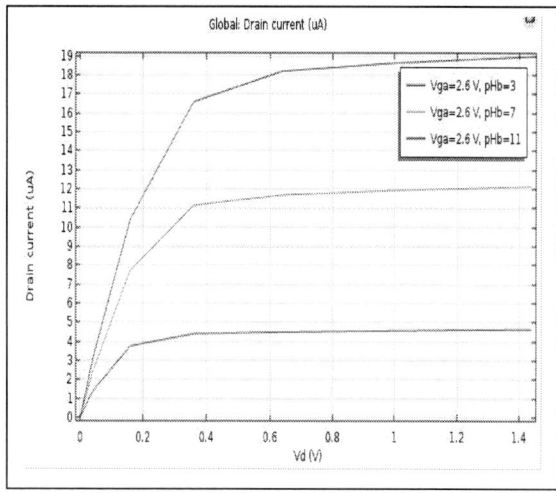

Fig. 6. Drain current-drain voltage (Id-Vd) curve for three different pH.

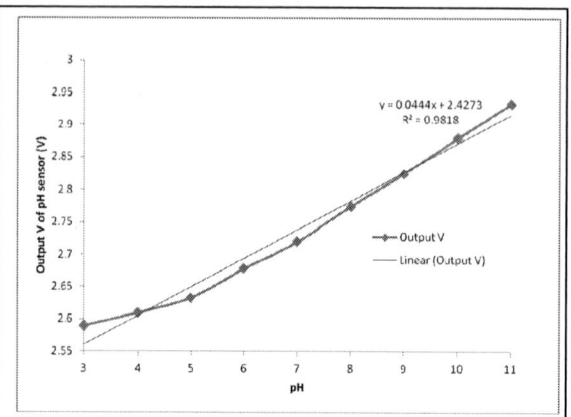

Fig. 7. Sensitivity curve of ISFET pH sensor.

Fig. 4 shows the electrolyte potential along center line of electrolyte domain comparing the simulation and approximation result. The trend of 1D and 2D match each other. Next. Fig. 5 present the Id-Vg curve. The drain voltage was kept constant to zero value while the gate voltage swept from 0V value to 3V value. Furthermore, the Id-Vd curve for three different pH value as shown in Fig. 6 has pH value as the control parameter with constant gate voltage, Vga = 2.6V.

The pH sensitivity of this ISFET which Si_3N_4 as sensing membrane can be shown in Fig. 7. The resulting gate voltage (the output sensor) as a function of the ph value (the input sensor) is plotted. As the results shown in the graph, it indicates that the value of pH sensitivity is 44.4 mV/pH which it poor sensitivity compared to previous research which is 52.93 mV/pH, as the sensitivity of Si_3N_4 vary with pH [7].

VI. CONCLUSIONS

The characterization of silicon nitride as sensing membrane based ISFET sensor by using simulation of COMSOL Multiphysics has been investigated. The pH sensitivity of the ISFET sensor with silicon nitride was 44.4 mV/pH which is very low sensitivity compared to Nernstian limit.

ACKNOWLEDGMENT

The author gratefully acknowledge the sponsor HICOE-AKU-95 grant that funded by The Ministry of Higher Education, Malaysia. The author also would like to thank the sponsor Dana Pembangunan Penyelidikan PTJ under grant DPP-2018-006 : Peranti bioerubatan dengan IOT, and also would like to thank Skim Geran Penyelidikan Pembangunan Prototaip (PRGS) for supporting under grant PRGS/1/2017/TK05/UKM/02/1 : Design and Fabrication of Silicon Membrane Filtration System for Artificial Kidney.

REFERENCES

[1] D. L. Polla, "BioMEMS applications in medicine," *MHS 2001 - Proc. 2001 Int. Symp. Micromechatronics Hum. Sci.*, pp. 13–16, 2001.

[2] C. Nordqvist, "All you need to know," vol. 18, no. 11. p. 4922, 2002.

[3] J. D. Petty, J. N. Huckins, and A. David, "(12) Patent Application Publication (10) Pub . No .: US 2002/0187020 A1," vol. 1, no. 19, 2002.

[4] M. Kaisti, "Detection principles of biological and chemical FET sensors," *Biosens. Bioelectron.*, vol. 98, no. April, pp. 437–448, 2017.

[5] P. Em, "ISFET, Theory and Practice," no. October, pp. 1–26, 2003.

[6] V. K. Khanna, "Remedial and adaptive solutions of ISFET non-ideal behaviour," vol. 3, pp. 228–237, 2013.

[7] S. Sinha, R. Rathore, S. K. Sinha, R. Sharma, R. Mukhiya, and V. K. Khanna, "Modeling and Simulation of Isfet Microsensor," 2014.

[8] D. Comsol, C. Books, and C. Consultants, "Microfluidic Design of neuron-MOSFET based on ISFET," *Comsol.Com*, 2010.

[9] R. A. Rani and O. Sidek, "ISFET pH sensor characterization: towards biosensor microchip application," *2004 IEEE Reg. 10 Conf. TENCON 2004.*, vol. D, pp. 660–663, 1970.

[10] S. Swaminathan, "Microsensor characterization in an integrated blood gas measurement system," pp. 15–20.

[11] V. K. Khanna, "Fabrication of ISFET microsensor by diffusion-based Al gate NMOS process and determination of its pH sensitivity from transfer characteristics," *Indian J. Pure Appl. Phys.*, vol. 50, no. 3, pp. 199–207, 2012.

[12] C. S. Fatt and M. Haron, "Design anD Fabrication oF n-isFet using si 3 n 4 as a sensing MeMbrane For pH MeasureMent," vol. 2, no. June 2011, pp. 23–30.

[13] A. A. Hamzah, J. Yunas, B. Y. Majlis, and I. Ahmad, "Sputtered Encapsulation as Wafer Level Packaging for Isolatable MEMS Devices: A Technique Demonstrated on a Capacitive Accelerometer," pp. 7438–7452, 2008.

[14] F. W. Yunus, A. A. Hamzah, M. R. Buyong, J. Yunas, and B. Y. Majlis, "Negative charge dielectrophoresis by using different radius of electrodes for biological particles," *Proc. 2017 IEEE Reg. Symp. Micro Nanoelectron. RSM 2017*, pp. 84–87, 2017.

[15] M. R. Buyong, F. Larki, M. S. Faiz, A. A. Hamzah, J. Yunas, and B. Y. Majlis, "A tapered aluminium microelectrode array for improvement of dielectrophoresis-based particle manipulation," *Sensors (Switzerland)*, vol. 15, no. 5, pp. 10973–10990, 2015.

[16] K. Ho, T. Lu, and C. Chang, "Sodium and Potassium Ion Sensing Properties of EIS and ISFET Structures with Fluorinated Hafnium Oxide Sensing Film," pp. 1128–1131, 2009.

978-1-5386-5284-8/18 $31.00 © 2018 IEEE

Characteristic Study of Doped ZnO Thin Film

S. Ishak, S. Johari, M. M. Ramli, A. Ibrahim, A. Arbae, M. S. Jaafar and A. S. Azlan
School of Microelectronic Engineering
Universiti Malaysia Perlis, Perlis, Malaysia
shazlinajohari@unimap.edu.my

Abstract— **In this paper, we report on the characterization of undoped and doped ZnO semiconductor for potential detection of formaldehyde gas, which involve the investigation of their morphological structures and optical properties. Four different dopants are used, namely Al, Ga, In and Sn with three different concentration which are 0.5 at%, 1.0 at% and 1.5 at%. Undoped and doped ZnO are prepared using sol gel method and annealed at 500°C for 5 hours. From scanning electron microscope (SEM), the result showed significant changes of surface properties for each dopant. The grain size exhibited from atomic force microscope (AFM) for undoped ZnO is greater compared to doped ZnO. Average surface roughness for all dopants type also increase as the doping concentration is increased. The obtained transmittance spectra shows that doping concentration affects the optical properties of ZnO as the transmittance percentage decreases as the doping concentration is increased.**

Keywords—ZnO, dopant, gas sensor

I. INTRODUCTION

Formaldehyde is an organic compound that exists in different forms of gas and solid. As a gas, formaldehyde is colorless and has a pungent characteristic with irritating odor. Formaldehyde can be easily found in outdoor environment as it is from automobile fumes, open burning of waste product, forest fires and burned of fossil fuels [1-2]. The most common symptoms that are related to over exposure of formaldehyde gas are irritation of nose, eyes and throat. In addition, the risk of asthma, allergy, nausea, edema and dyspnea may increase when formaldehyde is combined with protein [3]. Moreover, according to International Agency for Research on Cancer (IARC), the National Toxicology Program (NTP), and the US Environmental Protection Agency (US EPA), formaldehyde gas may also cause nasopharyngeal cancer and leukaemia [4]. Hence, the development of gas sensor to detect the presence of formaldehyde gas is crucial in order to avoid the risk of health threatening diseases. There are a few methods that have been reported to detect formaldehyde, which are spectrophotometry [1][5-6], gas chromatography [7], polarography [8], ion chromatography [9], high performance liquid chromatography [10] and fluorescene [11-12]. Unfortunately, these methods are unable to detect formaldehyde in real time basis and required strict conditions and large instrumentation. Metal oxides-based gas sensors for instance SnO_2 [13-14], TiO_2 [15] and ZnO [16-17] are of importance in the gas sensing area since they

have good responses and it is possible to modify their electrical characteristics. In particular, gas sensor based on thin film Zinc Oxide (ZnO) is the most chosen for sensing applications due to its tuneable surface morphology, very large surface-to-volume ratio, and superior stability due to better crystallinity. The mechanism of gas detection is by means of surface reaction, where the absorption of gas by ZnO thin film will affect its conductivity, hence reducing its electrical properties. ZnO has a wide bandgap of ~3.4 eV at 1.2 K. This has been proved to be an excellent gas sensing material for measuring both oxidative and reductive target gases at parts-per-million (ppm) level and above [18-19]. Even though great progress have been made for ZnO based gas sensor for formaldehyde detection, there are still room to improve the sensor performances such as sensitivity, response-recovery and working. One of the way to improve the performance of gas sensor is by doping ZnO with other elements. The function of dopants in the gas sensing process is still not well acknowledged. Based on the dopants loading concentration and their own intrinsic characteristics such as atomic/ionic size and charge, it will affect the thin film crystal structure and surface area. For instance, when group III elements such as Al, In and Ga are doped into the ZnO nanostructures, the dopant is expected to act as a single charged donors and supply the excess carriers to conductance band, thus increasing the conductance of ZnO and provide a path to improve the sensing properties of the thin film. Doping ZnO with different kinds of anions or cations is one of the best ways to modify or alter the morphology, carrier concentrations, density of states, transmittance, bandgap, and electrical conductivity. Moreover, the alteration of the thin film properties via doping leads to a change of attraction towards gases, and this can modify deeply the sensing properties of the thin film. Identifying the suitable dopant that could greatly enhance the sensing performance of ZnO towards formaldehyde gas is a challenge, as each dopant has their own unique properties. There are various methods that can be applied in order to produce doped ZnO thin film such as spray pyrolysis [20], pulse laser deposition [21], electron beam evaporation [22], magnetron sputtering [23] and sol gel [24]. Sol gel method is cheap and easy to be used. Moreover, it has a high uniformity and can be handled in a low temperature. Besides that, it also can give a great adhesion between the substrate and the top coating material and can help avoid corrosion by producing a thick coating on a substrate [25]. In this work, we studied the doping effect on the characteristics of ZnO thin film by observing its morphological and optical properties. Four different dopants are used, namely Sn, Al, Ga and In. These dopants will be doped with ZnO in order to see the effect of the doping performance.

II. EXPERIMENTAL PROCEDURE

In preparation of the undoped and doped ZnO solution, sol gel method was used. The concentration of zinc acetate used was 1.25×10^{-5} mol/L. The concentration of all dopants were set at 0.5 at%, 1.0 at% and 1.5 at%. Two types of samples were prepared, with and without dopants. The resultant solution was stirred at 80°C for 1 hour. Then, the solution was aged for 24 hours to yield a homogenous solution. Spin coating was used to prepare the ZnO and dopant solution at the rotation speed of 2000 rpm for 60s. Then, the thin film was soft baked for 10 minutes at 150°C. The process of spin coating and soft baked were repeated for 6 times. After that, the samples were annealed in a furnace at 500°C for 5 hours. The preparation flowchart of ZnO is shown in the Figure 1.

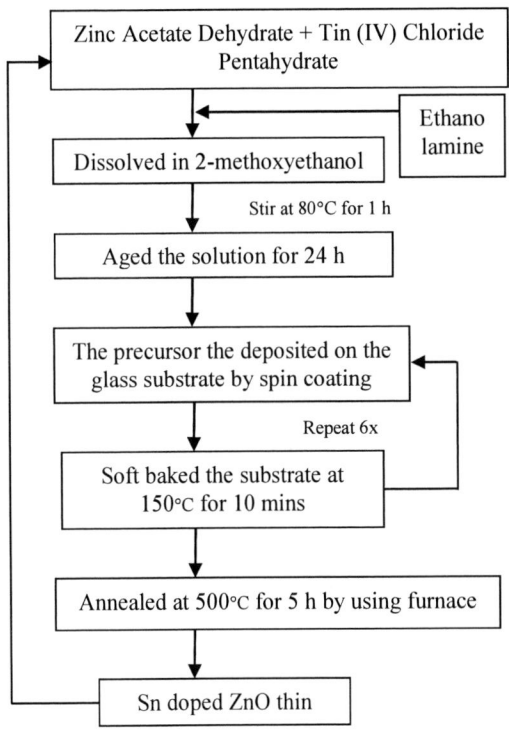

Fig. 1. The flow chart of preparation of Sn-doped ZnO solution

III. RESULTS AND DISCUSSION

SEM images of undoped and doped ZnO (1.0 at%) films are shown in Figure 2 (a-e). From the image, we can see that these micrograph show significant changes in morphology for different types of dopant. Based on the micrographs, undoped ZnO has a uniform nanocrystalline size and meanwhile, when dopant (Sn, Al, Ga and In) is added, the grains have some spongy clusters. There is some white cluster due to charging effect (accumulation of electron on

the film) on the image because SEM analysis is done without the metal coating. Based on SEM results, morphological surface is strongly affected by the different dopants used. When 1.0 at% of Al is used, the AZO surface looks more obvious compare to Ga, In and Sn. Meanwhile, for 1.0 at% Ga and In, particles with different shape and size are mixed together and the film is not fully distributed on the glass substrate. Next, for Al (0.054nm) and Sn (0.064nm), they have smaller ionic radii compare to Zn (0.074nm), thus, particles become smaller as doping occurred. This is due to the differences in ionic radii as compression stresses happened when the grain growth with the aid of dopant concentration [23].

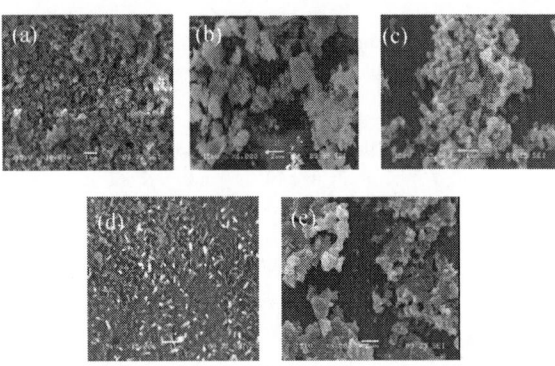

Fig.2 . SEM images of (a) undoped and 1.0 at.% (b) Al doped (c) Ga doped (d) In doped and (e) Sn doped ZnO thin films.

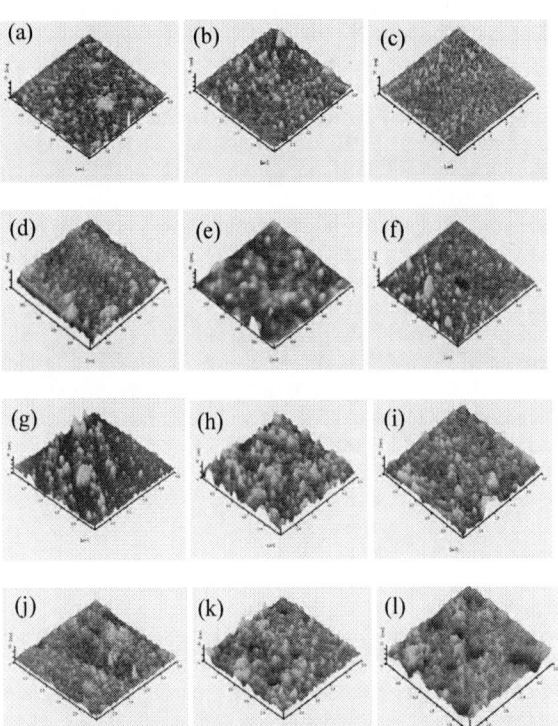

Fig. 3. 3D AFM top view image of (a) - (c) 0.5 at.%, 1.0 at.% and 1.5 at.% Al doped thin film respectively, (d) - (f) 0.5 at.%, 1.0 at.% and 1.5 at.% Ga doped thin film respectively, (g) - (i) 0.5 at.%, 1.0 at.% and 1.5 at.% In doped thin film respectively and (j) - (l) 0.5 at.%, 1.0 at.% and 1.5 at.% Sn doped thin film respectively

TABLE I.   DIAMETER (NM) GRAINS OF UNDOPED AND SN-DOPED ZNO

Doping Concentration (at. %)	Average Grain Size (nm)
0	73
0.5	65
1.0	72
1.5	61

Figure 3 (a-l) shows doped ZnO on AFM in 3D images with various dopants and concentrations. The glass substrate is uniformly covered by the grains on the surface. For Sn-doped ZnO, the surface of the glass shows the thickness of Sn-doped is thinner compared to the undoped ZnO. Undoped and 0.5 at. % dopant concentration indicate an evenly spread of grain to the surface of the film. Besides that, 1.0 at. % dopant concentration shows peak with sharp grain and combination of the grain on the certain area, whilst 1.5 at. % dopant concentration shows a smooth surface, less peak with sharp grain and also some area is not nicely covered. The grain size of the undoped is bigger than the doped grain size. This is due to the presence of Sn when the doping process occur with increase in concentration between ZnO and Sn. AFM imply that the Sn doped gives smaller grain size compared to the undoped. The grain size fluctuated as the concentration is increased. For 0.5 at. % dopant concentration, the grain size is 65 nm, then increase to 72 nm at 1.0 at. % dopant concentration, but decrease to 61 nm at 1.5 at. % dopant concentration (Table 1). These changes show that the concentration of Sn-doped will give effect to the grain size to compare with the undoped ZnO.

The size, shape and morphology are found to effect the gas sensing properties due to dependent of morphological properties on their shape and size distribution. Wang et al. [26] found that the smaller grain size will give a better gas sensing properties. The reason is that, the grain size is fully involved to the space-charge layer compared to a bigger size grain. Space-charge layer is known as electron depleted region where all the charge carriers movement take place. Similar results have been obtained by [27] and [28] which is as the grain size decrease, the sensitivity properties in terms of sensor will increase. However, it is not necessarily that a bigger grain size cannot display a good sensitive gas sensing properties. According to [26] an experiment with different surface area between octahedral, truncated octahedral, and 14-faceted polyhedral, it is found that 14-faceted polyhedral give a better sensitivity caused by a wider surface active area, thus more interaction between Zinc stannate-IV (ZnSnO3) and particular gases.

The transmittance spectra of undoped and Gallium doped ZnO is shown in Figure 4 as a function of wavelength. Based on the transmittance spectra, undoped ZnO shows the highest transmittance and the transmittance percentage decrease as the doping concentration increase. The reduction of the transmittance value after doping could be due to the increase of the film thickness. This result is similar as reported in [29] where ZnO is doped with Ti using sol-gel method, similar with this work.

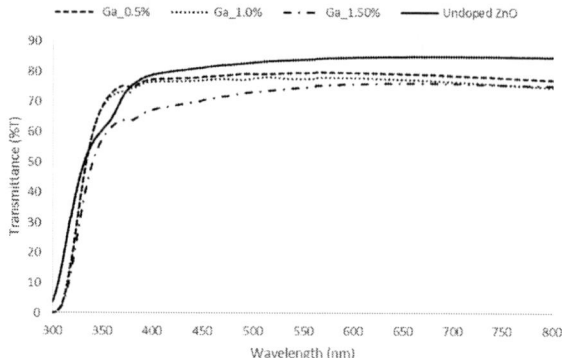

Fig. 4. Transmittance spectra of undoped and doped ZnO

The trend is similar for Sn doped ZnO, where undoped sample also shows an optical transmittance of 78%) compared to when Sn is added, as the transmittance decrease (<75%). This is similar as reported in [28], where transmittance of doped Sn always show the lowest percentage compared to other dopants. When ZnO is doped with Sn, the surface roughness of the film increases, hence lowering the transmittance percentage. When there is surface roughness, optical scattering may increase, thus lowering the optical transmittance. Thus, the optical band gap also may decrease as there is addition of Sn concentration. For gas sensing application, the sensing properties of ZnO depends on the reaction between dopant and gases in the atmosphere. These reaction is expected to change the electrical properties on the metal oxides. Other parameters like film thickness, crystallite size, porosity, amount and nature of dopants have been reported to be significant in improving gas sensors sensitivity [30-31].

The optical properties can be used to determine the band gap energy. The band gap energy is the difference in energy between valence band and conduction band, where the electrons are able to move freely through the material. Thus, when the band gap smaller, the electron transfer through the material become faster and the response in detecting the gas will become shorter. In other words, finding the band gap will lead to obtaining the time response of the gas. It is important to know the time response as it will determine the time taken for the gas to be detected [32-33]. In the future, this work will be used to measure formaldehyde sensing properties by measuring the changes of resistance of the sensors, before and after formaldehyde is introduced.

IV. CONCLUSION

Undoped and doped ZnO with four different dopants of Sn, Ga, In and Al, and three different concentration have been prepared by using sol gel and spin coating method. All samples are annealed at 500°C for 5 hours. The characteristics of the thin film have been investigated based on its morphological and optical properties. In terms of structural properties, the surface roughness of the AFM structure increase as the dopant concentration increases for all types of dopants. As for the optical properties, it can be observed that the optical transmittance spectra of doped ZnO thin films represent a strong dependence on the doping concentration. As the concentration increase, the thickness of the dopant on the film surface also increase. This will lead to reducing the optical transmittance as obtained in the results. Future plan include testing the doped thin film with the introduction of formaldehyde gas.

ACKNOWLEDGMENT

We wish to express our sincere thanks to School of Microelectronic technicians and staffs at Universiti Malaysia Perlis (UniMAP) for all their guidance and advice.

REFERENCES

[1] F. Lipari, S.J. Swarin, "Determination of formaldehyde and other aldehydes in automobile exhaust with an improved 2,4-dinitrophenylhydrazine method," Journal of Chromatography A, vol. 247, pp. 297-306, October 1982.

[2] T. Salthammer, S. Mentese, R. Marutzky, "Formaldehyde in the indoor environment," Chemical Reviews, vol. 110, pp. 2536-2572, January 2010.

[3] J. Xiong, P. Zhang, S. Huang, and Y. Zhang, "Comprehensive influence of environmental factors on the emission rate of formaldehyde and VOCs in building materials: Correlation development and exposure assessment," Environmental Research, vol. 151, pp. 734-741, 2016.

[4] R. Golden, "Identifying an indoor air exposure limit for formaldehyde considering both irritation and cancer hazards," Critical Reviews in Toxicology, vol. 41, pp. 672-721, 2011.

[5] O. Bunkoed, F. Davis, P. Kanatharana, P. Thavarungkul, and S. P. Higson, "Sol-gel based sensor for selective formaldehyde determination," Anal Chim Acta, vol. 659, pp. 251–257, Feb. 2010.

[6] Y.-L. Li, J. Liu, and W.-S. Guan, "Determination of trace formaldehyde in alcoholic beverage by chromotropic acid spectrophotometry," Proc. 3rd Int. Conf. Bioinformat. Biomed. Eng., pp. 1–4, June. 2009.

[7] H. Bagheri, M. Ghambarian, A. Salemi, and A. Es-Haghi, "Trace determination of free formaldehyde in DTP and DT vaccines and diphtheria–tetanus antigen by single drop microextraction and gas chromatography–mass spectrometry," J. Pharmaceutical Biomed.Anal., vol. 50, pp. 287–292, Oct. 2009.

[8] Z. Q. Zhang, H. Zhang, and G. F. He, "Preconcentration with membrane cell and adsorptive polarographic determination of formaldehyde in air," Talanta, vol. 57, pp. 317–322, May 2002.

[9] M. L. Chen, M. L. Ye, X. L. Zeng, Y. C. Fan, and Z. Yan, "Determination of sulfur anions by ion chromatography–postcolumn derivation and UV detection," Chin. Chem. Lett., vol. 20, pp. 1241–1244, Oct. 2009.

[10] J. C. P. Penteado, A. C. Sobral, and J. C. Masini, "Evaluation of monolithic columns for determination of formaldehyde and acetaldehyde in sugar cane spirits by high-performance liquid chromatography," Analytical Letters., vol. 41, pp. 1674–1681, July 2008.

[11] B. K. Wagner et al., "Small-molecule fluorophores to detect cell-state switching in the context of high-throughput screening," J. Amer. Chem. Soc., vol. 130, pp. 4208–4209, Apr. 2008.

[12] J.-Y. Lai and Y.-T. Li, "Functional assessment of cross-linked porous gelatin hydrogels for bioengineered cell sheet carriers," Biomacromolecules, vol. 11, pp. 1387–1397, May 2010.

[13] S.D. Kim, B.J. Kim, J.H. Yoon, J.S. Kim, J. Korean Phys. Soc., vol. 51, pp. 2069–2076, 2017.

[14] T. Pisarkiewicz, A. Sutor, P. Potempa, W. Maziarz, H. Thust, T. Thelemann, "Microsensor based on low temperature cofired ceramics and gas-sensitive thin film," Thin Solid Films, vol. 436, pp. 84–89, 2003.

[15] J.W. Lim, H.H. Kim, B.H. Kang, D.D. Lee, "Fabrication and characteristics of suspended-type micro gas sensor," J. Korean Phys. Soc. vol. 33, pp. 432-435, 1998.

[16] A. Nemeth, E. Horvath, Z. Labadi, L. Fedak, I. Barsony, Sensors Actuat. B, vol. 127, pp 157–160, 2007.

[17] E.K. Kim, H.Y. Lee, J. Park, J.H. Kwak, S.E. Moon, S. Maeng, K.H. Park, S.W. Kim, H.J. Ji, G.T. Kim, J. Korean Phys. Soc., vol. 51, 2007.

[18] R. A. Potyrailo and V. M. Mirsky, "Combinatorial and high-throughput development of sensing materials: The first 10 years," Chem. Rev., vol. 108, pp. 770–813, Feb. 2008.

[19] L. Schmidt-Mende and J. L. MacManus-Driscoll, "ZnO-Nanostructures, defects, and devices," Mater. Today, vol. 10, pp. 40–48, May 2007.

[20] Paudel, T. R., & Lambrecht, W. R. L., Growth and characterization of Sn doped ZnO thin films by pulsed laser deposition. Physical Review B, vol. 203, pp. 1383–1389, 2006 7

[21] Agarwal, D. C., Chauhan, R. S., Kumar, A., Kabiraj, D., Singh, F., Khan, S. A., Satyam, P. V. "Synthesis and characterization of ZnO thin film grown by electron beam evaporation," Journal of Applied Physics, vol. 99, 2006.

[22] Dave, P. Y., Patel, K. H., Chauhan, K. V., Chawla, A. K., & Rawal, S. K. "Examination of Zinc Oxide Films Prepared by Magnetron Sputtering," Procedia Technology, vol. 23, pp. 328–335, 2016.

[23] Hussein, H. F., Shabeeb, G. M., & Hashim, S. S. "Preparation ZnO thin film by using Sol-gel-processed and determination of thickness and study optical properties," Journal of Materials and Environmental Science, vol. 2 pp. 423–426, 2011.

[24] Park, C. O. and Akbar, S. A., "Ceramics for chemical sensing," Journal of Materials Science, vol. 38 pp. 4611–4637, 2003.

[25] Synthesis, optical and morphological studies of Sol-Gel derived ZnO/PVP one dimensional Nanocomposite. S. Ravichandran, G.Ramalingam, J. Nanosci & Nanotech, vol. 1, pp. 39-43, 2013.

[26] C. Wang, L. Yin, L. Zhang, D. Xiang, and R. Gao, "Metal oxide gas sensors: Sensitivity and influencing factors," Sensors, vol. 10, no. 3, pp. 2088–2106, 2010.

[27] A. Maldonado, "Particle Size Effect On Gas Sensing Properties Of ZnO Pellets," no. 31, 2016.

[28] S. Aksoy, Y. Caglar, S. Ilican and M. Caglar, "Effect of Sn dopants on the optical and electrical properties of ZnO films," Optical Applicata, vol.40, 2010.

[29] Hung-Peng Chang, Fang-Hsing Wang, Jen-Chi Chao, Chia-Cheng Huang, Han-Wen Liu, Effects of thickness and annealing on the properties of Ti-doped ZnO films by radio frequency magnetron sputtering, Curr. Appl. Phys. 11 (2011) S185–S190.

[30] Bahgat AA, Ibrahim FA, El-Desoky MM, Proceding AIP Proceedings of the Fifth Saudi Physical Society Conference AIP Conf Proc, pp. 61–67, 2011.

[31] Benkara S, Zerkout S, Ghamri H, Mater Sci Semi Process, vol. 16, pp.1271–127, 2013.

[32] C. S. Prajapati, A. Kushwaha, and P. P. Sahay, "Optoelectronics and formaldehyde sensing properties of tin-doped ZnO thin films," Appl. Phys. A Mater. Sci. Process., vol. 113, no. 3, pp. 651–662, 2013.

[33] J. Qi, H. Zhang, S. Lu, X. Li, M. Xu, and Y. Zhang, "High performance indium-doped ZnO gas sensor," J. Nanomater., vol. 2015, 2015.

2018 IEEE International Conference on Semiconductor Electronics (ICSE)

Characterization of Permittivity and Conductivity for ESKAPE Pathogens Detection

Muhammad KhairulAnwar Abdul Rahim, Muhamad Ramdzan Buyong, Nur Mas Ayu Jamaludin, Azrul Azlan Hamzah, Kim Shyong Siow, Burhanuddin Yeop Majlis

Institute of Microengineering and Nanoelectronics (IMEN),
Level 4, Research Complex,
Universiti Kebangsaan Malaysia (UKM),
43600 Bangi, Selangor Darul Ehsan.
Email: muhdramdzan@ukm.edu.my

Abstract — **This paper is to addresses the important diagnostic / detection technology gap by describing a rapid, portable, low-cost, and easy-to-use microfluidic, Dielectrophoresis Lab: On-A-Chip based system for detecting the Enterococcus faecium, Staphylococcusaureus, Klebsiella pneumoniae, Acinetobacter baumannii, Pseudomonas aeruginosa, and Enterobacter (ESKAPE) bacterial pathogens that are most commonly associated with antibiotic resistance. However, in this study was focus used the sizes of ESKAPE pathogens to identified the unique identification by translated it into dielectric properties. The MATLAB software was used to analyze and simulated the dielectric properties of each ESKAPE species based on their sizes. The MATLAB simulation was successfully conducted to identified the permittivity and conductivity crossover frequencies for all of Enterococcus faecium, Staphylococcusaureus, Klebsiella pneumoniae, Acinetobacter baumannii, Pseudomonas aeruginosa, Enterobacter species were 1.07 MHz, 0.91 MHz, 1.17 MHz, 0.62 MHz, 0.69 MHz and 0.76 Mhz with each average radius 0.59 μm, 0.687 μm, 0.545 μm, 0.962 μm, 0.875 μm and 0.81 μm respectively. The smallest size of ESKAPE species Klebsiella pneumonia, r of 0.545 μm have the higher crossover frequency (COF) of 1.17 MHz. In contrast the Acinetobacter baumannii species have the largest size, r of 0.962μm but it has lower COF of 0.62MHz.**

Keywords—Dielectrophoresis (DEP), ESKAPE pathogens,

I. INTRODUCTION

Basically, each of ESKAPE pathogens has unique identifications of phenotype and physiological state (interior or exterior morphology - size, shape and arrangements also surface roughness). Hence, in this research is focus and use the physiological state in determines a unique identification of each ESKAPE pathogens. More focus to the exterior morphology of ESKAPE sizes. Compared to interior ESKAPE pathogens composition which is not really significant due to in a group of bacteria. The unique identifications of size, shape and arrangements are simplified into dielectric properties and can be modeled to electrically equivalent circuit. The simple electrical equivalent circuit of an ESKAPE pathogens consisted connection of series and parallel resistance and capacitance. The outer surface of the bacteria is composed of parallel resistance and capacitance. The internal of bacteria also consist of series resistance and capacitance. Combination of series-parallel for resistance and capacitance represented their permittivity and conductivity. Theoretically the equivalent circuit can be illustrated the behavior of conductivity and permittivity due to the increasing frequencies magnitude. Normally at the smallest frequency caused the values of permittivity high. Otherwise the conductivity is rising caused by the frequency increase. In ESKAPE application, each of ESKAPE pathogens has different frequency response due to different permittivity and conductivity values that represent of phenotype and physiological state (interior or exterior morphology). This mechanism is use in detection of each of ESKAPE pathogens. Implementation of taper DEP electrode with two spot intensity electric field, as we exposed electric field to the sample based DEP working principle, this device able to manipulate by lateral attraction or vertical repulsion at region of interest. As integration DEP platform with microfluidics, dielectrophoresis lab: on-a-chip for each ESKAPE sample.

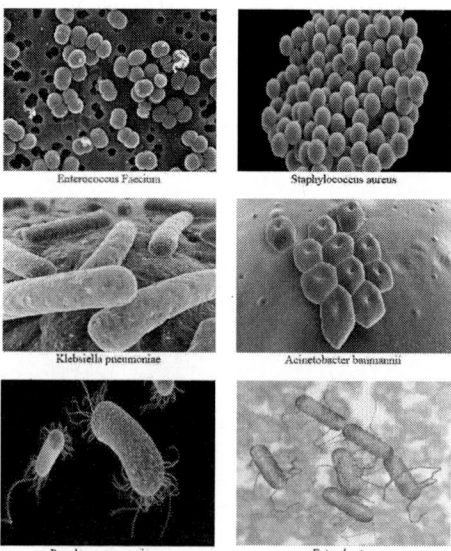

Fig. 1. Physical structure of ESKAPE pathogens [1]

978-1-5386-5284-8/18 $31.00 © 2018 IEEE   132

II. DEP CONCEPT

Dielectrophoresis has been widely used for the determination of cell, bacteria, and virus and protein dielectric characteristics and is applicable to the detection, manipulation, separation, and isolation of target cells from mixtures in suspension [1-11]. This phenomenon is useful to explain the behavior of DEP. Basically; it consists of two following these fundamental concept [4]. Firstly, the polarity properties are influenced by the external electric field acting on it [12]. It depends on the strength and direction of the electric field, also their polarizability dielectric of the particles and the medium that it suspended. Secondly, each particle and medium have a distribution of different charges. When it is revealed to a non-uniform electric field then the distribution of the charges will be interrupted. Dipolar moment will be induced. Particles or even more polarized mediums will be attracted to the region of high electric field intensity while the opposite will avoid to the region have high electric field intensity. Based on these two theories, Maxwell was introduced the relation of existed free charges on the particle and medium when electric field applied [4]. Principally, the system no needed to changed the particle but only used the non-uniform electric field to driven the particle based on the difference net charges between their particle and medium.

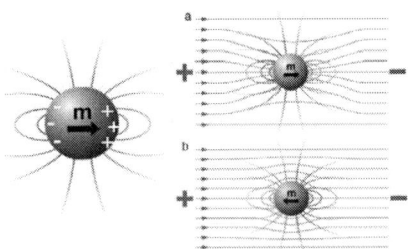

Fig. 2. Free electrical charge and electrical flux [4]

Basically the all of ESKAPE pathogens species have outer membrane and nucleus at inner part. By using these physical cell properties, the equivalent circuit can be determining resistance and capacitance of both outer and inner part of ESKAPE. The inner part (ESKAPE nucleus) consists of series resistance and capacitance, meanwhile the outer part (EKSAPE outer membrane) consists of parallel resistance and capacitance. The combination of both series and parallel (resistance and capacitance) represented the conductivity and permittivity. The fig. 3 illustrated equivalent circuit.

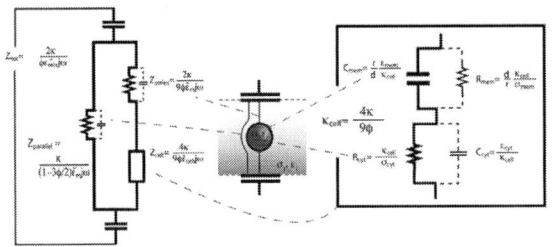

Fig. 3. Details of discrete electric equivalent circuit model of an ESKAPE [11]

Details of discrete electric equivalent circuit model of an ESKAPE pathogen in suspension, taking into account an ESKAPE pathogen in term of conductivity, σ_2 and permittivity ε_2, medium in term conductivity σ_1 and permittivity ε_1 and electrical double layer impedance of the microelectrodes. Based on unique identifications of each ESKAPE, which is physical structure exterior morphologies later can be converting those properties into dielectric properties for each ESKAPE. In term of dielectric properties, the permittivity and conductivity value are manipulating for each of ESKAPE in DEP applications [4].

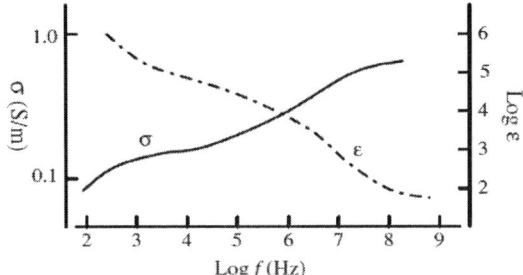

Fig. 4. The co-relation of permittivity and conductivity [4]

The co-relation of permittivity and conductivity value of ESKAPE pathogen will give the information about relationship between frequency and dielectric properties. Basically when manipulated the values of frequency from lower to higher, the conductivity also gaining. The greater value of permittivity is at the smallest frequency value. The ability of ESKAPE outer surface tends to store the more polarization charges can make these bacteria species act as insulator. More charges are stored in the ESKAPE surface because it allowed the greater out of flow charges stored at the low level of frequency. At in another hands, the rising of frequencies given lose of the resistance and capacitance. The permittivity values also became smaller. In this experiment MATLAB software were used to analyzed and simulated the permittivity and conductivity of ESKAPE pathogen versa their medium based on the ESKAPE sizes to identified each COF.

$$\langle F_{\text{DEP}} \rangle = \pi \varepsilon_m a^3 \text{Re}[\text{CMF}(\omega)] \nabla |E|^2 \quad \text{where } |E|^2 = E \cdot E* \quad (1)$$

$$\text{Re}[\text{CMF}(\omega)] = \frac{\tilde{\varepsilon}_p - \tilde{\varepsilon}_m}{\tilde{\varepsilon}_p + 2\tilde{\varepsilon}_m} \quad (2)$$

$$\tilde{\varepsilon}_{m/p} = \varepsilon_{m/p} - i\frac{\sigma_{m/p}}{\omega} \quad (3)$$

The minimum requirement of applied electric field to successfully throw into the inner part of ESKAPE is started from 100 MHz frequency or greater than that value [4]. Therefore, the polarizations disappear is occurring at the ESKAPE pathogen. At low frequencies the ESKAPE tend to extremely low of effectiveness conductivity to their solution. The gaining of frequencies value, also caused the effectiveness conductivity rising. With increasing frequency, the effective conductivity of the ESKAPE pathogens are increases. Furthermore, based on the better knowledge of dielectric properties of ESKAPE pathogen, the Clausius Mossotti factor (CMF) are useful to summarize their electrical properties. The polarization particle can be produced the DEP force (the unique identifications of ESKAPE in term dielectric properties) if it occurred under non-uniformly electric field. The CMF factor is transform the differ polarisability to ESKAPE pathogens and their medium. The CMF of ESKAPE pathogen is based on their sizes, medium and alternating current frequencies [13, 14, 15, 16].

The development of circulating tumor cells (CTC) detection, for example Triple Negative Breast Cancers (TNBC) CMF is opens new doors to studies the unique identification of ESKAPE pathogens [4, 17, 18]. This is because the ESKAPE pathogens have average sizes less than 1 μm meanwhile the TNBC has larger than 20 μm. The ESKAPE pathogens consist of six types of differences species but the TNBC only has one type of cell. The ESKAPE pathogens have higher frequencies response compared to TNBC because the average radius size of ESKAPE smaller than TNBC, thus the higher intensity of electric field can have penetrated EKSAPE at inner part of shell it gives good response of frequencies. Accordance to these behaviours the DEP response can completely act on outer and inner part in ESKAPE bacteria but in contrast the DEP response only acting on internally in TNBC cell.

III. RESULT AND DISCUSSION

The fig. 5 were summarized and listed the result simulated by MATLAB to determined each COF for all ESKAPE species. The conductivity value of the medium was determined 0.0055 S/m as red blood cell (RBC). This study was indicated the smallest size of ESKAPE species *Klebsiella pneumoniae*, radius (r) = 0.545 μm have the higher crossover frequency (COF) of 1.17 MHz. In contrast the *Acinetobacter baumannii* species with the largest size r of 0.962 μm have the lower COF of 0.62 MHz). Although the value of the radius is taken in the average form, it does not mean that the same size of ESKAPE pathogens for different species can be attracted or repelled together during the separation process. This is because each bacteria or particle has a different atomic configuration arrangement so it causes different polarization factors for each ESKAPE bacteria species or other particles.

Species	ESKAPE pathogens			
	Range diameter (μm)	Average diameter (μm)	Average radius (μm)	Crossover frequency (Hz)
Enterococcus faecium	0.36 - 2.0	1.18	0.59	1.07x10⁶
Staphylococcus aureus	0.75 – 2.0	1.375	0.687	9.14x10⁵
Klebsiella pneumonia	0.18 – 2.0	1.09	0.545	1.17x10⁶
Acinetobacter baumannii	1.35 – 2.5	1.925	0.962	6.22x10⁵

Species	ESKAPE pathogens			
	Range diameter (μm)	Average diameter (μm)	Average radius (μm)	Crossover frequency (Hz)
Pseudomonas aeruginosa	0.5 – 3.0	1.74	0.875	6.96x10⁵
Enterobacter	0.24 – 3.0	1.62	0.81	7.59x10⁵

Fig. 5. The average sizes of ESKAPE pathogens

The Fig. 6 was shown the CMF values with increased of input frequencies for all ESKAPE sized 0.59 μm, 0.687 μm, 0.545 μm, 0.962 μm, 0.875 μm and 0.81 μm respectively. From this real part CMF curve was indicated the gap unique dielectric properties between each ESKAPE species. It were useful to manipulated and separated the ESKAPE pathogens based on their sizes with used DEP force technique.

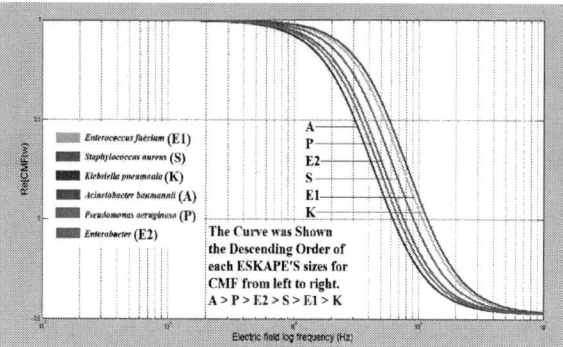

Fig. 6. CMF for ESKAPE pathogens

The Fig. 7 was shown the permittivity of ESKAPE pathogens and medium. This permittivity versus input frequencies curve was proved the permittivity of ESKAPE pathogens decreased with respond of rising input frequencies but their medium no significantly differences at intersection points.

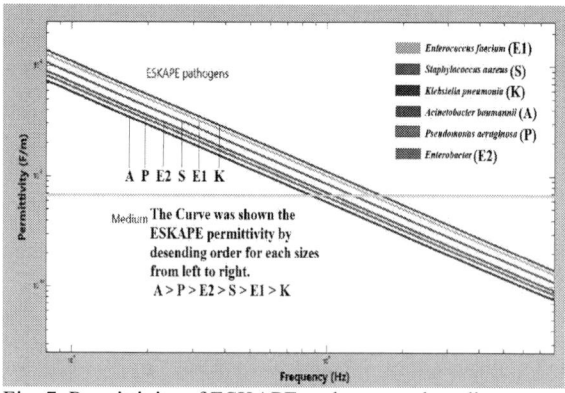

Fig. 7. Permittivity of ESKAPE pathogen and medium.

The fig. 8. was shown the conductivity of ESKAPE pathogens no significantly difference with rising of input frequencies but their medium extremely increased with respond of rising input frequencies.

978-1-5386-5284-8/18 $31.00 © 2018 IEEE

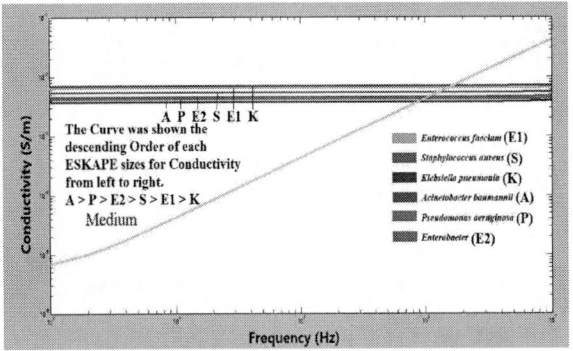

Fig. 8. Conductivity of ESKAPE pathogen and medium.

The Fig. 9 was shown the merger of permittivity and conductivity versus input frequencies curves. The first difference was the conductivity of ESKAPE species no significantly difference when the input frequencies rising but the conductivity of medium extremely increase when input frequencies rising. In contrast the permittivity of ESKAPE species are deceasing due to increasing of input frequencies but the permittivity of medium no significantly difference when increasing the input frequencies.

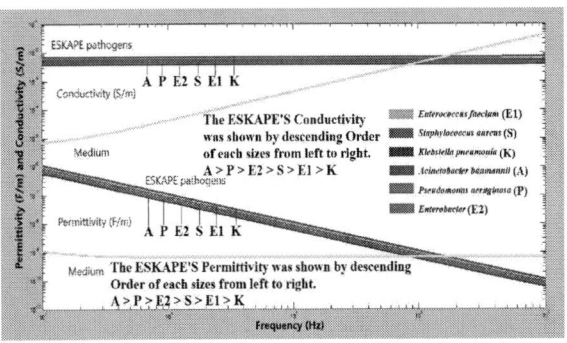

Fig. 9. Permittivity and conductivity for ESKAPE pathogens and medium.

IV. CONCLUSION

Based on the Matlab simulation, the CMF of ESKAPE pathogens can be determined by their sizes that reflect the dielectric properties of permittivity and conductivity values of each bacteria pathogens. These unique identifications can be used to manipulated and separated ESKAPE pathogens from their medium.

ACKNOWLEDGEMENT

Researcher wants to appreciate and thankful because of given the Dana Cabaran Perdana (DCP-2017-003/3) and Geran Galakan Penyelidik Muda (GGPM-2017-028) grants funded by Universiti Kebangsaan Malaysia (UKM) as a Research University.

REFERENCES

[1] S. Santajit, N. Indrawattana, 'Mechanisms of Antimicrobial Resistance in ESKAPE Pathogens", Biomed Res Int. 2016; 2016: 2475067. 10.1155/2016/2475067

[2] https://phys.org/news/2017-03-toolkit-rapid-acterial.html

[3] S. Mitsuhashi, K. Kryukov, S. Nakagawa, JS. Takeuchi, Y. Shiraishi, K. Asano, and T.Imanishi, "A portable system for rapid bacterial composition analysis using a nanopore-based sequencer and laptop computer", Scientific Reports 7, Article number: 5657(2017) doi:10.1038/s41598-017-05772-5.

[4] R. Pethig, "Dielectrophoresis: An assessment of its potential to aid the research and practice of drug discovery and delivery," Adv. Drug Deliv. Rev.,vol. 65, pp. 1589-1599, 2013.

[5] T. Honegger, David Peyrade, "Moving Pulsed Dielectrophoresis", Lab Chip 2013, 13, 1538 - 1545.

[6] K. Khoshmanesh, S. Nahavandi, S. Baratchi, A. Mictchell, K. Kalantar-Zadeh, "Dielectrophoretic Platforms for Bio-microfluidics Systems", Biosensors and Bioelectronics 26 (2011) 1800-1814.

[7] R. Pethig, "Review Article Dielectrophoresis: Status of the Theory, Technology and Applications", Biomicrofludics 4, (2010) 022811-1 – 022811-35.

[8] R. Pethig, A. Menachery, S. Pells, P. De Sousa, "Dielectrophoresis: A Review of Application for Stem Cell Research" Journal of Biomedic. and Biotech. (2010) 1-7.

[9] H. A. Pohl, 1978. Dielectrophoresis. Cambridge, UK: Cambridge University Press.

[10] K. F. Hoettges, "Dielectrophoresis as a Cell Characterization Tool. Microengineering in Biotech". Methods in Molecular Biology 583 (2010) 183-198.

[11] N. Piacentini, G. Mernier, P. Tornay, P. Renaud, "Separation of platelets from other blood cells in continuous-flow by dielectrophoresis field-flow-fractionation" Biomicrofludics 5, 034122 (2011)

[12] N. Marsi, B. Y. Majlis, A. A Hamzah and F. M. Yasin. The mechanical and electrical effects of MEMS capacitive pressure sensor based 3C-SiC for extreme temperature. Journal of engineering, (2014), pp.8.

[13] M. R. Buyong, F. Larki, M. S. Faiz, A. A Hamzah, J. Yunas and B. Y. Majlis. A tapered Aluminium microelectrode array for improvement of dielectrophoresis-based particle manipulation. Sensor, (2015), 15(5), 10973-10990.

[14] M. R. Buyong, F. Larki, Y. Takamura and B. Y. Majlis. Tapered microelectrode array system for dielectrophoretically filtration: fabrication, characterization, and simulation study. Journal of Micro/Nanolithography, MEMS, and MOEMS, (2017), 16(4), 044501.

[15] M. R. Buyong, F. Larki, C. E. Caille, Y. Takamura, A. A. Hamzah and B. Y. Majlis. Determination of lateral and vertical dielectrophoresis force using tapered microelectrode array. Micro & nano letters, (2018), 13(2): 143.

[16] M. R. Buyong, J. Yunas, A. A. Hamzah, B. Y. Majlis, F. Larki. N. A. Aziz. Design, fabrication and characterization of dielectrophoresis microelectrode array for particle capture. Microelectronics International (2015) 32 (2), 96-102.

[17] M. Karhade, C. Hall, P. Mishra, S. Krishnamurthy, and A. Lucci, "Circulating tumor cells in non-metastatic triple-negetive breast cancer," pp. 325-333, 2014.

[18] S. Shim, et al., "Dielectrophoresis has broad applicability to marker-free isolation of tumor cells from blood by microfluidic systems," Biomicrofluidics, vol. 7, p. 011808, 2013.

Effect of ZnO Composition on the Electrical Properties of MEH-PPV: ZnO Nanocomposites Thin film via Spin Coating

N.E.A. Azhar
NANO-ElecTronic Centre, Faculty of Electrical Engineering, Universiti Teknologi MARA,
Shah Alam, Selangor, Malaysia
najwaezira@yahoo.com

S.S. Shariffudin
NANO-ElecTronic Centre, Faculty of Electrical Engineering, Universiti Teknologi MARA,
Shah Alam, Selangor, Malaysia
sobihana@gmail.com

R. Abdul Rani
NANO-SciTech Centre, Institute of Science, Universiti Teknologi MARA,
Shah Alam, Selangor, Malaysia
rozina.abdulrani@yahoo.com

A.S. Zoolfakar
NANO-ElecTronic Centre, Faculty of Electrical Engineering, Universiti Teknologi MARA,
Shah Alam, Selangor, Malaysia
ahmad074@salam.uitm.edu.my

M.F. Malek
NANO-SciTech Centre, Institute of Science, Universiti Teknologi MARA,
Shah Alam, Selangor, Malaysia
mfmalek07@uitm.edu.my

Salman Alrokayan
Research Chair for Biomedical Applications of Nanomaterials, Department of Biochemistry, College of Science, King Saud University,
Riyadh, Saudi Arabia
dr.salman@alrokayan.com

Haseeb A. Khan
Research Chair for Biomedical Applications of Nanomaterials, Department of Biochemistry, College of Science, King Saud University,
Riyadh, Saudi Arabia
haseeb@ksu.edu.sa

M. Rusop
NANO-SciTech Centre, Institute of Science, NANO-ElecTronic Centre, Faculty of Electrical Engineering, Universiti Teknologi MARA,
Shah Alam, Selangor, Malaysia
nanouitm@gmail.com

Abstract—**Organic semiconductor have been commercialized for optoelectronic device application particularly in organic light emitting diodes (OLEDs). Poly [2-methoxy-5(2' – ethylhexyloxy)-1, 4- phenylenevinylene), MEH-PPV used in optoelectronic devices because it is easily synthesized and deposited in high molecular weight and good purity. The MEH-PPV: ZnO nanocomposites was prepared by spin coating method at room temperature. The MEH-PPV: ZnO nanocomposites thin film was investigated at different ZnO compositions. The electrical properties showed the ZnO composition at 0.2 wt% exhibits the highest conductivity of nanocomposites thin film (7.40×10^{-1} S. cm^{-1}) and suitable applied in optoelectronic devices.**

Keywords— MEH-PPV: ZnO nanocomposites thin film; ZnO composition; spin coating; electrical properties

I. INTRODUCTION

In the twenty-first century, the nanotechnology is comparable to the information and semiconductor technologies [1, 2]. Nanotechnology has developed a link between all fields of science and technology. The nanotechnology consists of the application in the chemical, physical, and biological field ranging from small atom to submicron dimension and evolving nanostructures into a larger system. The nanotechnology has been introduced in science and technology area such as nano-electronic, material and manufacturing, biotechnology and information technology.

In 1950, an organic light emitting diodes (OLEDs) was first observed by Bernanose and groups for the first time as the organic materials that can show electroluminescence

(EL). The OLEDs are effective light emitting diode (LED) made from semiconducting organic polymers consists of a single layer and multilayer of organic materials [3]. Basically, energy band gap of conjugated polymer is typically in the range between of 1.5 to 3.0 eV [4, 5]. There are various types of conjugated polymers such as poly [2-methoxy-5(2' –ethylhexyloxy)-1, 4-phenylenevinylene] (MEH-PPV), poly (3-hexylthiophene) (P3HT), and poly (methyl methacrylate) (PMMA). MEH-PPV is commonly used in fabrication devices such as OLED and photovoltaic cells due to its excellent spectroscopic properties and favorable electronic. Zinc oxide (ZnO) is a II-VI compound semiconductor and with a direct band gap of 3.37 eV [6, 7] at room temperature which it is suitable for short wavelength in application devices. The ZnO exhibits as an *n*-type material with attractive features. In recent years, it captured a lot of attention due to the direct band gap of 3.44 eV at low temperature.

A. Aleshin et al. [8] was deposited MEH-PPV polymer composite with the metal oxide material (ZnO nanoparticle) at different concentrations in the range between of 0 to 50 wt%. For the formation of an emission layer in photonic devices based on organic materials, the semiconducting polymer layers is commonly used and mixtures with different polymer molecules, as well as composite (inorganic–organic) materials, which, in comparison with pure polymer mixtures, are structurally and chemically more stable. But the problem in this field is associated with elucidation the mechanism of interaction between organic and inorganic phases in such nanocomposites systems.

In this work, the electrical properties MEH-PPV: ZnO nanocomposites thin film has been reported. We deposited

978-1-5386-5284-8/18 $31.00 © 2018 IEEE

the MEH-PPV: ZnO nanocomposites thin film by spin coating method. This work is very important to evaluate the highly conductive of MEH-PPV: ZnO nanocomposites thin film for OLED applications.

II. METHODOLOGY

A. Preparation MEH-PPV Thin Film

MEH-PPV powder was dissolved into a toluene solvent (5 mg/ml) to form a solution. The MEH-PPV solution was stirred onto a hot plate for 48 hours at room temperature. After 48 hours, the solution was deposited onto substrate (2 cm x 2 cm) by a spin coating method. The solution was deposited at 2000 rpm for 60 seconds. Then, MEH-PPV thin films were annealed at 50°C.

B. Preparation of ZnO Nanotetrapods Thin Film

The ZnO nanotetrapods was prepared by thermal chemical vapor deposition method. 1g of zinc (Zn) powder (99% purity; Sigma-Aldrich) was used as a precursor and placed in alumina boat at the furnace 1. Glass substrate was placed on the alumina boat in the furnace 2. Argon gas was fed into the tube for 15 minutes to eliminate all other gases. Temperature of furnace 1 and 2 was set at 750°C and 500°C, respectively. When the temperatures reached at a desired value, the oxygen gas with a flow rate of 5 sccm was fed into the tube. The growth of the nanotetrapods was deposited for 30 minutes under a constant argon flow of 100 sccm.

C. Preparation of MEH-PPV: ZnO Nanocomposites Thin Film

MEH-PPV: ZnO nanocomposites thin film was prepared at different weight compositions at 0.1 wt% to 0.3 wt% of ZnO. The nanocomposites thin film was deposited by the spin coating method. The ZnO nanotetrapods with different compositions was dissolved in MEH-PPV solution and sonicated using ultrasonic waterbath for 30 minutes. Then, the MEH-PPV: ZnO nanocomposites solutions were stirred for 60 minutes. After 60 minutes, the nanocomposites solution was deposited onto Indium Tin Oxide (ITO) coated glass and spun at 2000 rpm with 60 seconds. The nanocomposites thin film was dried at 50°C for 5 minutes onto hot plate.

D. Characterization

The thickness of MEH-PPV: ZnO nanocomposites thin film was measured by surface profiler (VEECO DEETAK 150). The electrical properties was characterized using 2-point probe solar simulator (CEP 2000). The metal electrode (60 nm) was deposited by thermal evaporator (Edward 306 Turbo, UK)

III. RESULT AND DISCUSSION

A. Electrical Properties for MEH-PPV Thin film

Current-voltage (I-V) measurement of MEH-PPV thin film was measured by 2-point probe solar simulator. The metal electrode was deposited using Al (60 nm) on top MEH-PPV thin films to measure the I-V characteristic as shown in Fig. 1.

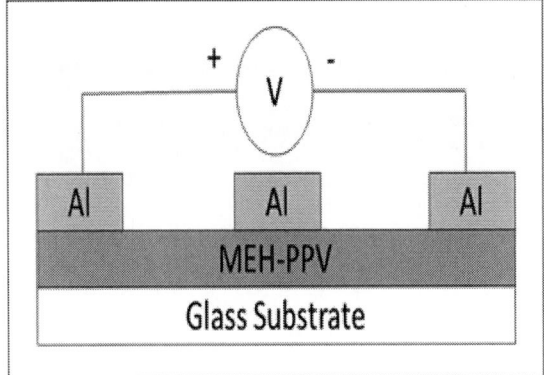

Fig. 1.   MEH-PPV thin film configuration

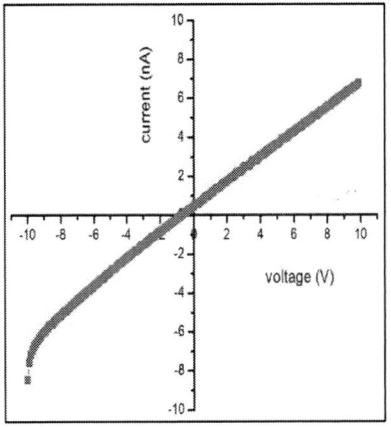

Fig. 2.   I-V curve of MEH-PPV thin film

Fig. 2 shows the I-V measurement of MEH-PPV thin film with applied voltage from -10 to 10V. The I-V curve indicates that polymer thin films give an ohmic behavior. From the previous study, Azhar et al. [9] reported the I-V measurement for polymer thin film deposited at weight powders of 5 mg/ml. The weight at 5 mg/ml had the highest current. This may be due to the MEH-PPV acts as p-type semiconductor material [10]. The resistivity and conductivity of MEH-PPV thin film can be defined using Equation 1 and 2 below:

$$\rho = R\,(wt/l) \qquad (1)$$

$$\sigma = 1/\rho \qquad (2)$$

where R is resistance, w is width of metal contact, t is thickness of film, l is the length between metal contact, ρ is resistivity and σ is conductivity of thin film.

The conductivity of MEH-PPV thin film at 5 mg/ml is 3.29×10^{-5} S. cm^{-1} and shows the polymer has good semiconductor properties as shown in Table I. The MEH-PPV thin film with the weight of 5 mg/ml was dissolved in toluene shows an excellent possibility to be used as the emissive layer in the nanocomposites thin film for OLEDs application.

B. Electrical Properties for ZnO Nanotetrapods Thin film

The ZnO nanotetrapods was prepared at temperatures of 750°C as shown in Fig. 3. The substrate was prepared using gold (Au) as a catalyst to grow the nanotetrapods onto the substrate. The result shows the current obtained at the temperature of 750°C which is 4.09×10^{-4} A as shown in Fig. 4.

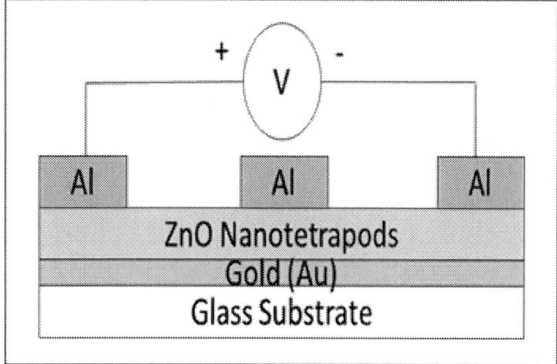

Fig. 3. ZnO nanotetrapods thin film configuration

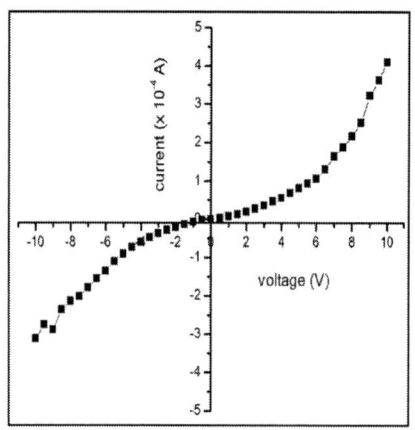

Fig. 4. I-V curve of ZnO nanotetrapods thin film

TABLE I.   RESISTIVITY AND CONDUCTIVITY OF MEH-PPV AND ZnO NANOTETRAPODS THIN FILM

Sample (°C)	Resistivity, ρ (Ω. cm)	Conductivity, σ (S. cm^{-1})
5	0.030×10^6	3.29×10^{-5}
750	2.07×10^1	4.84×10^{-2}

Table I shows the conductivity of ZnO nanotetrapods at temperature of 750°C ($\sim 4.84 \times 10^{-2}$ S. cm^{-1}) was calculated using Equation 1 and 2. For this reason, the energy will supply for the atoms rearrangement process at the higher conductivity. Therefore, the optimum current for ZnO nanotetrapods thin film was evaporated at 750°C.

C. Electrical Properties of MEH-PPV: ZnO Nanocomposites Thin Film

The Al was chosen and deposited onto MEH-PPV: ZnO nanocomposites thin film as the cathode metal electrode and the ITO substrate is often used as the anode electrode because it is transparent and allow out coupling of light. Fig. 5 shows the I-V characteristic of MEH-PPV: ZnO nanocomposites thin film deposited at the different ZnO compositions.

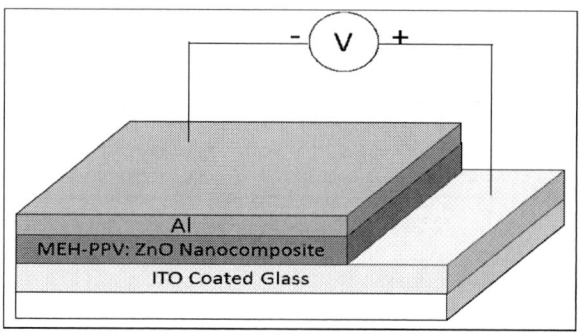

Fig. 5. MEH-PPV: ZnO nanocomposites thin film configuration

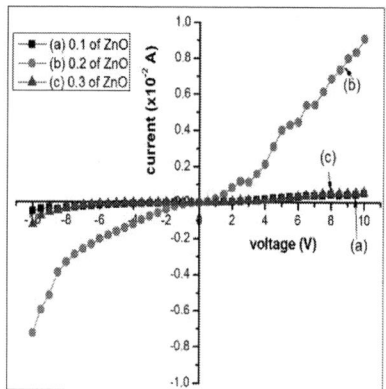

Fig. 6. I-V curve of MEH-PPV: ZnO nanocomposites thin film deposited at (a) 0.1, (b) 0.2, and (c) 0.3 wt% of ZnO

The resistivity of MEH-PPV: ZnO nanocomposites thin film deposited at 0.1 wt% (pure MEH-PPV) to 0.3 wt% were in the range of 1.351 to 1.47×10^1 Ω.cm. The resistivity of nanocomposites thin film decreased when ZnO weight composition increased to 0.2 wt%. When the weight composition increased to 0.3 wt%, the resistivity value increased.

The conductivity of nanocomposites film increased when the ZnO composition increased to 0.2 wt% (7.40×10^{-1} S. cm^{-1}) as tabulated in Table II. The conductivity of nanocomposites thin films increased contribute to the ability of nanotetrapods to fill in the pores in the polymer. The conductivity of nanocomposites thin film decreased when the composition of ZnO increased to 0.3 wt%. The decreasing of conductivity was related to the recombination of the process between the n-type of ZnO and the p-type of conjugated polymer (MEH-PPV).

Even though the conductivity of ZnO nanotetrapods is high but by adding the ZnO nanotetrapods with organic material, its enhanced the conductivity of MEH-PPV: ZnO nanocomposite thin film. Therefore, the 0.2 wt% of ZnO showed the high performance of nanocomposite thin film, thus it is suitable used for OLEDs application.

TABLE II.   RESISTIVITY AND CONDUCTIVITY OF MEH-PPV: ZNO NANOCOMPOSITES THIN FILM DEPOSITED AT DIFFERENT ZNO COMPOSITIONS

Sample (wt%)	Resistivity, ρ (Ω. cm)	Conductivity, σ (S. cm^{-1})
0.1	1.63	6.61×10^{-1}
0.2	**1.35**	**7.40×10^{-1}**
0.3	1.47×10^{1}	6.79×10^{-2}

Theoretically, in OLEDs device measurement, the electrons flow from ITO to the metal electrode is because of the work function difference between them. The ITO transparent was chosen and acted as the anode material that provides a good electrical conductivity, good energy-level matching for efficient injection of holes into organic layers [11, 12]. An organic layer such as MEH-PPV layer with the good hole blocking characteristic and good electron transport typically used between the emissive layer and the cathode. The influence of an external field, the holes and electrons migrated in opposite directions and driven to the emissive zone where they form excitons that decay radiatively to the output of light [13]. Due to the lower barrier at the cathode-polymer junction, the injected electron and holes are balanced, thus increasing the luminescence efficiency and reducing the turn-on voltage of the device.

IV. CONCLUSION

In this work, the MEH-PPV: ZnO nanocomposites thin film was successful deposited on the ITO coated glass at different ZnO composition by the spin coating method. The deposition of MEH-PPV was deposited at 5 mg/ml and annealed at 50°C exhibited the highest conductivity of 3.29 $\times 10^{-5}$ S. cm^{-1}. Hence, this judgement was made on the account that at the temperature of 750°C exhibits the highest conductivity, 4.84×10^{-2} S.cm^{-1} due to the energy will supply for the atoms rearrangement process at the higher conductivity. Based on the result, it can be concluded that the most suitable compositions of ZnO for MEH-PPV: ZnO for OLEDs application is 0.2 wt%. This judgement was made on the account that at 0.2 wt% it exhibits the highest

conductivity of 7.40×10^{-1} S. cm^{-1} which is the most important factor in improving the performance of polymer for OLEDs.

ACKNOWLEDGMENT

The author would like to express our gratitude Ministry Education Malaysia (MOE), Research Management Centre (RMC), Universiti Teknologi MARA (UiTM), Shah Alam, Selangor, Malaysia through the project 60-IRMI/PERDANA 5/3 BESTARI (101/2018) for financial support. This work also supported by the Research Chair for Biomedical Applications of Nanomaterials, Department of Biochemistry, College of Science, King Saud University, Riyadh, Saudi Arabia.

REFERENCES

[1] L. Filipponi and D. Sutherland, "Nanotechnology: A brief introduction," *Journal of Clinical and Experimental Neuropsychology*, vol. 31, no. 2, pp. 1–11, 2007.

[2] B. Bhushan, "Introduction to Nanotechnology," in *Springer Handbook of Nanotechnology*, Springer, 2010, pp. 1–15.

[3] V. K. Chandra, B. P. Chandra, and P. Jha, "Organic Light - Emitting Diodes and their Applications," *Defect Diffus. Forum*, vol. 357, pp. 29–93, 2014.

[4] Y. Li, "Conducting Polymer," in *Organic Optoelectronic Materials*, 1st ed., Spinger International Publishing, pp. 27 – 33, 2015.

[5] Liming Dai, "Intelligent Macromolecules for Smart Device: From Materials Synthesis to Dvice Applications (Engineering Materials and Processes)," in *Intelligent Macromolecules for Smart Devices From Material Synthesis to Device Applications*, 1st ed., vol. 1980, Springer, pp. 41–80, 2004.

[6] H. H. Lee, S.-H. Kim, and S. Fujita, "Catalyst-Free Sythesis of ZnO Nanorods on Metal Substrates by Using Thermal Chemical Vapor Deposition," *J. Korean Phys. Soc.*, vol. 53, no. 1, pp. 183–187, 2008.

[7] H. I. Abdulgafour, Z. Hassan, and F. K. Yam, "Growth of high-quality ZnO nanowires without a catalyst," *Phys. B Phys. Condens. Matter*, vol. 405, no. 19, pp. 4216–4218, 2010.

[8] A. N. Aleshin, I. P. Shcherbakov, and I. N. Trapeznikova, "Temperature and concentration dependences of the photoluminescence of MEH-PPV polymer composite films with ZnO nanoparticles," *Phys. Solid State*, vol. 56, no. 2, pp. 405–411, 2014.

[9] N. E. A. Azhar, S. . Shariffudin, I. H. . Affendi, S. A. H. Alrokayan, H. A. Khan, and M. Rusop, "Characteristic of Conjugated Polymer MEH-PPV Thin Films Deposited by Spin Coating Method," in *IEEE*, pp, 620–624, 2015.

[10] B. Kang, Y. Yang, L. Wang, and Y. Qiu, "Solvent induced semiconductor type conversion of MEH-PPV investigated by surface photovoltage spectra," *Displays*, vol. 25, no. 2–3, pp. 57–60, 2004.

[11] F. Nüesch, E. W. Forsythe, Q. T. Le, Y. Gao, and L. J. Rothberg, "Importance of indium tin oxide surface acido basicity for charge injection into organic materials based light emitting diodes," *J. Appl. Phys.*, vol. 87, no. 11, pp. 7973–7980, 2000.

[12] V. Singh, C. K. Suman, and S. Kumar, "Indium Tin Oxide (ITO) films on flexible substrates for organic light emitting diodes," *Proc. ASID*, pp. 388–391, 2006.

[13] A. Misra, P. Kumar, M. N. Kamalasanan, and S. Chandra, "*White organic LEDs and their recent advancements*," Semicond. Sci. Technol. Semicond. Sci. Technol, *vol. 21, no. 21, pp. 35–47, 2006*.

Synthesis of ZnO nanoflakes by 1064 nm Nd:YAG pulsed laser deposition in a horizontal tube furnace

* Kong Eng Ng[1], Teck Yaw Tiong[1], MohdAmbri Mohamed[1], Wei Sea Chang[2], Chang Fu Dee[1], BurhanuddinYeop Majlis[1]

[1]Institute of Microengineering and Nanoelectronics (IMEN),

UniversitiKebangsaan Malaysia (UKM),

43600, Bangi, Malaysia

[2]Advanced Engineering Platform, Mechanical Engineering Discipline,

School of Engineering, Monash University Malaysia,

Bandar Sunway, Selangor 47500, Malaysia

Email: kongeng12@gmail.com

Abstract— **In this research, a 1064 nm Nd:YAG laser with pulse duration of 8 ns was used to fire ZnO pellet in a furnace using 6 different energy densities. A substrate was placed above the ZnO target which was positioning at an angle of 45° within the frame of the target holder. ZnO nanoflakes were formed on the surface of ceramic, Al₂O₃ substrate was placed upside down on a substrate holder. This simple modification on the substrate orientation has greatly reduced the deposition of useless debris on the substrate. High crystalline ZnO nanoflakes with diameter ranging from 30 to 100 nm were characterized by Field Emission Scanning Electron Microscope (FESEM). Three sharp peaks (100), (002) and (101) characterized by X-Ray Diffraction (XRD) confirm the deposition of ZnO nanoflakes on the Al₂O₃ substrate. The depositions were remarkable as it required only low tube-furnace vacuum of 3.9 Torr (5.2 mbar) without extra thermal source other than the heat generated by the laser itself.**

Keywords— ZnO nanoflakes, pulsed laser deposition, nanostructured materials.

I. Introduction

ZnO has drawn huge attraction of researchers due to its large binding energy 60 meV and wide direct band gap of 3.37 eV applicable in various sectors such as optoelectronics[1], UV photodetectors[2], transistors, piezoelectric devices[3], solar cells[4] and sensors[5]–[7]. There are numerous methods used to fabricate ZnO nanostructured for example hydrothermal[8], wet chemical synthesis[9] and electrochemical deposition[10]. In this research, we reported the growth of high crystallinity ZnO nanoflakes using pulsed laser deposition on ceramic, Al₂O₃ substrate under low vacuum condition.One of the advantage of laser ablation is the light does not induce any contamination to the materials and the chemical process is stoichiometry under vacuum condition[11], [12]. This work is at its early stage which focus on conducting the experiment at room temperature environment to investigate the feasibility of fabrication high quality ZnO nanostructures.Unlike the conventional set up for substrate position in a furnace system[13], we placed the substrate above the ZnO target to avoid falling of side products of laser ablation such as debris or unwanted particles which

potentially contaminate the substrate surface. The surface morphologies of ZnO substrates ablated by different laser power are also demonstrated in this research characterized using FESEM. The deposition composition of ZnO on Al₂O₃ substrate is characterized using XRD.

II. Experimental

Cylindrical ZnO pellet of about 10 mm diameter and 5 mm height was placed in the middle of a furnace held 45° by the target holder. The ceramic, Al₂O₃ substrate was positioned above the target to avoid falling debris during laser ablation. The pressure within furnace was vacuumed to low pressure of 3.9 Torr (5.2 mbar). The experiment was conducted in room temperature to study the possibility of fabricating high crystallinity ZnO without the usage of high thermal source.

When the pressure in the furnace was stabilized, 1064 nm ND-YAG pulsed laser was directed toward the target for ablation process. The fluence on the target was kept at a focal spot of diameter 1mm with six variation of laser power intensity, which was 0.83 J, 2.62 J, 5.17 J, 7.56 J, 9.92 J and 12.04 J respectively. Schematic diagram in Figure 1(a) represents the experiment set up for the laser ablation of ZnO confined in horizontal tube furnace. Figure 1(b) illustrates the beam of pulsed laser focused on ZnO pellet. The surface morphology of deposited substrate was characterized using FESEM.

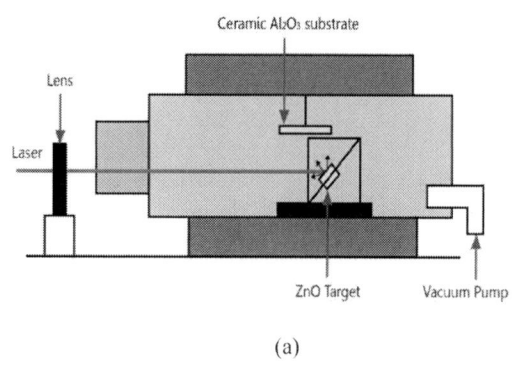

(a)

2018 IEEE International Conference on Semiconductor Electronics (ICSE)

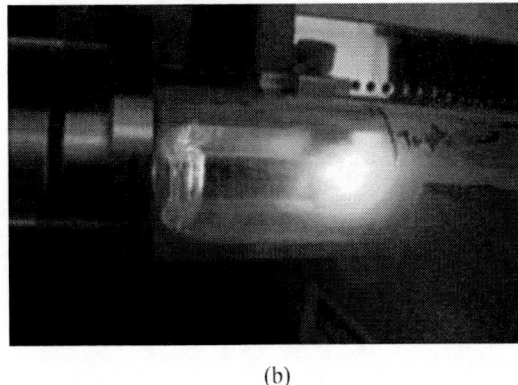

(b)

(b)

Fig. 1. (a) Schematic diagram of experiment set up (b) Laser firing on the ZnO target

(c)

III. RESULTS AND DISCUSSION

FESEM images of ZnO nanostructures fabricated by pulsed laser deposition in low vacuum condition at different laser power are shown in Figure 2. These figures demonstrate that different laser power intensity produces different surface structures. The laser ablation with power intensity of 0.83 J yields morphology of flatten nanoflakes (Figure 2a). In the case of 2.62 J, high porosity nanoparticles are produced compared to the case of 0.83 J (Figure 2b). This suggests the power intensity of laser can penetrate the ZnO target into tiny fragments during the deposition. This statement is evidently supported by the case of 5.17 J in which the morphology consists of highly aggregated nanoflakes with spiky wire-like structure as shown in Figure 2c. In Figure 2(d), (e) and (f), rounded nanoflakes with particles diameter ranging from 30 to 100 nm are observed. These images also suggest that the deposition is more uniform with increasing laser power intensity.

(d)

(a)

978-1-5386-5284-8/18 $31.00 © 2018 IEEE

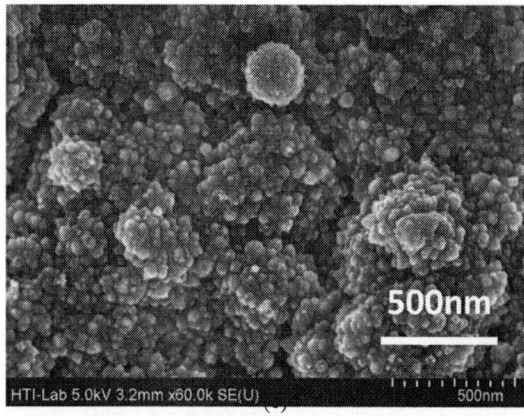

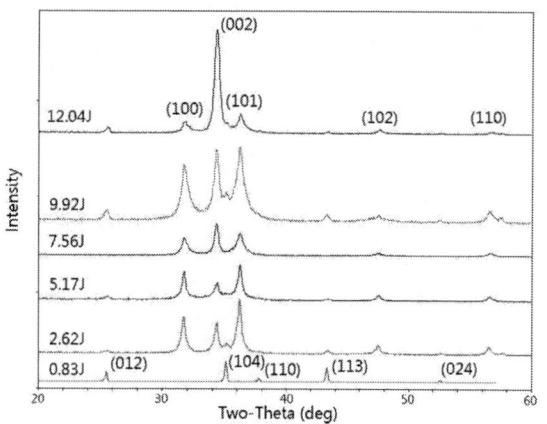

Fig. 3. XRD Image of ZnO nanoflakes by varying the laser power intensity (a) 0.83 J, (b) 2.6 2J, (c) 5.17 J, (d) 7.56 J, (e) 9.92 J and (f) 12.04J on Al_2O_3 substrate

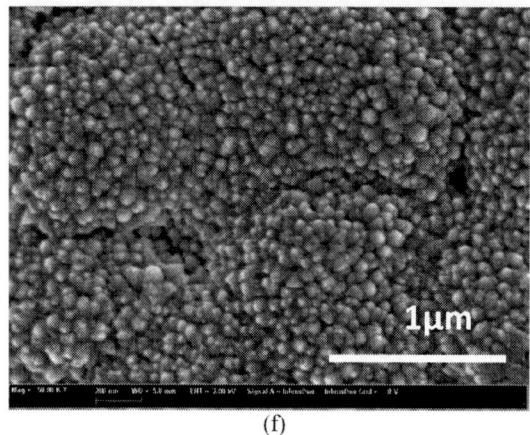

Fig. 2. FESEM Images of ZnO nanoflakes by varying the laser power intensity (a) 0.83 J, (b) 2.62 J, (c) 5.17 J, (d) 7.56 J, (e) 9.92 J and (f) 12.04J

Typical ZnO peaks of (100), (002), (101), (102) and (110) are observed in laser power intensity 2.62 J, 5.17 J, 7.56 J, 9.92 J and 12.04 J[8], [14], [15] in Figure 3. For the case of 2.62 J, 5.17 J, 7.56 J and 9.92 J, ZnO nanostructres were preferentially aligned on (100), (002) and (101) resulting in strong peaks on these planes. (002) plane are found strongest in the case of 12.04 J, implies that the laser intensity favors ZnO orientation on this specifically plane. However, the ZnO relevance peaks did not appear in laser power intensity of 0.83 J. This can be explained by further investigation on the peaks of (012), (104), (110), (113) and (024) where these peaks belong to the ceramic substrate, Al_2O_3 as discussed in previous study[16]. In other words, the power intensity of 0.83 J is insufficient to ablate ZnO target thus resulting in low deposition rate on the substrate. Therefore, ZnO peaks are not detectable under XRD.

IV. CONCLUSION

In conclusion, we observed a flatten nanoflakes at low laser power intensity. As the laser power intensity increases, the surface substrate is deposited with tiny fragment of ZnO nanoflakes which are rounded and more uniform at high laser power. Rounded high crystallinity nanoparticles of size ranging 30 to 100 nm are observed suggesting that the feasibility of fabrication of ZnO nanostructures under low vacuum and room temperature condition. From the XRD results, we deduce that high laser intensity favors the growth of ZnO nanostructure on (002) plane. This setup also has the ability for working on mixing, doping and alloying of different materials by using the hybrid method of vapor transport and pulsed laser deposition in our future experiment. Thermal source can also be provided by the furnace system for future study on the uniformity deposition of ZnO particles with sufficient kinetic movement ability during laser ablation process.

REFERENCES

[1] P. K. Samanta, S. Datta, S. Basak, and T. Kamilya, "Wet chemical growth of ultra-long ZnO nanoplates and their optical property," *Chem. Phys. Lett.*, vol. 584, pp. 155–158, 2013.

[2] Y. Cai *et al.*, "High performance ultraviolet photodetectors based on ZnO nanoflakes/PVK heterojunction," *Appl. Phys. Lett.*, vol. 109, no. 7, 2016.

[3] W. L. Hughes and Z. L. Wang, "Formation of piezoelectric single-crystal nanorings and nanobows," *J. Am. Chem. Soc.*, vol. 126, no. 21, pp. 6703–6709, 2004.

[4] C. O. Solar, "Cuprous Oxide Solution Preparation and Application to Cu2O-ZnO Solar Cells," *Thin Solid Films*, vol. 178, no. c, pp. 163502–163502, 2010.

[5] S. M. U. Ali, Z. H. Ibupoto, M. Kashif, U. Hashim, and M. Willander, "A potentiometric indirect uric acid sensor based on ZnO nanoflakes and immobilized uricase," *Sensors*, vol. 12, no. 3, pp. 2787–2797, 2012.

[6] I. Y. Y. Bu and C. C. Yang, "High-performance ZnO nanoflake moisture sensor," *Superlattices Microstruct.*, vol. 51, no. 6, pp. 745–753, 2012.

[7] B. Behera and S. Chandra, "An innovative gas sensor incorporating ZnO-CuO nanoflakes in planar MEMS technology," *Sensors Actuators, B Chem.*, vol. 229, pp. 414–424, 2016.

[8] B. Deka Boruah and A. Misra, "Energy-Efficient Hydrogenated Zinc Oxide Nanoflakes for High-Performance Self-Powered Ultraviolet Photodetector," *ACS Appl. Mater. Interfaces*, vol. 8, no. 28, pp. 18182–18188, 2016.

[9] P. K. Samanta and A. Saha, "Wet chemical synthesis of ZnO nanoflakes and photoluminescence," *Optik (Stuttg).*, vol. 126, no. 23, pp. 3786–3788, 2015.

[10] T. Pauporté, G. Bataille, L. Joulaud, and F. J. Vermersch, "Well-aligned ZnO nanowire arrays prepared by seed-layer-free electrodeposition and their cassie-wenzel transition after hydrophobization," *J. Phys. Chem. C*, vol. 114, no. 1, pp. 194–202, 2010.

[11] S. Besner and M. Meunier, "Laser Synthesis of Nanomaterials," in *Laser Precision Microfabrication*, vol. 135, 2010, pp. 163–187.

[12] T. Ohnishi, M. Lippmaa, T. Yamamoto, S. Meguro, and H. Koinuma, "Improved stoichiometry and misfit control in perovskite thin film formation at a critical fluence by pulsed laser deposition," *Appl. Phys. Lett.*, vol. 87, no. 24, pp. 1–3, 2005.

[13] G. R. Ashirovich *et al.*, "Nanomaterials: Laser-Based Processing in Gas Phase," in *Nanomaterials*, Wiley-Blackwell, 2012, pp. 105–201.

[14] K. Mahmood, S. Bin Park, and H. J. Sung, "Retracted Article: Enhanced photoluminescence, Raman spectra and field-emission behavior of indium-doped ZnO nanostructures," *J. Mater. Chem. C*, vol. 1, no. 18, pp. 3138–3149, 2013.

[15] S. Gadag and M. Gupta, "Laser Synthesis of ZnO Nanostructures.," *Lasers Eng. Old City Publ.*, vol. 17, no. 3/4, pp. 239–250, 2007.

[16] V. S. K. G. Kelekanjeri, W. B. Carter, and J. M. Hampikian, "Deposition of α-alumina via combustion chemical vapor deposition," *Thin Solid Films*, vol. 515, no. 4, pp. 1905–1911, 2006.

Low Transmittance of Anatase Titanium Dioxide (TiO₂) Prepared via Doctor Blade Technique

M.S.P Sarah
Faculty of Electrical Engineering,
Universiti Teknologi MARA
40450 Shah Alam, Selangor
puterisarah@salam.uitm.edu.my

N. Norizan
Faculty of Electrical Engineering,
Universiti Teknologi MARA
40450 Shah Alam, Selangor
nurallif.norizan@gmail.com

S.S. Shariffudin
Faculty of Electrical Engineering,
Universiti Teknologi MARA
40450 Shah Alam, Selangor
sobihana@salam.uitm.edu.my

H. Hashim
Faculty of Electrical Engineering,
Universiti Teknologi MARA
40450 Shah Alam, Selangor
hashimah655@salam.uitm.edu.my

Abstract— The effect of deposition layer on the physical, optical and electrical properties of TiO₂ thin film that were deposited on glass substrates by doctor blade was studied. X-ray diffraction pattern showed no disruption of crystal and the material is confirmed anatase TiO₂. The thicknesses of the thin films increased as the deposition layer increased. FESEM images show crack-free film when the deposition layer was increase. The optical transmittance showed a very high transparency thin films where the readings decreases from 5% to 0% in the visible region. I-V results indicate that the current does not shows any trend of increasing or decreasing from 1 layer to 5 layers' deposition. However, layer 1 exhibited the highest current with 300 x 10⁻⁹ A.

Keywords— Titanium Dioxide (TiO₂); Doctor-Blade; deposition layer; thickness

I. INTRODUCTION

Nanocystalline Titanium Dioxide (TiO₂) is a well-known semiconductor because of its physical, chemical and optical characteristics. TiO₂ can be found in its three polymorphous (anatase, rutile and brukite). However, anatase phase is usually used in applications due to its excellent stability and photoactivity.

TiO₂ is also sensitive to light and suitable to be use in photo-electrochemical devices such as solar module or photovoltaic module. When it comes to dye sensitized solar cell (DSSC), the composition of TiO₂ and selected dyes could produce chemical reaction as same as redox reaction. Once the light energy is absorbed, oxidation process is occurred and the dye molecule helps to inject electron to the semiconductor conduction band [1]. Although, TiO₂ in DSSC has been study for quite a long time, there are still contingencies that need to be investigated. One of the contingencies is how the thickness of the TiO₂ thin film can manipulates the absorption of light in order to give a better photocurrent conductivity [2]. This work is mainly investigating either a thicker film or a thinner film could give a better performance in application like DSSC [3]. The strong absorption peak in the visible region (between 350 nm to 800 nm wavelength) can be used as natural sensitizer in the visible light range [4, 5]. Thus, thickness of the thin film is a main parameter as it provides larger or smaller internal surface area [6, 7].

There are many methods on preparing TiO₂ such as electrospinning [8], hydrothermal treatment [9], sol-gel method [10] and doctor blade [11]. This work mainly focused on the preparation of TiO₂ thin film by using Doctor

Blade's. Doctor-blade was chosen due to the simple procedure and low cost while the method is famously used compare to the others. Plus, the technique can give a uniform coating of TiO₂ paste. A desired thickness can be control by applying a few layers of TiO₂ paste on top of the glass substrates.

II. METHODOLOGY

The experimental methodology was divided into several sections. Each of the main process are discussed detail in every section.

A. Glass substrate cleaning

The glass was cut into 2 cm x 2 cm sample glass. The cut glasses were then cleaned with acetone. The process is repeated with methanol and distilled water. After the cleaning process is done, all the glass substrates were rinsed using nitrogen blower to ensure no more water droplets on the sample glass.

B. Preparation of TiO₂ by using Doctor-blade's method

For this process, 3 grams of TiO₂ nano-powder (anatase) is place in a beaker weighted by using digital scale. While, 7.5 ml of 99.6% acetic acid glacial and 3 ml of distilled water are prepared in two different measuring cylinders. Then, all of the materials are mixed in the beaker that contained TiO₂ powder. The mixture is gently stirred by using spatula. Lastly, half drop of Triton X-100 is added in the mixture. A smooth white paste is formed as in Fig. 1(a). Before the process ended, the mixture is placed in ultrasonic cleaner for 20 minutes under ambient temperature. Fig. 1(b) shows the TiO₂ paste is applied to the glass and flatten it using glass rod until the TiO₂ films became a uniform layer. The process is then repeated for several times.

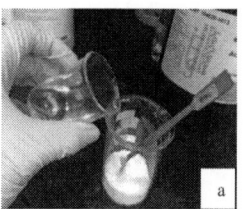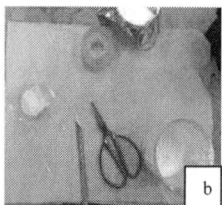

Figure 1: Fabrication process; a) TiO₂ nano-powder is mixed with 99.6% of acetic acid glacial, b) Deposition of TiO₂ paste onto the substrate glass

C. Anealling process

While preparing the TiO_2 paste, the furnace was set to 450 °C. Once the preparation of samples is done, all the samples are placed in the furnace for 30 minutes.

D. Characterizations

After all the process are done, the samples undergo characterization and measurements process. The thickness of the thin films can be measured using surface profiler. The surface morphologies were obtained from FESEM. The crystallinity of the samples was shown from XRD. While, transmittance can be measured by using UV-vis spectrophotometer. Current-voltage are obtained from 2 probes I-V measurement as can be seen in Fig. 2. The parameter is set up and the sample is placed on the stage. Two probes were then attached on the metal contact (Au). The current is measured. Before measuring the current, the metal contact was deposited onto the thin films by using thermal evaporator.

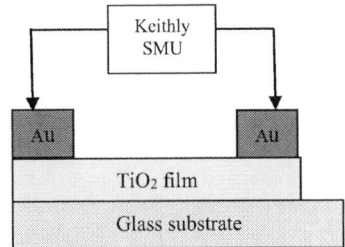

Figure 2: Illustration on *I-V* measurement

III. RESULTS AND DISCUSSION

A. Physical properties

The structural properties were discussed in several section which include XRD pattern, thicknesses of the thin film and FESEM images for all the samples.

I. XRD Pattern

Fig. 3 presents the XRD pattern of TiO_2. As shown in Fig. 1, both layer clearly show the thin films are polycrystalline. From the graph we can see that the diffraction peaks for both layers are at (101) plane due to the electronegativity property of the TiO_2 itself [11]. Within the thin film, some unidentified minerals are also found. This result may be explained by the fact that the lattice structure allows the replacement of oxygen position in TiO_2. However, no disruption of crystal structure and the thin films are confirmed purely composed of TiO_2. It is possible to hypothesis that this conditions are less likely to occur due to annealing process. According to this data, we can infer that TiO_2 thin films was annealed under optimum temperature, 450 °C and very suitable to make TiO_2 became crystalline with anatase structure.

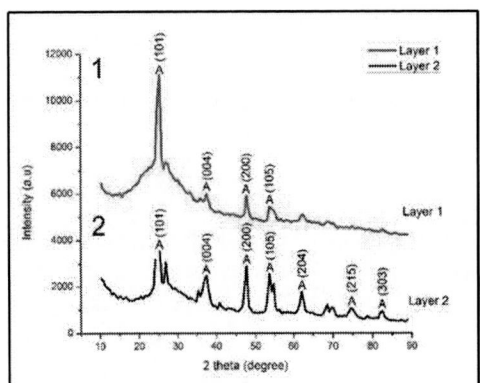

Fig. 3: XRD pattern for 1 and 2 layer

II. Thickness of the thin film

Table 1 provides the experimental data on the thickness of TiO_2 thin film. What stands out in the table is that layer 1 has the minimum thickness which is 16.71 μm, while layer 2, 3 and 4 have thickness of 21.79 μm, 28.62 μm and 39.13 μm respectively. Layer 5 gives the maximum thickness which is 46.16 μm. It is apparent from this table that very few significant changes on the thickness for the whole prepared samples. The result, as shown in Table 1, indicate that increment in thickness is due to the high concentration of TiO_2 paste. As been reported by [7] optimum thickness for a DSSC is around 13.5 μm. This is somehow a significant result outcome.

Table 1: Average thickness for each sample

Layer	Reading 1 (μm)	Reading 2 (μm)	Reading 3 (μm)	Average (μm)
1	16.26	17.05	16.83	16.71
2	21.04	22.11	22.23	21.79
3	28.03	29.01	23.83	28.62
4	38.72	39.45	39.23	39.13
5	46.33	45.93	46.21	46.16

III. FESEM

Fig. 4 compares an overview of the morphological obtained for all 5 samples. The magnification for all the samples were taken at 5000*x*. The most striking result to emerge from the data is that all of the samples are crack-free. This is an interesting outcome since the method used to prepare these samples produced quite thick film. Further analysis showed that the morphological surface of TiO_2 thin film is not abundance with TiO_2 as can be seen in Fig. 4a. While Fig. 4e shows the most abundance of TiO_2 among the other samples. It might be due to this sample is the thickest among the other samples. *Fig.* 4b, 4c and 4d suggested that the growth of the TiO_2 is quite uniform with less agglomeration of TiO_2 particles. This experiment did not detect any evidence of agglomeration of TiO_2 particles for all samples even though technique used to prepare samples produce quite thick film. According to these data, we can conclude that thick films do not show agglomeration of TiO_2 at any point. There are, however, other possible explanations. It seems possible that these results are due to all samples are

978-1-5386-5284-8/18 $31.00 © 2018 IEEE

annealed at optimum temperature as discussed in the XRD pattern.

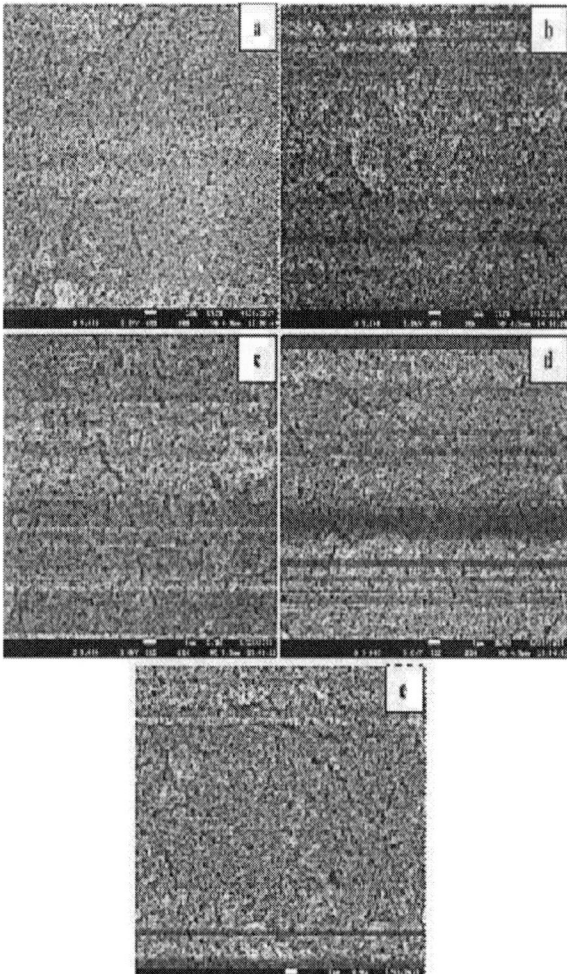

Fig. 4: FESEM images for a) layer 1; b) layer 2; c) layer 3; d) layer 4; e) layer 5

B. Optical properties

Fig. 5 shows the experimental data on transmittance measurement for wavelength between 400 to 800 nm. From the figure we can see no increment or decrement in transmittance. Layer 1 presents transmittance value around 5% at 450 nm. Layer 2 and layer 3 shows the transmittance value decrease to almost 0% respectively. While, layer 4 shows increment up to around 4% and at layer 5 it decreases back to almost 0%. These findings are rather disappointing. An implication of this is the possibility that layer 4 sample is not properly prepared. This inconsistency may be due to the surface morphology as shown in Fig. 4. However, these results are in line with those of previous studies; thicker film produced less transparency thin film.

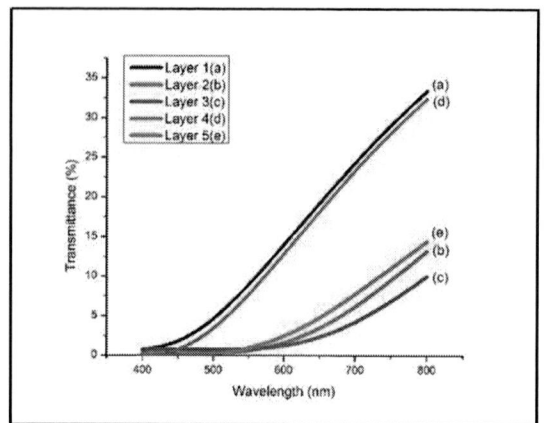

Fig. 5: Transmittance spectra for all samples

C. I-V measurement

Fig. 6 presents the experimental data on I-V measurement. The voltage was applied from -10 V to 10 V. From the graph below we can see that layer 1 shows the highest current among other samples. Layer 1 shows value of current other with 300 nA. Followed by layer 3, 4 and 5 with 50 nA, 15 nA and 10 nA respectively. From layer 1 to layer 3,4 and 5 the current shows trend of decreasing. But, layer 2 shows the lowest current with almost 0 nA. The inconsistency is likely due to the adhesion of the metal contact (Au) with the thin film were not properly done. Due to that reason, there are challenges in the flow of the electron. Also, this result suggested that roughness of the surface contribute to the inconsistency of the current obtained. These results are likely related to transmittance value which also do not demonstrate trends of decreasing for all the samples.

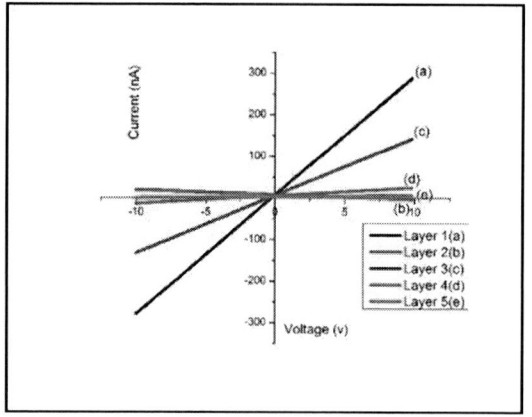

Fig. 6: I-V readings for a) layer 1; b) layer 2; c) layer 3; d) layer 4; e) layer 5

IV. CONCLUSION

In this investigation, the aim was to assess the effect of deposition layer to the physical, optical and electrical properties. The nanocrystalline TiO_2 thin films have been successfully fabricated on the glass substrate by using Doctor-Blades method for five different layers. An additional uncontrolled factor is the possibility that the transmittance and current do not give consistent reading. Whilst this study did not confirm the effect of deposition

978-1-5386-5284-8/18 $31.00 © 2018 IEEE

layer could influence the optical and electrical properties, it successfully produced a crack-free film.

ACKNOWLEDGMENT

The authors would like express their deepest appreciation to all the staffs of NANO-ElecTronic Centre (NET) and NANO-SciTech Centre (NST) for their assistance in completing this work. Special thanks also go to the Faculty of Electrical Engineering, UiTM Shah Alam for the funding.

REFERENCES

[1] M. Grätzel, "Dye-sensitized solar cells," *Journal of Photochemistry and Photobiology C: Photochemistry Reviews,* vol. 4, pp. 145-153, 2003.

[2] A. Sedghi and H. N. Miankushki, "Influence of TiO2 electrode properties on performance of dye-sensitized solar cells," *Int J Electrochem Sci,* vol. 7, pp. 12078-12089, 2012.

[3] A. Nikfarjam, R. Mohammadpour, A. Kasaeian, and Z. Zebhi, "Improving transparency in dye-sensitized nanostructured solar cells by optimizing nano-porous titanium dioxide photo-electrode," *Journal of Materials Science: Materials in Electronics,* vol. 28, pp. 7811-7818, 2017.

[4] J. Wu, J. Wang, J. Lin, Z. Lan, Q. Tang, M. Huang, *et al.,* "Enhancement of the Photovoltaic Performance of Dye-Sensitized Solar Cells by Doping Y0. 78Yb0. 20Er0. 02F3 in the Photoanode," *Advanced Energy Materials,* vol. 2, pp. 78-81, 2012.

[5] W. Wang, H. Yuan, J. Xie, D. Xu, X. Chen, Y. He, *et al.,* "Enhanced efficiency of large-area dye-sensitized solar cells by light-scattering effect using multilayer TiO2 photoanodes," *Materials Research Bulletin,* vol. 100, pp. 434-439, 2018.

[6] R. Keshavarzi, A. Jamshidvand, V. Mirkhani, S. Tangestaninejad, M. Moghadam, and I. Mohammadpoor-Baltork, "The effect of the number of calcination steps on preparing crack free titania thick templated films for use in dye sensitized solar cells," *Materials Science in Semiconductor Processing,* vol. 73, pp. 99-105, 2018.

[7] D. L. Domtau, J. Simiyu, E. O. Ayieta, L. O. Nyakiti, B. Muthoka, and J. M. Mwabora, "EFFECTS of TiO2 FILM THICKNESS and ELECTROLYTE CONCENTRATION on PHOTOVOLTAIC PERFORMANCE of DYE-SENSITIZED SOLAR CELL," *Surface Review and Letters,* vol. 24, 2017.

[8] N. Qin, J. Xiong, R. Liang, Y. Liu, S. Zhang, Y. Li, *et al.,* "Highly efficient photocatalytic H2 evolution over MoS2/CdS-TiO2 nanofibers prepared by an electrospinning mediated photodeposition method," *Applied Catalysis B: Environmental,* vol. 202, pp. 374-380, 2017/03/01/ 2017.

[9] Y. Qiu, F. Ouyang, and R. Zhu, "A facile nonaqueous route for preparing mixed-phase TiO2 with high activity in photocatalytic hydrogen generation," *International Journal of Hydrogen Energy,* vol. 42, pp. 11364-11371, 2017/04/20/ 2017.

[10] J. Reszczyńska, T. Grzyb, Z. Wei, M. Klein, E. Kowalska, B. Ohtani, *et al.,* "Photocatalytic activity and luminescence properties of RE3+–TiO2 nanocrystals prepared by sol–gel and hydrothermal methods," *Applied Catalysis B: Environmental,* vol. 181, pp. 825-837, 2016/02/01/ 2016.

[11] T. Azizi, A. E. Touihri, M. Ben Karoui, and R. Gharbi, "Comparative study between dye-synthesized solar cells prepared by electrophoretic and doctor blade techniques," *Optik,* vol. 127, pp. 4400-4404, 2016/05/01/ 2016.

2018 IEEE International Conference on Semiconductor Electronics (ICSE)

Performance Comparison on Current Consumption Between Arduino Nano and Arm Coretex M3 for Portable Dialysis System

Farrah Masyitah Bt Mohd Shuib[1,2], Shafii B Abd Wahab[1], Abdul Hafiz B Mat Sulaiman[1], Muammar Nur Iman B Ibrahim[1], Marinah Othman[2], Sawal Hamid Md Ali[1], Azrul Azlan Hamzah[1].

Corresponding Author : farrah_5782@yahoo.com

[1] Microengineering and Nanoelectronic System Laboratory (MAINS), Institute of Microengineering and Nanoelectronics (IMEN), Level 4, Research Complex, Universiti Kebangsaan Malaysia, 43600, Bangi, Selangor, Malaysia.

[2]Faculty of Engineering and Built Environment, Level 2, Faculty of Science and Technology Buiding, University Sains Islam Malaysia, Bandar Baru Nilai, 71800, Nilai, Negeri Sembilan, Malaysia.

Abstract—The need of a long battery life span is crucial in portable devices. Following that, this study is conducted to extensively monitor the current performance drawn from the Arduino Nano and ST Arm Core ST32L152RE used as the controller in a portable dialysis system, specifically at the Active and Low Power Mode. The results obtained showed that the ST Arm Core ST32L152RE managed to achieve 99.99% current saving as opposed to the Arduino which gave a rate of 99.74%, despite the fact that the former runs at twice the frequency of the latter.

Keywords—ST Arm Core, Arduino Nano, Current Consumptions

I. Introduction

The Internet of Things (IoT) is a concept reflecting a connected set of anyone, anything, anytime, anyplace, any service, and any network, a megatrend in next-generation technologies that can impact the whole business spectrum and can be thought of as the interconnection of uniquely identifiable smart objects and devices within today's internet infrastructure with extended benefits. Benefits typically include the advanced connectivity of these devices, systems, and services that goes beyond machine-to-machine (M2M) scenarios [1]. Therefore, introducing automation is conceivable in nearly every field. The IoT provides appropriate solutions for a wide range of applications such as smart cities, traffic congestion, waste management, structural health, security, emergency services, logistics, retails, industrial control, and health care [1]. Medical care and health care represent one of the most attractive application areas for the IoT [2].

With the advent of technology, comes the challenge of safeguarding the data being both transmitted and received [3]. It has been found previously that an increase in the number of total bits alone gave the best hardware performance overall [4]. The challenge therefore lies on the ability to maintain the security of the data which will generally involve an increase in computational process while at the same time, trying to either maintain or reduce the

current consumption. While the algorithms in use during the encryption process have the same level of randomness, and hence strength in terms of security, at the hardware level, it has been shown that for certain designs, when the total bits in use is increased, the total power consumed actually reduces [5].

Some of the IoT devices have constrained capabilities and limited access to power. Battery capacity also limits the performance of the wireless communication channel and lifetime of the device. Therefore reduction of cost and power consumption of IoT devices is an important issue [6]. The IoT revolution in the e-health will improve the quality of healthcare services as well as reduce the healthcare cost [7]. Indirectly, it is crucial to sustain the battery duration along with time operation especially in wearable devices such as in portable dialysis system, hence the need to look at the current consumption factor in order to increase the device performance as a whole.

The conventional Dialysis System makes use of gravity to control the flow rate of the Dialysate and waste fluid. The introduction of a pump will help to ease and monitor the flow rate during the process, while enabling the portability of the device.

In this study, the Arduino Nano Board and ST Arm Core (ST32L152RE Nucleo) mini processor are used. The Arduino Nano, fabricated by Arduino Incorporated, is the miniaturized version of Arduino Uno, and has similar functions to the ST Arm Core (ST32L152RE Nucleo) fabricated by ST Microelectronics.

Portable Dialysis System

In coming up with an efficient Portable Dialysis System various factors needs to be taken into account. It was reported that proper encapsulation to achieve complete isolation of the sensor or actuator elements would increase device performance as well as its lifetime [8]. In addition to this, a new supercapacitor design [9] has been shown to enhance the charging mechanism, an important factor, as the biomedical and bioMEMS systems makes use of energy

978-1-5386-5284-8/18 $31.00 © 2018 IEEE   148

storage devices in order to power the system up. Yunas et. al., in turn, looked at the effect of the substrate and observed a significant improvement in the performance of micro-transformers fabricated a highly resistive glass substrate as opposed to that fabricated on a silicon substrate. The newly developed fabrication process which utilizes a simple micromachining technique that includes an additional adhesive bonding between glass and silicon substrate using conductive epoxy material to integrate the device with the silicon CMOS devices, enable the fabrication of a fully integrated circuit block with arbitrary shapes of coil structures over silicon substrate [10].

This paper presents the study of two mini processor which have been used as controller to establish function of Portable Dialysis System which are the Arduino Nano and ST Arm Core (ST32L152RE Nucleo). Comparison is done in terms of the curent consumed during the operation.

II. METHODOLOGY

The basic component making up the Portable Dialysis System is shown in Figure 1.

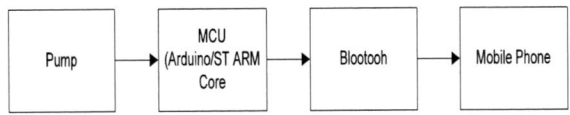

Figure 1: Block Diagram of Components in Portable Dialysis Machine

A. Coding Step for Portable Dialysis System

To establish the function of the Portable Dialysis System, both the mini processor boards are coded as shown in the following flow charts.

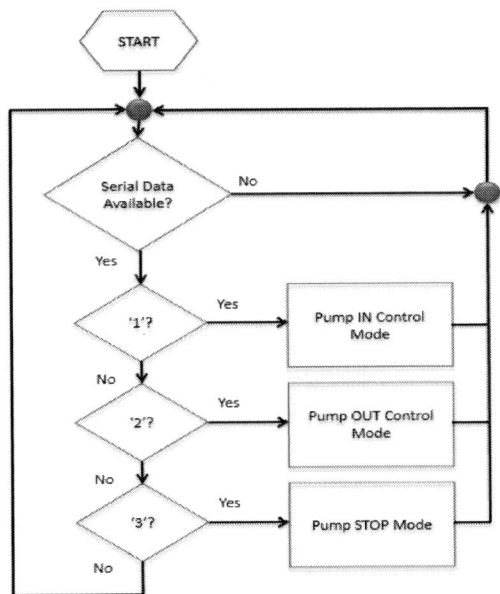

Figure 2: Flowchart of Main Program in Portable Dialysis Machine

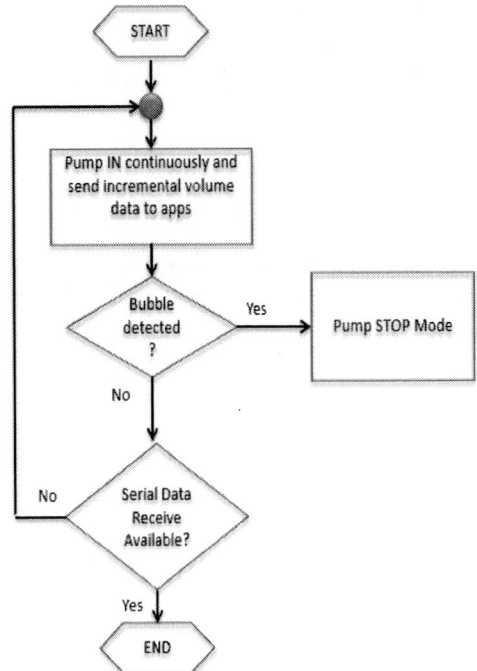

Figure 2(a): Flowchart of Subprogram (Pump In Mode) in Portable Dialysis Machine

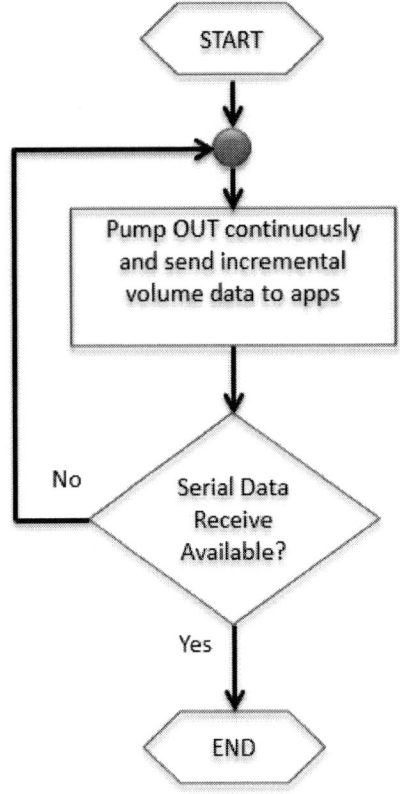

Figure 2(b): Flowchart of Subprogram (Pump Out Mode) in Portable Dialysis Machine

978-1-5386-5284-8/18 $31.00 © 2018 IEEE

B. Low Power Mode Execution for Arduino and ST Arm Core Board.

Once set up, to allow for a better comparison, both boards have been set to Low Power Mode when looking at the current consumption by both.

The steps taken to activate the Low Power Mode are as follows:

1.1) Arduino Nano Board
To establish the Low Power Mode, a new library for executing Low Power Mode has been added to the compiler, to force sleep the ADC and BOD Timer while running at half of the original board's oscillator speed.

1.2 ST Arm Core (ST32L152RE Nucleo)Board
Likewise for the ST Arm Core board, where upon the addition of another set of program in order to perform the Low Power Mode, the WDT and clock are automatically forced to run at half of the original processor's speed.

The experimental setups for both boards are shown in Figure 3(a) and 3(b):

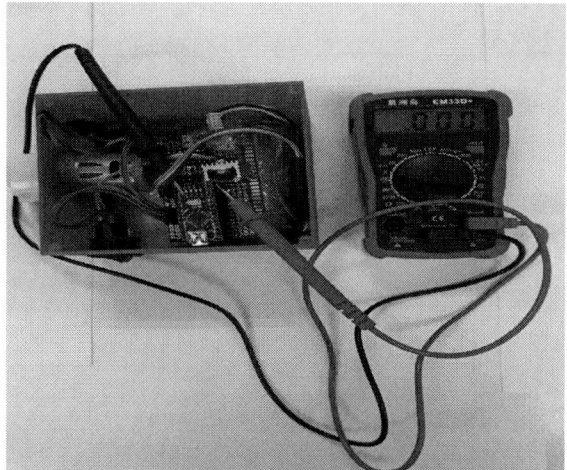

Figure 3(a): Experiment Setup for Arduino Nano Board.

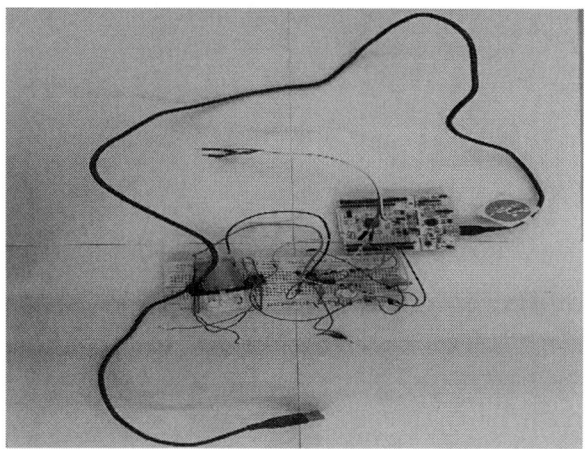

Figure 3(b): Experiment Setup for ST Arm Core (ST32L152RE) Board.

III. RESULT AND DISCUSSION

In this section, the performances in terms of the current consumption in both the boards used will be presented in order to compare the efficiency of the tested mini processors in both Active and Low Power Mode, with the latter at half the original oscillator frequency.

Boards	Mode	Oscillator Frequency	Current	Percentage
Arduino Nano	Active	16Mhz	10mA	99.74
	Low Power	8Mhz	26μA	
ST Arm Core (ST32L152RE Nucleo)	Active	32Mhz	16mA	99.99
	Low Power	16Mhz	1.7μA	

Table 1: Result tested on Active and Low Power Mode for both processor.

From Table 1, it can be seen that the ST Arm Core (ST32L152RE) boards shows a 99.99% reduction of current consumed when running at half of the original clock speed, and likewise the Arduino board. In terms of the total current reduction, the St Arm core board gave a better performance in saving current n sleep mode even when running at double frequency as compared to the Arduino board.

IV. CONCLUSION

It has been observed that the ST Arm Core (ST32L152RE) has shown better performance in terms of current reduction despite running at twice of Arduino Nano Low Power Mode's frequency thus indirectly indicating its ability to achieve the desired response time as it is known that lower the frequency might affect the processor speed in in terms of receiving and transmitting dialysis data.

ACKNOWLEDGMENT

Sincere regards to React Native from MAINS group IMEN UKM for cooperative effort given as well goes to FKAB, USIM as to accomplish this project. Authors also specifically honored the gratitude to these respective funds as follows:

1) DPP-2018-006

Dana Pembangunan Penyelidikan PTJ.

Peranti bioperubatan dengan IOT.

2) PRGS/1/2017/TK05/UKM/02/1

Skim Geran Penyelidikan Pembangunan Prototaip (PRGS).

Design and Fabrication of Silicon Membrane Filtration System for Artificial Kidney.

REFERENCES

[1] S. M. R. Islam, D. Kwak, H. Kabir, M. Hossain, and K.-S. Kwak, "The Internet of Things for Health Care : A Comprehensive Survey," *Access, IEEE*, vol. 3, pp. 678–708, 2015.

[2] Z. Pang, *Technologies and Architectures of the Internet-of-Things (IoT) for Health and Well-being*, vol. Doctoral, no. ISBN 978-91-7501-736-5. 2013.

[3] S. Y. A. M. Fauzi, M. Othman, F. M. M. Shuib, and K. Seman, "Modified A5/1 Stream Cipher for Secured Global System for Mobile (GSM) Communication," *Proceeding 3rd Int. Conf. Artif. Intell. Comput. Sci.*, no. October 2015, pp. 12–13, 2015.

[4] S. Yohana, M. Fauzi, M. Othman, F. Masyitah, M. Shuib, and K. Seman, "The Effect of Total Bits and Number of LFSRs on the Hardware Performance of Modified A5 / 1 Stream Ciphers," pp. 1–5, 1994.

[5] S. Y. A. M. Fauzi, M. Othman, F. M. M. Shuib, and K. Seman, "Hardware Implementation of Modified A5/1 Stream Cipher," *Int. J. Comput. Theory Eng.*, vol. 9, no. 5, pp. 385–389, 2017.

[6] O. Horyachyy, "Comparison of Wireless Communication Technologies used in a Smart Home: Analysis of wireless sensor node based on Arduino in home automation scenario," no. June, 2017.

[7] S. Niranjana and A. Balamurugan, "Intelligent E-Health Gateway Based Ubiquitous Healthcare Systems in Internet of Things," *Int. J. Sci. Eng. Appl. Sci.*, vol. 1, no. 9, pp. 284–290, 2015.

[8] A. A. Hamzah, J. Yunas, B. Y. Majlis, and I. Ahmad, "Sputtered encapsulation as wafer level packaging for isolatable MEMS devices: A technique demonstrated on a capacitive accelerometer," *Sensors*, vol. 8, no. 11, pp. 7438–7452, 2008.

[9] H. E. Z. Abidin, A. A. Hamzah, B. Y. Majlis, J. Yunas, N. A. Hamid, and U. Abidin, "Electrical characteristics of double stacked ppy-pva supercapacitor for powering biomedical mems devices," *Microelectron. Eng.*, vol. 111, pp. 374–378, 2013.

[10] J. Yunas, A. A. Hamzah, and B. Y. Majlis, "Fabrication and characterization of surface micromachined stacked transformer on glass substrate," *Microelectron. Eng.*, vol. 86, no. 10, pp. 2020–2025, 2009.

Silver-Graphene Oxide Nanocomposite Film-based SPR Sensor for Detection of Pb^{2+} Ions

Wan Maisarah Mukhtar
Faculty of Science and Technology
Universiti Sains Islam Malaysia (USIM)
Nilai, Malaysia
wmaisarah@usim.edu.my

Razman Mohd Halim
National Metrology Institute of Malaysia
(NMIM)
Sepang, Malaysia
razmanmh@sirim.my

Karsono Ahmad Dasuki
Faculty of Science and Technology
Universiti Sains Islam Malaysia (USIM)
Nilai, Malaysia
drkarsono@usim.edu.my

Affa Rozana Abdul Rashid
Faculty of Science and Technology
Universiti Sains Islam Malaysia (USIM)
Nilai, Malaysia
affarozana@usim.edu.my

Nur Athirah Mohd Taib
Faculty of Science and Technology
Universiti Sains Islam Malaysia (USIM)
Nilai, Malaysia
athirahtaib@usim.edu.my

Abstract—**This study discussed the effect of nanocomposite film's thicknesses on the generation of SPR for detection of Pb^{2+} ions. Ag-GO nanocomposite film was deposited on the hypothenuse side of the triangular prism via Kretschmann configuration. The thicknesses of Ag-GO were varied between 50nm and 60nm by manipulating the Ag's thicknesses from 30nm to 50nm and GO's between 2nm and 10nm. The optimum SPR signal was obtained as thickness of Au-GO was fixed at 50nm with $t_{Ag}=48nm$ and $t_{GO}=2nm$ resulting 75.77% of SPP excitation. As thicknesses of GO increased above 2nm and thicknesses of Ag were set less or more than 48nm, the SPP excitation became weaker. The sensitivity of SPR sensor shows a good agreement with the generation of SPP. The sensor exhibits the optimum sensitivity of $s=0.452°/RIU$ with angle shifting of 0.905° as nanocomposite film under configuration of $t_{Ag}=48nm$: $t_{GO}=2nm$ and $t_{Ag}=50nm$: $t_{GO}=2nm$ were employed. In conclusion, the amplification of SPR sensor's sensitivity can be achieved by introducing 50nm nanocomposite films consist of $\pm2nm$ thicknesses of GO.**

Keywords—nanocomposite film, Ag-GO, SPR sensor, sensitivity, Kretschmann configuration

I. INTRODUCTION

The development of water quality monitoring system is a critical element in the assessment of polluted water's level. Water pollution usually occurs due to the rapid urbanization and industrialization activities [1]. Heavy metal ions such as Cd, nickel (Ni), Cu, As, Hg and Pb are the main contributor in water pollution due to the emission from industrial wastewater [2-3]. The exposure of heavy metal ions to the human results numerous hazardous illnesses such as kidney damage, hallucinations, reproductive effect, carcinogenicity and neurological effects [4-5]. Thus, it is important to evaluate the concentrations and distribution of heavy metals in our ecosystem. Until now, many studies have been

conducted to cure and prevent water pollution issues by introducing numerous remediation method [5-7] and employing sensor to detect the presence of heavy metals ion such as tapered fiber optics sensors, electronic sensors and electrochemical sensors [8-10].

Recently, surface plasmon resonance (SPR) sensor has earned more attention in environmental issues due to its privilege in detecting diverse types of hazardous chemicals either in a form of liquid or gas [11-12]. Apart from its simple optical setup which consists of prism that acts as light coupler, p-polarized laser and photodetector, the crucial requirement for the development of SPR sensor is the presence of noble metal which allowed the generation of surface plasmon polaritons (SPP) [13]. The existence of plasmons is characteristic of the interaction metal structures with light. Most common examples of noble metals which have been used in SPR are gold (Au), silver (Si) and platinum (Pt). Despite the shortcoming of Ag which is easily oxidized than Au, the Ag based SPR sensor shows better sensitivity and selectivity than Au [14]. Ag have been attracting attention because of their excellent optical and electronic properties, high catalytic activity and biocompatibility [15].

Lately, the application of graphene has emerged in SPR based sensors [16-17]. Graphene is a smart novel material for sensing due to its large specific surface area for molecular adsorption, high aspect ratio and outstanding electrical properties and high carrier mobility [18]. An extensive research on graphene leads to the studies on graphene oxide (GO) [19-20]. The difference between graphene and GO is the addition of oxygen atoms bound with the carbon scaffold which results opposite properties between these two chemicals. Graphene is hydrophobic, meanwhile GO is hydrophilic which is easily dispersible in water [18].

The authors would like to acknowledge the support of Malaysian Ministry of Higher Education (MOHE) through Universiti Sains Islam Malaysia (USIM) under grant USIM/FRGS/FST/32/51514 and Knoll Group from Max Planck Institute Research for the Winspall 3.02 simulation software.

Owing to the rapid development of SPR sensor, increasing the sensitivity of this optical sensor become the top priority. The optimum thickness of noble metal to generate strong SPR is 50nm as reported elsewhere [21-24]. Too thin metal results electron damping issue, meanwhile the too thick leads to the photon absorption that prohibits the oscillation of SPP [14]. The introduction of nanocomposite materials which consist of two or more materials such as hybridization of Au-GO and Au-GO-chitosan etc is found possible to enhance the excitation of SPP [25-26]. However, total thicknesses of nanocomposite film must be take into consideration since light can penetrate a metal only to a very small extent. This issue become the motivation of this study in which the influence of nanocomposite films consist of Ag-GO with numerous thicknesses is investigated. Throughout this study, the range of total thickness are set between 50nm to 60nm by varying both thicknesses of Ag and GO. At the end of this paper, we discuss the influence of nanocomposite thicknesses in detecting the presence of Pb^{2+} ion. We believe that the output of this work will provide a new perspective on the employment of nanocomposite films in SPR sensing application.

II. METHODOLOGY

Surface plasmon polaritons (SPP) was excited by incident p-polarized red laser (λ=633nm) on the Ag-GO coated triangular prism (RI=1.51) under Kretschmann configuration as displays in Fig. 1(a) using Winspall 3.02 simulation software [20]. RI values for Ag and GO were set as $0.0585+4.2665k$ and $3.000+1.1491k$ respectively. By tuning the angle of light incidence of the totally reflected beam inside the prism, the resonance condition for excitation of SPPs can be fulfilled [13]. The values of incident angles were modulated from 39° to 52° with an increment of 0.3° per reading. In this work, thicknesses of Ag and GO were controlled to investigate the effect of nanocomposite film thicknesses on the generation of SPR as depicted in Table 1. The characteristics of SPR were investigated by studying the value of optical minimum reflectance R_{min}, SPR curve depth (Q-factor) and full-width-half-maximum (FWHM). The value of Q factor represents the percentage of incident light which is absorbed by nanocomposite film resulting the generation of SPP. The greater the SPP generation, the stronger the SPR signal [18]. Large FWHM indicated by wide SPR curve defined poor SPP excitation [11].

TABLE I.   VARIOUS THICKNESSES OF AG-GO NANOCOMPOSITE FILM RANGING FROM 50NM TO 60NM

Thin films configuration	Thickness	
	Layer 1, Ag (nm)	Layer 2, GO (nm)
Ag: GO	30	20
	35	15
	40	10
	45	15
	48	2
	50	2
	50	5
	50	10

Fig.1(b) illustrates the SPR experimental setup for detection of the Pb^{2+} ions. Pb^{2+} ions were introduced on top of the nanocomposite film coated prism. The sensitivity of

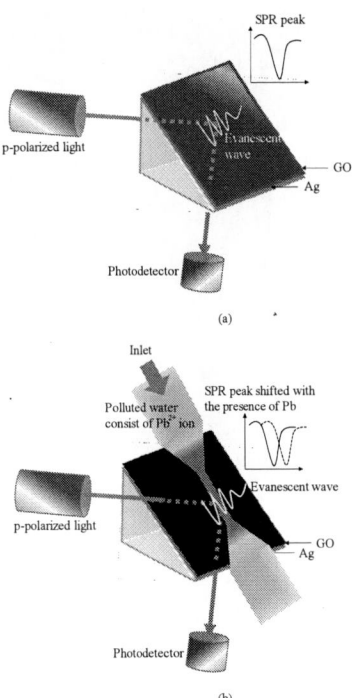

Fig. 1. (a) SPR setup consists of nanocomposite film Ag-GO coated prism (b) SPR peak was shifted as Pb^{2+} ion metal was introduced on top of the coated prism

SPR, s, is determined in degree (°) per refractive index unit (RIU). It was calculated by using equation 1 [27]:

$$s = \frac{\Delta\theta}{\Delta n} \quad \text{(unit: °/RIU)} \tag{1}$$

To evaluate the performance of SPR sensor, analyses on the amount of angle shifting, $\Delta\theta$ and sensitivity, s between with the absence and the presence of Pb^{2+} were performed.

III. RESULTS AND DISCUSSIONS

Fig. 2 displays the SPR curves as p-polarized light was incident through the nanocomposite film Ag-GO coated prism. Previous works proved that the optimum thickness of noble metal for maximum SPR signal can be obtained as the thickness was set at t=50nm [24]. It is clearly seen that maximum SPR signal was obtained as thicknesses of Ag and GO were set at 48nm and 2nm respectively with R_{min}=0.1443 a.u. A sharp and deep SPR curve represented an excellent criterion of SPP excitation. As thickness of Ag was slightly increased to 50nm with thickness of GO remain fixed at 2nm, R_{min} was decreased about 39.50% to 0.2014 a.u. By maintaining t_{AG} at 50nm, we found that the SPR signal became weaker as GO was raised to 5nm where the value of R_{min} was obtained as R_{min}=0.4768 a.u. SPR was almost diminished which was indicated by a very shallow curve as the total thickness of nanocomposite film was raised to 60nm with t_{Ag}=50nm and t_{GO}=10nm. When thicknesses of Ag and GO were reduced to 45nm and 5nm respectively which resulted total thickness of 50nm, R_{min} was obtained as 0.3306 a.u. Poor SPR signals of R_{min}=0.4319 a.u represented by small Q-factor and large FWHM were achieved as

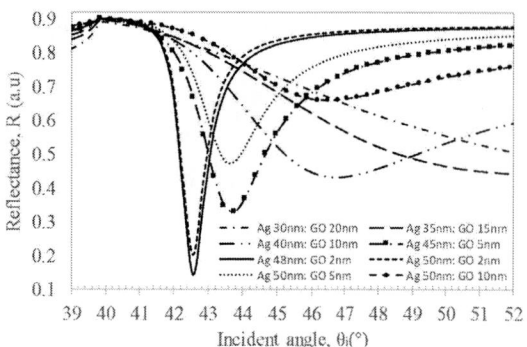

Fig. 2. Influence of various thicknesses of Ag-GO nanocomposite film on SPR curves

thicknesses of nanocomposite were set at $t_{Ag}=40$nm and $t_{GO}=10$nm. The excitation of SPP became unobvious in which the SPR peak are less apparent as thickness of Ag was reduced below 35nm and GO were increased from 15nm to 20nm.

Strong SPR portrays an excellent plasmon absorption through the noble metal [13]. SPP excitation can be determined by considering the SPR curve depth or Q factor of the SPR peak as illustrated in Fig. 3. The weakest SPP excitation about 23.80% was resulted as the thicknesses of films were set at $t_{Ag}=50$nm and $t_{GO}=10$nm. At $t_{Ag}=30$nm and $t_{GO}=20$nm, the sensor experienced 39.22% of plasmon excitation which resulted better enhancement than $t_{Ag}=50$nm: $t_{GO}=10$nm configuration. Maximum excitation up to 75.77% was obtained when total thickness of 50nm ($t_{AG}=48$nm; $t_{GO}=2$nm) of films were deposited on the prism surface. Plasmon excitation was decreased about 5.74% as 52nm thickness of nanocomposite film consisting of $t_{AG}=50$nm and $t_{GO}=5$nm were introduced. Almost 57% of plasmon was generated through the films at $t_{AG}=45$nm and $t_{GO}=5$nm. Note that strong excitation of SPP was produced by employing nanocomposite film with t=50nm where only a very thin layer of GO was employed.

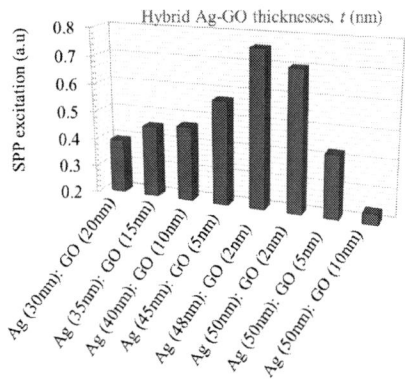

Fig. 3. Various thicknesses of nanocomposite film Ag-GO resulted numerous percentages of SPP excitation ranging from 23.80% to 75.77%

To date, most of the SPR sensors have been developed by depositing a monolayer of noble metal with 50nm thickness. It is noteworthy to highlight that high sensitivity SPR sensor

can be achieved by appointing either 50nm of monolayer metal or composite layer with total thickness of (50±2) nm where ±2nm is referring to the GO thickness. According to the analyses, we obviously observed that thicknesses combination of 96% Ag and 4% GO provide excellent enhancement of SPP excitation by setting $t_{Ag}=48$nm: $t_{GO}=2$nm or $t_{Ag}=50$nm: $t_{GO}=2$nm. If we focus on the influence of composite film thickness in SPR, it is less accurate to conclude that 50nm of thickness provide impressive plasmon excitation. This condition can be proved by considering the condition in which 90% of the composite film thickness consists of Ag ($t_{Ag}=45$nm: $t_{GO}=5$nm) resulted the percentage drop of SPP excitation to 42.67%. This situation occurred due to the limitation coverage part of evanescent field amplitude which depends on metal's thickness.

Next, the influence of nanocomposite film thicknesses of Ag and GO on the sensitivity of SPR sensor was investigated by introducing Pb^{2+} ions metal on top of the coated prism as portrayed in Fig.4. In this analysis, we only consider four significant conditions where SPR were obviously occurred (Fig. (4(a): $t_{Ag}=45$nm: $t_{GO}=5$nm, Fig. (4(b) $t_{Ag}=48$nm: $t_{GO}=2$nm, Fig. (4(c) $t_{Ag}=50$nm: $t_{GO}=2$nm and Fig. (4(d) $t_{Ag}=50$nm: $t_{GO}=5$nm)). The sensitivity of SPR sensor showed a good agreement with SPP excitation. The SPR sensor exhibited excellent sensitivity of s=0.452°/RIU with angle shifting of 0.905° as nanocomposite film under configurations of $t_{Ag}=48$nm: $t_{GO}=2$nm and $t_{Ag}=50$nm: $t_{GO}=2$nm were introduced. The sensitivity of SPR sensor consisting of $t_{Ag}=45$nm: $t_{GO}=5$nm and $t_{Ag}=50$nm: $t_{GO}=5$nm experienced 0.693° of angle shifting which leads to the sensitivity of s=0.426°/RIU for both. Note that for configurations which consists of 96% Ag and 4% GO with different thicknesses ($t_{Ag}=48$nm: $t_{GO}=2$nm and $t_{Ag}=50$nm: $t_{GO}=2$nm) result the similar values of sensitivity. Identical situation was observed for $t_{Ag}=45$nm: $t_{GO}=5$nm and $t_{Ag}=50$nm: $t_{GO}=5$nm configurations. This condition happened due to the amount of SPP excitations between both configurations which were nearly the same with very small percentage difference of ≈7.58%.

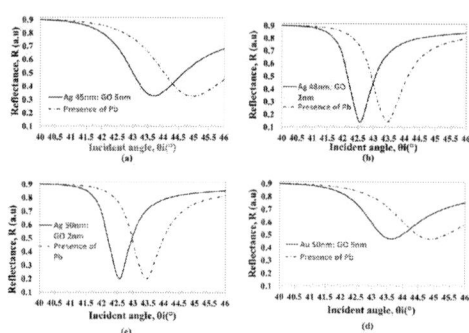

Fig. 4. SPR angle was shifted as Pb2+ ion was introduced on the hypothenuse side of the coated prism (a) $t_{Ag}=45$nm: $t_{GO}=5$nm (b) $t_{Ag}=48$nm: $t_{GO}=2$nm (c) $t_{Ag}=50$nm: $t_{GO}=2$nm (d) $t_{Ag}=50$nm: $t_{GO}=5$nm

IV. CONCLUSION

In conclusion, the thickness of nanocomposite films for each layer is very important in developing a high sensitivity SPR sensor. Although the introduction of GO is proven able to enhance the performance of SPR sensor, its thickness should be set within ±2nm. Maximum SPP excitation can be

978-1-5386-5284-8/18 $31.00 © 2018 IEEE

generated by maintaining the total thickness of nanocomposite film at 50nm. By considering these factors, the development of high efficient SPR sensor is possibly to be achieved. For future works, the influence of other types of noble metals such as gold(Au) and platinum (Pt) in nanocomposite films on generation of SPP will be investigated.

ACKNOWLEDGMENT

The authors would like to acknowledge the support of Malaysian Ministry of Higher Education (MOHE) through Universiti Sains Islam Malaysia (USIM) under grant USIM/FRGS/FST/32/51514 and Knoll Group from Max Planck Institute Research for the Winspall 3.02 simulation software.

REFERENCES

[1] Y. Lu, S. Song, R. Wang, Z. Liu, J. Meng, A. J. Sweetman, A. Jenkins et al. "Impacts of soil and water pollution on food safety and health risks in China." *Environment international, vol.* 77, pp. 5-15, (2015).

[2] Y. Tian, M. Wu, R. Liu, Y. Li, D. Wang, J. Tan, R. Wu and Y. Huang, "Electrospun membrane of cellulose acetate for heavy metal ion adsorption in water treatment", *Carbohydrate Polymers*, vol. 83, no. 2, pp.743-748, 2011.

[3] U. Förstner and G. T. Wittmann, "Metal pollution in the aquatic environment". Springer Science & Business Media. 2012.

[4] P.B Tchounwou, C. G.Yedjou, A. K. Patlolla and D. J. Sutton, "Heavy metal toxicity and the environment", Basel: Springer, 2012.

[5] S. A. Hashim, F. N. D Samsudin, C. S. San Wong, K. A. Bakar, S. L. Yap and M. F. M. Zin, "Non-thermal plasma for air and water remediation", *Archives of biochemistry and biophysics*, 605, pp.34-40. 2016.

[6] G. Mambaand A. K. Mishra, "Graphitic carbon nitride (g-C3N4) nanocomposites: a new and exciting generation of visible light driven photocatalysts for environmental pollution remediation", *Applied Catalysis B: Environmental*, vol. 198, pp.347-377, 2016.

[7] R. Prabhu, F. Hewitt, G. Georgieva, L. A. Lawton, H. N. Meenakshi, and P. K. J. Robertson, "Energy efficient operation of photocatalytic reactors based on UV LEDs for pollution remediation in water". *In IEEE OCEANS 2017-Aberdeen*, pp. 1-5, 2017.

[8] B. Højris, S. C. B. Christensen, H. J. Albrechtsen, C. Smith, C. and M. Dahlqvist, "A novel, optical, on-line bacteria sensor for monitoring drinking water quality", *Scientific reports*, vol. 6, pp.23935, 2016.

[9] F. Attivissimo, C. G. C Carducci, A. M. L. Lanzolla, A. Massaro and M. R. Vadrucci, "A portable optical sensor for sea quality monitoring", *IEEE Sensors Journal,* vol. 15, no. 1, pp.146-153, 2015.

[10] W. M. Mukhtar, P. S. Menon and S. Shaari, "Microfabricated fiber probe by combination of electric arc discharge and chemical etching techniques", *In Advanced Materials Research*, vol. 462, pp. 38-41, 2012.

[11] W. M. Mukhtar, R. M. Halim, K. A. Dasuki, A. R. A. Rashid and N. A. M. Taib, "SPR sensor for detection of heavy metal ions: Manipulating the EM waves polarization modes", *Malaysian Journal of Fundamental and Applied Sciences*, vol. 13, no.4, 2017.

[12] S. Han, X. Zhou, Y. Tang, M. He, X. Zhang, H. Shi and Y. Xiang, "Practical, highly sensitive, and regenerable evanescent-wave biosensor for detection of Hg^{2+} and Pb^{2+} in water", *Biosensors and Bioelectronics*, vol. 80, pp.265-272, 2016.

[13] W. M. Mukhtar, S. Shaari, A. A. Ehsan and P. S. Menon, "Electro-optics interaction imaging in active plasmonic devices", *Optical Materials Express*, vol. 4, no. 3, pp.424-433. 2014

[14] L. Novotny and B. Hecht, "Principles of Nano-Optics," USA: Cambridge University Press, 2006.

[15] S. A. Zynio, A. V. Samoylov, E. R. Surovtseva, V. M. Mirsky and Y. M. Shirshov, "Bimetallic layers increase sensitivity of affinity sensors based on surface plasmon resonance", *Sensors*, vol. 2, no. 2, pp.62-70, 2002.

[16] K. Z. Kamali, A. Pandikumar, G. Sivaraman, H. N. Lim, S. P. Wren, T. Sun and N. M. Huang, "Silver@ graphene oxide nanocomposite-based optical sensor platform for biomolecules", *RSC Advances*, vol. 5, no. 23, pp.17809-17816, 2015.

[17] J. N. Dash R. and Jha, "Graphene-based birefringent photonic crystal fiber sensor using surface plasmon resonance", *IEEE Photonics Technology Letters*, vol. 26, no. 11, pp.1092-1095, 2014.

[18] X. Yang, Y. Lu, B. Liu and J. Yao, "Analysis of graphene-based photonic crystal fiber sensor using birefringence and surface plasmon resonance", *Plasmonics*, vol. 12, no. 2, pp.489-496, 2017.

[19] S. K. Mishra, S.N. Tripathi, V. Choudhary and B. D. Gupta,. "SPR based fibre optic ammonia gas sensor utilizing nanocomposite film of PMMA/reduced graphene oxide prepared by in situ polymerization", *Sensors and Actuators B: Chemical*, vol. 199, pp.190-200, 2014

[20] N. F. Murat, W. M. Mukhtar, A. R. A. Rashid, K. A. Dasuki and A. A. R. A. Yussuf, "Optimization of gold thin films thicknesses in enhancing SPR response", *In Semiconductor Electronics (ICSE), 2016 IEEE International Conference on* , pp. 244-247, 2016

[21] Q. Wang, Q. Li, X. Yang, K. Wang, S. Du, H. Zhang and Y. Nie, "Graphene oxide–gold nanoparticles hybrids-based surface plasmon resonance for sensitive detection of microRNA", *Biosensors and Bioelectronics*, vol. 77, pp.1001-1007, 2016.

[22] W. M. Mukhtar, R. M. Halim and H. Hassan, "Optimization of SPR signals: Monitoring the physical structures and refractive indices of prisms", *In EPJ Web of Conferences*, vol. 162, pp. 01001, 2017.

[23] N. F. Murat, W. M. Mukhtar, P. S. Menon, A. R. A. Rashid, K. A. Dasuki and A. A. R. A. Yussuf, "Influence of electromagnetic (EM) waves polarization modes on surface plasmon resonance", *In EPJ Web of Conferences*, vol. 162, pp. 01008, 2017.

[24] S. Venkatesh, S. Mandava, L. Nayak, S. S. Ramamurthyand S. Neeleswar, "Novel Synthesis of Nanoparticles for Enhancements in Surface Plasmon Coupled Emission", *In International Conference on Fibre Optics and Photonics Optical Society of America*, pp. M4A-44, 2014..

[25] N. F. Chiu, S. Y. Fan, C. D. Yang and T. Y. Huang, "Carboxyl-functionalized graphene oxide composites as SPR biosensors with enhanced sensitivity for immunoaffinity detection", *Biosensors and Bioelectronics*, vol. 89, pp.370-376, 2017.

[26] N. F. Lokman, A. A. A. Bakar, F. Suja, H. Abdullah, W. B. W. A. Rahman, , N. M. Huang and M. H.Yaacob, "Highly sensitive SPR response of Au/chitosan/graphene oxide nanostructured thin films toward Pb (II) ions", *Sensors and Actuators B: Chemical*, vol. 195, pp.459-466, 2014.

[27] B. Lahiri, S. G. McMeekin, R. M. De la Rue, and N. P. Johnson, "Enhanced Fano resonance of organic material films deposited on arrays of asymmetric split-ring resonators (A-SRRs)," Opt. Express, vol. 21, no. 8, pp. 9343–52, 2013.

Optimisation of Pattern Transfer in Fabrication of GFET for Biosensing Applications

Reena Sri Selvarajan
Institute of Microengineering
and Nanoelectronics (IMEN)
Universiti Kebangsaan Malaysia
Bangi, Malaysia
sreenat90@gmail.com

Azrul Azlan Hamzah
Institute of Microengineering
and Nanoelectronics (IMEN)
Universiti Kebangsaan Malaysia
Bangi, Malaysia
azlanhamzah@ukm.edu.my

Siti Aisyah Zawawi
Institute of Microengineering
and Nanoelectronics (IMEN)
Universiti Kebangsaan Malaysia
Bangi, Malaysia
aisyahzawawi@gmail.com

Burhanuddin Yeop Majlis
Institute of Microengineering
and Nanoelectronics (IMEN)
Universiti Kebangsaan Malaysia
Bangi,Malaysia
burhan@ukm.edu.my

Abstract— **This work presents optimized pattern transfer technique for fabrication of GFET by using a conventional microfabrication technique. Although there are many advanced methods are available to produce nanostructures such as ion-beam lithography (IBL), electron beam lithography (EBL), nano-imprint lithography and focused ion beam milling, these techniques often requires high maintenance costs, time and very complicated processes compared to conventional photolithography. This conventional technique is good option to pattern feature size more than 1 micron. In this work, microgap design from chrome mask are transferred on silicon wafer via conventional photolithography. The final stage of fabrication which is developing technique is optimized to the pattern the desired dimension of electrodes. Therefore, developing time is carefully monitored; drawbacks and optimized techniques are reported in this work.**

Keywords— pattern transfer, electrodes, developing time, GFET, photolithography, biosensors

I. INTRODUCTION

Biosensors are defined by International Union of Pure and Applied Chemistry (IUPAC) as 'a device which uses specific biochemical reactions mediated by isolated enzymes, immunosystems, tissues, organelles or whole cells to detect chemical or biological compounds, usually by the use of electrical, thermal or optical signals'[1]. Biosensors are distinguished analytical tools in healthcare, environmental screening and monitoring of food toxins and pathogens due to affordability and simplicity. Incorporating graphene as detecting platform in biosensors provides significant advantages over current available standards due to its unique properties which are ultrahigh sensitivity and excellent stability [2]. Graphene exhibits excellent transducing properties by transmitting signals to indicate presence of analytes in the environment with the assistance of bioreceptors which makes it favorable to be used as

This research is sponsored by AKU-95 HICoE MEMS for Artificial Kidney grant funded by the Ministry of Higher Education (MOHE) Malaysia, Dana Pembangunan Penyelidikan PTJ (peranti Bioperubatan dengan Peranti), and PRGS.

biosensors [3]. Among the myriad biosensing architectures reported, biosensors based on field effect transistors architecture have attracted high attention to be implemented in biosensing applications.

A conventional planar field effect transistor (FET) comprised of three metallic contact conducting electrodes which are the source(S), drain (D) and gate (G) electrodes, thin insulating layer (dielectric) and semiconductor, latter being the active part where charge carriers flow [4]. The current carrying channel is exposed to environment and is in direct contact with the surrounding. Thus, this provides exceptional control on the surface charge of graphene [5]. Therefore, surface based FET biosensors are more sensitive as it directly perceive interactions of biological molecules on its surface and translated it into readable electrical signals [6]. Introducing carbon nanomaterial especially graphene as a sensing material opens up pathway for ultrasensitive, time efficient, low cost, low-noise and portable electrical sensors for biosensing application [7].

Photolithography is an established fabrication method used to transfer patterns and designs onto the surface of photoresist materials [8]. During photolithography process, light source penetrates through a mask with transparent and opaque regions. The patterns or designs on the mask will be transferred on the substrate which is coated with a light sensitive photoresist. Usually positive photoresist dissolves in the light and negative photoresist gets cured as the light passes through, eventually it leaves behind the pattern on the mask [9]. Photolithography resolution is influenced by diffraction of light. As light projects an image onto the wafer, its angle changed. The amount of change of angle is based on the size of the features on the substrate and the wavelength of the light. Too large of an angular deviation can attribute to image distortion of pattern transferred [10]. Therefore, the critical dimension of the feature size is

$$CD=k1*\lambda NACD=k1*\lambda NA$$

where NA is the numerical aperture and k1 is a process related coefficient. Hence, it can be concluded that shorter wavelength attributes to lesser diffraction of light and allows clean fine pattern transferring process.

Over the years, this process has been refined and miniaturized [11]. Currently microlithography is used to produce items such as semiconductors for computers and arrays of different biosensors. Up to date, photolithography has become one of the most successful technologies in the field of microfabrication [12]. Photolithography involves several generalized steps, describes as follows: cleaning of the substrate, coating of the photoresist material, soft baking, exposure, developing, and hard baking [13].

Therefore, this work demonstrates the optimization of pattern transfer process in GFET fabrication using photolithography microfabrication technique. This step is crucial in order to produce source and drain electrodes with desired dimension [14]. Therefore, microgap pattern from chrome mask is transferred on substrate which is silicon wafer via conventional lithography process. Hence, optimized techniques in pattern transfer process and drawbacks are reported in this work.

II. METHODOLOGY

The fabrication process of GFET is illustrated in Fig. 1. Silicon wafer with oxide layer of 300nm was used as a substrate [15]. The wafer was cleaned with acetone, IPA, and DI water. Then, RCA 1 cleaning method was employed to remove any remaining organic particles staggered on top of the substrate [16].Then, the wafer was spin coated with HDMS (hexamethyldisilazane) as an adhesion layer [17] and then followed by positive photoresist (AZ 1500). Subsequently, the wafer was exposed under UV rays for pattern transfer. The UV exposed sample was developed using AZ 300 MIF (developer) [18]. The development of resist is monitored closely to ensure all the unwanted resist is removed. Based on the result obtained, the time taken to develop resist, achieved gap size and quality of developing (underdeveloped or overdeveloped) are monitored and recorded using optical microscope.

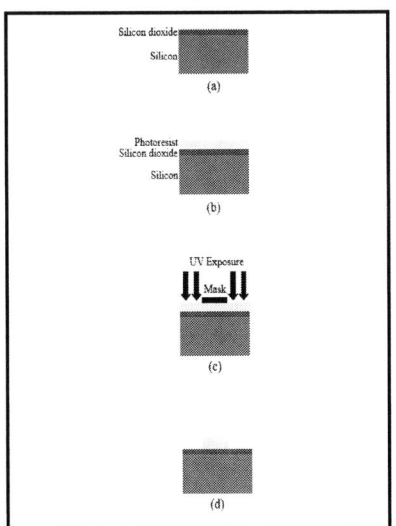

Fig.1: Fabrication steps of GFET. (a) Silicon wafer with oxide layer of 300nm used as substrate. (b) Photoresist coating. (c) UV exposure (d) Pattern transfer after developing process.

III. RESULT AND DISCUSSION

This experiment is carried out to transfer electrodes pattern with 5um gap for GFET based biosensing applications. In order to attain this desired featured dimension, parameters such as resist thickness, PEB time, UV exposure time and developer concentration were carefully observed in this work. The factor that had high implication on the pattern transfer process is development time of the device. In order to investigate the time taken to fully develop the transferred pattern, both normal and abnormal conditions during the development process were given special attention. The samples were immersed in AZ 300 MIF developer and time taken for development of the transferred pattern was recorded.

(a)

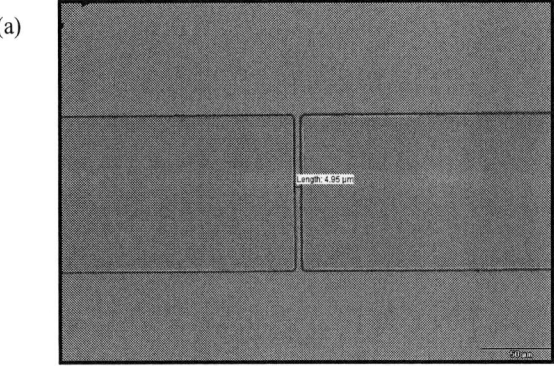

(b)

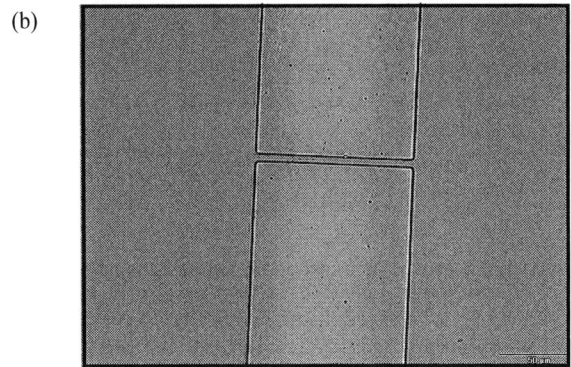

978-1-5386-5284-8/18 $31.00 © 2018 IEEE

(c)

Fig.2: (a) Image of well-developed electrodes.
(b) Under-developed electrodes. (c) Over-developed electrodes.

Fig. 2 (a) shows the image of well-defined image of pattern transferred via photolithography process. The gap between the two electrodes is well attained. The edges or outline of the transferred pattern is well defined. Hence, development process removes the UV exposed and softened region to convert the latent image into usable pattern. It's also worthy to take note that the pattern is fully transferred and this proves that the UV exposure dosage is at optimum.

However, Fig. 2 (b) represents image of an under developed device. This is because the positive photoresist is not fully removed and it remains in the area it should be removed. This result shows that development time should be increased in order to attain desired pattern. However, Fig. 2 (c) illustrates the overdeveloped structures of the transferred pattern. This image is the result of increase in developer beyond the optimum developing time. It's vital to take note that the patterned region turned into rainbow colour complexion. Therefore, optimized developing time is essential in order to transfer and produce pattern with desired dimensions.

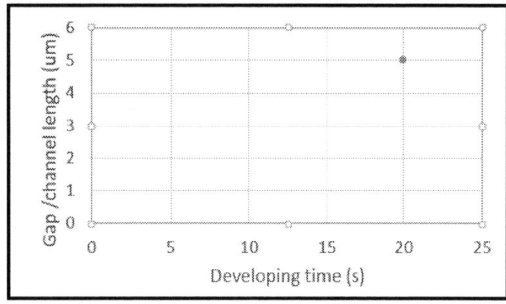

Fig.3: Optimum developing time to pattern electrodes with 3um gap for GFET based biosensing applications.

Fig. 3 illustrates the graphical form of the optimized developing time to transfer pattern with 5um gap for fabrication of GFET device. The developing time for an excellent pattern transfer is at 20 seconds. Immersion of sample in developer (AZ 300MIF) for 20 seconds and rinsed using DI water gives the best outcome. For developing time more than 20 seconds leads to over-development whereas developing time less than 20 seconds attributes to underdevelopment of electrodes.

IV. CONCLUSION

An elementary and conventional method to transfer pattern on substrate is outlined in this work. Process optimization especially for developing coated photoresist is illustrated to transfer pattern with 5um gap. Optimized developing time to transfer pattern to fabricate electrodes for GFET is presented in this work.

Therefore, it can be concluded that developing time to transfer pattern of 5um gap size electrodes are 20seconds. Additional parameters such as UV exposure, post exposure bake time and resist thickness were optimized and kept constant for all the samples throughout the entire work.

ACKNOWLEDGMENT

This research is sponsored by AKU-95 HICoE MEMS for Artificial Kidney grant funded by the Ministry of Higher Education (MOHE) Malaysia, Dana Pembangunan Penyelidikan PTJ (peranti Bioperubatan dengan Peranti), and PRGS.

REFERENCES

[1] Amine, A., Mohammadi, H., Bourais, I. & Palleschi, G. 2006. Enzyme inhibition-based biosensors for food safety and environmental monitoring. Biosensors and Bioelectronics 21(8): 1405-1423.

[2] Pumera, M. (2011). Graphene in biosensing. Materials Today, 14(7–8), 308–315.

[3] Selvarajan, R. S., Hamzah, A. A., & Majlis, B. Y. (2017). Transfer Characteristics of Graphene based Field Effect Transistor (GFET) for Biosensing Application, 88–91.

[4] Zhang, A. & Zheng, G. 2015. Semiconductor nanowires for biosensors. In Semiconductor Nanowires: Materials, Synthesis, Characterization and Applications. Cambridge: Woodhead Publishing. pp. 471-490.

[5] Chaudhary, R., Sharma, A., Sinha, S., Yadav, J., Sharma, R., Mukhiya, R. & Khanna, V.K. 2016. Fabrication and characterisation of Al gate n-metal-oxide-semiconductor field-effect transistor, on-chip fabricated with silicon nitride ion-sensitive field-effect transistor. IET Computers and Digital Techniques 10(5): 268-272.

[6] Hamzah, A. A., Selvarajan, R. S., & Majlis, B. Y. (2017). Graphene for biomedical applications: A review. Sains Malaysiana, 46(7), 1125-1139.

[7] He, Q., Wu, S., Yin, Z., & Zhang, H. (2012). Graphene-based electronic sensors. Chemical Science, 3(6), 1764.

[8] P. Van Zant, Microchip Fabrication: A Practical Guide to Semiconductor Processing. McGraw-Hill, 2000.

[9] McCord, M. a, & Rooks, M. J. (2000). SPIE Handbook of Micromachining and Microfabrication Volume 1. SPIE Handbook of Microlithography, Micromachining and Microfabrication, 1.

[10] Marsi, N., Majlis, B. Y., Mohd-Yasin, F., & Hamzah, A. A. (2014). The fabrication of back etching 3C-SiC-on-Si diaphragm employing KOH + IPA in MEMS capacitive pressure sensor. Microsystem Technologies, 21(8), 1651–1661. https://doi.org/10.1007/s00542-014-2267-8

[11] Postnikov, S., Hector, S., Garza, C., Peters, R., & Ivin, V. (2003). C ritical dimension control in optical lithography. Microelectronic Engineering, 69, 452–458.

[12] B. Ziaie, "Hard and soft micromachining for BioMEMS: review of techniques and examples of applications in microfluidics and drug delivery," Advanced Drug Delivery Reviews, vol. 56, no. 2, pp. 145–172, Feb. 2004.

[13] T. H. S. Dhahi, U. D. A. B. I. N. Hashim, N. M. Ahmed, and A. M. A. T. Taib, "A review on the electrochemical sensors and biosensors composed of nanogaps as sensing material," Journal of Optoelectronics and Advanced Materials, vol. 12, no. 9, pp. 1857–1862, 2010.

[14] Rao, B. S., & Hashim, U. (2014). Pattern Transfer of 1μm Sized Microgap and Microbridge Electrode for Application in Biomedical Nano-Diagnostics. Advanced Materials Research, 925, 533–537. http://doi.org/10.4028/www.scientific.net/AMR.925.533

[15] A. A. Hamzah, B. Y. Majlis, and I. Ahmad, "HF Etching of Sacrificial Spin-on Glass in Straight and Junctioned Microchannels for MEMS Microstructure Release," J. Electrochem. Soc., vol. 154, no. 8, p. D376, 2007.

[16] N. A. Aziz, B. Bais, A. A. Hamzah, and B. Y. Majlis, "Characterization of HNA etchant for silicon microneedles array fabrication," IEEE Int. Conf. Semicond. Electron. Proceedings, ICSE, pp. 203–206, 2008.

[17] Y. Li et al., "Fully integrated graphene electronic biosensor for label-free detection of lead (II) ion based on G-quadruplex structure-switching," Biosens. Bioelectron., vol. 89, no. October, pp. 758–763, 2017.

[18] J. Johari, J. Yunas, A. Hamzah, and B. Majlis, "Piezoelectric Micropump with nanoliter per minute flow for drug delivery systems," Sains Malaysiana, vol. 40, no. 3, pp. 275–281, 201

978-1-5386-5284-8/18 $31.00 © 2018 IEEE

A Study on the Atomic Topography of Nanostructured TiO₂ Thin Films: Effect of Annealing

S. Munirah
NANO-SciTech Centre, Institute of Science, Faculty of Applied Science, UniversitiTeknologi MARA, 40450 Shah Alam, Selangor, Malaysia.
munirahsafiay@gmail.com

Rozina Abdul Rani, N. A. M. Asib, M. Robaiah
NANO-SciTech Centre, Institute of Science, Faculty of Applied Science, UniversitiTeknologi MARA, 40450 Shah Alam, Selangor, Malaysia.
rozina.abdulrani@yahoo.com

Z. Khusaimi, Saifollah Abdullah
Faculty of Applied Science, UniversitiTeknologi MARA, 40450 Shah Alam, Selangor, Malaysia.
zurai142@salam.uitm.edu.my

Fazlena Hamzah
Faculty of Chemical Engineering, UniversitiTeknologi MARA, 40450 Shah Alam, Selangor, Malaysia.
fazlena@salam.uitm.edu.my

Salman Alrokayan, Haseeb Khan
Research Chair for Biomedical Applications of Nanomaterials Department of Biochemistry College of Science King Saud University, Riyadh
salrokayan@ksu.edu.sa
haseeb@ksu.edu.sa

M. Rusop
NANO-SciTech Centre, Institute of Science, Faculty of Applied Science, UniversitiTeknologi MARA, 40450 Shah Alam, Selangor, Malaysia.
nanouitm@gmail.com

Abstract—**Titanium dioxide (TiO₂) is one of the most investigated metal oxides due to the wide range of applications such as photocatalysts. Photocatalytic activity will increased when the surface area is higher. Hence, this research focus on the surface topography and roughness of nanostructured TiO₂ films characterized by atomic force microscopy (AFM) in order to obtain thin films with the optimum roughness for photocatalytic activity. TiO₂ thin films were prepared by spin coating method at room temperature. The TiO₂ solutions of 0.1 - 0.2 M were synthesized from titanium butoxide in ethanol. TiO₂ films were deposited on the silicon substrates and annealed at 450°C. The results shown when the film was annealed, the grain were clearly observed. The grain size and the roughness increased when the film were annealed at high temperature. 0.2 M of TiO₂ thin film exhibit the higher roughness with Ra and RMS values were 51.29 and 78.90 nm, respectively.**

Keywords— TiO₂; Molarity; Topography; Roughness; AFM

I. INTRODUCTION

Recently, metal oxides have received a lot of interest and widely studied due to their electrical and optical properties [1]. Among the metal oxides, titanium dioxide (TiO₂) is one of the most important semiconductors that has attracted attention from researchers due to the excellent chemical stability [2], biocompatibility [3], non-toxic [4], high photocatalytic activity [5], and low cost [6]. Nowadays, a lot of applications of TiO₂ in various field have been explored such as solar cells [6], sensors [7], semiconductors [8], photocatalyst [6] and waste water treatment [9]. TiO₂ is an inexpensive semiconductor with a wide band gap of 3.2 eV [6]. TiO₂ exist in three different crystalline forms namely anatase (tetragonal), rutile (tetragonal) and brookite (orthorhombic) [10]. At temperature below 800°C, TiO₂ exist in anatase phase while at higher temperature, it transformed to the more stable rutile phase [11]. Various technique of synthesizing TiO₂ have been done such as sol-gel method [6,10], atomic layer deposition [5], chemical vapour deposition [12], spray pyrolysis [7] and electrodeposition method [10]. Among these methods, sol-gel method has been known to have several advantages including low synthesis temperature, low cost and produce thin, transparent and uniform thin film structure [5].

According to Kaewwiset et al. [11], the growth of crystalline structures of TiO₂ nanoparticles was controlled by the annealing temperature. Hence, it is essential to study the effect of annealing on the nanostructured TiO₂ thin films. The surface morphology was analyzed by using atomic force microscopy (AFM) [13]. The aim of this research is to study the effect of TiO₂ thin film annealing process (before and after annealing) by analyzing the surface morphology and roughness of TiO₂ thin films.

II. METHODOLOGY

A. TiO₂ Sol-gel Solution Preparation

First, To prepare TiO₂ thin films by sol-gel method, titanium butoxide (Ti(OBu)₄), ethanol and glacial acetic acid were used as the precursor, solvent and stabilizer, respectively. The solution was prepared by dissolving the precursor, (Ti(OBu)₄) into the solvent, ethanol at room temperature. Then, glacial acetic acid, distilled water and Triton X-100 were added and this solution was stirred for 2 hours to yield a clear and homogenous solution [14]. After 2 hours, the solution were left ageing for 24 hours before spin coating process.

978-1-5386-5284-8/18 $31.00 © 2018 IEEE

B. Preparation of Silicon Substrates

The silicon substrates were treated with acetone and distilled water and were repeated three times [15]. Lastly, the substrates was blown with Argon gas to dry and clean the substrates from any impurities. This step is important so that the water particles will bring the impurities with them and blown away by argon gas.

C. TiO₂ Thin Films Deposition by Spin Coating Technique

TiO₂ solution was deposited on silicon substrates by spin coating. The coating and drying process was repeated five times on each substrates. The TiO₂ thin films then were annealed at temperature of 450°C for 1 hour.

D. Characterization of TiO₂ Thin Films

Atomic Force Microscopy (AFM) analysis was done using XE-100 Park Systems AFM to study the effect of annealing on the topography properties of TiO₂ thin films with different molarities. The flowchart of the overall synthesis process is shown in Figure 1.

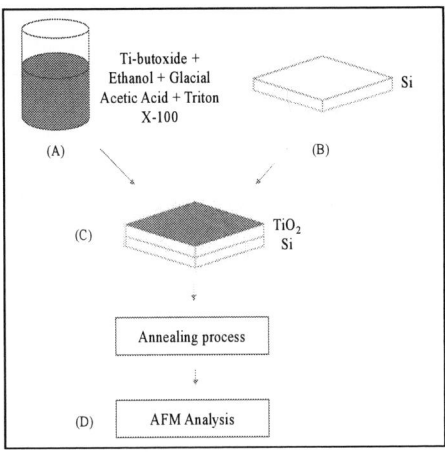

Fig. 1. Flowchart of the TiO₂ thin films preparation.

III. RESULT AND DISCUSSION

The surface topography of TiO₂ thin films of different molarities were observed by AFM and the effect of annealing has been studied. Figure 2 shows two and three dimensional AFM images of the 0.1 M TiO₂ thin films deposited on silica substrates before and annealing taken at 10 μm x 10 μm scan area with non-contact mode configuration. According to the quantitative analysis of the roughness from the AFM analysis in Table 1, the values of average roughness (Ra), root mean square (RMS) and coefficient of kurtosis (RKU) changed between the two samples.

The AFM surface morphology of the as-deposited TiO₂ thin film revealed that the film surface is rather smooth, compact and not uniform [15]. The Ra value obtained was 16.07 nm and RMS value was 26.95 nm. From Table 1, the values of Ra and RMS of this sample are lower than the sample after annealing at 450°C. The photocatalytic efficiency of TiO₂ can be improved by controlling the distribution and nanoparticles size of TiO₂. A significant difference can be seen in Figure 2(b) where the surface of the film shows a higher roughness and clearer grains can be seen. The after annealing sample Ra value was 24.54 nm and RMS value was 32.40 nm. According to Muaz et al. [16], this changes were due to the grains of the as-deposited TiO₂ thin film still in amorphous and surface topography and roughness were highly dependent on the annealing. This shows that the grain growth and regular grain shapes were formed on the surface of the film [9]. Annealing process also helps the grain growth of TiO₂ [17]. A roughness analysis from AFM confirmed that the annealing increased the Ra and RMS values of the sample. Figure 3 and 4 also shows the same results with Figure 2 where the Ra and RMS values of the samples increased after the annealing process. Figure 3 shows the as-deposited and after annealing samples of 0.15 M TiO₂ thin films. The Ra and RMS values of as-deposited sample were 19.78 and 33.57 nm, respectively. After annealing, these Ra and RMS values have increased to 40.51 nm and 50.06 nm. The same pattern was observed in Figure 4 where the as-deposited 0.2 M of TiO₂ thin films Ra was 22.51 nm and RMS was 42.24 nm while after annealing the values increased to 51.29 nm and 78.90 nm. It was clearly observed that the values of Ra and RMS increased with increasing molarities of TiO₂ thin films. This probably due to the scattering mechanism that influence the size of the particles [4]. Increasing RMS value in after annealing thin film also due to the increasing of the grain sizes. During annealing, high temperature resulted in increasing mobility of atoms [18] and the mobility of atoms will causes agglomeration of particles as the particles are compacted with each other. Agglomeration will form larger grains [19] and leads to an increase in the film roughness [16]. For the photocatalytic activity applications, thin films with higher surface roughness will be more photocatalytically active as rough surface provides higher surface area and more photocatalytic sites, leads to a higher photocatalytic activity [20].

TABLE I.   THE AVERAGE ROUGHNESS (RA), ROOT MEAN SQUARE (RMS) AND COEFFICIENT OF KURTOSIS (RKU) OF TiO₂ THIN FILM BEFORE AND AFTER ANNEALING AT 450°C.

Samples		Ra (nm)	RMS (nm)	RKU (nm)
0.1 M	Unannealed	16.07	26.95	13.65
	Annealed	24.54	32.40	4.74
0.15 M	Unannealed	19.78	33.57	14.80
	Annealed	40.51	50.06	3.87
0.2 M	Unannealed	22.51	42.24	17.01
	Annealed	51. 29	78.90	3.07

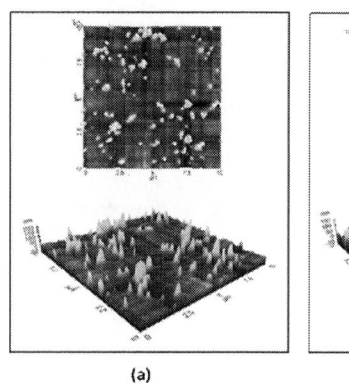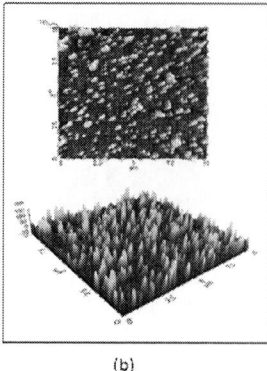

Fig. 2. AFM images of 0.1 M TiO$_2$ thin film on silica substrate (a) unannealed (b) annealed at 450°C.

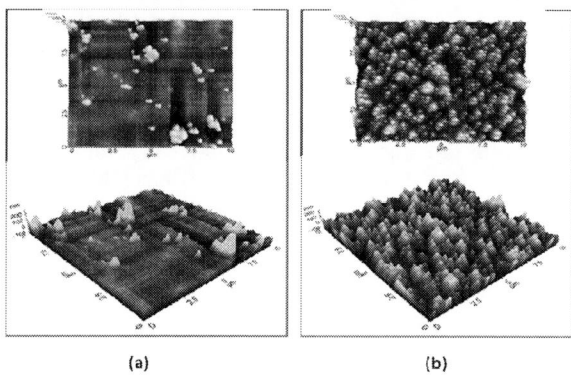

Fig. 3. AFM images of 0.15 M TiO$_2$ thin film on silica substrate (a) unannealed (b) annealed at 450°C.

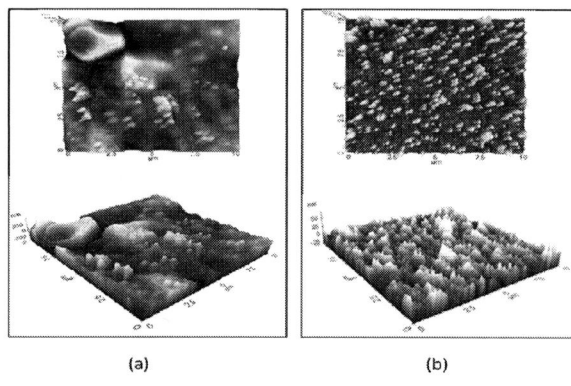

Fig. 4. AFM images of 0.2 M TiO$_2$ thin film on silica substrate (a) unannealed (b) annealed at 450°C.

From the unannealed sample, the value of coefficient of kurtosis (RKU) is higher, which means that the sample has infrequent extreme deviations of the measured height [21]. From the 3D image of the sample, this clearly can be seen as

spikes. The spikes suppressed when the film were annealed at high temperature, hence the RKU value of after annealing sample decrease.

AFM results clearly show that there is a change in the surface topology and roughness of TiO$_2$ thin films before and after annealing process. The most likely reason is due to the grain growth when increasing temperature of which is in good agreement with that reported by Corekci et al. [22]. Our observation also agrees with the previous work of Hadjoub et al. [23] where it was reported that the roughness values of TiO$_2$ thin films increased from 2.93 nm to 5.63 nm, before and after annealing at 400°C. In addition, increasing molarity also causes the roughness value to be increased. Molarity is one of the key for nanoparticles synthesis. The significance to study on the molarity of TiO$_2$ is because it affects the nanoparticles distribution, hence their size.

IV. CONCLUSION

Atomic Force Microscopy (AFM) was used to study the effect of annealing on the surface morphology and roughness of TiO$_2$ thin films synthesized by sol-gel method by using titanium butoxide as a precursor. The different molarities TiO$_2$ thin films were deposited on the silica substrates at room temperature by spin coating technique. In the 0.1 M, 0.15 M and 0.2 M TiO$_2$ thin films, the AFM topography of the unannealed film shows that the film surface is non-uniform and the grains were not clearly visible because it has not been annealed. A remarkable difference occurs after the thin film annealed at temperature 450°C, where the particles are formed more closely and dense. The thin film surface exhibits a higher roughness and clear grains can be observed. The higher molarity thin films gave the higher roughness where the 0.2 M of TiO$_2$ thin films Ra and RMS values were 51.29 and 78.90 nm, respectively.

ACKNOWLEDGEMENT

This work is financially supported by Bestari Perdana grant (600-IRMI/PERDANA 5/3 BESTARI (100/2013)), Institute of Research Management & Innovation (IRMI), Universiti Teknologi MARA (UiTM), and Research Chair for Biomedical Applications of Nanomaterials. Special thanks to Research Chair for Biomedical Applications of Nanomaterials, Department of Biochemistry, College of Science, King Saud University, Riyadh 11451, Saudi Arabia.

REFERENCES

[1] H. Li, J. Wang, H. Liu, H. Zhang, and X. Li, "Zinc oxide films prepared by sol-gel method," *J. Cryst. Growth*, vol. 275, no. 1–2, pp. 943–946, 2005.

[2] H. Dong *et al.*, "An overview on limitations of TiO$_2$ based particles for photocatalytic degradation of organic pollutants and the corresponding countermeasures," *Water Res.*, vol. 79, pp. 128–146, 2015.

[3] M. Hasan Nia *et al.*, "Stabilizing and dispersing methods of TiO$_2$ nanoparticles in biological studies," *J. Paramed. Sci. Spring*, vol. 6, no. 2, pp. 2008–4978, 2015.

[4] P. Sarah, M. Saad, H. B. Sutan, and S. S. Shariffudin, "TiO$_2$ Thin Film via Sol-Gel Method : Investigation on Molarity Effect," *IOP Conf. Ser. Mater. Sci. Eng.*, vol. 12006, 2015.

[5] N. S. Zulkiflee *et al.*, "Characterization of TiO$_2$, ZnO , And TiO$_2$ / ZnO Thin Films Prepared By Sol-Gel Method," *Arpn J. Eng. Appl. Sci.*, vol. 11, no. 12, pp. 7633–7637, 2016.

[6] N. Damsyik, B. Mohd, J. K. Elektrik, and P. Merlimau, "A Study Of TiO$_2$ Thin Film Using Sol-Gel Method," vol. 1, pp. 77–84, 2016.

[7] B. Guo, Z. Liu, L. Hong, and H. Jiang, "Sol gel derived photocatalytic porous TiO$_2$ thin films," *Surf. Coatings Technol.*, vol. 198, no. 1–3 SPEC. ISS., pp. 24–29, 2005.

[8] X. Wang, F. Shi, X. Gao, C. Fan, W. Huang, and X. Feng, "A sol-gel dip/spin coating method to prepare titanium oxide films," *Thin Solid Films*, vol. 548, pp. 34–39, 2013.

[9] M. Alzamani, A. Shokuhfar, E. Eghdam, and S. Mastali, "Influence of catalyst on structural and morphological properties of TiO$_2$ nanostructured films prepared by sol-gel on glass," *Prog. Nat. Sci. Mater. Int.*, vol. 23, no. 1, pp. 77–84, 2013.

[10] M. M. Byranvand, A. N. Kharat, L. Fatholahi, and Z. M. Beiranvand, "A Review on Synthesis of Nano-TiO$_2$ via Different Methods," *J. Nanostructures*, vol. 3, pp. 1–9, 2013.

[11] W. Kaewwiset, W. Onreabroy, and P. Limsuwan, "Effect of Annealed Temperatures on the Morphology of TiO$_2$ Films Sol-gel Dip-coating Annealed at different temperatures," *Nat. Sci.*, vol. 42, pp. 340–345, 2008.

[12] R. S. Sonawane, S. G. Hegde, and M. K. Dongare, "Preparation of titanium (IV) oxide thin film photocatalyst by sol-gel dip coating," *Mater. Chem. Phys.*, vol. 77, no. 3, pp. 744–750, 2003.

[13] O. Ceh, R. Baca, A. I. Oliva, R. D. Maldonado, and S. Debnath, "Understanding the thermal annealing process on metallic thin films," *IOP Conf. Ser. Mater. Sci. Eng.*, vol. 45, no. 12013, pp. 1–6, 2013.

[14] R. A. Rahman, M. A. Zulkefle, W. Fazlida, H. Abdullah, and S. H. Herman, "Effect of Post Deposition Annealing Process on the pH Sensitivity of Spin-coated Titanium Dioxide Thin Film," *Appl. Mech. Mater.*, vol. 749, pp. 197–201, 2015.

[15] I. H. H. Affendi, N. E. A. Azhar, M. S. P. Sarah, S. A. H. Alrokayan, H. A. Khan, and M. Rusop, "Annealing temperature and spin speed effect on TiO$_2$ nanostructured topology and electrical properties," *2015 IEEE Student Conf. Res. Dev. SCOReD 2015*, pp. 517–521, 2016.

[16] A. K. M. Muaz, U. Hashim, and F. I. K. L. Thong, "Effect of annealing temperatures on the morphology , optical and electrical properties of TiO$_2$ thin films synthesized by the sol – gel method and deposited on Al / TiO$_2$ / SiO$_2$ / p -Si," 2015.

[17] S. Nadzirah and U. Hashim, "Synthesis and Film Investigation of Titania," *Adv. Mater. Res.*, vol. 832, pp. 128–131, 2014.

[18] F. Hajakbari and M. Ensandoust, "Study of Thermal Annealing Effect on the Properties of Silver Thin Films Prepared by DC Magnetron Sputtering," vol. 129, no. 4, pp. 680–682, 2016.

[19] I. Senain, N. Nayan, and H. Saim, "Structural and electrical properties of TiO$_2$ thin film derived from sol-gel method using titanium (IV) butoxide," *Int. J. Integr. Eng.*, no. Issue on Electrical and Electronic Engineering, pp. 29–35, 2010.

[20] J. Moore, R. Louder, and C. Thompson, "Photocatalytic Activity and Stability of Porous Polycrystalline ZnO Thin-Films Grown via a Two-Step Thermal Oxidation Process," *Coatings*, vol. 4, no. 3, pp. 651–669, 2014.

[21] H. Ivan, A. Pullmannov, P. Martin, and H. Juraj, "Communications structural and morphological investigations of TiO$_2$ sputtered thin films," *Communications*, vol. 60, no. 6, pp. 354–357, 2009.

[22] S. Çörekçi and K. Kizilkaya, "Effects of Thermal Annealing and Film Thickness on the Structural and Morphological Properties of Titanium Dioxide Films," *Pol. A*, vol. 121, no. 1, pp. 247–248, 2012.

[23] I. Hadjoub *et al.*, "Post-deposition annealing effect on RF-sputtered TiO$_2$ thin-film properties for photonic applications," *Appl. Phys. A Mater. Sci. Process.*, vol. 122, no. 2, pp. 1–8, 2016.

Platinum (Pt) doped Nb₂O₅ for Enhancing Ultraviolet Photodetector

M. Anas A. Basir, A. Manut*, M. Hafiz Mamat,
Rosmalini Abdul Kadir. & A. Sabirin Zoolfakar
Faculty of Electrical Engineering, Universiti
Teknologi MARA, 40450 Shah Alam, Malaysia.
E-mail: azrifmanut@salam.uitm.edu.my

Rozina Abdul Rani & M. Rusop
Nano-SciTech Centre, Institute of Science, Universiti
Teknologi MARA, 40450 UiTM Shah Alam,
Malaysia.

Abstract— This work presents the development of Ultraviolet (UV) Photodetector based on platinum doped niobium oxide (Nb_2O_5) for enhancing UV Photodetector performance. The anodization duration was varied by using Ethylene Glycol and Ammonium Fluoride and observed by UV radiation with different voltage supply. The morphology of Nb_2O_5 nanoporous was characterized by using FESEM. Metal contact with 30 nm thickness of platinum was used using Thermal Evaporator process. Different duration of anodized produces different sizes of nanoporous diameter that affected UV performance. Lastly, I-V characteristic was used to record the reading of UV performance for every sample substrate. The thinner the thickness of the substrate, the higher the performance of UV characterized.

Keywords- Nb_2O_5; nanoporous; anodization; UV performance.

I. INTRODUCTION

Functional integration of Ultraviolet (UV) photodetector with low-dimensional semiconducting nanoporous is an emerging requirement.[1] In the past decade, the researchers have extended their efforts to develop nanostructure-based nanodevices such as gas sensors, solar cells, UV photodetector and others. Ultraviolet (UV) light is an electromagnetic radiation that exists around the surrounding but could not be detected using naked eyes. UV radiation consists 10% of the total light produced by the sun and present in sunlight. UV light consists of higher frequencies of an electromagnetic wave than those that can be detected by the human eyes and the sun emitted UV radiation with the range 200-400 nm. [1] A depletion of the ozone layer will increase UV radiation on earth. The atmosphere of earth is able to absorb the most of UV-B light with range 290-320 nm and the UV-C radiation at range 200-290 nm, but the longer wavelength of UV-A with range 320-400 nm can reach the surface of earth and the exposure to UV-A can cause skin cancer, premature aging of the skin, eye damage and others [1], [2]. However, UV also has a positive aspect where were commonly used in hospitals. UV lamp in hospitals use to sterilize surgical equipment in the operating room and for food and medicine companies use to sterilize their product[3], [4].

The quest of UV photodetector with high sensitivity, better selectivity, and long-term stability has increased in a few years ago [5], [6]. In particular, UV photodetector is receiving significant attention given its technology potential and requirement in environmental control. UV photodetector also widely used in many commercial and military application such as ozone layer monitoring, missile warning systems, and flame detection [7], [8]. Photodetectors is one of the most important semiconductor devices which is convert optical signals into electrical that have wide applications used in light-wave communication, optoelectronic circuits, and imaging [1], [9].

Recently, there is a huge interest to develop UV photodetector based on nanostructured metal oxide such as SnO_2, ZnO, TiO_2, WO_3, CuO, NiO, and Ga_2O_3 [5], [10]. ZnO is commonly used as nanostructure-based device in fabrication of UV photodetector. ZnO have a wide direct band-gap energy, 3.37eV and a larger exciton-binding energy, 60meV that used by sol-gel process [7]. However, there are small number of papers that address the feasibility of Nb_2O_5 for UV photodetector [2], [11].

Niobium pentoxide (Nb_2O_5) are known as a wide bandgap n-type of metal oxide and have desirable properties such as low firm stress, high refractive index, and good chemical stability [1][10]. Nb_2O_5 has been used in various application such as a biocompatible material and catalysis, batteries, solar cell and in electrochromic coating [10]. Nb_2O_5 with 3.4eV of bandgap is good for visible-blind UV-light sensor and optoelectronic circuits, those operating in UV-A [1],[12]. Nb_2O_5 also have excellent optical properties, corrosion resistance and coloration changes [13]. The crystalline of Nb_2O_5 has low conductivity which is approximately 10^{-6} S cm⁻¹ and exhibit a color change from transparent to light blue meanwhile amorphous Nb_2O_5 change color into brownish-gray.[13],[14]. Nb_2O_5 is the most thermodynamically stable between others niobium oxides [14]. Due to the excellent properties in this project, its attracted many researchers to develop UV photodetector based on Nb_2O_5 because of its crystallinity and nanostructure [5], [12].

There are many research published on fabrication of ultraviolet photodetector but the difference of this work is the anodization process of Nb_2O_5 UV photodetector. Anodization is most widely used in nanofabrication due to its capability to produce highly porous and ordered oxide morphologies [11], [15]. The anodized oxides obtained have many advantages such as large surface area, film thickness, self-organized formation, and controllable pores [11]. These properties shown the

978-1-5386-5284-8/18 $31.00 © 2018 IEEE

potential anodized in various application including sensing, solar cell, and others [10], [11], [13].

II. METHODOLOGY

A. Preparation of cleaning niobium foil.

The niobium foil was cut into dimension of 2.5cm x 1.0cm and then immersed into the acetone and treated in ultrasonic bath for 10 minutes at 50 ˚C. The treated foils were then rinsed with distilled water. Next, the foil was blown dry with Nitrogen gas (N_2) to eliminate any contaminant. After that, the foil was placed with ethanol solution and immersed in ultrasonic bath for 10 minutes at 50 ˚C. The foil then rinsed with distilled water for several times and blown dry with Nitrogen gas (N_2) to eliminate any contaminant left.

B. Anodization process and Annealing.

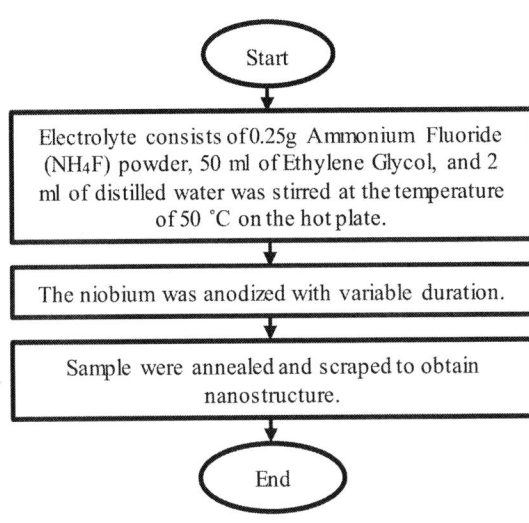

Figure 1: Flowchart of Nb_2O_5 anodization and annealing process.

Figure 1 shows the flowchart of Nb_2O_5 anodization process. For anodization process, the electrolyte was prepared with 0.25g of Ammonium Fluoride (NH_4F) powder, 50 ml of Ethylene Glycol, and 2 ml of distilled water. The Ammonium Fluoride (NH_4F) was weighed using electronic balance before. After weighing the substrate, the electrolyte was stirred on the hot plate and kept the temperature to about 50 ˚C. Usually, there were two terminals or two electrodes used in anodization process set-up which consist of cathode and anode. For anode, niobium metal was placed while for cathode, platinum was placed. Then, the anodizing duration was varied from 15, 30, 60, and 90 minutes in this process. By minimizing the induced defect states during the anodization process, the result produced anodized oxide that can be developed for various application from sensing to photoconversion [11], [16]. The fabrication of anodized Nb_2O_5 on foil without contaminant is by optimizing electrolyte and

anodization condition [11], [17]. The previous fabrication also included the anodization process but the study did not focus on the effect of anodized duration on UV photodetector of Nb_2O_5 nanoporous. Variation of anodized duration caused changes in the size and thickness of Nb_2O_5 nanoporous. Therefore, the UV performance of Nb_2O_5 nanoporous is determined by the photodetector thus the exposure UV radiation.

Next, the samples were annealed using furnace at the temperature of 440 ˚C. The process took 30 minutes to anneal niobium foil to improve crystallinity of porosity. Then, a cleaned razor blade was used to scrape the oxide nanostructure from the Nb foil substrates in Nb_2O_5 nanostructure.

III. RESULT AND DISCUSSION

A. Effect of duration of anodization process to Nb_2O_5 nanoporous size.

The anodization process of Nb_2O_5 layer started by supplying a direct current to niobium foil and platinum under chemical electrolyte. The anode side contains a niobium metal and oxygen produced. For cathode side which is the negative electrode, where the platinum is placed and hydrogen is formed. When the oxygen and hydrogens combines, it will convert niobium (Nb) metal to niobium oxide (Nb_2O_5). The electrolyte that were used in anodization process contains fluorine ions and chemical dissolution happens during anodization process. The voltage connected to niobium metal was 10V and chemical dissolution reacted as anode placed in Nb_2O_5 thin film. Thus, nanopores were produced in Nb_2O_5 thin film.

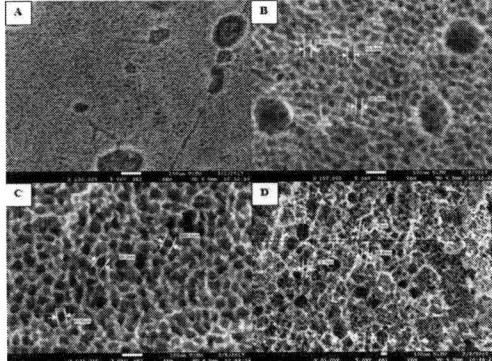

Figure 2: FESEM images of Nb_2O_5 nanopores that were produced in different times of anodization process; (A) 15 minutes, (B) 30 minutes, (C) 60 minutes and (D) 90 minutes.

Figure 2 shows the FESEM images of nanopores of Nb_2O_5 that were produced in different time of anodized process such as 15, 30, 60 and 90 minutes. The left top of image FESEM shown the structure of the Nb_2O_5 nanoporous for 15 minutes. As shown in the figure, a highly uniform pore distribution can be seen in nanosized pores. The size of nanopores depends on the duration of the anodization process. The longer duration of anodized caused more chemical dissolution on the thin film and the size of nanopores will increase until it reached the optimum period of anodization. The size of nanopores for the 90 minutes-sample is the largest compared to the others sample of 15, 30, and 60 minutes anodized. The sample of 90 minutes anodization shows the largest size of nanopores and are highly oriented. The electrolyte that contains fluoride ions has a crucial part on the surface morphology of the porous. According to *Gong et al.*, fluoride ion concentration changes the formation and morphology of the sample after undergoing anodization process [18]. Niobium that dissolution with fluoride ion affects the formation of surface morphology. It can be observed here that the duration of 90 minutes of anodization is the optimum condition to gain the largest size of nanopores.

B. Effect of Nb_2O_5 thickness.

The thickness of Nb_2O_5 corresponds directly to the duration of the anodizing process. If the duration of anodization of the sample is longer, the Nb_2O_5 is taking a time to detect the UV radiation. It shows that the thickness slows down the performance of the UV sensor. UV light that transmitted to the sample has to penetrate deeper into the oxide layer to detect the radiation.

In order to carry out the electrical testing, a voltage needs to be supplied to the device on the Nb_2O_5 surface. Metal contact with 30 nm thickness of platinum was placed on Nb_2O_5 surface. The thickness of nanopores is one of the factors for absorption surface area of the material. Absorption surface area which is nanoporous play important role in UV performance because the absorption and diffusion of molecule occur in nanoporous. The thinner the thickness of Nb_2O_5 nanoporous increase the efficiency of absorption and diffusion of a molecule of UV radiation, so that a better UV photodetector performance can be achieved. The thickness of nanoporous reduces the efficiency of absorption molecule from UV radiation so that the device takes a long time to detect the radiation. Therefore, the higher the number of nanopores can improve the device's performance in detecting UV radiation.

When supplying low voltage to the thicker sample of 90 minutes, the detection of UV radiation is low because UV radiation takes time to react with the Nb_2O_5. Unlike sample of 15 minutes anodization where the reaction of UV radiation is faster at a lower voltage because of thinner nanopores have better electron movement. For a good sensor, lower voltage is needed for the best performance. Based on observation, the sample 15 minutes shows the best performance compare to the 90 minutes of anodization.

The good results obtained when 10 V was supplied for each sample. For sample 90 minutes, the result

obtained from I-V graph is consistent. The high voltage supply triggered the quickest reaction of Nb_2O_5. Next, for the Sample 15 minutes anodization also shows a good respond. The sample 15 minutes have shown a higher reading compared to 90 minutes. Therefore, the thickness of the sample produces difference performance for each sample. Based on the result obtained, the best UV photodetector is with a thin layer of Nb_2O_5 nanopores. This feature can be concluded as the best performance for UV photodetector.

C. UV photodetector performance.

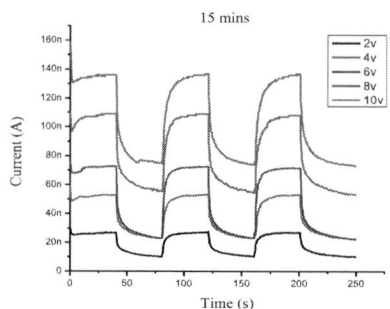

Figure 3: Radiation of UV light for 15 minutes anodization of Nb_2O_5 nanoporous.

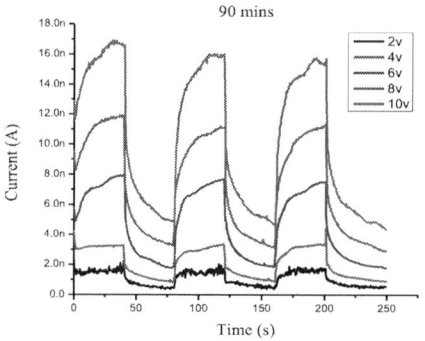

Figure 4: Radiation of UV light for 90 minutes anodization of Nb_2O_5 nanoporous.

Figure 3 and 4 shows the UV light radiation detected by the photodetector for different thickness of Nb_2O_5 nanopores which are 15 and 90 minutes. The graph presented the current gain over the time for different voltages of UV characterization of the anodization process. This characterization studied the effect of voltage Nb_2O_5 nanopores with different thickness. When the substrate supplied with high voltage density, the performance of Nb_2O_5 nanoporous via UV photodetector was also enhanced. The electrode reaction faster and active with the higher voltage. When the thickness of Nb_2O_5 nanoporous is thinner, so that the quantity of nanoporous obtains is bigger and increase the mobility of molecule to react

978-1-5386-5284-8/18 $31.00 © 2018 IEEE

with the voltage supply. If the thickness of nanoporous is thicker, the mobility of the molecule decrease due to the nanoporous size is bigger. The performance of the photodetector will be decreased due to the reduced quantity of nanoporous. The thinner of nanoporous size increases the absorption of electron from UV radiation. Therefore, based on the observation sample of 15 minutes anodize shows the higher current obtains when the voltage was supplied.

Figure 3 and 4 show each sample of niobium oxide has been tested with a different voltage such as 2V, 4V, 6V, 8V, and 10V. Based on Figure 3, the voltage at 2V starts detecting the radiation with a stable condition. It shows the good performance of niobium oxide with the lower voltage supply. After increasing the voltage into 4V, 6V, 8V, and 10V, the currents start to increase the reading with the time respectively.

Based on Figure 4, the sample shows that the current gain for 2V is not detecting the UV radiation very well compared to the sample 15 minutes which is the sample start increases the reading when supply the

voltage. This is because of the different thickness and size of Nb_2O_5 nanoporous. Based on the observation, when supply the higher voltage, it can produce the higher current with the time increase. The current gain for the thicker sample is lower than the thinner one. Thus, the sample 15 minutes is more sensitive than the sample 90 minutes due to the thickness and voltage supply.

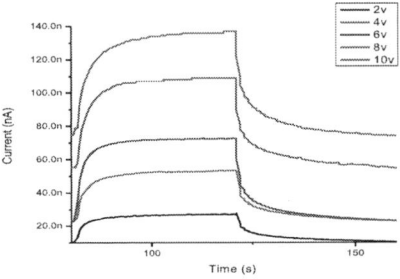

Figure 5: Detection of UV radiation for sample 15 minutes

Figure 5 shows the detection of UV radiation for sample 15 minutes with the different voltage supply. The current produced is higher over the time. This sample highly reactive when UV radiation occurs which is the current produce higher when the voltage supply is higher. Based on the calculation, the rise time of this sample is 27s meanwhile fall time is 32s. The time reaction when UV radiation occurs is a bit slow compared to the fall time.

Figure 6 shows the detection of UV radiation for sample 90 minutes with the different voltage supply. The rise time for sample 90 minutes is 30.2s meanwhile the fall time is 32s.

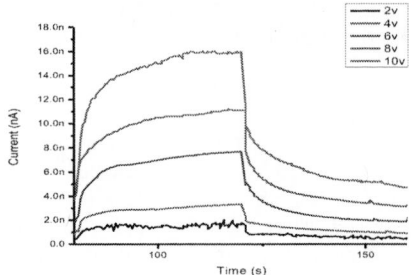

Figure 6: Detection of UV radiation for sample 90 minutes.

Table 1: Rise time and fall time.

UV/Thickness	Rise time, T_r (s)	Fall time, T_f (s)
15 mins	27.0	32
30 mins	30.5	35
60 mins	27.5	28
90 mins	30.2	32

This sample shows the time for the current increase is longer but for the fall time, it is the same as the sample 15 minutes. Based on observation, The sample 60 minutes shows good response time compared to others. It shows surface area to volume should be optimized to gain good performance of photodetector.

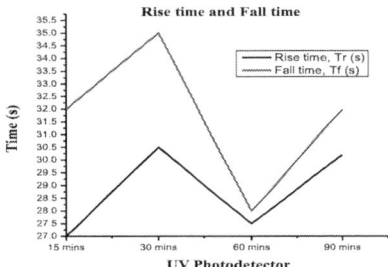

Figure 7: Rise time and fall time.

Figure 7 shows the rise time and fall time for each sample based on results in Table 1. Rise time required for a pulse to rise taken from 10 to 90 percent of the rising value of graph. While fall time is taken from 10 to 90 percent of dropped value. The thinner sample shows the faster rise time compared to the thicker. But, the thicker size shows the faster fall time compared to the thinner.

D. UV photodetector mechanism.

The photodetector activity takes place under UV light simulator. The UV photodetector used to test the response signal of UV radiation and comparing the response output at the different condition such as the thickness of Nb_2O_5 nanoporous and the voltage supply. The processes that involved in this activity are absorbent molecule from UV radiation. In this reaction, the photons have higher energy than the bandgap energy ($h\nu \geq E_g$).

When the Nb_2O_5 exposed to the UV lamp or UV light simulator, nanoporous will absorb the molecule from the radiation and allow the electricity to flow on the substrate, so that it will increase the current gain over the time. Meanwhile, if UV light is switched off, the nanoporous will release the molecule and there wil be no electricity flow. The process happens when the UV light exposed on the nanostructure surface, electron-hole pairs are photogenerated according to the following reaction[19]:

$$h\nu \longrightarrow h^+ + e^-$$

Based on electrolyte in the anodization process, the following shows the chemical reactions as the growth mechanism for these Nb_2O_5 nanostructure[19]

$$NH4F + C_2H_6O_2 + H_2O + Nb \longrightarrow Nb_2O_5$$

There are significant impacts of the nanoporous surface area toward photodetector performance. The nanoporous size presenting the absorption molecule of surface area. The small size of nanoporous can increase the absorption molecule by Nb_2O_5 nanostructure compare to the bigger size.

IV. CONCLUSION

The effects of anodized duration towards UV photodetector performance of Nb_2O_5 nanoporous have been successfully investigated. When the anodized duration is increased to 90 minutes, the size of Nb_2O_5 nanoporous is bigger. The increase of duration anodized process caused the performance of UV photodetector decrease. UV photodetector performance also depends on the thickness of Nb_2O_5 nanoporous and the voltage supply. When the thickness of Nb_2O_5 is thinner, the size of nanoporous is small but in large quantity. The ideal size of nanoporous and thickness is reflected in sample 60 minutes where the T_r and T_f are the shortest compared with other samples. In this sample, the nanoporous can absorb more UV light and show better performance. Overall, the anodization process at 60 minutes duration shows the best performance of UV photodetector.

ACKNOWLEDGMENT

We would like to thank Faculty of Electrical Engineering, Universiti Teknologi MARA (UiTM) Shah Alam for the facilities and Research Management Centre (RMC), UiTM for the grant 600-IRMI/MyRA5/3/LESTARI(081/2017).

REFERENCES

[1] X. Fang *et al.*, "New Ultraviolet Photodetector Based on Individual Nb2O5 Nanobelts," pp. 3907–3915, 2011.

[2] D. Zheng, "Investigation of Ultraviolet Transmission Characteristics of Detecting Window in Ultraviolet Fire Detector ∗," vol. 4, no. 2, 2006.

[3] L. Hu, J. Yan, M. Liao, L. Wu, and X. Fang, "Ultrahigh External Quantum Efficiency from Thin SnO2 Nanowire Ultraviolet Photodetectors," no. 8, pp. 1012–1017, 2011.

[4] C. Chen *et al.*, "Novel fabrication of UV photodetector based on ZnO nanowire / p-GaN heterojunction," *Chem. Phys. Lett*, vol. 476, no. 1–3, pp. 69–72, 2009.

[5] R. Fiz *et al.*, "Synthesis, Characterization, and Humidity Detection Properties of Nb2O5 Nanorods and SnO2 / Nb2O5 Heterostructures," no. Cvd, 2013.

[6] N. U. V Photoresponse, "Semi-Transparent ZnO-CuI / CuSCN Photodiode Detector with," 2015.

[7] K. J. Chen, F. Y. Hung, S. J. Chang, and S. J. Young, "Optoelectronic characteristics of UV photodetector based on ZnO nanowire thin films," vol. 479, pp. 674–677, 2009.

[8] S. M. Hatch, J. Briscoe, and S. Dunn, "COMMUNICATION A Self-Powered ZnO-Nanorod / CuSCN UV Photodetector Exhibiting Rapid Response," pp. 867–871, 2013.

[9] B. Mukherjee, Y. Cai, H. R. Tan, Y. P. Feng, E. S. Tok, and C. H. Sow, "NIR Schottky Photodetectors Based on Individual Single-Crystalline GeSe Nanosheet," 2013.

[10] R. A. Rani, A. S. Zoolfakar, J. Z. Oua, M. R. Field, M. Austin, and K. Kalantar-Zadeh, "Nanoporous Nb2O5 hydrogen gas sensor," *Sensors Actuators, B Chem.*, vol. 176, pp. 149–156, 2013.

[11] R. A. Rani *et al.*, "Reduced impurity-driven defect states in anodized nanoporous Nb2O5: the possibility of improving performance of photoanodes," pp. 6349–6351, 2013.

[12] A. Le Viet, R. Jose, M. V Reddy, B. V. R. Chowdari, and S. Ramakrishna, "Nb2O5 Photoelectrodes for Dye-Sensitized Solar Cells: Choice of the Polymorph," pp. 21795–21800, 2010.

[13] R. A. Kadir *et al.*, "Optical Gas Sensing Properties of Nanoporous Nb2O5 Films," 2015.

[14] J. Liu, D. Xue, and K. Li, "Single-crystalline nanoporous Nb2O5 nanotubes," pp. 2–9, 2011.

[15] S. G. Chen, S. Chappel, Y. Diamant, and A. Zaban, "Preparation of Nb2O5 Coated TiO2 Nanoporous Electrodes and Their Application in Dye-Sensitized Solar," no. 10, pp. 4629–4634, 2001.

[16] F. Lenzmann, V. Shklover, K. Brooks, and M. G. R. Atzel, "Mesoporous Nb2O5 Films: Influence of Degree of Crystallinity on Properties," pp. 175–180, 2000.

[17] J. Xia, N. Masaki, K. Jiang, and S. Yanagida, "Sputtered Nb2O5 as a Novel Blocking Layer at Conducting Glass / TiO2 Interfaces in Dye-Sensitized Ionic Liquid Solar Cells," pp. 8092–8097, 2007.

[18] M. Rahman, A. Rani, A. Z. Sadek, A. S. Zoolfakar, M. R. Field, and T. Ramireddy, "A vein-like nanoporous network of Nb2O5 with a higher lithium intercalation discharge cut-off voltage †," 2013.

[19] M. H. Mamat, Z. Khusaimi, M. Z. Musa, M. F. Malek, and M. Rusop, "Sensors and Actuators A: Physical Fabrication of ultraviolet photoconductive sensor using a novel aluminium-doped zinc oxide nanorod – nanoflake network thin film prepared via ultrasonic-assisted sol – gel and immersion methods," *Sensors Actuators A. Phys.*, vol. 171, no. 2, pp. 241–247, 2011.

Activation energy of thermal oxidation germanium oxide on germanium substrates

N. A. A. Halim
Faculty of Engineering
Universiti Malaysia Sarawak
IMEN, Universiti Kebangsaan
Malaysia
nurulatiqahabdulhalim@gmail.c
om

S. K. Sahari
Faculty of Engineering
Universiti Malaysia Sarawak
Kota Samarahan, Malaysia
sskudnie@unimas.my

A. A. Hamzah
Institute of Microengineering
and Nanotechnology
Universiti Malaysia Kebangsaan
Bangi, Malaysia
azlanhamzah@ukm.edu.my

B. Y. Majlis
Institute of Microengineering
and Nanotechnology
Universiti Malaysia Kebangsaan
Bangi, Malaysia
burhan@ukm.edu.my

S. Marini
Faculty of Engineering
Universiti Malaysia Sarawak
Kota Samarahan, Malaysia
smarini@unimas.my

L. Hasanah
Faculty of Mathematics and
Natural Sciences
Institut Teknology Bandung
Bandung, Indonesia
azlanhamzah@ukm.edu.my

M. Kashif
Faculty of Engineering
Universiti Malaysia Sarawak
Kota Samarahan, Malaysia
kmuhammad@unimas..my

Abstract— **The activation energy of GeO₂ was studied by determining the oxide thickness versus temperature from lower to higher oxidation temperature in the range between 450˚C and 600°C. It was found that a linear relationship can be obtained between oxidation time and oxide thickness for the oxidation temperature between 450 and 575°C while for the 600°C oxidation, a linear relationship can be obtained for the shorter oxidation time. The rate of oxidation increased until 0.55 and abruptly decreased after increasing oxidation temperature to 600°C which implies that the oxygen intermixing occurs during higher oxidation (600°C) rather that diffusion mechanism that leads to the lower activation energy.**

Keywords—activation energy, Ge oxidation, GeO₂, GeO desorption

I. INTRODUCTION

Over the past 40 years, impressive progress has been made in Silicon (Si) technology by continual scaling of devices to smaller size. This action limits the performance, power consumption, current and short-channel effects which have the trade-off relationship with each other. Therefore, device structures and materials with high carrier mobility without associated leakage current in gate capacitance are needed. Germanium (Ge) is a promising candidate and become a great current research interest that has higher electron and hole mobility to than Si. [1-2]. The lower melting point of Ge (938˚C) compared to Si (1414˚C) gives an advantage to Ge metal-oxide-semiconductor (MOS) field effect transistors (FETs) with much lower thermal budget processes [3]. However, with the problem of inferior properties of Ge compared to Si annihilate it in semiconductor foundry [13-16]. Faster oxidation just after cleaning compared to Si oxidation reflects a difficulty in integrating novel materials as a metal gate electrode and high-k dielectric [4]. Another big issue in implementing Ge as a metal gate electrode and high-k dielectric into advanced transistor is volatilization of Ge monoxide (GeO) during thermal process, which can degrade the performance of MOSFETs. However, previous study shown that the combination of Aluminium Oxide (Al₂O₃) and Ge formed Ge oxide (GeOx) after post anneal deposition lower the interface trap density [5]. From this finding, the GeOx can be a potential interfacial layer between high-k and Ge. Recent study shows that the surface cleaning of Ge and anneal temperature influences the growth of interfacial layer between GeOx and Al₂Ox that cause the intermixing between GeOx and Al₂O₃ [6]. In addition, theoretical study on the GeO₂/Ge interface has been performed based on the calculations of SiO₂/Si interfaces, but the reaction mechanism of Ge oxidation remains unsolved [7]. Therefore, fundamental understanding of the mechanism of Ge oxidation is essential to form a good quality of oxide on Ge surface. Hypothetically, the change of activation energy of Ge oxide growth also can influence the mechanism of Ge oxidation. From the activation energy, the reaction between Ge-O can be discussed in detail. In this work, the thickness of Ge oxide is evaluated with temperature dependence. The activation energy is calculated based on the Arrhenius plot. The value of activation energy will be compared with the reported diffusivity of oxygen in GeO₂ From experimental data, the mechanism of Ge oxide growth will be discussed in detail

II. EXPERIMENTAL PROCEDURES

A p-type Ge(100) with a resistivity of ~10 ohm cm were used in this study. The substrates were unfirmly cut into 1cm² pieces. Before dipping into HCl, Ge wafers were dipped in deionized pure water for 90 s and H_2O_2 for 60 s. These steps are conducted to ensure that all the native oxide and other contaminants are removed completely. Finally, the wafers were rinsed shortly in deionized pure water to minimize the Cl atom on the Ge surface. All the samples were prepared in clean room air at room temperature. In this work, hydrochloric acid (HCl) was used to remove the native oxide. As reported in our previous study, the less surface micro roughness has been determined after HCl cleaning [8]. In addition, the Cl termination has been observed after HCl cleaning that can slow the native oxidation of Ge surface before the next fabrication process [9]. After wet chemical cleaning process, a cleaned wafer was brought quickly to other areas of the fab to oxidize the surface. The wafers were directly kept on the quartz boat then transferred to the mouth of quartz furnace. After 5 minutes, the quartz boat was transferred to the middle of the furnace slowly in order to prevent thermal stress, which cause the wafers to fracture. Then, the sample was leaved for 5 minutes with 2l/min of N_2 flowing before start the oxidation. This process is to remove the contaminant attached on the Ge wafer during the transfer's process. Following the step, the O_2 gas was flowed with flowing rate of 2l/min and the N_2 gas was reduced to zero. After that, the oxidation process was performed at temperature of 375 to 600C. The spectroscopic ellipsometry and thin film mapper (TFM) measured the oxide thickness.

III. EXPERIMENTAL RESULTS

Figure 1 shows GeO_2 thickness versus oxidation time as a function if temperature ranging from 450°C to 600°C. It shows that the linear relationship was obtained between oxidation temperature and oxide thickness between 450°C and 575°C. However, for the 600°C oxidation, the linear relationship only can be obtained for the shorter time oxidation while the curve saturated after further increased of oxidation temperature. This result indicates that the different mechanism of Ge oxidation between two range of oxidation temperature; (between 450 and 575°C) and 600°C). To discuss the mechanism in detail, a slope, m was extracted by using linear extrapolation of log-scale plot as shown in Figure 2. It can be seen that the slope of oxidation increases with increasing oxidation temperature from 450°C to 575°C. Further increase in oxidation temperature to 600°C decreases the slope of oxidation. These result indicates that the rate of oxygen diffusion through Ge oxide is around 0.55. The reduction of slope at temperature of 600°C may be due to the intermixing of oxygen as discussed in the previous work which is difficult to explain the diffusion mechanism. [10].

To determine the activation energy, the oxidation rate is plotted versus reciprocal temperature as shown in Figure 3. In Figure 3, the logarithm of the oxidation rate, m is plotted against the reciprocal of the absolute temperature. It shows that a good straight line is obtained not in all cases. Here, the relationship between ln M and reciprocal temperature (1000/K) can be roughly described by [11]

$$k = Ae^{(-Ea/RT)} \qquad (1)$$

$$ln\ k = ln\ A - Ea/R\ (1/T) \qquad (2)$$

where k is a rate constant, A is a pre-exponential factor, Ea is an activation energy, R is a gas constant which is 8.314

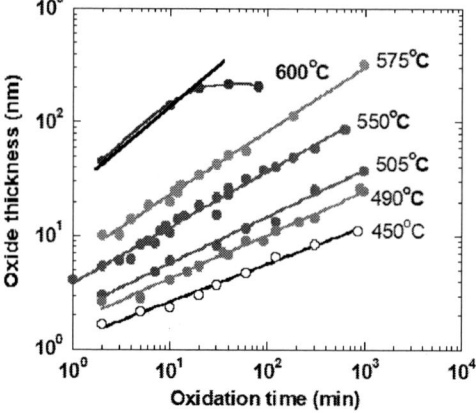

Fig 1. GeO_2 thickness at temperature between 450°C and 600°C,

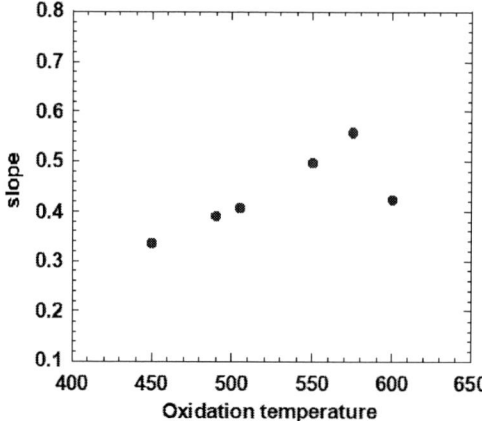

Fig 2. Slope versus temperature after 2 minutes of oxidation

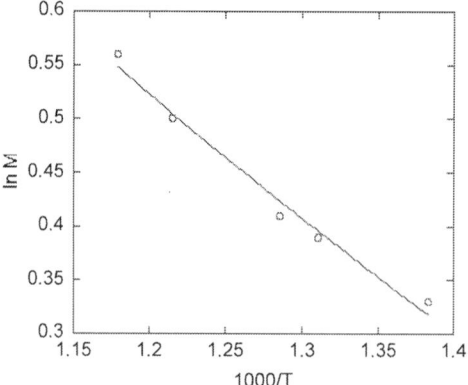

Fig 3. Plot of the oxidation rate versus reciprocal temperature

J/mol.k and T is an absolute temperature in Kelvin. The activation energy can be calculated by this equation; where m is the slope

$$m = -Ea/R \qquad (3)$$

An Arrhenius plot is often used to analyze the effect of temperature on the rates of chemical reaction. The Arrhenius plot was used to investigate the effect of temperature on the rates of interfacial layer growth. From the graph, the activation energy was determined to be 0.22eV which is not consistent with the reported activation energy of diffusivity of oxygen in GeO_2 [12]. The lower activation energy may be due to the GeO desorption occurs during Ge oxidation. The Ge atom diffused from the Ge substrate through GeO_2 and combined with oxygen near the surface and desorbs as Ge monoxide (GeO). After that the GeO recombined with the oxygen on the surface and form the GeO_2 on the top layer. The model of GeO desorption and Ge oxidation are illustrated in Figure 4. To check the validity of the proposed model, the experimental data was fitted with the mathematical equation for the thermally-grown GeO_2 on Ge(100) at 550° oxidized in dry oxygen ambience as shown in Figure 5. In the figure, the oxidation rate expressed by linear parabolic model (equation (1)) is also plotted as a function of thickness [11].

$$x^2 + Ax = B(t+\tau) \qquad (4)$$

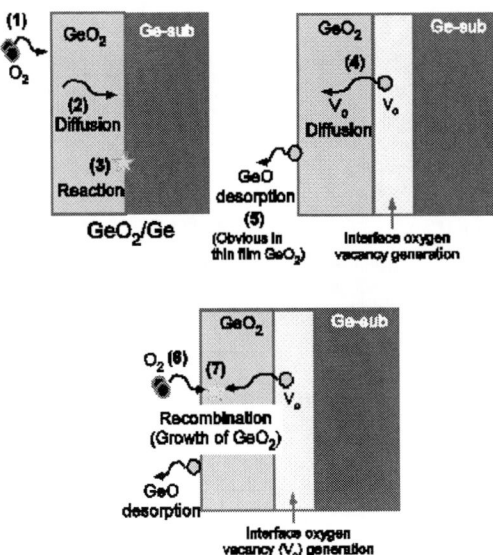

Figure 4 Model of Ge oxidation and GeO desorption

As in the figure, the thinner region (less than 70 nm) can be divided in to two phase; initial and intermediate phase. The rate rapidly decreases with thickness (oxide thickness less than 10nm) and at the intermediate phase, the rate decreases asymptotically to the onset linear-parabolic kinetics. The expression of the total growth is then written as:

$$dx/dt = B/(A+2x) + K1\ exp\ (-x/L1) + K2exp\ (-x/L2) - C/x \qquad (5)$$

where initial phase; $K1exp(-x/L1)$, intermediate phase: $K2exp(-x/L2)$, GeO desorption: $-C/x$

Consequently, by applying equation (2), the oxide growth rate as the function of oxide thickness can be fitted quite well which can represent the oxidation and desorption simultaneously as shown in Figure 5.

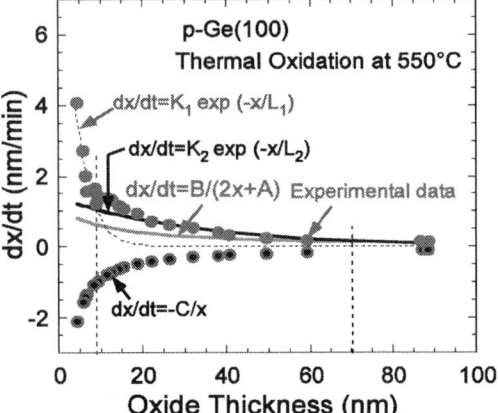

Figure 5 Oxide thickness dependence of growth rate for thermally-grown GeO_2 on Ge(100) at 550°C.

IV. CONCLUSION

The oxidation of Ge has been investigated from temperature of 450°C to 600°C. The mechanism of Ge oxidation between 450°C and 575C is different with Ge oxidation at 600°C. The oxidation of Ge between 450°C to 575°C is controlled by oxygen diffusion through oxide layer while intermixing of oxygen occurs during 600°C oxidation which leads to the difficulty to distinguish the diffusion mechanism.

ACKNOWLEDGMENT

The authors would like to acknowledgement the Universiti Malaysia Sarawak and Ministry of higher education for funding this study via Research Acculturation Collaborative Effort (RACE) grant -: Race/c (1)/1330/2016(3). A part of this works was supported by DPP-2018-006, Dana Penyelidikan PTJ, Peranti Bioperubatan dengan IOT and Prototype Research Grant Scheme (PRGS); PRGS/1/2017/TK05/UKM/02/1 by Design and Fabrication of Silicon Membrane Filtration System for Artficial Kidney.

REFERENCES

[1]   Hennessy, J., and D.A. Antoniadis, "High electron mobility germanium MOSFETs: Effect of n-type channel implants and ozone surface passivation", Device Research Conference (2009) 257

[2]   Sze S M and Ng K K, Physics of Semiconductor Devices (New York, Wiley 3rd ed) pp.786, 2007

[3] Littlejohns, C. G., Dominguez Bucio, T., Nedeljkovic, M., Wang, H., Mashanovich, G. Z., Reed, G. T., & Gardes, F. Y. "Towards a fully functional integrated photonic-electronic platform via a single SiGe growth step", Scientific Reports, 6, pp. 19425, 2016

[4] S.K.Sahari, H. Murakami, T. Fujioka, T. Bando, A. Ohta, K. Makihara, S. Higashi and S. Miyazaki , "Native Oxidation Growth on Ge(111) and (100) Surfaces", Japanese Journal of Applied Physics, Volume 50, pp 4S, 2011

[5] R. Zhang, T. Iwasaki, N. Taoka, M. Takenaka, and S. Takagi, "Al2O3/GeOx /Ge gate stacks with low interface trap density fabricated by electron cyclotron resonance plasma post oxidation", IEDM 2011, p. 56, 2011

[6] S. K. Sahari, M. Kashif, N. M. Sultan, N. A. F. N. Z. Fathi, A. A. Hamzah, R. Sapawi, B. Y. Majlis, I. Ahmad, "Growth kinetic and composition of the interfacial for RF sputtering Al_2O_3 layer on germanium", Microelectronics International 34(2), 64, 2017

[7] M. Yang, R. Q. Wu, Q. Chen, W. S. Deng, Y. P. Feng, J. W. Chai, J. S. Pan and S. J. Wang, "Impact of oxide defects on band offset at GeO_2/Ge interface", Appl. Phys. Lett. Vol. 94 (14). (2009)

[8] S. K Sahari, N. A. F. N. Z Fathi, N. M. Sutan, R Sapawi, AA. Hamzah, B. Y. Majlis "Wet chemical cleaning effect on the formation of ultrathin interfacial layer between Germanium (Ge) and high-k dielectric" 2015 IEEE Regional Symposium on Micro and Nanoelectronics (RSM) , pp. 1-3, 2015

[9] Siti Kudnie Sahari, Muhammad Kashif, Marini Sawawi, Nik Amni Fathi Nik Zaini Fathi, Azrul Azlan Hamzah, Burhanuddin Yeop Majlis, Norsuzailina Muhammad Sutan, Rohana Sapawi, Kuryati Kipli, Nurul Atiqah Abdul Halim, Nazreen Junaidi, Sharifah Masniah Wan Masra, "Stability of Chlorine Termination on Ge (100) and Ge (111) Surfaces", MATEC Web of Conferences 87, 05005, 2017

[10] Shengkai Wang1 , Koji Kita1,2 , Tomonori Nishimura1,2 , Kosuke Nagashio1,2 and Akira Toriumi1 "18O isotope isotope isotope isotope tracing tracing tracing tracing study of GeO Desorption Desorption Desorption Desorption from GeO_2/Ge Structure" Extended Abstracts of the 2009 International Conference on Solid State Devices and Materials, Sendai, pp1002-1003, 2009

[11] Deal B E and Grove A S, "General Relationship for the thermal oxidation of silicon", J. Appl. 36 (12), pp. 3770-3778, 1965

[12] C. Haas," The diffusion of oxygen in silicon and germanium", Journal of Physics and Chemistry of Solids 15(1-2), 108, 1960

[13] N. Marsi, B. Y. Majlis, F. Mohd-Yasin and A. A. Hamzah, "The fabrication of back etching 3C-SiC-on-Si diaphgram employing KOH+ IPA in MEMS capacitive pressure sensor", Microsystem Technologies 21 (8), 1651-1661, 2015

[14] N. Marsi, B. Y. Majlis, A. A. Hamzah and F. Mohd-Yasin, "Comparison of mechanical deflection and maximum stress of 3C SiC-and-Si-based pressure sensor diaphragms for extreme environment", Semiconductor Electronics (ICSE), 2012 10th IEEE International Conference on, 2012

[15] N. Marsi, B. Y. Majlis, A. A. hamzah and F. Mohd-Yasin, "The mechanical and electrical effects of MEMS capacitive pressure sensor based 3C-SiC for extreme temperature", Journal of engineering 2014

[16] A. A. hamzah, J. Yunas, B. Y. Majlis and I. Ahmad,"Sputtered encapsulation as wafer level packaging for isolatable MEMS devices: A technique demonstrated on a capacitive accelerometer", Sensors 8 (11), 7438-7452, 2008

978-1-5386-5284-8/18 $31.00 © 2018 IEEE

Optimization of wurtzite GaN-based Gunn diode as terahertz source

Wen Zhao Lee, Duu Sheng Ong, Kan Yeep Choo
Faculty of Engineering
Multmedia University
Cyberjaya, Malaysia
leewenzhao1@gmail.com, dsong@mmu.edu.my, kychoo@mmu.edu.my

Abstract— The performance of GaN-based Gunn diode as a terahertz source has been investigated using Monte Carlo particle simulations. The 4-valley analytical band Monte Carlo model developed in the work for modelling electron transport in GaN consists of various scattering mechanisms including impact ionization. The conventional Gunn diode n^+-n^--n-n^+ device structure with a notch is optimized to achieve current oscillation in the THz range by studying the effects of transit region length and bias levels. We found that a Gunn diode with 550 nm transit length is capable to achieve a 500 GHz signal of 2.61 W with 2.27% efficiency under 22 V DC and 5V RF condition.

Keywords—wurtzite GaN, Gunn diode, terahertz, Ensemble Monte Carlo

I. INTRODUCTION

Terahertz radiation, which has the non-destructive and non-invasive characteristics, spark interest in fields such as medical imaging, astronomy and security [1]. Utilization of terahertz technology has always been a challenge, especially in developing terahertz source and receiver operating at room temperature. An $In_{0.23}Ga_{0.77}As$ planar Gunn diode is reported to have 60.1 GHz with multiple channels [2]. It is also reported that a 2-domain GaN Gunn diode has operating frequency up to 525 GHz with output power of 87 mW [3].

Theoretical work [4]-[8] and experimental data [9][10] from various research groups show great disparity in the velocity-field characteristics for wurtzite GaN. In general, experimental results from previous work [9][10] show that the peak drift velocity and the drift velocity drop in negative differential resistance (NDR) region is much lower than that predicted from the theoretical simulations. These theoretical works are based on the Monte Carlo particle method using the analytical-band approximation or full band from first principle calculation. In this work, material parameters used in these analytical-band models are reviewed to guide the selection of parameters in our Monte Carlo model in order to reproduce the velocity-field characteristics from the full band Monte Carlo model of [6].

In section II, the Monte Carlo model developed for this work and the simulated device structure are presented. In section III, the results for velocity-field characteristics and Gunn diode performance such as operating frequency and

power efficiency are shown and discussed. The findings and the best performance device structure are concluded in section IV.

II. MODEL AND DEVICE STRUCTURE

A 4-valleys Monte Carlo model based on [11] is used for the simulation of electron velocity-field characteristics in a bulk GaN semiconductor. The scattering mechanisms considered for this model including polar optic phonon, acoustic deformation potential, intervalley scattering, and impact ionization. This Monte Carlo model is self-consistently coupled with a Poisson solver to simulate the Gunn diode device performance using the ensemble particle method. The Gunn diode device structure studied in this work is the standard structure with a notch (n^+ - n^- - n - n^+) as shown in Table 1. The structure has a doping density of $2.0 \times 10^{23} \text{ m}^{-3}$ in the n-region. For a device with 550 nm transit region, $nL = 1.1 \times 10^{13} \text{ cm}^{-2}$, which fits the criteria of $nL > 1 \times 10^{12} \text{ cm}^{-2}$ for Gunn effect to occur, where L is the active region. While a higher doping density can be used in a shorter transit length device in order to achieve a higher frequency signal, the heating effect would potentially break the device.

The simulation of electron velocity-field characteristics used 10,000 super particles with a time interval of 5 fs. Over 10^5 super particles are used to represent the electrons in device for the simulations of Gunn diode performance. To capture the dynamics of electron transport in the Gunn diode, the self-consistent potential distribution in the one-dimensional structure is derived by the Poisson solver at a smaller time step of 1 fs on a uniform spatial mesh with spacing of 1 nm. The simulation of the Gunn diode device is used to study the effects of transit region length and bias voltage on the resulting output current density and operating frequency.

TABLE I.   DEVICE STRUCTURE CHOSEN FOR PERFORMANCE ANALYSIS

	n^+	n^-	n	n^+
Doping density, m^{-3}	2.0×10^{24}	5.0×10^{22}	2.0×10^{23}	2.0×10^{24}
Length, nm	100	100	550 - 800	100

TABLE II. WURTZITE GaN PARAMETERS CHOSEN VALUE

Material parameters	Value				
	General	Γ1	Γ3	U	M
Effective mass, m_* [m_e]	-	0.24	0.6	0.4	0.57
Band gap, E_g [eV]	3.44	-	-	-	-
Energy band minimum [eV]	-	0.0	2.4	2.33	3.01
Number of equivalent valley	-	1	1	6	3
Static dielectric constant	8.9	-	-	-	-
High frequency dielectric constant	5.35	-	-	-	-
Material density [kg/m^3]	6150	-	-	-	-
Longitudinal sound velocity [m/s]	6560	-	-	-	-
Lattice constant [Å]	5.185	-	-	-	-
Scattering parameters	Value				
	General	Γ1	Γ3	U	M
Non-parabolicity [eV^{-1}]	-	0.168	0.029	0.065	0.028
Acoustic deformation potential [eV]	-	11.0	11.0	11.0	11.0
Intervalley deformation potential [eV/m][a]	1.0	1.0	1.0	1.0	1.0
Polar optic phonon energy [eV]	0.98	-	-	-	-
Intervalley optic phonon energy [eV][a]	0.98	0.98	0.98	0.98	0.98
Valley cut-off energy [eV]	-	5.96	7.71	8.33	10.71
Softness coefficient for impact ionization[b]	-	0.5	60.0	60.0	130.0
Power exponent for impact ionization[b]	-	4.8	1.4	1.5	2.4

[a]. Value across all intervalley scattering are set to be the same.

[b]. From the impact ionization rate equation $P = P_0\lambda (E - E_{th})^\gamma$, where $P_0\lambda$ is the softness coefficient and γ is the power exponent.

III. RESULTS AND DISCUSSIONS

A. Velocity-field Characteristic

The velocity-field characteristics are simulated to guide the selection of material parameters of GaN in the Monte Carlo model. The previous theoretical works predicted peak drift velocity ranging from 2.48×10^7 cm/s [1] to 3.1×10^7 cm/s [12] and the critical field ranging from 150 kV/cm [1][13] to 220 kV/cm [5][14]. Both of the experimental work [9][10] reported a peak drift velocity of 1.89×10^7 cm/s, which differ a lot from all the theoretical work. They did measure a critical field around 220 kV/cm, although they were not able to observe clearly the negative differential resistance effect.

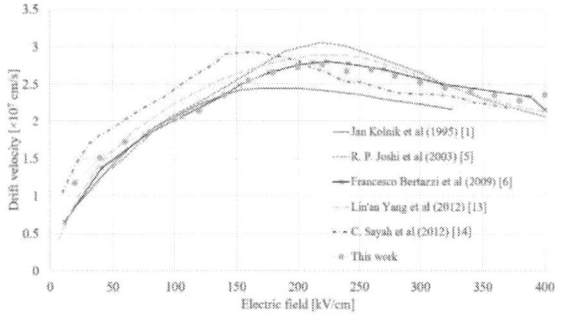

Fig. 1. Simulated electron velocity-field characteristics in comparison with other results. Temperature is set to be 300 K.

The work from [6] is chosen as the reference and guidance for material parameters fitting because they employed a full band structure derived from nonlocal empirical pseudopotential calculations and the ab initio techniques to determine the phonon dispersion relation in their Monte Carlo model. Energy bandgap and band minimum for this work are chosen based on [6]. The impact ionization parameters of this work are also fit to their impact ionization scattering rates. The material parameters used in this work are tabulated in Table 2 and the velocity-field characteristic is shown in Fig. 1. The peak drift velocity is calculated as 2.76×10^7 cm/s and associated to a critical field of 220 kV/cm.

B. Gunn Diode Device Model Simulation

Our Monte Carlo model for GaN is first used to compare with the work of [5] to study the effect of velocity field characteristics on Gunn diode performance. The device structure is a conventional Gunn diode structure of n$^+$-n$^-$-n-n$^+$, where the doping density for the cathode and anode is 2×10^{24} m^{-3}. The notch is 200 nm long with 4×10^{22} m^{-3} and the transit region is 1000 nm long with 2×10^{23} m^{-3}. They have simulated the current oscillation with external circuit built with the Gunn diode. In Fig. 6 of [5], the operating frequency is predicted to be 135 GHz, with peak-to-peak current oscillation of 0.13 A, whereas with material parameters chosen in this work, the device oscillates at 234 GHz with peak-to-peak current oscillation of 1.61 A. The result is expected as we predicted a higher critical electric field than [5], shown in Fig. 1.

978-1-5386-5284-8/18 $31.00 © 2018 IEEE

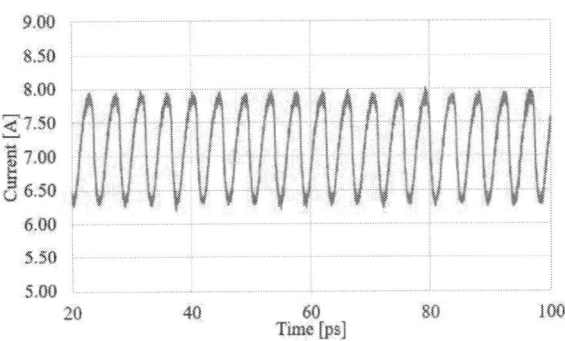

Fig. 2. Current oscillation as a function of time simulated using our chosen material parameters for the device structure from [5]. Bias voltage of 50 V and 300 K temperature were used.

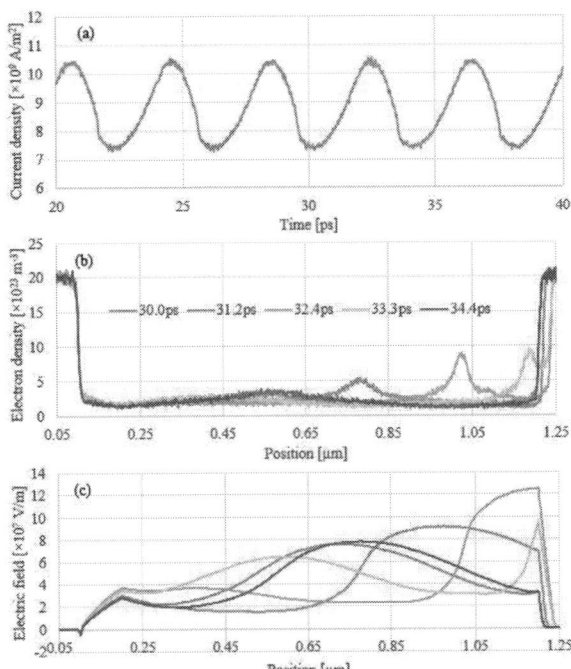

Fig. 3. Device performance of Gunn diode with 100 nm notch length and 1000 nm transit length with bias of 50 V. (a) Current density waveform. (b) Electron density within the device. (c) Electric field distribution within the device.

The device structure proposed in this work has a notch length of 100nm and doping density of 5×10^{22} m^{-3} instead, with the transit length kept the same at 1000 nm is simulated under the same condition. From Fig. 3 (a), the estimated operating frequency is around 250 GHz and peak-to-peak current density is 3.18×10^9 A/m^2. With cross-sectional area of the device same as used in above, at 1×10^{-9} m^2, the peak-to-peak current is 3.189 A. This is a good improvement as the operating frequency is increased slightly, while the peak-to-peak current is increased by nearly 2 times.

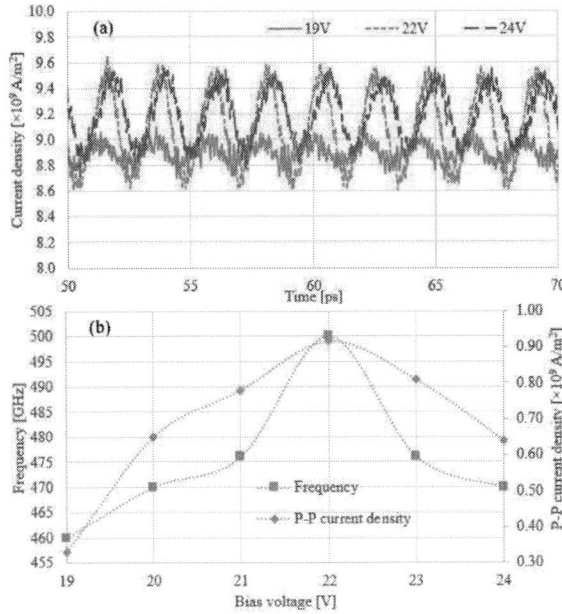

Fig. 4. Comparison of current oscillation of 550 nm transit length device with varying bias from 19 V to 24 V. (a) Current density waveforms for 19 V, 22 V and 24 V. (b) Frequency and peak to peak current density for 19 V to 24 V biases.

C. Device Optimization

The device performance is simulated starting from 800nm of transit length. The transit length is reduced for each simulation setup to compare their operating frequency and peak-to-peak current density. The criteria are having stable current density oscillation and low impact ionization scattering.

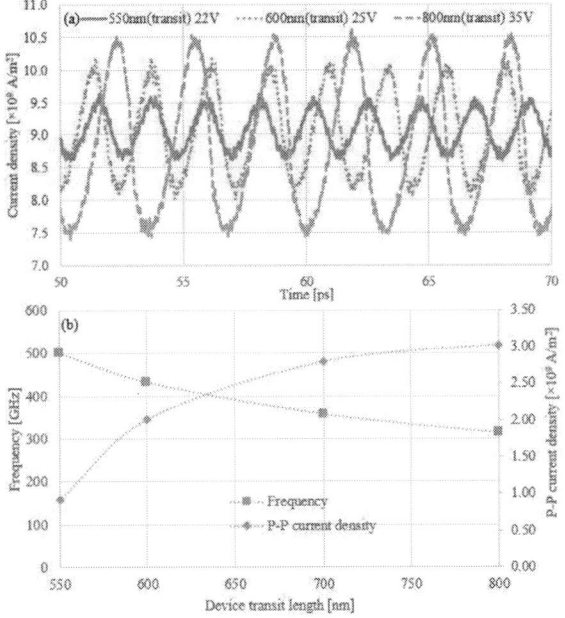

Fig. 5. Comparison of current oscillation of varying transit length with their respective best bias voltage. (a) Current density waveform for 550 nm, 600 nm and 800 nm. (b) Operating frequency and peak-to-peak current density for 550 nm, 600 nm, 700 nm and 800 nm.

978-1-5386-5284-8/18 $31.00 © 2018 IEEE

The Gunn diode with 800 nm transit length is simulated for different biases, and found that at 35 V it has the highest peak-to-peak current density of 3.02×10^9 A/m^2 oscillating at 313 GHz. When the bias voltage is increased to 40 V, a slight increase in frequency (323 GHz) but decrease in peak-to-peak current density (2.95×10^9 A/m^2) is observed. Bias of 41 V and above would generate too high number of impact ionization scattering, thus in this work they will not be taken into consideration.

For the device with 700 nm transit length, a bias of 30 V produces peak-to-peak current density of 2.8×10^9 A/m^2 at 357 GHz. The shorter device with 600 nm transit length at bias of 25 V produces smaller peak-to-peak current density of 2.01×10^9 A/m^2 but oscillating at a higher frequency of 432 GHz. For device with 550 nm transit length, a 22 V bias voltage produces peak-to-peak current density of 0.919×10^9 A/m^2 at 500 GHz, as shown in Fig. 4. We found that the oscillation of current density in devices shorter than 550 nm, could not be sustained. Fig. 5 shows the peak-to-peak current density and their corresponding oscillation frequency for the Gunn diode with different transit lengths. It is obvious that while shorter transit length can generate higher frequency signal, it will suffer for the lower output peak to peak current.

D. Power Estimation

For the power and efficiency estimation of a Gunn diode, we follow the standard practice of [15], where an AC bias was imposed on the device and the cross sectional area of 6×10^{-10} m^2 for the device was used the simulations. With input AC voltage of 5V to simulate as the outer circuit, the RF power and power efficiency are calculated for the device with 550 nm transit length. It is estimated to have 2.61 W RF power and efficiency of 2.27% at 500 GHz. This shows that the wurtzite GaN Gunn diode is a promising room temperature THz source.

IV. CONCLUSION

We have developed an analytical band Monte Carlo for modelling wurtzite GaN Gunn diode using material parameters fitted to a full band Monte Carlo model results. A device structure is proposed for GaN Gunn diode capable of generating stable current oscillation in the THz range. For a device with the transit length of 550 nm, the model predicted THz signal of 500 GHz with RF power of 2.61 W with efficiency of 2.27% at room temperature.

ACKNOWLEDGMENT

The authors wish to acknowledge the support of Motorola Foundation Scholarship and Fund.

REFERENCES

[1] R A Lewis, "A review of terahertz sources," J. Appl. Phys. 47, 374001 (2014).

[2] Bin Li, Hongxia Liu, Yasaman Alimi and Aimin Song, "Simulation investigation of multiple domain observed in In$_{0.23}$Ga$_{0.77}$As planar Gunn diode," International Journal of Hydrogen Energy, Vol. 41, No. 35, September 2016. 15772-15776.

[3] Smita Francis and Robert van Zyl, "Evaluating the microwave performance of a two domain GaN Gunn diode for THz applications," Terahertz Science and Technology, ISSN 1941-7411, Vol.8, No.1, March 2015.

[4] J Ján Kolník, İsmail H. Oğuzman, Kevin F. Brennan, Rongping Wang and P. Paul Ruden, "Electronic transport studies of bulk zincblende and wurtzite phases of GaN based on an ensemble Monte Carlo calculation including a full zone band structure," J. Appl. Phys. 78, 1033 (1995).

[5] R. P. Joshi, V. Sridhara, P. Shah and R. D. del Rosario, "Monte Carlo analysis of GaN-based Gunn oscillators for microwave power generation," J. Appl. Phys. 93, 4836 (2003).

[6] Francesco Bertazzi, Michele Moresco and Enrico Bellotti, " Theory of high field carrier transport and impact ionization in wurtzite GaN. Part I: A full band Monte Carlo model," J. Appl. Phys. 106, 063718 (2009).

[7] Lin'an Yang et al, "Temperature effect on the submicron AlGaN/GaN Gunn diodes for terahertz frequency," J. Appl. Phys. 109, 024503 (2011).

[8] A. Reklaitis and L. Reggiani, "Monte Carlo study of hot-carrier transport in bulk wurtzite GaN and modeling of a nearterahertz impact avalanche transit time diode," J. Appl. Phys. 95, 7925 (2004).

[9] M. Wraback et al, "Time-resolved electroabsorption measurement of the electron velocity-field characteristic in GaN," Appl. Phys. Lett. 76, 1155 (2000).

[10] O. Yilmazoglu et al, "Measured negative differential resistivity for GaN Gunn diodes on GaN substrate," Electronics Letters, 12th April 2007, Vol. 43, No.8.

[11] K Y Choo and D S Ong, "The applicability of analytical-band Monte Carlo for modelling high field electron transport in GaAs," Semicond. Sci. Technol. 19 (2004) 1067–1073.

[12] A. Reklaitis and L. Reggiani, "Monte Carlo study of hot-carrier transport in bulk wurtzite GaN and modeling of a near terahertz impact avalanche transit time diode," J. Appl. Phys. 95, 7925 (2004).

[13] Lin'an Yang, Shuang Long, Xin Guo, and Yue Hao, "A comparative investigation on sub-micrometer InN and GaN Gunn diodes working at terahertz frequency," J. Appl. Phys. 111, 104514 (2012).

[14] C. Sayah, B. Bouazza, A. Guen-Bouazza and N. E. Chabane-Sari, "Simulation of Electron Transport in GaN Based MESFET Using Monte Carlo Method," World Applied Programming, Vol (2), No (2), February 2012. 97-103

[15] G.M. Dunn and M.J. Kearney, "A theoretical study of differing active region doping profiles for W-band (75–110 GHz) InP Gunn diodes," Semicond. Sci. Technol. 18 (2003) 794–802.

Bandgap Engineering of GaAsBi Alloy for Emission of up to 1.52 μm

Abdul Rahman Mohmad
Institute of Microengineering and Nanoelectronics
National University of Malaysia (UKM)
Bangi, Malaysia
armohmad@ukm.edu.my

John P. R. David
Department of Electronic and Electrical Engineering
University of Sheffield
Sheffield, United Kingdom
j.p.david@sheffield.ac.uk

Abstract—**Band gap engineering by incorporating bismuth (Bi) into GaAs to form ternary $GaAs_{1-x}Bi_x$ alloy was investigated. Series of GaAsBi samples with different Bi concentrations were grown by molecular beam epitaxy. Based on the high resolution X-ray diffraction (HR-XRD) measurements, Bi concentration of up to 0.108 was successfully incorporated into the lattice. Sample with the highest Bi concentration, $GaAs_{0.892}Bi_{0.108}$, show room temperature photoluminescence (PL) emission with a peak wavelength of 1.52 μm and full-width-at-half-maximum (FWHM) of 89 meV. It was found that the incorporation of Bi into GaAs lattice affected both the conduction band as well as the valence band. The conduction band minimum reduces linearly by 23 meV/%Bi while the valence band maximum was best fitted by using the valence band anti-crossing (VBAC) model with coupling parameter, C_{Bi} of 1.65 eV.**

Keywords—band gap engineering, GaAsBi, photoluminescence, X-ray diffraction, molecular beam epitaxy

I. INTRODUCTION

The optical fiber communication networks require optical devices to operate at specific wavelengths which are 1.3 and 1.55 μm. This is because the 1.3 μm wavelength has zero dispersion in fiber while the 1.55 μm window exhibits the lowest optical loss of 0.2 dB/km [1]. However, choices of semiconductor materials that operates in these wavelengths and has lattice parameter which is almost lattice matched with the commercially available substrates (typically InP and GaAs) are limited. Therefore, band gap engineering or controlling the band gap via alloying process is necessary.

GaAs based semiconductor alloys are more attractive compared to InP based materials due to its low cost. Various GaAs based alloys have been reported to operate at 1.3 and 1.55 μm wavelengths including InGaAs, GaAsN and InGaAsN [2,3]. However, the incorporation of indium into GaAs is limited by the formation of dislocations while high concentration of nitrogen generally degrades the optical quality of the grown alloy [4]. Recently, increasing interest has been shown towards bismuth (Bi) containing semiconductors due to the relatively large band gap reduction with small addition of Bi and the large spin orbit bowing making it promising for long wavelength

optoelectronic and spintronic devices [5-8]. Besides, other substrates and structures were also reported including $GaAs_{1-x}Bi_x$/GaAs nanowires grown on silicon substrate and $GaAs_{1-x}Bi_x$ on germanium substrate [9,10].

II. EXPERIMENTAL DETAILS

The GaAsBi samples studied in this work were grown on undoped GaAs substrate by molecular beam epitaxy. The sample structure consists of a GaAs buffer layer, followed by 25-50 nm of $GaAs_{1-x}Bi_x$ layer and then capped with 50 nm of GaAs. In order to increase the incorporation of Bi into GaAs, the growth temperature was varied between 400 – 320 °C while the temperature of the Bi cell was increased from 500 – 550 °C. The (004) high resolution X-ray diffraction (HR-XRD) θ-2θ scans were carried out to verify the incorporation of Bi into GaAs and to assess the structural quality of the samples. Then, HR-XRD data fitting using RADS Mercury software was carried out to determine the Bi concentration in the samples. The optical properties of the samples were assessed by room temperature PL. The excitation source is a continuous wave 532 nm green laser and the PL emission was detected by a germanium detector operating at 77 K. A standard phase sensitive lock-in method was also used to reduce the background light and to increase the signal to noise ratio.

III. RESULTS AND DISCUSSIONS

A. Bandgap Simulation

The introduction of Bi into GaAs lattice results to the presence of Bi states which is located close to the valence band (VB) edge of GaAs [11]. Based on the valence band anti-crossing (VBAC) model, the interaction between these two states leads to the splitting of the VB to two subs bands which is [11]

$$E_{\pm}(GaAsBi) = \frac{E_v(GaAs) + E_{Bi} \pm \sqrt{(E_v(GaAs) - E_{Bi})^2 + 4xC_{Bi}^2}}{2},$$ (1)

where $E_v(GaAs)$ is the energy of GaAs valence band maximum (VBM), E_{Bi} is the energy of Bi defect state, x is Bi concentration and C_{Bi} is the coupling parameter. The value of

$E_v(GaAs)$ is referenced to zero while E_{Bi} is 0.4 eV below the GaAs VBM and $C_{Bi} = 1.6$ eV [11].

Besides, the incorporation of Bi into GaAs also may reduce the energy of the conduction band minimum (CBM) and can be estimated by the virtual crystal approximation (VCA) which is

$$E_{CB\text{-}VCA} = E_g(GaAs) - \Delta E_{CBM}x, \qquad (2)$$

where $E_{CB\text{-}VCA}$ is the VCA conduction band minimum and ΔE_{CBM} is 2.3 eV (the CB offset between GaBi and GaAs) [11,12]. Therefore, the band gap of $GaAs_{1-x}Bi_x$ is

$$E_g(GaAsBi) = E_{CB\text{-}VCA} - E_+(GaAsBi). \qquad (3)$$

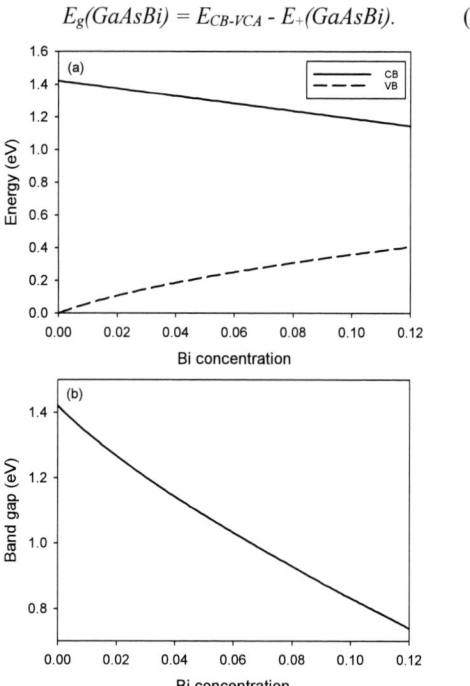

Fig. 1. Evolution of the (a) conduction band minimum and the valence band maximum and (b) band gap of GaAsBi for Bi concentration between 0 to 0.12.

Fig. 1(a) shows the calculated CBM and VBM of $GaAs_{1-x}Bi_x$ alloy for Bi concentrations between 0 (GaAs) to 0.12. The results show that the energy of the VBM increases with the increase of Bi concentration while the CBM reduces linearly by 23 meV/% Bi. This results to a significant reduction of the band gap. Fig. 1(b) shows that the band gap of $GaAs_{1-x}Bi_x$ reduces linearly by 64 meV/%Bi for x up to 0.06 but becomes non-linear for higher Bi concentrations. Based on these results, PL peak emission of 1.3 and 1.5 μm may be obtained with Bi concentration of 0.075 and 0.1, respectively.

B. HR-XRD and PL Results

In order to assess the structural properties of the grown samples, HR-XRD measurements were carried out and the data are shown in Fig. 2. The sharp and narrow peak located at 0 arcsec originated from the GaAs while the broad peak located at negative peak splitting corresponds to the $GaAs_{1-x}Bi_x$ layer. Based on data fitting using RADS Mercury software (results not shown), Bi concentrations in the samples are 0.022, 0.048, 0.081 and 0.108. Sample with the highest Bi concentration ($x = 0.108$) has a peak splitting of 3160 arcsecs which correspond to 1.27% of compressive strain. Despite the relatively high strain, well-defined fringes still can be observed which indicate a smooth and coherent interface without any dislocations.

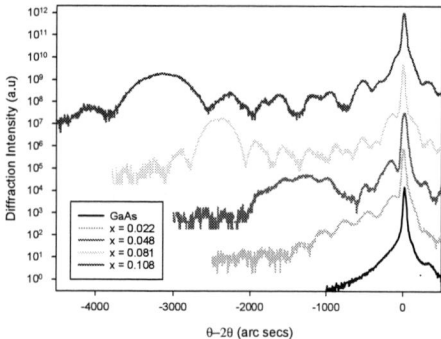

Fig. 2. HR-XRD spectra of GaAs and GaAsBi samples with different Bi concentrations. The spectra were vertically offset for clarity

Fig. 3(a) shows the PL spectra of GaAs and $GaAs_{1-x}Bi_x$ samples measured at room temperature. The GaAs control sample has a PL peak wavelength of 870 nm but redshifted to 980 nm, 1123 nm, 1342 nm and 1520 nm for samples with $x = 0.022$, 0.048, 0.081 and 0.108, respectively. The results show that higher Bi concentration results to a smaller band gap, as expected. Then, band gap versus Bi concentration is plotted, as shown in the inset of Fig. 3(a). A nice fit was obtained between the experimental data and the calculated values from (3). This shows that the incorporation of Bi into GaAs affect both the conduction band as well as the valence band.

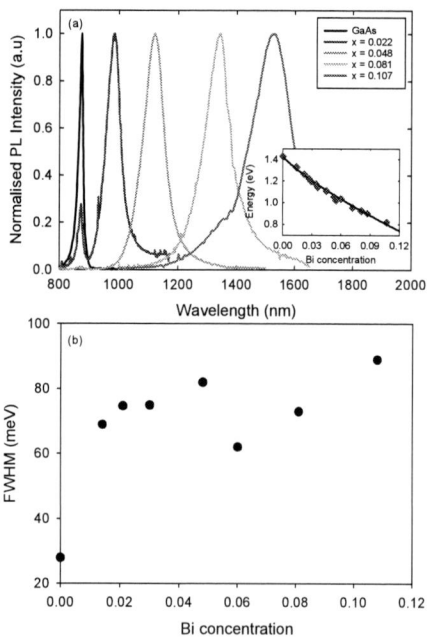

Fig. 3. (a) Room temperature PL spectra and (b) PL FWHM of GaAs and GaAsBi samples with different Bi concentrations. The inset in (a) is band gap versus Bi concentration.

At room temperature, the GaAs sample has a PL FWHM of 28 meV, as shown in Fig. 3(b). However, the linewidth increases significantly to 69 meV for GaAs$_{1-x}$Bi$_x$ sample with $x = 0.014$. The FWHM remains between 60 - 90 meV when Bi concentration is further increased to 0.108. The large FWHM indicates significant alloy fluctuation and Bi clustering which typically result to PL localization effects at low temperatures [13-15]. Despite the large FWHM, our reported values are much smaller compared to other reported values in the literature [16].

IV. CONCLUSION

In summary, the incorporation of Bi into GaAs has been shown to reduce the band gap of GaAs$_{1-x}$Bi$_x$. This is attributed to the reduction of the CB minimum and the increase of the VB maximum due to the interaction between the Bi defect state and the valence band edge of GaAs. Room temperature PL peak wavelength of up to 1.5 μm was reported for GaAs$_{1-x}$Bi$_x$ sample with $x = 0.108$. These results show that GaAs$_{1-x}$Bi$_x$ alloy is a promising material for long wavelength optoelectronic applications.

ACKNOWLEDGMENT

A. R. Mohmad thanks F. Bastiman, the Ministry of Higher Education Malaysia and the University of Sheffield – EPSRC.

REFERENCES

[1] G. Keiser, Optical Fiber Communications, 2nd ed., McGraw-Hill: Singapore, 1991, pp. 8.

[2] N. Tansu, Y. Jeng-Ya, and L.J. Mawst, "High-performance 1200-nm InGaAs and 1300-nm InGaAsN quantum-well lasers by metalorganic chemical vapor deposition", IEEE J. Selected Topics in Quantum Electronics, vol. 9, pp. 1220-1227, 2003.

[3] L. Bellaiche, "Band gaps of lattice-matched (Ga,In)(As,N) alloys", Appl. Phys. Lett., vol. 75, pp. 2578-2580, 1999.

[4] S. Spruytte, C. Coldren, J. Harris, W. Wampler, P. Krispin, K. Ploog, and M. Larson, "Incorporation of nitrogen in nitride-arsenides: Origin of improved luminescence efficiency after anneal", J. Appl. Phys., vol. 89, pp. 4401, 2001.

[5] A. R. Mohmad, F. Bastiman, C. J. Hunter, R. D. Richards, S. J. Sweeney, J. S. Ng, J. P. R. David and B. Y. Majlis, "Localization effects and band gap of GaAsBi alloys", Physica Status Solidi B Basic Research, vol. 251, pp.1276-1281, 2014.

[6] M. Masnadi-Shirazi, R. B. Lewis, V. Bahrami-Yekta, T. Tiedje, M. Chicoine, P. Servati, "Bandgap and optical absorption edge of GaAsBi alloys with 0<x<17.8%", J. Appl. Phys. vol. 116, pp. 223506, 2014.

[7] B. Fluegel, S. Francoeur, A. Mascarenhas, S. Tixier, E. C. Young, and T. Tiedje, "Giant spin-orbit bowing in GaAsBi," Phys. Rev. Lett. vol. 97, pp. 067205, 2006.

[8] A. R. Mohmad, F. Bastiman, C. J. Hunter, J. S. Ng, S. J. Sweeney and J. P. R. David, "The effect of Bi composition to the optical quality of GaAsBi", Appl. Phys. Lett. vol. 99, pp. 042107, 2011.

[9] S. J. Zelewski, J. Kopaczek, W. M. Linhart, F. Ishikawa, S. Shimomura, R. Kudrawiec, "Photoacoustic spectroscopy of absorption edge for GaAsBi/GaAs nanowires grown in Si substrate", Appl. Phys. Lett. vol. 109, pp. 182106, 2016.

[10] P. Wang, W. Pan, X. Wu, C. Cao, S. Wang, Q. Gong, "Heteroepitaxy growth of GaAsBi on Ge(100) substrate by gas source molecular beam epitaxy", Appl. Phys. Express. vol. 9, pp. 45502, 2016.

[11] K. Alberi, O.D. Dubon, W. Walukiewicz, K.M. Yu, K. Bertulis, and A. Krotkus, "Valence band anticrossing in GaBiAs", Appl. Phys. Lett., vol. 91, pp. 051909-3, 2007.

[12] K. Alberi, J. Wu, W. Walukiewicz, K. M. Yu, O. D. Dubon, S. P. Watkins, C. X. Wang, X. Liu, Y. J. Cho, and J. Furdyna, "Valence-band anticrossing in mismatched III-V semiconductor alloys," Phys. Rev. B vol. 75, pp. 045203, 2007.

[13] A. R. Mohmad, F. Bastiman, C. J. Hunter, F. Harun, D. F. Reyes, D. L. Sales, D. Gonzalez, R. D. Richards, J. P. R. David and B. Y. Majlis, "Bi concentration inhomogeneity in GaAsBi bulk and quantum well structures", Semicond. Sci. Tech. vol. 30, pp. 094018, 2015.

[14] T. Wilson, N. P. Hylton, Y. Harada, P. Pearce, D. Alonso-Alvarez, A. Mellor, R. D. Richards, J. P. R. David, N. J. Ekin-Daukes, "Assessing the nature of the distribution of localised states in bulk GaAsBi", Sci. Rep. vol. 8, pp. 6457, 2018.

[15] J. Kopaczek, W. M. Linhart, M. Baranowski, R. D. Richards, F. Bastiman, J. P. R. David, R. Kudrawiec, "Optical properties of GaAsBi/GaAs quantum wells: photoreflectance, photoluminescence and time-resolved photoluminescence study" Semicond. Sci. Technol. vol. 30, pp. 094005, 2015.

[16] A. R. Mohmad, F. Bastiman, J. S. Ng, S. J. Sweeney and J. P. R. David, "Photoluminescence investigation of high quality GaAsBi on GaAs", Appl. Phys. Lett. vol. 98, pp. 122107, 2011.

Photonic crystal embedded waveguide for compact C-band band-pass filter

Mohd Nuriman Nawi
Institute of Microengineering and Nanoelectronics (IMEN)
Universiti Kebangsaan Malaysia (UKM)
Bangi, Malaysia
mnuriman.nawi@siswa.ukm.edu.my

Nurulhani Diana Rashid
Institute of Microengineering and Nanoelectronics (IMEN)
Universiti Kebangsaan Malaysia (UKM)
Bangi, Malaysia
haniidiana@siswa.ukm.edu.my

Dilla Duryha Berhanuddin
Institute of Microengineering and Nanoelectronics (IMEN)
Universiti Kebangsaan Malaysia (UKM)
Bangi, Malaysia
dduryha@ukm.edu.my

Burhanuddin Yeop Majlis
Institute of Microengineering and Nanoelectronics (IMEN)
Universiti Kebangsaan Malaysia Malaysia (UKM)
Bangi, Malaysia
burhan@ukm.edu.my

Mohd Adzir Mahdi
Wireless and Photonics Networks Research Centre
Faculty of Engineering Universiti Putra Malaysia (UPM)
Serdang, Malaysia
adzir@ieee.org

Ahmad Rifqi Md Zain
Institute of Microengineering and Nanoelectronics (IMEN)
Universiti Kebangsaan Malaysia (UKM)
Bangi, Malaysia
rifqi@ukm.edu.my

Abstract— We report the modelling of a band-pass filter at conventional band (c-band) and the effect of different shapes of photonic crystal (PhC) holes embedded on silicon-on-insulator (SOI) waveguide. The designed embedded waveguides was simulated with LUMERICAL finite different time domain (FDTD) and a filter response with a bandwidth of approximately 30 nm complying with international telecommunication unit (ITU-T) standard was observed. The simulated bandwidth observed was sufficient for guarding against other band interference in telecommunication applications such as wavelength division multiplexing (WDM). The waveguide was designed with a dimension of 600 nm width x 260 nm height and embedded with PhC of 4 mirrors and 3 cavities. 2 mirrors at both end of the whole structure were designed with less number of holes for obtaining the band-pass filter profile. With a value of lattice constant a, hole radius r and cavity distance c of 370, 115 and 315 nm respectively, the simulated device spectrum complimented the erbium doped fiber amplifier (EDFA) spectrum to obtain wavelength profile flatness. The PhC embedded waveguide was tailored to give a 70 percent value of transmission. A flat profile was observed by reducing the photonic crystal hole radii in the middle mirrors. The wavelength band and the bandwidth of the band can be tuned by manipulating the number of mirrors and cavities in waveguide. A different types of PhC hole shapes were also studied and compared. The transmission quality and band-pass quality with different types of hole shapes show that the circular PhC shape are superior in comparison with square hole shapes.

Keywords— *photonic crystal; cavity; mirror; waveguide; silicon on insulator; c-band; EDFA; WDM; FDTD*

I. INTRODUCTION

Photonic crystal (PhC) embedded waveguide is the future solution for compacting device in replacing electrical interconnect in electronic microprocessors due to the Moore's limitation [1-3]. Not only for replacing the electrical interconnect, the application of PhC has been researched for various range application such as photodetectors and biosensors [4-6]. With the advancement of the fabrication technique, a device at a nanometer scale can be fabricated without jeopardizing the performance of the device [7, 8]. A 1-dimensional (1D) PhC silicon waveguide is preferable due to the simplicity and good confinement of light that gives high selectivity in the designed waveguide [9, 10]. PhC is a medium with different refractive index to be arranged in a manner where the index differences are violated for wavelength transmission dictation. Silicon based waveguide offers a clear window at near-infrared region that makes it suitable for telecommunication [11]. The size of the waveguide is optimized to 2 (width): 1 (height) on silicon-on-insulator (SOI) platform depending on the polarization of light to be guided for the optimum light confinement [12].

In wavelength division multiplexing (WDM), the conventional band (c-band) and other wavelength bands are utilized to deliver information. The international telecommunication unit (ITU-T) has rule out a grid for WDM for the industrial standardization where the range of wavelength for c-band falls from approximately 1.53 to 1.56 µm [13]. In order to protect the specific band, a compact band-pass filter is required.

II. DEVICE SIMULATION MODEL

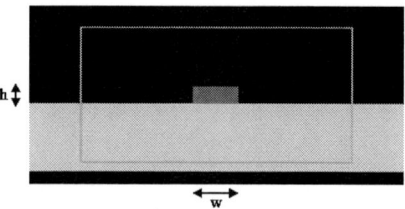

Fig. 1. The MODE Solution interface with the cross section of a waveguide with a dimension of 600 nm width x 260 nm height on SOI platform. Silicon, air and insulator (silicon dioxide) refractive indexes used were 3.48, 1 and 1.45 respectively.

The authors would like to thank the Ministry of Higher Education (MOHE) under the grant no: LRGS (2015)/NGOD/UM/KPT for the financial support of this project.

978-1-5386-5284-8/18 $31.00 © 2018 IEEE

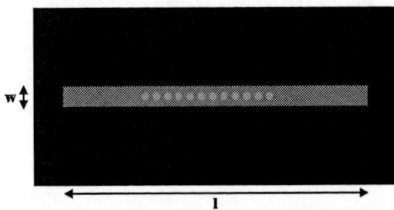

Fig. 2. A 12 periodic holes with a radius **r**, 115 nm and a lattice constant **a**, 370 nm are arranged in a 12 μm waveguide to give a photonic band gap (forbidden region) to forbid the light from passing through.

The device was modeled with a dimension of 600 nm width x 260 nm height on an SOI platform. The structure was modelled by using MODE solution from Lumerical as shown in Fig.1 and the light confinement was observed.

The waveguide was embedded with a PhC structure that consists of mirrors and cavities. As shown in Fig. 2, a row of holes with a radius **r**, 115 nm and a lattice constant **a**, 370 nm (for periodicity arrangement) were designed to give a sufficient photonic band-gap (PBG) to block the wavelength [4].

The PhC was then arranged with 4 mirrors and 3 cavities as shown in Fig. 3 to give the desired band-pass filter spectrum. The cavity **c**, 315 nm was introduced in between the 4 mirrors. 3D-FDTD solution from Lumerical was used for the simulation. The simulation time was set according to the device dimension and the refractive index to ensure that the simulation will stop after all the light has go through the distance set before leaving the simulation area.

$$t = \frac{simulation\ region\ distance}{c\,n} \qquad (1)$$

where t, c and n are the simulation time, velocity of light and material refractive index respectively where the refractive index of silicon at 3.48 was used in the simulation. The mesh of the area to be simulated was configured to be as automatic non-uniform mesh due to the different sizes and patterns involved in the structure. The minimum size of the mesh setting was set to be 0.25 nm for simulation accuracy purposes. The boundary condition for the simulation region was configured to be as perfectly matched layer (PML) to avoid any light reflected disturbing the results obtained.

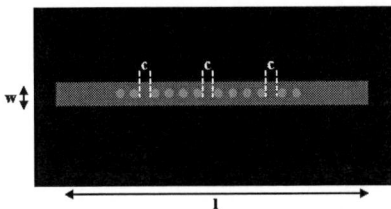

Fig. 3. A periodic holes arranged in a waveguide with 4 mirrors and 3 cavities. The cavity **c**, 315 nm was introduced in between mirrors. Both mirror at the end of PhC comprises of 2 holes instead of 4 like the middle mirrors.

III. RESULTS AND DISCUSSION

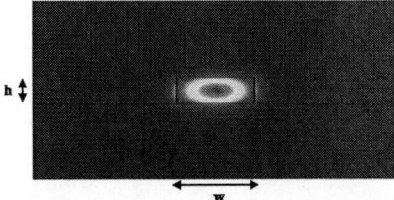

Fig. 4. The light confinement for a silicon waveguide with a dimension of 600 nm (width) x 260 nm (height) that shows minimum loss transmission for transverse electric (TE) mode light transmission where the highest intensity of light was confined in the center of the structure.

Fig. 4 shows 1.5 μm light source confined in the waveguide simulated. The light confined was the transverse electric (TE) mode light with nearly 100 percent confinement shows the structure was giving a minimal loss with the selected dimension of waveguide. PBG obtained shows a prominent wavelength blocking function for the waveguide as shown in Fig. 5 where the forbidden region was from 1.34 to 1.68 μm.

The introduction of a single cavity results in a single wavelength excitation as shown in Fig. 6. The excitation wavelength was determined by the cavity distance [4].

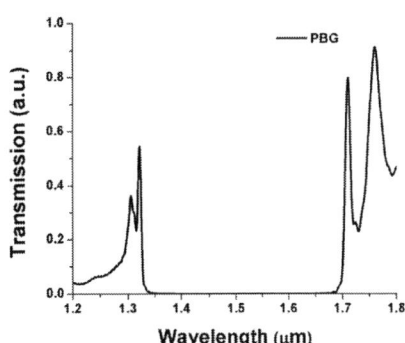

Fig. 5. A photonic band-gap (PBG) obtained with a range of 1.34 to 1.68 μm for a periodic structure of PhC embedded without any cavity.

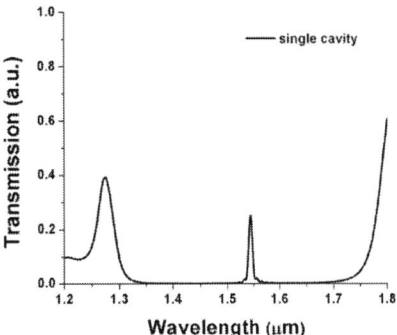

Fig. 6. A single wavelength excitation at the middle of the band for single cavity. The excitation wavelength was govern by the cavity distance in the periodicity of the PhC [4].

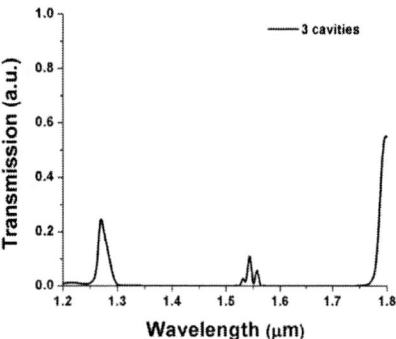

Fig. 7. 3 cavities produces 3 excitation wavelengths at the center of the forbidden region.

By adding more cavities in the PhC periodic structure (in a symmetry manner in conjunction with the mirrors), the excitation wavelengths was excited additively as shown in Fig. 7. By controlling the number of holes in the mirrors, the multiple resonant wavelengths will merge and a single wavelength excitation with broader wavelength complies with ITU-T standard in c-band was produced as observed in Fig. 8. The spectrum observed complimented the erbium doped fiber amplifier (EDFA) output response to obtain flatness in WDM system. A smooth light transmission was obtained and the flatness of the spectrum was observed when the radii of the 2 holes in the 2 middle mirrors each were reduced from 115 nm to 110 nm as shown in Fig. 9. The PhC hole shape was replaced with a square shape instead of circular. With a similar properties; cavity distance, width and length of the square dimension complying with the diameter of circular structure and periodicity, the results obtained was as shown in Fig. 10. The results obtained shows that the square shape of PhC will reduce the transmission of light in the waveguide. This was due to the sharp edges of the square holes PhC that introduces an abrupt changes in the flow of light in the structure. The abrupt change in the structure will influence the transmission of light due to the some portion of light traveled in the waveguide will get reflected and scattered.

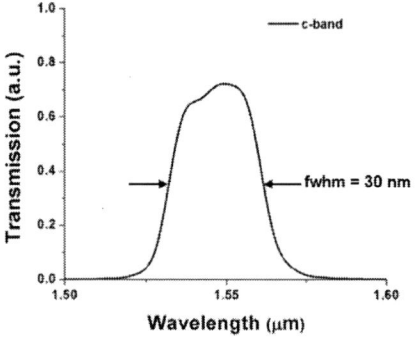

Fig. 8. C-band wavelength band-pass filter response with 30 nm bandwidth complying with ITU-T G.694.1 standard and 70 percent transmission quality to compliment EDFA.

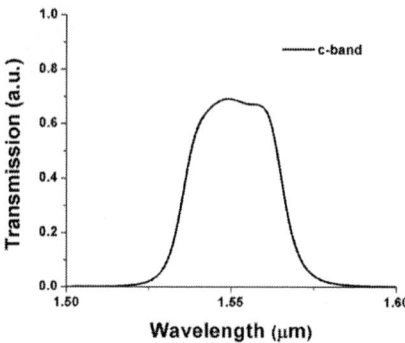

Fig. 9. C-band wavelength band-pass filter with a flat response and 70 percent transmission suitable for next generation WDM application where the flatness of the channels is important in the long haul fiber optic transmission.

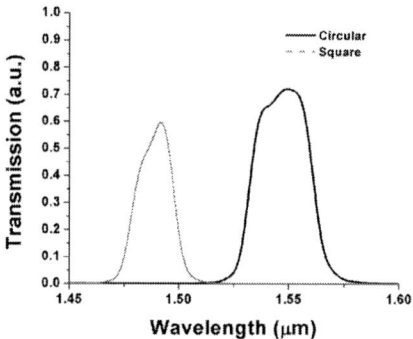

Fig. 10. A comparison between the circular embedded PhC and square embedded PhC where the wavelength downshift and reduced transmission peak with non-flat response were observed.

IV. CONCLUSION

The PhC waveguide simulated has shown a good response in conjunction with a compact nanowire c-band guard in WDM channel to be fabricated. It was shown that by manipulating the structure of the PhC; the wavelength selection can be easily selected for it to be blocked or to be pass-through. From the results shown, the mirror radius size, then number of holes, the shape of holes and the cavity distances play a major role in the bandwidth and wavelength selection. The number of cavity in the periodic structure also will gives a good control on the number of excitation wavelengths in the device. The circular shape PhC gives a very good and ease of wavelength control in comparison with the square shape structure and in term of fabrication, due to the etching technology, a sharp edge in nanostructure for PhC is very difficult to be realized that gives the circular PhC structure the best option for fabrication. It is good to note that in order to simulate a device with a simulation tool such as FDTD, a good meshing selection is necessary in order to obtain good results without stressing the computer power and memory. With a bandwidth of 30 nm, higher output transmission and more compact design as compared to the components available in the market [14], this will open up a new path for future compact photonics devices.

REFERENCES

[1] G. E. Moore, "Cramming more components onto integrated circuits (Reprinted from Electronics, pg 114-117, April 19, 1965)," (in English), *Proceedings of the Ieee,* Reprint vol. 86, no. 1, pp. 82-85, Jan 1998.

[2] Y. Yu, G. Chen, C. Sima, and X. Zhang, "Intra-chip optical interconnection based on polarization division multiplexing photonic integrated circuit," *Optics Express,* vol. 25, no. 23, pp. 28330-28336, 2017/11/13 2017.

[3] A. R. M. Zain, N. P. Johnson, M. Sorel, and R. M. D. L. Rue, "High Quality-Factor 1-D-Suspended Photonic Crystal/Photonic Wire Silicon Waveguide Micro-Cavities," *IEEE Photonics Technology Letters,* vol. 21, no. 24, pp. 1789-1791, 2009.

[4] M. N. Nawi, N. D. Rashid, D. D. Berhanuddin, A. R. M. Zain, B. Y. Majlis, and M. A. Mahdi, "Photonic crystal (PhC) nanowires for infrared photodetectors," in *2016 IEEE International Conference on Semiconductor Electronics (ICSE),* 2016, pp. 153-156.

[5] M. H. Haron, A. R. M. Zain, and B. Y. Majlis, "Sensitivity increment of one dimensional photonic crystal biosensor," in *2017 IEEE Regional Symposium on Micro and Nanoelectronics (RSM),* 2017, pp. 116-118.

[6] N. D. M. Zamani, A. R. M. Zain, and B. Y. Majlis, "Modelling of 2-D Gallium Nitride (GaN) photonic crystal," in *2016 IEEE International Conference on Semiconductor Electronics (ICSE),* 2016, pp. 54-56.

[7] N. Nawi, B. Y. Majlis, M. A. Mahdi, R. M. De La Rue, M. Lonĉar, and A. R. Md Zain, "Enhancement and reproducibility of high quality factor, one-dimensional photonic crystal/photonic wire (1D PhC/PhW) microcavities," *Journal of the European Optical Society-Rapid Publications,* journal article vol. 14, no. 1, p. 6, February 20 2018.

[8] A. R. M. Zain and R. M. De La Rue, "Control of coupling in 1D photonic crystal coupled-cavity nano-wire structures via hole diameter and position variation," (in English), *Journal of Optics,* Article vol. 17, no. 12, p. 5, Dec 2015, Art. no. 125007.

[9] A. R. M. Zain, N. P. Johnson, M. Sorel, and R. M. D. L. Rue, "Ultra high quality factor one dimensional photonic crystal/photonic wire micro-cavities in silicon-on-insulator (SOI)," *Optics Express,* vol. 16, no. 16, pp. 12084-12089, 2008/08/04 2008.

[10] M. Zain and A. Rifqi, "One-dimensional photonic crystal/photonic wire cavities based on silicon-on-insulator (SOI)," University of Glasgow, 2009.

[11] E.-K. Tien *et al.,* "Discrete parametric band conversion in silicon for mid-infrared applications," *Optics Express,* vol. 18, no. 21, pp. 21981-21989, 2010/10/11 2010.

[12] D. X. Xu *et al.,* "Silicon Photonic Integration Platform - 2014;Have We Found the Sweet Spot?," *IEEE Journal of Selected Topics in Quantum Electronics,* vol. 20, no. 4, pp. 189-205, 2014.

[13] International Telecommunication Union, "Spectral grids for WDM applications: DWDM frequency grid," International Telecommunication Union, T-REC-G.694.1-201202. [Online]. Available: https://www.itu.int/rec/T-REC-G.694.1-201202-I/en [Accessed: April 1, 2018].

[14] Thorlabs, "B1550-30 - Ø1" Bandpass Filter, CWL = 1550 ± 6 nm, FWHM = 30 ± 6 nm" drawing no 20453-E01, part no FB1550-30, Feb. 2011. [Online], Available: https://www.thorlabs.com [Accessed: April. 1, 2018].

Optical Properties of Multilayer Porous Silicon with Different Fabrication Conditions for Application along Telecom Band

Ahmad Afif Safwan Mohd Radzi
Solar Energy Research Institute (SERI)
Universiti Kebangsaan Malaysia
Bangi, Malaysia
afifseri@ukm.edu.my

Mohamad Rusop
NANO-SciTech Centre, Institute of Science
Universiti Teknologi MARA
Shah Alam, Malaysia
nanouitm@gmail.com

Saifollah Abdullah
School of Physics and Material Studies, Faculty of Applied Sciences
Universiti Teknologi MARA
Shah Alam, Malaysia
saifollah@salam.uitm.edu.my

Abstract—**Multilayer porous silicon structure was fabricated to study its optical response along the telecommunication band wavelength. The structure was fabricated using electrochemical etching technique. By modulating the etching parameters, different porosity layers can be achieved which will determine the refractive index value of each layer. Reflectance study was conducted to analyze the optical response for different fabrication thickness. Also the number of layers were set consistent at 40 layers for each sample for optimization. High degree of accuracy from thickness measurements achieved when comparing results from FESEM and Ellipsometer. Thicker layer films were determined to have higher reflectance along the C-band region which is very useful for communication application. Multilayer porous silicon has been studied for multiple optical devices and applications including Bragg grating waveguide, Fabry-Perot interferometers, photonic integrated circuit and optical gas sensor.**

Keywords—*Porous silicon, multilayer thin films, optical properties, refractive index, Bragg mirror*

I. INTRODUCTION

Porous silicon (PSi) was first discovered by Uhlir [1] in 1956 during the procedure of electro-polishing silicon wafer. The concentrated acidic electrolyte etches the surface of silicon under electrochemical etching process and accidentally produced a dark, low reflective, unpolished surface. It was until 1990 that PSi really gains the popularity when Canham et al. discovered the photoluminescence (PL) and electroluminescence from its surface [2]. PSi is a form of nanocrystalline structure with complex orientation that possesses high porosity resulted from the silicon's chemical properties. It has introduced perpendicular nanoporous holes in its structure producing large surface area to volume ratio [3]. Modulation of electrochemical etching current densities

with a specific layer etching time will produce the multilayer PSi (MPSi) structure. The etching process is selective to each alternating layers meaning the etching parameters for new etched layer does not influence or disturb the structural properties of the previously etched layers [4]. By this, a consistent alternating thickness and porosity multilayer structure could be fabricated. This would allow the realization of multilayer thin films structure with optical properties that can behave as distributed Bragg reflectors, optical microcavities, waveguides, Fabry-Perot interferometer, and Bragg grating waveguide [1-2, 5]. In order for these MPSi to behave as such, it must first obey the primary condition for reflective multilayer thin film, the Bragg wavelength condition. In any Bragg reflector, the reflected wavelength, λ_B, is defined by the Bragg condition,

$$\lambda_B = 2n_{eff}\Lambda \qquad (1)$$

where n_{eff} is the effective refractive index and Λ is the grating periodicity [6]. Several studies have been conducted for implementations of multilayer structure in filtering specific wavelength especially within the telecom band [7-8]. Telecom band is a range of wavelength that is used for long distance radio telecommunication. The range for its signal utilization which follows the IEEE standard is from short wavelengths S-band into conventional C-band microwaves. The wavelength range is from 1460–1530 nm of S-band to 1530–1565 nm of the C-band [5, 9]. Recent studies emphasized that a proper optimization of fabrication parameters is required to achieve and maintain high reflectance across the telecom band [10-11]. The thickness, porosity and refractive index contrast of MPSi plays an important role to attain desired reflectance. Telecom band is very useful in signal and data transfer especially in satellite, radar, and aviation technology, both military and commercial sectors.

II. EXPERIMENTAL DETAILS

Electrochemical etching was performed in a Teflon cell using anodic reaction to produce porous silicon surface. The electrolyte used for this work is a mixture of hydrofluoric (HF 48%) rich acid-based solution and absolute ethanol (C_2H_5OH 99%). The mixture were composed with HF:Ethanol ratio of 1:1. Crystalline silicon substrate used was a highly doped p-type silicon wafer, <1 0 0> oriented and 525μm thickness. In order to produce multilayer porous silicon, the etching current density and etching time were varied during etching process so that the realization of thin films with different porosity can be achieved. Multilayer number for MPSi fabrication in this study is optimized at 40 layers to compare the optical properties especially reflectance spectroscopy corresponds to the thickness. According to literature, high number of layers will produce a higher reflectance which corresponds to the multilayer thickness and also incident light wavelengths [12]. The alternating etching current density sequences are 15.38 mA/cm^2 for low current density, J_1 and 46.15 mA/cm^2 for high current density, J_2 with constant 100 Volts of electrical potential supply. Three etching time-per-layer of 10 seconds, 20 seconds and 30 seconds were used to show thickness variation effects. Sample's nomenclature can be referred with Λ as sum of alternating layers. Symbol T_{10} indicates that the etching time of 10 seconds was applied for every layer, T_{20} for 20 seconds and T_{30} for 30 seconds. Table 1 summarize the etching parameters for each sample.

TABLE I. ETCHING PARAMETERS TO FABRICATE EACH MPSI SAMPLE

Sample	First Sequence J_1 (mA/cm^2)	Second Sequence J_2 (mA/cm^2)	No. of Layers (Λ)	Etching Time per Layer T (s)
40Λ T_{10}	15.38	46.15	40	10
40Λ T_{20}	15.38	46.15	40	20
40Λ T_{30}	15.38	46.15	40	30

For cross section view, MPSi samples were characterized by Field Emission Scanning Electron Microscopy (FESEM JEOL JSM 7600F). For ellipsometric measurement, Variable Angle Ellipsometer L 116S (Gaertner Scientific Corp.) with fix wavelength He-Ne laser source of 632.8 nm was used. Reflectance spectra is measured using UV-Vis (PerkinElmer Lambda 750).

III. RESULTS AND DISCUSSION

A. Determination of thickness and refractive index of MPSi

Figure 1 shows the FESEM cross section view of 40Λ sample with its thickness consisted of layers with alternating high and low refractive index. The realization of multilayer structure is coherent with the physical properties of the porous structure. Higher porosity layer will possess larger void area and vice versa [13]. Due to different crystal composition, high contrast layers of alternating porosity was observed and thicknesses were able to be verified [14-16]. Thickness estimated from FESEM for 40Λ T_{10} multilayer structure is between 105-119nm per layer. From this measurement, the thickness of the whole 40 layers structure produced is estimated at 4.46 μm. Higher porosity layer will possess larger void area and vice versa [13]. Due to different crystal composition, high contrast layers of alternating

porosity was observed and thicknesses were able to be verified [14-16]. Figure 1 (b) shows the FESEM cross section of sample 40Λ T_{20} and measured whole thickness is 10.2 μm. Average single layer thickness calculation from the whole multilayer of 40Λ T_{10} can be referred to the equation below:

$$\text{Average thickness} = \frac{\text{Thickness of the whole multilayer}}{\text{No. of Multilayer}} \quad (2)$$

From FESEM whole thickness measured, using equation 2, calculated thickness of a single layer is estimated at 255 nm. Figure 1 (c) shows the thickness of the whole MPSi structure for 40Λ T_{30} which is 15,500 nm. The average thickness of each layer for 40Λ T_{30} calculated using equation 2 is 387.5 nm.

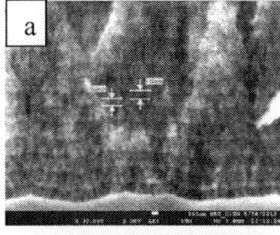

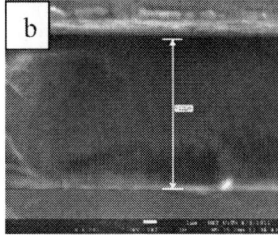

Fig. 1. FESEM Cross section view of (a) 40Λ T_{10}, (b) 40Λ T_{20}, (c) 40Λ T_{30}

Table 2 summarize the average thickness per layer and average thickness of the whole MPSI structure for each sample estimated from the FESEM images.

TABLE II. THICKNESS PER LAYER AND TOTAL THICKNESS ESTIMATED FROM FESEM IMAGE

Sample	No. of Layers (Λ)	Etching Time per Layer, T (s)	Average Thickness per Layer (nm)	Total Average Thickness (μm)
40Λ T_{10}	40	10	112	4.46
40Λ T_{20}	40	20	255	10.2
40Λ T_{30}	40	30	388	15.5

The whole thickness determined for 40Λ samples are 15.5 μm for T_{30}, 10.2 μm for T_{20} and 4.85 μm for T_{10}. The approximation of average thickness of each layer for T_{30} sample is 387.5 nm, T_{20} sample is 255 nm and T_{10} sample is 120 nm. It is obvious that T_{30} samples, which has longer etching time, is slightly thicker than T_{10} and T_{20} samples. The realization of multilayer stacking of PSi structure is achievable complete with thickness estimations.

B. Ellipsometer analysis of MPSi samples

Table 3 shows the refractive index and thickness values measured by ellipsometer for all samples. The refractive index for high porosity layer is referred to n_H while for lower porosity is n_L.

TABLE III.   REFRACTIVE INDEX AND THICKNESS VALUES FOR ALL SAMPLES

Sample	Refractive index, n		Thickness, t (nm)	
	n_H	n_L	t_1	t_2
40Λ T_{10}	1.412	1.538	163	166
40Λ T_{20}	1.453	1.847	284	293
40Λ T_{30}	1.366	1.528	337	372

It is known from literature [17] and [18] that lower refractive index value corresponds to high porosity structure since the structure consisted of larger void area [18. Refractive index values for high porosity layer, n_H is consistent for all samples within range of 1.358 to 1.495. This range is resulted from the second sequence of higher etching current density, 46.15 mA/cm^2 which produced high porosity layer [19]. For lower porosity layer, n_L the refractive index range is higher but with inconsistent values between 1.528 to 1.847. This range is resulted from lower first sequence etching current density, 15.38 mA/cm^2 which produced lower porosity layer [13,20]. The average thickness per layer measured is within 1~10% varied from each of them. For 30 seconds etching, thickness measured is 337~372 nm and from FESEM is 387.5 nm. For 20 seconds, average thickness measured from ellipsometer is 284~293 nm and from FESEM is 255 nm. Average layer thickness for 10 seconds etching from ellisometer is 163~166 nm and 120 nm from FESEM. Table 4 shows the whole thickness for all samples measured using ellpisometer and FESEM. Using ellipsometer, thickness measured for 40Λ T_{10} is 6.58 um 40Λ T_{20} for 11.54 um and 14.18 um for 40Λ T_{30}.

TABLE IV.   THICKNESS FOR EACH ALTERNATING LAYER OF EACH SAMPLE ESTIMATED FROM ELLIPSOMETER

Sample	Thickness, t_1 (nm)	Thickness, t_2 (nm)	Total Average Thickness (μm)
40Λ T_{10}	163	166	6.58
40Λ T_{20}	284	293	11.54
40Λ T_{30}	337	372	14.18

These thickness values are comparable with FESEM estimation from Figure 1. Thickness values obtained by ellipsometer are almost identical with FESEM estimation giving a good agreement between both measurements.

C. UV-Vis Reflectance analysis of MPSi

Fig. 2 shows reflectance spectra from 40Λ samples in the Infra-Red region. The reflectance values are clearly high in the C-band region. Highest peak is measured in the regions between S-band and C-band. Along the C-band region, reflection for thicker layer thickness sample is highest followed by less thick layer samples due to both thickness and reflectivity are directly related [21-22].

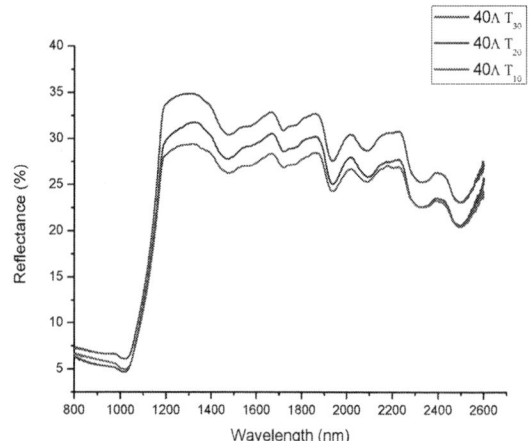

Fig. 2. Reflectance spectra from 40Λ samples

Another reason that caused the reflectance to increase is the Bragg wavelength, λ_B condition of MPSi itself [16, 23, 24]. MPSi's thickness and refractive index contrast produced a λ_B condition which gives the high reflectance across the pass band spectrum. 40Λ T_{30} sample has the highest reflectance value which indicates that the alternating of high and low refractive index with thicker layer and deeper thickness has produced a better mirror effect within the pass band [21]. Also from Fig. 2, the λ_B of all samples starts across the same wavelength of between 1000-1100 nm and continued high across the infra-red region. Reflectance at telecommunication band range (1550 nm) with reflectance of 30 % achieved [7,11]. These reflectance values are quite low due to multilayer thickness errors-effect [23]. These errors are affected by the low number multilayer thickness and also refractive index contrast between layers. Refractive index contrast, Δn is division of low refractive index layer, n_{low} over high refractive index layer, n_{high} of a sample as shown in the formula below [23]:

$$\Delta n = \frac{n_{low}}{n_{high}} \qquad (3)$$

When refractive index between layers is almost similar in values, the contrast will be higher approaching 1. Higher refractive index contrast will strengthen the thickness error effects. Refractive index contrast calculated using equation 3 for MPSi samples are high between 0.80 to 0.91. Due to the high refractive index contrast, and thickness deviation of each sample, the thickness error-effect becomes large. However, by fabricating higher multilayer number and thicker thickness, thickness error-effect can be reduced as proven from measured reflectance spectra. From FESEM and ellipsometer results, the thicknesses of each sample of MPSi are not uniform with slight deviation between t_1 and t_2. This deviation increases the thickness error effect reducing reflected light from MPSi structure. However this effect reduced as the number of multilayer and its thickness increased [23]. This explains why reflectance for 40Λ T_{30} is the highest compared to other samples. Fig. 3 shows the comparison between reflectance of 40Λ samples from their reflectance percentage values and maximum reflectance values at 1550 nm.

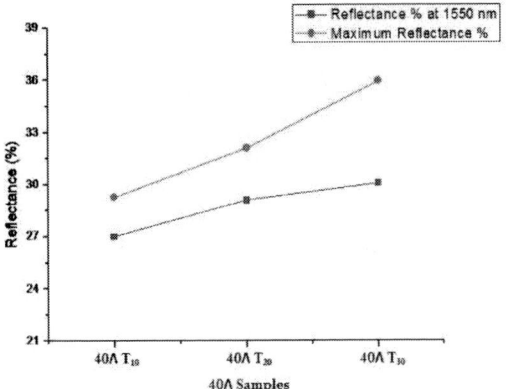

Fig. 3. Reflectance properties of 40Λ samples

It is clear that 40Λ T_{30} sample with longer etching time produced better reflectivity both for maximum reflectance value and also at 1550 nm wavelength position. This indicates that higher etching time produces better reflectivity compared to less etching time samples [21, 25].

CONCLUSION

Structural and optical properties of MPSi were characterized by several methods and a good agreement between those results obtained. From FESEM images, MPSi was successfully fabricated with alternating thickness and refractive index. The thickness of each layer was determined and measurement from ellipsometer is in a good agreement with the estimation from FESEM. Alternating refractive indexes measured are co-relating with the porosity of MPSi and stimulates the reflectance of MPSi. From reflectance spectra, reflectivity of around 30 % was achieved at 1550 nm within the desired telecommunication C-band wavelength region. However some errors could not be avoided which led to low reflectance value such as thickness error and low refractive index contrast. These errors led to degradation of output reflectance and also shifting the reflectance spectra. Also from the reflectance study it can be concluded that the prepared MPSi has the ability as a dielectric mirror to reflect light in the telecommunication pass band.

ACKNOWLEDGMENT

The author would like to thank for the help from officers of NOR Laboratory USM Penang and NanoSciTech UiTM Shah Alam.

REFERENCES

[1] M. Archer, M. Christophersen, and P. M. Fauchet, "Electrical porous silicon chemical sensor for detection of organic solvents," Sensors and Actuators B: Chemical, vol. 106, pp. 347-357, 2005.

[2] L. T. Canham, "Silicon quantum wire array fabrication by electrochemical and chemical dissolution of wafers," Applied Physics Letters, vol. 57, pp. 1046-1048, 1990.

[3] C. Wongmanerod, S. Zangooie, and H. Arwin, "Determination of pore size distribution and surface area of thin porous silicon layers by

spectroscopic ellipsometry," Applied Surface Science, vol. 172, pp. 117-125, 2001.

[4] L. Pavesi, "Influence of dispersive exciton motion on the recombination dynamics in porous silicon," Journal of Applied Physics, vol. 80, pp. 216-225, 1996.

[5] N. A. Olsson, "Lightwave systems with optical amplifiers," Lightwave Technology, Journal of, vol. 7, pp. 1071-1082, 1989.

[6] H. Zhang and P. R. Herman, "Bragg Grating Waveguides: Extending Bragg Technology into Three Dimensions optical filters in laser-written 3-D waveguides," Photonics Spectra, vol. 42, p. 98, 2008.

[7] A. Ramdane and A. Ougazzaden, "Quantum well bandgap engineering for 1.5 µm telecom applications," Materials Science and Engineering: B, vol. 74, pp. 66-69, 2000.

[8] K. Nandy, S. Biswas, R. Bhattacharyya, S. N. Saha, and A. Deyasi, "Novel Band-reject Filter Design using Multilayer Bragg Mirror AT 1550 NM," in National Conference on Advancement of Computing in Engineering Research (ACER 2013), West Bengal India, 2013, pp. 419–425.

[9] D. O. Caplan, "Laser communication transmitter and receiver design," Journal of Optical and Fiber Communications Reports, vol. 4, pp. 225-362, 2007.

[10] H.Y. Seba, T. Hadjersi, "Bragg mirrors porous silicon back reflector for the light trapping in hydrogenated amorphous silicon" Applied Surface Science, vol. 350, pp. 57–61, 2015.

[11] Mi-Ae Park and Honglae Sohn "Fabrication and optical characterization of Bragg resonance luminescence porous silicon" Semicond. Sci. Technol., vol. 31, 014013, 2016.

[12] R. Dubey and D. Gautam, "Synthesis and characterization of nanocrystalline porous silicon layer for solar cells applications," Journal of optoelectronic and biomedical materials, vol. 1, pp. 8-14, 2009.

[13] E. Xifré Pérez, "Design, fabrication and characterization of porous silicon multilayer optical devices," 2007.

[14] D. D. H. M. Ng, E. Iliopoulos, and T. D. Moustakas, "Distributed Bragg reflectors based on AlN/GaN multilayers," Appl. Phys. Lett. , vol. 74, p. 3, 1999.

[15] M. Achtenhagen, N. V. Amarasinghe, and P. Young, "Spectral Properties of High-Power Distributed Bragg Reflector Lasers," Lightwave Technology, Journal of, vol. 27, pp. 3433-3437, 2009.

[16] X.-y. Furu Zhong, Lv Zhen-hong, Jia Jiaqing Mo, "Fabrication of porous silicon-based silicon-on-insulator photonic crystal by electrochemical etching method," Optical Engineering vol. 51, pp. 1-3, 2012.

[17] O. Bisi, S. Ossicini, and L. Pavesi, "Porous silicon: a quantum sponge structure for silicon based optoelectronics," Surface Science Reports, vol. 38, pp. 1-126, 2000.

[18] P. B. Nair, V.B.Justinvictor, V. Ramakrishnan, D. D. Kumar, et al., "Structural, optical, photoluminescence and photocatalytic investigations on Fe doped Tio2 thin films," Thin Solid Films, vol. 550, pp. 121-127, 2014.

[19] Y. Tamar, M. Tzabari, C. Haspel, and Y. Sasson, "Estimation of the porosity and refractive index of sol–gel silica films using high resolution electron microscopy," Solar Energy Materials and Solar Cells, vol. 130, pp. 246-256, 2014.

[20] V. Torres-Costa, R. J. Martin-Palma, and J. M. Martinez-Duart, "Optical characterization of porous silicon films and multilayer filters," Applied Physics A, vol. 79, 2004.

[21] S. Chan, "Porous Silicon Multilayer Structures: from Interference Filters to Light," University of Rochester, 2000.

[22] A. Dodabalapur, L. Rothberg, R. Jordan, T. Miller, R. Slusher, and J. M. Phillips, "Physics and applications of organic microcavity light emitting diodes," Journal of applied physics, vol. 80, pp. 6954-6964, 1996.

[23] S. M. Feng, T. Chen, and Y. Q. Lin, "Random thickness errors-effect on reflectance of multilayer," Optik - International Journal for Light and Electron Optics, vol. 121, pp. 934-937, 2010.

[24] D. Ariza-Flores, J. S. Pérez-Huerta, and V. Agarwal, "Design and optimization of antireflecting porous silicon dielectric multilayers," Solar Energy Materials, vol. 123, pp. 144-149, 2014.

[25] S. Chan, Y. Li, L. J. Rothberg, B. L. Miller, and P. M. Fauchet, "Nanoscale silicon microcavities for biosensing," Materials Science and Engineering: C, vol. 15, pp. 277-282, 2001.

978-1-5386-5284-8/18 $31.00 © 2018 IEEE

Design and simulation of tapered optical fiber by enhancing the evanescent field region for sensing application

Nurulhani Diana Rashid
Institute of Microengineering & Nanoelectronics (IMEN)
Universiti Kebangsaan Malaysia (UKM)
Bangi, Selangor
haniidiana@siswa.ukm.edu.my

Dilla Duryha Berhanuddin
Institute of Microengineering & Nanoelectronics (IMEN)
Universiti Kebangsaan Malaysia (UKM)
Bangi, Selangor
dduryha@ukm.edu.my

Mohd Adzir Mahdi
Wireless and Photonics Network Research Centre, Faculty of Engineering
Universiti Putra Malaysia (UPM)
Serdang, Selangor
mam@upm.edu.my

Burhanuddin Yeop Majlis
Institute of Microengineering & Nanoelectronics (IMEN)
Universiti Kebangsaan Malaysia (UKM)
Bangi, Selangor
burhan@ukm.edu.my

Ahmad Rifqi Md Zain
Institute of Microengineering & Nanoelectronics (IMEN)
Universiti Kebangsaan Malaysia (UKM)
Bangi, Selangor
rifqi@ukm.edu.my

Abstract— We report the design and simulation of the tapered optical fiber with large presence of the evanescent field. The evanescent field of the optical fiber is strongly affected by the surrounding environment which will be exploited into fabricating variety of photonic-based devices such as photodetectors, optical sensors and ultra-high Q resonators. The simulation results show that by adiabatically tapered the waist region, there is a fairly large amount of evanescent field intensity observed at the air-cladding region. The smooth transition region of the tapered fiber has also minimized the multimode interference in the waist and transition region thus reducing the energy loss and contributing to the higher output power.

Keywords— *tapered fiber, evanescent field, mode profile, sensor*

I. Introduction

Tapered fiber has been introduced in various sensing devices especially for optical sensing detection [1-2], nanofiber-based evanescent wave spectroscopy [3], nonlinear optics, and cold atom physics [4].The increasing demand is due to high sensitivity of tapered fiber to the changes of refractive index as evanescent field generated at this region is able to interact with the surrounding.

Following the success of optical fibers in the telecommunication sector, they have shown great potentials to be utilized in sensing application. Evanescent wave fiber optic sensor (EWFOS) have been studied and accelerated with the progress and demand for chemical sensors. The interest on EWFOS is due the several advantages over conventional sensors such as low attenuation. compact size, flexibility and immunity to electromagnetic interference (EMI).

By exposing the evanescent field, it will enhance the potential of optical fiber to be utilized as a sensor. When light is transmitted through a tapered fiber, the evanescent field interacts with the surrounding in the tapered region. If the surrounding of the tapered fiber absorbs the transmission

wavelength, the transmission decreases [5]. Tapering has also been proved to increase the magnitude of evanescent field [6]. A tapered fiber can be categorized as adiabatic if most of the power remains in the fundamental mode and does not couple to higher order modes as it propagates along the taper as the radius changes gradually [7]. The beam propagation is strongly affected by the surrounding environment thus making it a suitable parameter for optical refractometric or absorption sensors. In most devices, tapered fiber waist diameter affect the evanescent field and coupling coefficient. Since fiber is designed for low loss transmission, adiabatically tapered fiber with uniform waist diameter and low surface roughness are among the properties of the tapered fiber that contributes to a high quality tapered fiber with low loss and strong evanescent field [8].

In this paper, we have designed and simulated the tapered optical fiber to enhance the formation of the evanescent field at the waist region of fiber. The tapered fiber will be used in the next step of designing a functional evanescent field-based optical sensor.

II. Device Simulation Model

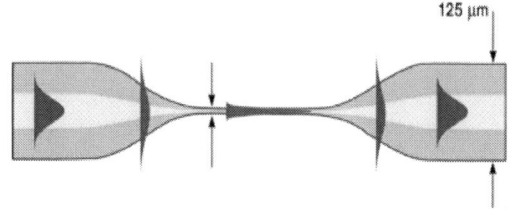

Fig. 1. Schematic of a tapered optical fiber (TOF) showing the light propagation and conversion of the fundamental mode in the original fiber, tapered region and fiber output [8].

The tapered fiber has been designed as Single-Mode Fiber (SMF) with cladding and core diameter of 125 μm and

8 µm respectively. The fiber is tapered to cladding and core diameter of 7 µm and 0.448 µm for 5 mm length and 30 mm for both down and up taper length as shown in Fig 1.

Light propagation in fiber can be described using the concept of an effective refractive index;

$$n_{eff} = \frac{c}{v_{phase}} = \frac{\beta}{k_0} \qquad (1)$$

Where c is the speed of light in vacuum, v_{phase} is the phase velocity of the mode, β is the mode propagation constant, and k_0 is the vacuum wavenumber. For this purpose, we have used Lumerical Tools MODE Solution that allows the observation of propagated light in the tapered fiber and field profile of the guided modes. Simulation model of tapered optical fiber in MODE solution is shown in Fig 2. The light source for this model is defined at a wavelength of 1550 nm.

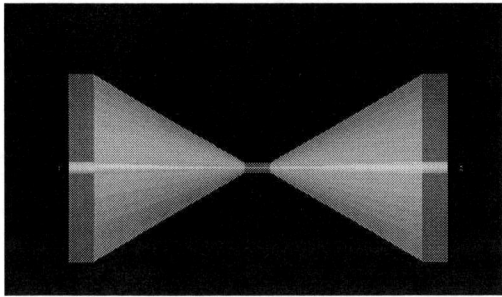

Fig. 2. Tapered optical fiber as simulated in Lumerical MODE.

III. RESULTS AND DISCUSSION

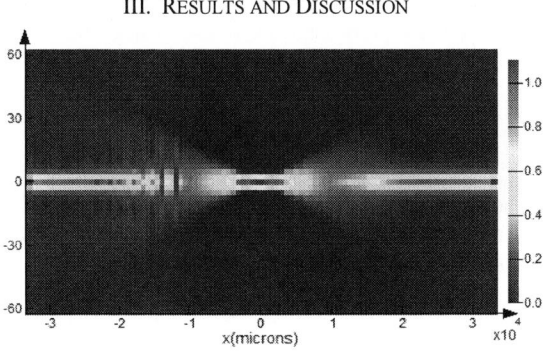

Fig. 3. Light propagation in tapered optical fiber in the untapered fiber to the fiber output through taper waist.

Fig. 3 shows the propagation of light in the tapered fiber through original fiber region to tapered region via down-taper transition region. The light is guided inside the core by total internal reflection principal at the core-cladding boundary. As the core diameter decreases in the transition region, the guided mode is compressed until it reaches taper waist diameter. At this point, cladding acts as new guiding modes thus forming strong evanescent field due to the mode expansion in the air.

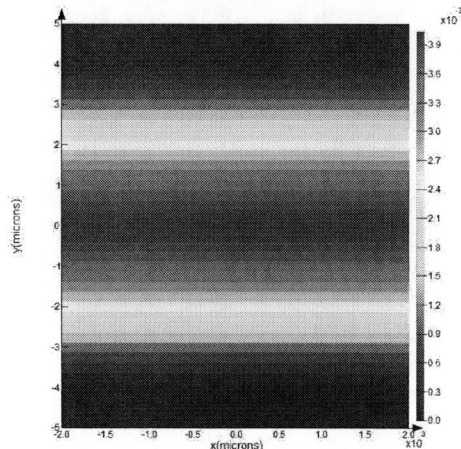

Fig. 4. The air-cladding region at the taper waist shows the evanescent field formed outside of the fiber.

Fig. 4 shows the evanescent field that is formed on the air-cladding region. The intensity of the evanescent field depends on the ratio of wavelength λ, propagating in the fiber and radius r. Higher λ/r ratio and n_{surr} (refractive indices of the medium surrounding the optical fiber) will increase the magnitude of evanescent field. This effect is important criteria in designing and realizing sensor, based on the evanescent field of the tapered fiber [9].

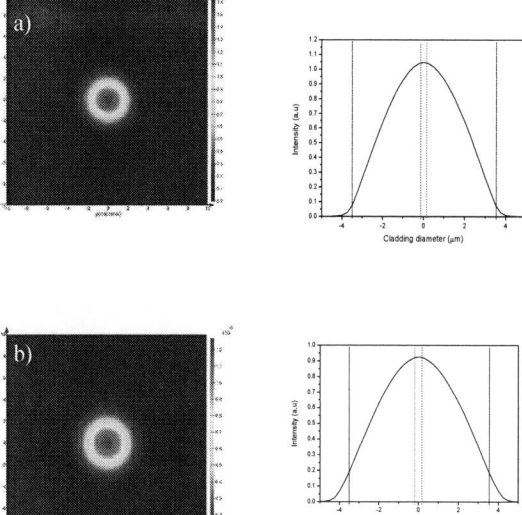

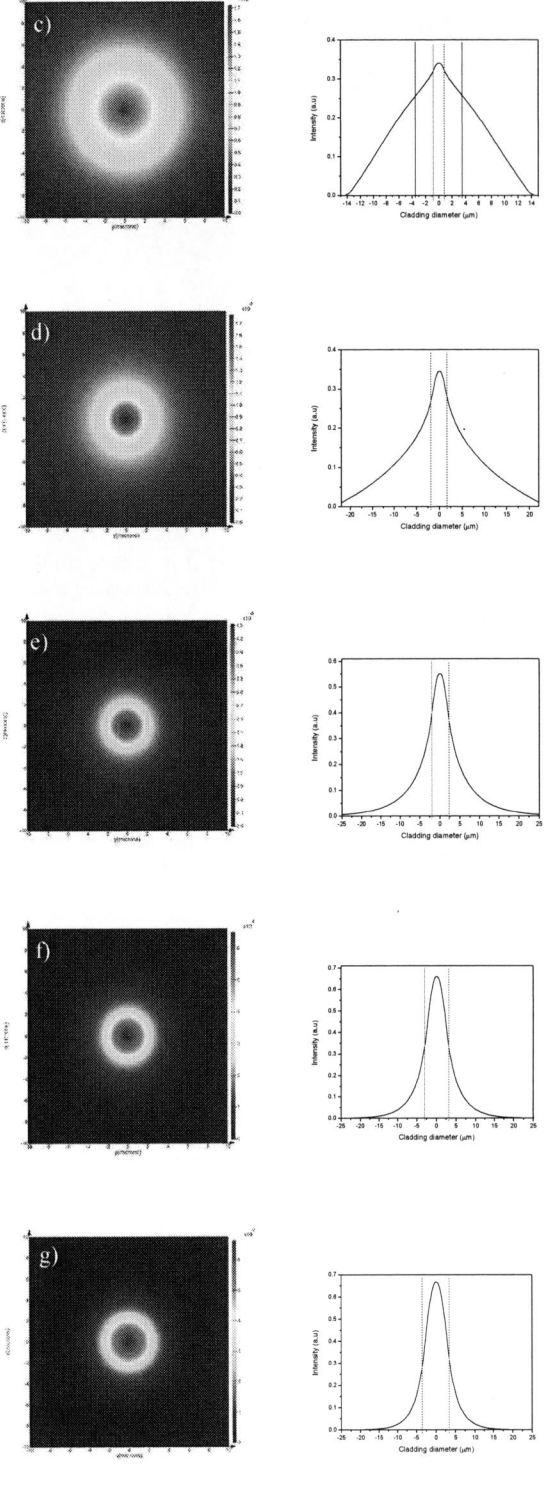

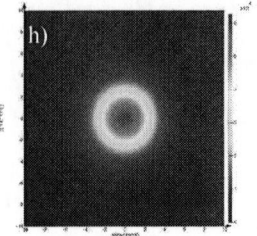

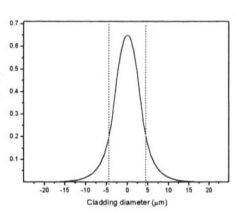

Fig. 5. Field profiles at different region of a),b) taper waist c)-g) transition region and h) fiber output spectra.

Next, we study the mode profile in different region of the fiber. Fig. 5(a) shows the modes is confined in the taper waist region while from the graph, the dotted line marks the air-cladding region of the taper waist and it shows the evanescent field is formed outside of the waist cladding. Fig. 5(c) to (h) shows the modes in the up-taper region. After the modes were confined in the taper waist and reached the up-taper region, the mode transformation is reversed to the down-taper region.

This was achieved by adiabatically tapered the transition region and all the guided modes are continuously transformed into cladding guided modes in the down-taper region which were then couple back into the fiber at the up-taper region (fiber output) [9]. The smooth transition region of the tapered fiber minimizes the multimode interference in the taper waist and transition region thus reducing the energy loss and contributing to the higher output [10-11].

IV. CONCLUSION

We have successfully designed and simulated tapered optical fiber with strong presence of the evanescent field at the air-cladding region. This was done with adiabatically tapered optical fiber which was then coupled back into the fiber output. The results are important in determining the next step which is to utilize the evanescent field as a component in realizing the photodetector and consequently to build an optical sensor.

ACKNOWLEDGMENT

The authors would like to thank the ministry of Higher Education (MOHE) under grant no: LRGS (2015)/NGOD/UM/KPT/KPT and LRGS/2015/UKM-UKM/NanoMiTe/04/01 for the financial support of this project.

REFERENCES

[1] Y. Mustapha Kamil, M. H. Abu Bakar, M. A. Mustapa, M. H. Yaacob, A. Syahir, and M. A. Mahdi, "Sensitive and Specific Protein Sensing Using Single-Mode Tapered Fiber Immobilized With Biorecognition Molecules," *IEEE Photonics J.*, vol. 7, no. 6, 2015.

[2] L. Zhang, F. Gu, J. Lou, X. Yin, and L. Tong, "Fast detection of humidity with a subwavelength-diameter fiber taper coated with gelatin film.," *Opt. Express*, vol. 16, no. 17, pp. 13349–53, 2008.

[3] A. Stiebeiner, O. Rehband, R. Garcia-Fernandez, and A. Rauschenbeutel, "Ultra-sensitive fluorescence spectroscopy of isolated surface-adsorbed molecules using an optical nanofiber.," *Opt. Express*, vol. 17, no. 24, pp. 21704–21711, 2009.

[4] G. Sagué, A. Baade, and A. Rauschenbeutel, "Blue-detuned

evanescent field surface traps for neutral atoms based on mode interference in ultrathin optical fibres," *New J. Phys.*, vol. 10, 2008.

[5] A. Leung, P. M. Shankar, and R. Mutharasan, "A review of fiber-optic biosensors," *Sensors Actuators, B Chem.*, vol. 125, no. 2, pp. 688–703, 2007.

[6] R. A. Yotter and D. M. Wilson, "A review of photodetectors for sensing light-emitting reporters in biological systems," *IEEE Sens. J.*, vol. 3, no. 3, pp. 288–303, 2003.

[7] T. K. Yadav, R. Narayanaswamy, M. H. Abu Bakar, Y. M. Kamil, and M. A. Mahdi, "Single mode tapered fiber-optic interferometer based refractive index sensor and its application to protein sensing," *Opt. Express*, vol. 22, no. 19, p. 22802, 2014.

[8] S. W. Harun, K. S. Lim, C. K. Tio, K. Dimyati, and H. Ahmad, "Theoretical analysis and fabrication of tapered fiber," *Optik (Stuttg).*, vol. 124, no. 6, pp. 538–543, 2013.

[9] R. Garcia-Fernandez *et al.*, "Optical nanofibers and spectroscopy," *Appl. Phys. B Lasers Opt.*, vol. 105, no. 1, pp. 3–15, 2011.

[10] G. Brambilla, "Optical fibre nanowires and microwires: a review," *J. Opt.*, vol. 12, no. 4, p. 43001, 2010.

[11] H. Ahmad, M. J. Faruki, M. Z. A. Razak, Z. C. Tiu, and M. F. Ismail, "Evanescent field interaction of tapered fiber with graphene oxide in generation of wide-bandwidth mode-locked pulses," *Opt. Laser Technol.*, vol. 88, no. September 2016, pp. 166–171, 2017.

J_{sc} And V_{oc} Optimization of Perovskite Solar Cell With Interface Defect Layer Using Taguchi Method

Mohd Shaparuddin Bahrudin
URND Sdn. Bhd.
Universiti Tenaga Nasional
Kajang, Malaysia
shaparuddin@yahoo.com

Siti Fazlili Abdullah
Department Of Electronics And
Communication Engineering
Universiti Tenaga Nasional
Kajang, Malaysia
siti@uniten.edu.my

Ibrahim Ahmad
Department Of Electronics And
Communication Engineering
Universiti Tenaga Nasional
Kajang, Malaysia
Aibrahim@uniten.edu.my

Ahmad Wafi Mahmood Zuhdi
Department Of Electronics And
Communication Engineering
Universiti Tenaga Nasional
Kajang, Malaysia
wafi@uniten.edu.my

Azri Husni Hasani
URND Sdn. Bhd.
Universiti Tenaga Nasional
Kajang, Selangor, Malaysia
Azri.Husni@uniten.edu.my

Fazliyana Za'abar
URND Sdn. Bhd.
Universiti Tenaga Nasional
Kajang, Selangor, Malaysia
fazliyana@uniten.edu.my

Mazin Malik
Center Of Graduate Studies
Universiti Tenaga Nasional
Kajang, Selangor, Malaysia
Abatshy91@hotmail.com

M. Najib Harif
Faculty of Applied Sciences
Universiti Teknologi Mara
Shah Alam, Malaysia
najib@ns.uitm.edu.my

Abstract—This paper is a study on Perovskite Solar Cell to optimize open circuit voltage, V_{oc} and short circuit current density, J_{sc} for maximum efficiency at variation depth of interface defect layer (IDL). The Perovskite Solar Cell structure is simulated with combinations of IDL at 6nm, 8nm and 10nm of thickness sandwiches on both side of the solar cell absorber layer. Taguchi Method using L9 Orthogonal Array with Larger-The-Better (LTB) was used on finding most effective value on three material parameters: Cadmium Sulfide (CdS) as an electron transport layer (ETL), Perovskite absorber layer (CH3NH3PbI3) and Copper Telluride (CuTe) as hole transport layer (HTL) in order to achieved best V_{oc} and J_{sc} values. The works was done by simulating a numerical model using Analysis Of Microelectronic and Photonic Structures (AMPS-1D) software. Using ANOVA, it was discovered the Perovskite absorber layer thickness is vital in affecting the increasing and decreasing on both V_{oc} and J_{sc}. Taguchi predicted a 200nm of thickness for best J_{sc} but predicted 300nm for best V_{oc}. The thickness of 200nm is selected for cost effectiveness. Taguchi method also predicted CdS and CuTe are considered slightly significant on improving the efficiency. Post Taguchi optimization approach shows Perovskite Solar Cell with CH3NH3PbI3 absorber layer has average power conversion efficiency of 20.7% on any combination of mentioned IDL thickness. With the aid of Taguchi method, a stable Perovskite Solar Cell efficiency with variation IDL thickness is achieved.

Keywords— Device simulation, Perovskite, Interface Defect Layer, Taguchi Method

I. INTRODUCTION

This Perovskite Solar Cell is a one of thin film solar cell type that is thriving in its development [1]. With power conversion efficiency of over 20% made the solar cell performance on par with the Silicon-based solar cell that is currently dominating the market as well on par with its other second generation thin film cell counterparts.

The most commonly studied Perovskite absorber is Methylammonium lead Trihalide (CH3NH3PbX3, where X is a halogen ion such as I−, Br−, Cl−), with an optical band gap between 1.6eV and 2.3eV depending on halide content [2].The combination of material thus far has been the efficient and low cost to produce.

The aim of this paper is to optimize Perovskite Solar Cell's J_{sc} and V_{oc} to achieve maximum fill factors (FF) and consequently better power conversion efficiency. According to Roldan-Carmona et al, an increase of thickness raised the J_{sc} from 18.5(mA/cm^2) to 23.6(mA/cm^2) but only achieved 1.04V of V_{oc} from 1.06V initially [4]. Feng et al, also reported as the thickness of Perovskite absorber layer increases, the J_{sc} also progressed until it saturated when the thickness reaches 1μm. V_{oc} however will slightly decrease due to more charge recombination activity occurred in thicker film [5].

Another factor affecting the V_{oc} is Interface Defect Layer (IDL). It is one of the main factors affecting the electrical properties of Perovskite include the active layer and their interface [6]. The IDL's are expressed into two separate additional layers, an interface IDL$_1$ between the buffer and absorber layers, and another interface IDL2 between the absorber and HTM layers. The existence of the IDL in Perovskite film play a critical role in determining the performance of the device, as the photoelectrons generated in this layers and the charge recombination behaviors here can become dominant in determining the V_{oc}, J_{sc}, FF and as well as the Efficiency of the device.

The optimization is done utilizing Taguchi Optimization Method. It is an orthogonal array based statistical method developed by Genichi Taguchi in the 60's originally used for improving the quality of manufactured goods in industry. However, in the recent years, the method also applied in engineering researchers as a robust design method [6].

II. METHODOLOGY

The physical device structure of Perovskite Solar Cell is as shown in FIGURE 1.The absorber used for this study is Methylammonium Lead Triiodide (CH3NH3PbI3)

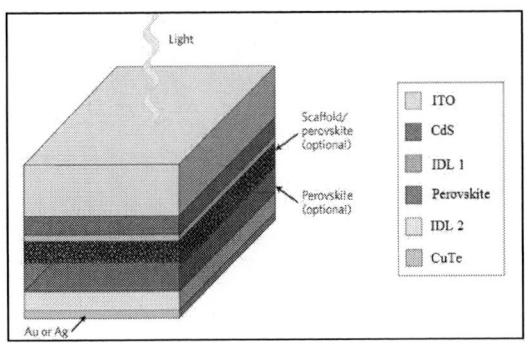

Fig. 1. Pervoskite solar cell structure

A. Material Parameters

The sample structure of the solar cell profile is a modified mathematical model calculated by Takashi Minemoto et al with IDL layers added to its structure [7]. There are three main materials, Cadmium Sulfide (CdS) buffer layer, Perovskite Absorber layer and Cuprum Telluride (CuTe) Hole Transport layer are respectively assigned with three different thickness values. These values are later to be optimized by Taguchi. Both IDLs are varied into three level of thickness, 6nm, 8nm and 10nm. The experiments were done by AMPS-1D simulator.

B. Taguchi Optimization Method

Taguchi method assisted in identifying the most dominant factors that influence the both J_{sc} and V_{oc} when a variation of main parameters and variation of IDL thickness applied in the experiments. This is done by forming all the parameters values into an Orthogonal Array (OA). There are several types of OA tabulated in Taguchi Method [8]. In this paper, L9 Orthogonal Array is utilized.

C. L9 Orthogonal Array

The basis for using L9 Orthogonal array (L9 OA) is because there are three parameters involved that became the Control Factors (CF) and L9 OA is capable to take up to four maximum numbers of CF [9]. These factors are CdS layers, Perovskite Absorber layers, and CuTe layers. Each CF is defined with 3 levels of values as shown in TABLE 2.

TABLE 1. CONTROL FACTORS PARAMETERS AND THEIR LEVEL

Symbol	Parameters (Thickness)	Unit	Level 1	Level 2	Level 3
A	Cadmium Sulfide (CdS)	nm	60	80	100
B	Perovskite (CH₃NH₃PbI₃)	nm	200	300	400
C	Copper Telluride (CuTe)	nm	100	150	200

These level of values are technically determined a minimum, medium and maximum configuration of Perovskite Solar Cell that produced good efficiency but not fundamentally optimized or economically fitting. Taguchi Method is used to identify the optimum configuration values from mentioned combinations for best efficiency result.

D. Interface Defect Layers as Noise Factor

In Taguchi Method, Noise factor (NF) is defined as an external parameter that cannot be controlled. The values of NF can be defined statistically but the actual values in a specific situation cannot be known [9].

In this study, IDL is chosen as NF since it uncontrollably occurs on unspecified condition during Perovskite solar cell operation. The L9 OA support maximum up to three noise factors [9], but in the case of this study, only two are utilized since there only two NF's became the focus of the study. The first noise factors is designated as IDL1, an interface defect layer occurred on the top part of Perovskite Absorber layer. The second noise factor is IDL2, another interface defect layer occurred at the bottom part of Perovskite Absorber layer. Both IDL1 and IDL2 layer configuration are as in TABLE 3.

TABLE 2. NOISE FACTORS AND THEIR RESPECTIVE LEVEL

Symbol	Parameters (Thickness)	Unit	Level 1	Level 2	Level 3
D	IDL1	nm	6	8	10
E	IDL2	nm	6	8	10

E. Experiment layout on L9 OA

L9 OA systematically organized the experiment that needs to be done and reduced their numbers. The results are then studied using Signal-to-Noise (S/N) ratio and ANOVA analysis. As its name implied, the L9 OA required nine experiments. TABLE 4 showed the experimental layout of the L9 Orthogonal Array. This layout is important to be used later for ANOVA.

TABLE 3. EXPERIMENTAL LAYOUT USING L9 ORTHOGONAL ARRAY

. Exp. No	Control Factor Level		
	Cadmium Sulfide (CdS)	Perovskite (CH₃NH₃PbI₃)	Copper Telluride (CuTe)
1	A₁	B₁	C₁
2	A₁	B₂	C₂
3	A₁	B₃	C₃
4	A₂	B₁	C₂
5	A₂	B₂	C₃
6	A₂	B₃	C₁
7	A₃	B₁	C₃
8	A₃	B₂	C₁
9	A₃	B₃	C₂

TABLE 4 shows how the configuration of NF for each experiments. The NF configuration is arranged on repetition basis to ensure all variation IDL thickness is covered. Given that there are three level of NF's as shown in TABLE 3, the configuration of NF's is varied nine times in each nine experiments [9]. That mean there are total of 81 experiments run with the configuration provided by both Tables.

TABLE 4. NOISE FACTORS CONFIGURATION FOR EACH REPETITION

Exp. Repetition	Noise Factors Configuration	
	IDL 1	IDL 2
1	1	1
2	1	2
3	1	3
4	2	1
5	2	2
6	2	3
7	3	1
8	3	2
9	3	3

III. RESULT AND DISCUSSION

TABLE 5 and TABLE 6 show the result for current density and open circuit voltage respectively in Perovskite solar cell run on the L9 OA with the NFs.

TABLE 5 CURRENT DENSITY (J_{sc}) FROM RESPECTIVE EXPERIMENT

	Repetitions								
	1	2	3	4	5	6	7	8	9
1	25.94	25.86	25.73	25.82	25.68	26.02	25.64	25.98	25.90
2	26.31	26.18	25.99	26.14	25.96	26.37	25.92	26.34	26.21
3	25.94	25.81	25.63	25.78	25.60	26.01	25.56	25.98	25.85
4	26.25	26.16	26.01	26.12	25.97	26.33	25.93	26.29	26.20
5	26.54	26.40	26.20	26.37	26.17	26.60	26.14	26.57	26.43
6	25.72	25.61	25.44	25.57	25.41	25.79	25.37	25.76	25.64
7	26.56	26.46	26.30	26.42	26.26	26.63	26.22	26.59	26.49
8	26.20	26.08	25.90	26.05	25.87	26.27	25.84	26.24	26.11
9	25.88	25.76	25.59	25.72	25.55	25.94	25.52	25.91	25.79

TABLE 6 OPEN CIRCUIT VOLTAGE (V_{oc}) FROM RESPECTIVE EXPERIMENTS

	Repetitions								
	1	2	3	4	5	6	7	8	9
1	0.99	0.99	0.98	0.99	0.99	0.99	0.99	0.99	0.99
2	0.99	0.98	0.98	0.98	0.98	0.98	0.98	0.98	0.98
3	0.98	0.98	0.98	0.98	0.98	0.98	0.98	0.98	0.98
4	0.99	0.99	0.98	0.99	0.99	0.99	0.99	0.99	0.99
5	0.99	0.98	0.98	0.98	0.98	0.98	0.98	0.98	0.98
6	0.98	0.98	0.98	0.98	0.98	0.98	0.98	0.98	0.98
7	0.99	0.99	0.98	0.99	0.99	0.99	0.99	0.99	0.98
8	0.98	0.98	0.98	0.98	0.98	0.98	0.98	0.98	0.98
9	0.98	0.98	0.97	0.98	0.98	0.98	0.98	0.98	0.98

As mentioned in methodology, there are nine experiments and nine repetitions with different NF configuration as explained in TABLE 3 and TABLE 4.

The values conceived from the experiments is used to calculate the signal to noise (S/N) ratio. Analysis on S/N ration performance determined which CF affecting J_{sc} and V_{oc} the most.

Analysis on S/N ratio performance determined which CF affecting. There are three categories of S/N ratio performance characteristic, i.e. lower-the-best, nominal-the-best and larger-the-best [9]. For this study, since largest Jsc and Voc is the objective so larger-the-best characteristic is selected. The S/N ratio for higher-the-best can be expressed as:

$$\eta = -10 \log_{10} \left(\frac{1}{n} \sum (Y_1^2 + Y_2^2 + \cdots + Y_n^2) \right) \qquad (1)$$

Where Y is the experimental values and n is the number of experiments. By applying equation (1), the S/N ratio (η) for the efficiency is calculated and listed in TABLE 8.

TABLE 7. (S/N) RATIO RETRIEVED FROM ALL EXPERIMENTS

Experiment No.	Signal-To-Noise (S/N) ratio (dB) for Larger-The-Better	
	Current Density (Jsc)	*Open Circuit Voltage (Voc)*
1	28.25	-0.11
2	28.35	-0.15
3	28.23	-0.18
4	28.35	-0.12
5	28.43	-0.16
6	28.16	-0.20
7	28.44	-0.12
8	28.32	-0.17
9	28.21	-0.20

From the results, individual value of S/N ratio (Larger-The-Better) for each Level of Control Factors and the Overall Mean is calculated . Those S/N ratio values are summarized in TABLE 9 for both J_{sc} and V_{oc}.

TABLE 8 (S/N) RATIO FOR JSC AND VOC

	Control Factors Parameters	Signal to Noise (S/N) Ratio (dB) for Larger-The-Better			Overall Mean (dB)
		Level 1	*Level 2*	*Level 3*	
J_{sc}	CdS	28.28	28.31	28.32	28.30
	Perovskite	28.35	28.37	28.20	
	CuTe	28.24	28.30	28.37	
V_{oc}	CdS	-0.15	-0.16	-0.17	-0.16
	Perovskite	-0.12	-0.16	-0.20	
	CuTe	-0.16	-0.16	-0.15	

From the summarized results in TABLE 9 , two Factor Effect graphs are plotted base on each Control Factor's level S/N Ratio values on both tables. These two plots are represented as Figure 2(a) and Figure 2(b).

Both Figures shows the selection for the best level on each factor effects that yield the best result can be decided. Fundamentally, the larger the S/N ratio the better the result would be. However, there is a disparity between factor effect points that influence the J_{sc} and V_{oc}. In Figure 2(a) stated that while increasing the CdS thickness have little improvement on the J_{sc}, but cause a small reduction on the V_{oc}. On the thickness of Perovskite layer, Figure 2(a) mentioned that 300nm yield a better result than 200nm thickness for J_{sc}, however, yield the worst result for V_{oc} in Figure 2(b). Figure 2 also asserted that increasing in CuTe thickness is better for J_{sc} but produce insignificant improvement on V_{oc}.

TABLE 9. RESULT OF ANOVA FOR BOTH JSC AND VOC

	Control Factors	DF	SS	MS	Factor Effect on SNR (%)
J_{sc}	CdS	2.0	0.00	0.00	5.02
	Perovskite	2.0	0.05	0.02	63.26
	CuTe	2.0	0.02	0.01	30.37
V_{oc}	CdS	2.0	0.00	0.00	4.64
	Perovskite	2.0	0.01	0.00	95.04
	CuTe	2.0	0.00	0.00	0.32

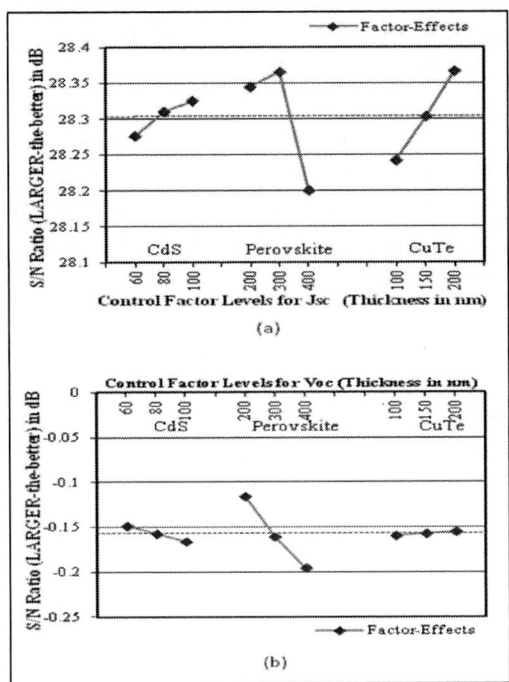

FIG. 2 FACTOR EFFECT PLOT FOR (A) J_{sc} (B) V_{oc}

A. Analysis of variance (ANOVA)

Analysis of Variance or ANOVA is a common statistical technique to determine the percentage the contribution of each factor for the result [10]. ANOVA is capable to verify which control factors that significantly affect the performance of the solar cell. It is done by calculating the Degree of Freedom (DF) given by

$$DF = t - 1 \qquad (2)$$

Where t is the total sum of square for each CF given by

$$SS_t = \sum_{i=n}^{n}(\eta_i - m)^2 \qquad (3)$$

Where n is a number of the experiment repeated, η_i is the mean of S/N ratio for the i-th experiment and m is the average of nine η_i values. In order to find the ratio percentage, the sum of square of each factor has to be determined. For example, CF for CdS thickness, labeled by symbol A in TABLE 1, its SS_{cf} is given by

$$SS_{cf} = 3[(mA_1 - m)^2 + (mA_2 - m)^2 + (mA_3 - m)^2] \quad (4)$$

Where m is the average of nine η_i values. Equation (4) is repeated for the rest of control factors for both J_{sc} and V_{oc} to determine their respective SS_{cf}. Then, the variance is determined by calculating the mean square (MS) given by

$$MS = \frac{SS}{DF} \qquad (5)$$

Next step is to determined the percentage of each factor effect. It is given by

$$Factor\ Effect\ (\%) = \frac{SS_{cf}}{SS_t} \qquad (6)$$

Equation (6) is repeated for the all factors effect also for both J_{sc} and V_{oc}. Finally the results on the ANOVA is summarized in TABLE 11.

According to ANOVA, The percentage of the S\N ratio indicates the priority of the factor level for best result [9]. Thus, the factor with the highest percentage and with the smallest means square among all factors will have great contribution [10]. First CF representing CdS thickness layer affecting only 5.02% on J_{sc} improvement for its best level and only 4.64% improvement for the V_{oc}. This is considered negligible by ANOVA as any selection on its factor levels insignificantly improve both J_{sc} and V_{oc}. So 80nm thickness (Level 2) is selected as it's in the middle that balance the performance between two results.

Perovskite layer thickness is found to be the major factor that affecting both J_{sc} and V_{oc} with factor effect on SNR at 63.26% and 95.04% respectively. The percentage of factor effect on SNR for V_{oc} is regarded higher than the one in J_{sc}, so for level selection, level 1 (Perovskite thickness layer of 200nm) is selected. For the last CF, at only 0.32%, the CuTe thickness is obviously posing almost zero influence on V_{oc}. But ANOVA asserted that it is the second most dominant factor that influences the performance of J_{sc} (at 30.37%). It is obvious for this third CF, level 3 (CuTe thickness of 200nm) is selected.

The ANOVA predicted the SNR range for J_{sc} is between 28.16dB and 28.35dB while SNR range for V_{oc} is between -0.17dB and -0.06dB. The best level selection of each control factors is then finalized in TABLE 12.

TABLE 10. BEST SETTING SELECTION FOR RESPECTIVE PARAMETER

Control Factor Parameters	Best Level	Best Value (nm)
CdS thickness	2	80
Perovskite thickness	1	200
CuTe	3	200

B. Confirmation run

A confirmation experiment is done to retrieve final efficiency base on parameters value on TABLE 12 in order to validate the analysis [10]. The experiment is again repeated nine times, similar to OA L9 experiments mentioned in the methodology to cater variation of NF's arranged in TABLE 4. The results on the Jsc and Voc from the experiment is laid out in TABLE 13 along with the efficiency and calculated S/N Ratio (Larger-The-Better) using equation (1).

978-1-5386-5284-8/18 $31.00 © 2018 IEEE

TABLE 11. RESULT OF THE CONFIRMATION EXPERIMENTS

Noise Factors (NF) combination									Avg
1 (1,1)	2 (2,2)	3 (3,3)	4 (1,2)	5 (2,3)	6 (3,1)	7 (1,3)	8 (2,1)	9 (3,2)	
J_{sc} (mA/cm^2)									
26.56	26.42	26.23	26.39	26.20	26.62	26.16	26.59	26.45	26.41
V_{oc} (V)									
0.99	0.99	0.98	0.99	0.99	0.99	0.99	0.99	0.99	0.99
Efficiency (%)									
20.71	20.53	20.29	20.56	20.33	20.63	20.36	20.67	20.49	20.51

The results show the J_{sc} is averaging at 26.40 mA/cm2 despite the variation of IDLs. These shows at 200nm of thickness, the current density of the Solar Cell can be maintained despite variations of IDL's. It is clear that due to the thickness of IDL's layers and the selection of thinner absorber layer by Taguchi Optimization Method, the V_{oc} is relatively very low, achieving lower than 1.0V. But the V_{oc} of the Solar Cell is maintained at stable 0.99V despite being exposed to a variation of IDL's thickness situations. The SNR for the J_{sc} and V_{oc} are 28.30 and -0.16 respectively. Both values are within the ANOVA predicted values mentioned previously. As far as efficiency is concerned, the result of the Taguchi optimization approach yield an impressive average power conversion efficiency of 20.51% even with IDL's layers up to 10nm of thickness on both side of CH3NH3Pbl3 Absorber layer.

C. Result comparison

TABLE 12. RESULT COMPARISON (IDL AT 10NM THICKNESS)

Paper Author	Bizuneh-Genene	Feng Juan et. Al.	Taguchi Method result
J_{sc} (mA/cm^2)	20.3	24.03	26.23
V_{oc} (V)	1.05	1.04	0.99
Efficiency (%)	18.5	20	20.53

TABLE 14 outlined the Taguchi Method result and its comparison with two most recent result from two others sources. Both using numerical simulation method. Each result represent the J_{sc} and V_{oc} of the solar cell and their respective efficiency when the IDL thickness occurred at 10nm on both top and bottom of the Perovskite absorber layer. The results concluded optimization with Taguchi Method yield best J_{sc} value with only 5 to 6% loss of V_{oc}. Perhaps, better V_{oc} can be achieved with slightly increased of the absorber layer for future development.

IV. CONCLUSION

In conclusion, the optimum solution in achieving maximum efficiency of Perovskite solar cell with a variation of IDL is a success by utilizing Taguchi Method. The method accurately predicted that Perovskite absorber layer is critical that influence both J_{sc} and V_{oc} that lead to best power conversion efficiency regardless the thickness of IDL that will otherwise drag the efficiency down. It also revealed that both CdS and CuTe present a very minor to none role that influence on the V_{oc} and increase thickness on CuTe bring an improvement on the J_{sc}. By using Taguchi optimization approach, a better understanding of parameters that govern the efficiency of CH3NH3PbI3-based solar cell devices is

realized. With this knowledge, further improvements in device performance can be accomplished in future.

ACKNOWLEDGMENT

The authors would like to thanks to Professor Dr. Nowshad Amin from Universiti Kebangsaan Malaysia (UKM) for his expertise and experience on Perovskite Solar Cell as well providing the AMPS software that was used in the project and Dr. Fauziyah binti Salehuddin from Universiti Teknikal Melaka (UTeM) for her knowledge sharing on Taguchi Optimization method.

REFERENCES

[1] Huang, Like, et al. *Electron transport layer-free planar perovskite solar cells performance enhancement perspective from device simulation.* s.l. : Elsivier B.V, 2016. pp. 1038-1047. Vol. 157.

[2] Askari, Mohammad Bagher, Mirzaei, Abadi Vahid Mahmoud and Mirhabibi, Mohsen. *Types of Solar Cells and Application.* Tehran : Science Publishing Group, 2015. Vol. 3. 2330-8494.

[3] Qi Chena, Nicholas De Marcoa, Yang (Michael) Yanga, Tze-Bin Songa, Chun-Chao Chena, Hongxiang Zhaoa, Ziruo Honga, Huanping Zhoua, Yang Yanga. Under the spotlight: The organic–inorganic hybrid halide perovskite for optoelectronic applications. 2015, Vol. 10, 3.

[4] K.S. Rahman, M. A. Islam, M. S. Hossain, M. M. Alam, K. Sopian, N. Amin. The Effect of Cu2Te as the back Surface field Layer in CdS/CdTe Solar Cells. August 2013, pp. 61-68.

[5] C. Rolda´n-Carmona, P. Gratia, I. Zimmermann, G. Grancini,a P. Gao, M. Graetzelb, Mohammad Khaja Nazeeruddin. High efficiency methylammonium lead triiodide perovskite solar cells: The relevance of non-stoichiometric precursors. 2015, Vol. 8, 2015.

[6] *Numerical Simulation: Towards the design of high-efficiency planer perovskite solar cells.* Feng, Liu, et al. 104, Beijing : AIP Publishing LLC, 2014.

[7] Si Fengjuan, Tang Fuling, Xue Hongtao,Qi Rongfei. Effects of defect states on the performance of perovskite solar cells. 2016, Vol. 37, 7.

[8] J.L. Rosaa, A. Robina, M.B. Silvab, C.A. Baldana, M.P. Peresd. Electrodeposition of copper on titanium wires: Taguchi experimental design approach. Brazil : Journal of Materials Processing Technology, 2009, Vol. 209.

[9] Genichi, Taguchi. *System Of Experimental Design (Jikken Keikakuho).* 3rd Edition. Tokyo : Maruzen, 1978.

[10] Phadke, Madhav. *Quality Engineering using Robust Design.* s.l. : Pearson Education, 2008.

[11] F. Salehuddin, I. Ahmand, F. A. Hamid, A. Zaharim, H. A. Elgomati, B. Y. Majlis, P. R. Apte. *Optimization of HALO Structure Effect In 45 nm P-type MOSFETs Device Using Taguchi Method.* Kuala Lumpur : s.n., 2011. pp. 296-302. Vol. 5.

[12] *The effect of skin-depth interfacial defect layer in perovskite solar cell.* Difer, Bizuneh Gebremichael and Mola, Genene Tessema. 215, Berlin : Springer-Verlag, 2016, Vol. 122.

[13] Islam, Mohammad Aminul, Amin, Nowshad and Sulaiman, Yusuf. *A Comparative Study Of Bsf Layers For Ultra-Thin Cds:O/Cdte Solar Cells.* Bangi : Chalcogenide Letters, 2011. pp. 65-75. Vol. 8.

[14] M. A. Matin, M. U. Tomal, A. M. Robin, N. Amin. Numerical Analysis of Novel Back Surface Field for High Efficiency Ultrathin CdTe Solar Cells. s.l. : Hindawi Publishing Corporation, 2013, Vol. 2013.

[15] Minemoto, Takashi and Murata, Masashi. Impact Of Work Function Of Back Contact Of Perovskite Solar Cell without Hole Transport Material Analyzed By Device Simulation. November 2014, Vol. 14, 11, pp. 1428-1433.

[16] Mohammad I. Hossain, Fahhad H. Alharbi, Nouar Tabet. Copper oxide as inorganic hole transport material for lead halide perovskite based solar cells. August 2015, Vol. 120, 2015, pp. 370-380.

[17] Yijie Xia, Kuan Sun, Jingjing Changa, Jianyong Ouyang. Effects of organic inorganic hybrid perovskite materials on the electronic properties and morphology of poly(3,4-ethylenedioxythiophene):poly(styrenesulfonate)and the photovoltaic performance of planarperovskite solar cells. 2015, Vol. 3, 2015.

Flexible Photoanode on Titanium Foil for Back-Illuminated Dye Sensitized Solar Cells

Suraya Shaban*
Institute of Advanced Technology
Universiti Putra Malaysia
Serdang, Selangor
surayashaban@gmail.com

Suhaidi Shafie
Institute of Advnaced Technology / Faculty of Engineering
Universiti Putra Malaysia
Serdang, Selangor
suhaidi@upm.edu.my

Yusran Sulaiman
Institute of Advanced Technology / Faculty of Science
Universiti Putra Malaysia
Serdang, Selangor
yusran@upm.edu.my

Fauzan Ahmad
Malaysia Japan International Institute of Technology
Universiti Teknologi Malaysia
Kuala Lumpur
fauzan.kl@utm.my

Muhammad Quisar Lokman
Malaysia-Japan International Institute of Technology
Universiti Teknologi Malaysia
Kuala Lumpur
muhammadquisar@gmail.com

Noor Fadzilah M. Sharif
Faculty of Engineering
Universiti Putra Malaysia
Serdang, Selangor
noorfadzilah@upnm.edu.my

Abstract—This paper reports the fabrication and analysis of flexible photoanode on titanium (Ti) foil for back-illuminated dye sensitized solar cells (DSSCs). Performance comparison with the solid state FTO glass based DSSC using the back-illumination and frond-illuminated techniques were also carried. During the fabrication process, the surface of Ti foil, had been treated with H_2O_2 and doctor blade method was applied for deposition of the photoanode on substrates. The measurement results show that the fabricated DSSC with flexible photoanode has power conversion efficiency of 1.00% under back illuminated solar radiation of 1.5 A.M while DSSCs with solid state photoanodes have power conversion efficiency of 0.53% (back-illuminated) and 2.22% (front-illuminated), respectively. The DSSC with flexible photoanode has better power conversion efficiency than the DSSC with solid-state photoanode under back-illumination condition. However, it is comparatively low from front illumination DSSC due to platinized counter electrode partially reflects light, while iodine in the electrolyte absorbs photons affects the performance.

Keywords—*DSSC, flexible Solar Cell, Ti foil, back-illumination, Photoanode.*

I. Introduction

Nowadays renewable energy experiencing an increase in attention worldwide due to many long term benefits such as low maintenance cost and environmental friendly. Renewable energy is generally originated from renewable resources such as bioenergy, geothermal, hydroelectric, hydrogen, ocean, solar and wind. Among them, solar energy are the most popular and well utilized in large scale such as solar farm till small scale such as in calculator. Solar energy can be converted to electricity via solar modules which convert the photon to direct current. This form of energy relies on the nuclear fusion power from the core of the Sun [1].

Photovoltaic (PV) Solar cell technologies are divided into three generations. First generation are mainly based on mono-crystalline and poly-crystalline silicon, typically demonstrate about 15-25 % of efficiency. The benefits of this solar cell technology lie in their good performance, as well as their high stability. However, they are rigid and require a lot of energy and cost in production. The second generation are based on thin film such as amorphous Silicon, Copper Indium Gallium Selenide (CIGS) and Cadmium Telluride (CdTe), where the typical performance is 10 - 15%. It has a lower material consumption since it avoid use of silicon wafers. However, as it still includes vacuum processes and high temperature treatments, there is still a large energy consumption associated with the production. Furthermore, it is based on scarce elements and this is a limiting factor in the price. The Cadmium used in thin film solar cell is also toxic for the human being therefore the production must be handle carefully. Third generation solar cells uses organic materials such as small molecules or polymers. It covers expensive high performance experimental multi-junction solar cells which hold the world record in solar cell performance. This type has only to some extent a commercial application because of the very high production price. A new class of thin film solar cells currently under investigation are Perovskite solar cells and show huge potential with record efficiencies beyond 20% on very small area [2]. However, Perovskite solar cell suffers from instability and the process is rigid since it must be done in very clean environment with controlled humidity.

Dye Sensitized Solar Cells (DSSC) are the third generation PV Solar Cell that converts photon into an electrical energy based on absorption of dye molecules associated in wide bandgap semiconductor. DSSC has received worldwide attention for its high-efficiency of energy conversion and low production cost [3,16,17].

The major key components of the conventional DSSC are the semiconductor photoanode, sensitizer (dye), redox mediator, counter electrode, and mechanical support. The photoanode which is usually a dye molecule coated

978-1-5386-5284-8/18 $31.00 © 2018 IEEE

nanoporous metal oxide film semiconductor deposited on a Transparent Conductive Oxide (TCO) substrate as photoanode electrode, Platinum (Pt) coated on TCO substrate as counter electrode and an electrolyte containing I-/I3- as redox couple [4].

Metal oxide consists of semiconductor such as TiO2, ZnO, and SnO. The semiconductor tasks as photon absorption and carrier for charge separation and transport, the two functions are different in a DSSC. In a conventional DSSC, light will be absorbed by a synthetic sensitizer dye typically N719, which led to the surface of a wide band gap n-type semiconductor [5]. The sensitizer (dye) is the photoactive component of DSSC as it converted to electricity once it is sensitized by visible light. The dye catches photons of incoming visible light and excites the electrons using its energy. Then, the dye injects the excited electron into the TiO2 semiconductor. The electron is conducted away by nanocrystalline TiO2 semiconductor. Pt coated on TCO glass performs a catalytic activity for the reduction of I-/I3- [6]. Electrolyte contains redox mediator serves to meet and restore the dye after oxidation of its molecules. The most mediators are the I-/I3- couple because of its slow recombination rate with injected electrons [7]. A chemical electrolyte in the cell then closes the circuit so that the electrons are returned back to the dye. These electrons create energy and harvested into a rechargeable battery or any electrical device.

Flexible DSSC is one of modification of conventional DSSC based on its substrate. Metal materials was chosen for having excellent electrical conductivity, good flexibility and ductility, thermal stability, could withstand high temperature treatment compared to the plastic, lower sheet resistance compared to ITO, as well as low cost in production. The use of metal materials as photo electrode substrates of DSSCs could lower the cost of the devices, and yet it helps to improve the performance of the solar cell by reducing the internal resistance. Nowadays, various metals are tested as substrates in DSSCs including StSt, W, Ti, Co, Ni, Pt, Al, Cu, Zn, etc. [8].

Previous paper have reported that pre-treatment of hydrogen peroxide (H2O2) performed on titanium (Ti) foil as an efficient photoanode substrate for dye-sensitized solar cell (DSSC). The H2O2-treated Ti shows high surface area because of the formation of networked TiO2 nanosheets, which enhances electrical contact between screen-printed TiO2 nanoparticles and Ti foil [9]. The surface property of the Ti wire, especially oxidized over layer and surface treatments have been found to greatly influence the adhesion as well as optimum electrical contact between coated nanoporous titanium oxide (TiO2) and Ti-wires. [10]

In this paper, the detail process of photoanode fabrication on flexible Titanium foil and the evaluation results are discussed.

II. EXPERIMENT

A. Material

The following materials were purchased from commercial suppliers: Fluorine Doped Tin Oxide (FTO) glass (< 15ohm/sq.; > 83 % ; Zhuhai Kaivo Optoelectronic Technology Co., Ltd.), Pure Metals (99.5% Titanium Sheet: The Nilaco Corporation), Glacial Acetic Acid (99.7% for

ACS Analysis; J.T Baker), Acetone, Ethanol, Isopropyl Alcohol (Quality Reagent Chemical), Di-tetrabutylammoniumcis-bis(isothiocyanato)bis(2,2′-bipyridyl-4,4′-dicarboxylato)ruthenium(II), Hexachloroplatinic Acid Hexahydrate, Alpha Terpineol, Titanium (IV) Oxide anatase-nanopowder, Ethyl Cellulose, Hydrogen Peroxide, H2O2 (Sigma Aldrich), Tri-iodide electrolyte (Solaronix)

B. Preparation of Glass Substrates

Fluorine doped tin oxide (FTO) glass with a surface resistivity of > 15 ohm/sq was used as the substrate, the glass was cut into 2 cm x 2.5 cm from the non-coated surface with the usage of a Glass Cutter System (ZHKV-DT300). The cut pieces were first washed with detergent and sonicated. The obtained substrates were placed in a beaker containing acetone (solvent) and then placed in an ultrasonic bath for 15 minutes. The acetone was subsequently replaced with Isopropanol alcohol and then subjected to sonication for another 15 minutes in order to remove organic and inorganic contaminants. Finally, the glass substrates were rinsed with deionized water (DI) water and then dried using a blower to remove excess liquid.

C. Preparation of Flexible Substrates

Firstly Ti foil was cut into 2cm x 2.5cm, then scratch the oxidize surface of Ti foil using fine sandpaper. Then cleaned with detergent, acetone and IPA for 10 minutes by sonication respectively. The Ti foil was then dipped in H2O2 solution at (30 wt. %, 20 ml) 500rpm at 70°C for 30 minutes. During dipping, the Ti foil reacted with H2O2 and formed networked TiO2 nanosheets on the surface. To enhance crystallization of TiO2 nanosheets, the Ti foil then sintered at 450°C, ramp 1hrs, bake 30 minutes. [9]

D. Preparation of Titanium Dioxide Photoanode

The commercial TiO2 powder was heated up to 400°C for 30 minutes to remove any absorbed moisture and organic impurities. Then the preheated nanopowder was added with 0.2ml glacial acetic acid and 0.2mL DI water. This mixture was grinded well in an agate mortar. The dispersion of TiO2 nanopowder into the dilute acetic acid was controlled by mortar grinding. This assisted in reducing the mean particle size by increasing the dispersion time. Transparency of the coated film also improved on increasing the grinding time. Powder which stuck inside the mortar was removed using a plastic spatula, to obtain a smooth homogenous paste. A well dispersed TiO2 particles in homogenous paste were then transferred using ethanol to a beaker. This solution was then subjected to ultra sonication about 15 minutes for homogenous mixing of particles. Followed by sonication, the obtained homogenous white liquid was subjected to magnetic stirring for 2 hours at a speed of 300rpm. The homogenous white liquid was evaporated for 15 minutes at 40°C produced a viscous paste. The viscous paste was again ultra sonicated for another 15 minutes to get a fine homogenous paste. This prepared paste was protected from moisture before coating on the substrates [11].

E. Preparation of Photoanode Deposition

This method is applied to both solid and flexible substrates. TiO$_2$ paste prepared with ethyl cellulose was sintered in furnace at 450°C for 30 minutes to crystalline the

TiO₂. A doctor blading technique was used to coat the glass substrate. TiO_2 paste was coated on the FTO (fluorine-doped tin oxide) conductive glass at 1 cm x 1 cm exposed area, in order to remove the solvents and other binding materials present in the pastes, the films were sintered at 450° C for 30 min in atmospheric air. After cooling down to 80° C, the films were dipped into a dye solution of 5×10−4 M cis–bis (isothiocyanato)-bis(2,2_-bipyridyl-4,4_-dicarboxylato) ruthenium(II) bis-tetrabutyl-ammonium (N719) dissolved in acetone and ethanol (volume ratio = 1:1) and kept at room temperature for 24 hrs.

F. Preparation of Counter Electrode

The FTO glass was used as the counter electrode for both photoanode. The glass were coated with Platinum solution for the catholic reduction of trioxide ion. 2 mM platinum solution was prepared using Hexachloroplatinic acid in 20 mL mixed with Isopropanol alcohol (IPA). Counter electrode was sintered at 450Ç for 30 minutes.

G. DSSC Assembly and IV Measurement

The dye-loaded photoanode was rinsed and dried. The Pt-counter electrodes were assembled by sandwich with the photoanode and 60 µm Surlyn polymer sheet. The polymer film were cut into 5mm x 15mm as spacer. Applied around the coated TiO2 film. Finally, liquid electrolyte was administered into the device. Solar energy conversion efficiency was measured under solar simulators used to mimic the solar spectrum of desired intensity at 1.5 air mass densities (AM1.5). A beam homogenizer which maintained evenly distributed incident power densities across the irradiation area was also incorporated into the device. Keithley4200 source meter was used to measure the output voltages and currents The plotted graph of I-V was then analyzed to calculate the field factor (FF) and energy conversion efficiency (ECE, η).

III. RESULTS AND DISCUSSIONS

The performance of fabricated flexible photoanode is investigated. The performance is compared with solid state photoanode on glass substrates, both with front illuminated and back illuminated test condition.

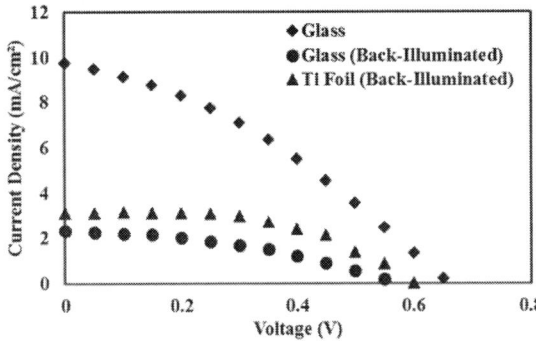

Fig. 1. Measured IV characteristics for i) glass (front-illuminated), ii) glass (back-illuminated) and iii) Ti foil (back-illumination).The irradiation was AM 1.5 (100 mW cm22).

TABLE I. DETAILS OF DSSC PERFORMANCE

	Substrates		
	Glass		Ti Foil
	Front-Illumination	Back-Illumination	
V_{oc} (V)	0.65	0.55	0.60
J_{sc} (mA/cm²)	9.77	2.32	3.12
FF	0.35	0.41	0.51
Efficiency η (%)	2.22	0.53	1.00

Fig. 1 shows I–V characteristics of DSSC based on flexible Titanium photoanode in comparison with solid state photoanode. From the graph, it shows that the back illuminated DSSC with flexible Titanium photoanode has better performance in term of current density, open circuit voltage and fill factor as compared with back illuminated solid state DSSC. However, the performance of front illuminated solid state DSSC is much better overall. This is due to the transparency of the counter electrode greatly influences the performance of DSSCs [12]. The back electrode consists of FTO and Pt layers which reduce the transparency. However, since the metal substrate photoanode based DSSCs have the only option of back-illumination. To increase the performance, the transparency of back electrode need to be improved.

Table 1 show the details performance of fabricated DSSC with flexible photoanode. The measured V_{oc}, I_{sc}, FF and η are 0.6 V, 3.12 mA/cm², 0.51 and 1.00%, respectively. As for comparison, the efficiency of back illuminated solid state DSSC halved the Flexible DSSC and front illuminated solid state DSSC doubled the Flexible DSSC.

The main drawback of this configuration not only relates to the transmission losses due to the Pt-based catalyst but also the I−/I3− liquid electrolyte [13]. The I32 electrolyte cuts the incident light significantly (from 400 nm to 600 nm) in the flexible DSC because of the back side illumination [14]. Despite this limitation the nanotube array DSSCs exhibit consistently larger values of Voc compared to typical published values for (front side illuminated) nanocrystalline cells. [15]

As reported in [3], the networked TiO2 formed after the treatment of H2O2 are the caused for the dye loading and better adsorption for the photocurrent to be improved on the Ti Foil. As it only sits in the H2O2 for 20 minutes at 95'C instead of 30 minutes at 70'C. Method used from this paper was using a screen print method instead of doctor blading method. The doctor blading method used was not suitable for the flexible Ti Foil and caused any cracks during assembly due to any bends during fabrication which detached it from the surface as doctor blading method gives a high density of paste applied.

Some recommendations on improving the performance of efficiency for the back illumination are replacing the counter electrode with a more transparent substrate and the catalyst itself from liquid electrolyte to solid electrolyte.

IV. CONCLUSION

Flexible photoanode on Titanium foil for back-illuminated DSSC has been successfully fabricated and measured. It shows a promising results with efficiency of 1.00% measured using back illuminated test setup. The lower efficiency of the back illumination was attributed to the higher recombination loss of photo-generated electrons at the front layer of the TiO_2 film in the shorter wavelength region of solar spectrum. Backside illumination is not optimal in DSSCs since the platinized counter electrode partially reflects light, while iodine in the electrolyte absorbs photons at lower wavelengths (400 – 600 nm).

ACKNOWLEDGMENT

This work was completely supported by Universiti Putra Malaysia grant (UPM/800-3/3/1/9629800)

REFERENCES

[1] Ellabban, O., Abu-Rub, H., &Blaabjerg, F. (2014). Renewable energy resources: Current status, future prospects and their enabling technology. Renewable and Sustainable Energy Reviews, 39, 748-764.

[2] Conibeer, G., Green, M., Corkish, R., Cho, Y., Cho, E. C., Jiang, C. W., ...&Trupke, T. (2006). Silicon nanostructures for third generation photovoltaic solar cells. Thin solid films, 511, 654-662.

[3] Tsai, T. Y., Chen, C. M., Cherng, S. J., &Suen, S. Y. (2013). An efficient titanium‐based photoanode for dye‐sensitized solar cell under back‐side illumination. Progress in Photovoltaics: Research and Applications, 21(2), 226-231.

[4] Fan, K., Li, R., Chen, J., Shi, W., & Peng, T. (2013). Recent development of dye-sensitized solar cells based on flexible substrates. Science of Advanced Materials, 5(11), 1596-1626.

[5] Cavallo, C., Di Pascasio, F., Latini, A., Bonomo, M., & Dini, D. (2017). Nanostructured semiconductor materials for dye-sensitized solar cells. Journal of Nanomaterials, 2017.

[6] Tsekouras, G., Mozer, A. J., & Wallace, G. G. (2008). Enhanced performance of dye sensitized solar cells utilizing platinum electrodeposit counter electrodes. Journal of the Electrochemical Society, 155(7), K124-K128.

[7] Gu, P., Yang, D., Zhu, X., Sun, H., Wangyang, P., Li, J., & Tian, H. (2017). Influence of electrolyte proportion on the performance of dye-sensitized solar cells. AIP Advances, 7(10), 105219.

[8] Wu, J., Lan, Z., Lin, J., Huang, M., Huang, Y., Fan, L., ...& Wei, Y. (2017). Counter electrodes in dye-sensitized solar cells. Chemical Society Reviews, 46(19), 5975-6023.

[9] Tsai, T. Y., Chen, C. M., Cherng, S. J., &Suen, S. Y. (2013). An efficient titanium‐based photoanode for dye‐sensitized solar cell under back‐side illumination. Progress in Photovoltaics: Research and Applications, 21(2), 226-231.

[10] Kapil, G., Pandey, S. S., Ogomi, Y., Ma, T., &Hayase, S. (2014). Titanium wire engineering and its effect on the performance of coil type cylindrical dye sensitized solar cells. Organic Electronics, 15(11), 3399-3405.

[11] Valsaraj D., Subramaniam MR2, Baiju G and Kumaresan D. (2016). Effect of Organic Binders of TiO2 Pastes in the Photoanodes of Cost-Effective Dye Sensitized Solar Cells Fabrication. Austin ChemEng 3(5).

[12] Balasingam, S. K., Kang, M. G., & Jun, Y. (2013). Metal substrate based electrodes for flexible dye-sensitized solar cells: fabrication methods, progress and challenges. Chemical communications, 49(98), 11457-11475.

[13] Jin, S., Shin, E., & Hong, J. (2017). TiO2 Nanowire Networks Prepared by Titanium Corrosion and Their Application to Bendable Dye-Sensitized Solar Cells. Nanomaterials, 7(10), 315.

[14] Ito, S., Rothenberger, G., Liska, P., Comte, P., Zakeeruddin, S. M., Péchy, P., ...&Grätzel, M. (2006). High-efficiency (7.2%) flexible dye-sensitized solar cells with Ti-metal substrate for nanocrystalline-TiO 2 photoanode. Chemical Communications, (38), 4004-4006.

[15] Paulose, M., Shankar, K., Varghese, O. K., Mor, G. K., Hardin, B., & Grimes, C. A. (2006). Backside illuminated dye-sensitized solar cells based on titania nanotube array electrodes. Nanotechnology, 17(5), 1446.

[16] Buda, S., Shafie, S. Rashid, SA., Jaafar, H., Sharif, NFM. (2017). Enhanced visible light absorption and reduced charge recombination in AgNP plasmonic photoelectrochemical cell. Results in Physics 7, 2311-2316

[17] Buda, S., Shafie, S. Rashid, SA., Jaafar, H., Khalifa, A. (2017). Response surface modeling of photogenerated charge collection of silver-based plasmonic dye-sensitized solar cell using central composite design experiments, Results in physics 7, 493-497

Simulation Analysis on CIGS Solar Cell On Different Absorber Layer Thickness Subject To Temperature Change Using SCAPS 1-D Software

M. Najib Harif
Faculty of Applied Sciences
Universiti Teknologi MARA
Shah Alam, Malaysia
najib@ns.uitm.edu.my

Siti Fazlili Abdullah
Dept of Electronics and
Communication Engineering
Universiti Tenaga Nasional
Kajang, Malaysia
siti@uniten.edu.my

Ahmad Wafi Mahmood Zuhdi
Dept of Electronics and
Communication Engineering
Universiti Tenaga Nasional
Kajang, Malaysia
wafi@uniten.edu.my

Fazliyana Za'abar
URND Sdn. Bhd
Universiti Tenaga Nasional
Kajang, Malaysia
fazliyana@uniten.edu.my

Mohd Shaparuddin Bahrudin
URND Sdn. Bhd
Universiti Tenaga Nasional
Kajang, Malaysia
shaparuddin@yahoo.com

Azri Husni Hasani
URND Sdn. Bhd
Universiti Tenaga Nasional
Kajang, Malaysia
azri.husni@uniten.edu.my

Abstract— The Copper Indium Gallium Selenide (CIGS) solar cell due to temperature change has been studied and the analysis was simulated using SCAPS 1-D software with the operating temperature varied from 25°C to 50°C. There are 3 parameters of thickness of CIGS absorber layer had been taken into account. The output characteristics were analyzed through a graphical method. The results showed a decreasing performance on the temperature change with respect to the efficiency, fill factor (FF) and open circuit voltage (V_{oc}) except for short circuit current density (J_{sc}). The average degradation of efficieny, fill factor and open circuit voltage with respect to temperature are 7.87%, 1.61% and 6.81% respectively while short current density increase 0.50%. It can also be suggested that there are significant direct correlations between band gap energy and temperature change.

Keywords—CIGS solar cell, SCAPS 1-D, Efficiency, Fill Factor, Open Circuit Voltage, Short Circuit Current, Band Gap Energy, Temperature Coefficient

I. Introduction

Thin film solar cell, also referred as the second generation photovoltaic technology, is made from compound material that is much thinner compared to first generation silicon solar cell technology due to its direct band-gap properties. It could be adapted with various engineering design to improve the efficiency performance. Moreover, the simple fabrication and deposition process for develop these solar cell surely would spark the idea of cost effectiveness durability [1]. Among the thin film is Copper Indium Galium diSelenide (CIGS) solar cell also known as chalcopyrite material solar cell. CIGS solar cell currently shows the highest efficiency among all thin film technologies. According to previous research that had been conducted, the highest efficiency of CIGS solar cell has achieved was 22.6% on lab scale. This progress performance has been improvised in terms of deposition method and designs [2].

Nowadays CIGS solar cell still ongoing research and development for pushing the current achievement for further increase in efficiency and reduce in cost. The CIGS layer acts as p-type absorber material with a direct band gap for potentially high coefficient absorption and simpler fabrication process from silicon solar cells are reasons why CIGS become perfect choice [3]. Despite higher band gap energy, another aspect that needs attention is the CIGS absorber layer thickness. It is important mainly because these materials would be the key ingredients in terms of performance and cost effective [4]. In other research also been done to find related behavior between band gap energy and temperature coefficient by NREL team reported an interesting finding of the correlation between thermal behaviour and temperature coefficient of the fill factor in CIGS solar cell modules. Accordingly, the temperature coefficient of the fill factor of CIGS modules is found to be increased after light soaking. It is suggested that this is due to light induced reduction in the conduction band offset between the buffer and the absorber [5]. Further investigation regarding temperature effect and behaviour are vital in order to develop more robust CIGS solar cells. In this paper, simulation study on the temperature change on CIGS absorber layer is demonstrated.

II. Simulation Approach Using SCAPS 1-D Software

The output characteristics of CIGS thin film solar cell were simulated using Solar Cell Capacitance Simulator (SCAPS) software version 3.3.0.7 which allows users enter the parameters to define the materials and interfaces of the solar cell [6]. The CIGS/CdS/ZnO cell structure that was designed in SCAPS is shown in Fig 1. The structure consist of window layer ZnO, buffer layer CdS, p-type absorber layer CIGS back contact layer (Molybdenum) and glass substrate layer. The thickness for both window layer and buffer layer are set to 0.05 µm. For this study, all materials properties for the buffer layer and window layer are kept constant in order to optimize the output characteristic of the

978-1-5386-5284-8/18 $31.00 © 2018 IEEE

absorber layer CIGS with regards to the temperature change as previously studied [7]. The thickness and material properties for the back contact and glass substrate are pre-defined in SCAPS software.

The absorber layer CIGS thickness on the other hand, was varied at 1 μm, 2 μm and 3 μm respectively. However the band gap energy, E_g was kept constant at 1.20 eV as well as the electron affinity, χ_e at 4.25 eV. These value were selected based on the optimum Ga/(Ga + In) composition as $Cu(In_{0.7},Ga_{0.3})Se_2$ which has been reported in [8]. Table 1 summarized the parameters used in the simulation design.

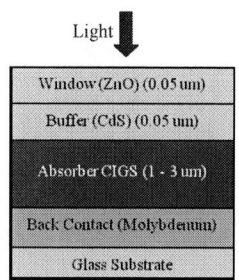

Fig 1. Schematic design for CIGS solar cell

TABLE 1. BASELINE PARAMETER FOR CIGS SOLAR CELL USING SCAPS SIMULATION

Parameters	p-CIGS	n-CdS	n-ZnO
Band gap, E_g (eV)	1.20	2.40	3.30
Electron affinity, χ_e (eV)	4.25	4.20	4.45
Thickness (μm)	1.0, 2.0, 3.0	0.05	0.05

In order to understand the trend of the CIGS cell output characteristics, each absorber layer CIGS thickness as 1μm, 2 μm and 3 μm were used to simulate at variant temperature. In this study, the operating temperature was setup from 25 °C to 50 °C. While both CdS and ZnO layer thicknesses were kept constant. The simulation focused on how temperature affects the efficiency and fill factor. These two parameters play an important part in solar cell which indicates the performance of a solar cell.

III. RESULTS AND DISCUSSION

Firstly, the efficiency for CIGS solar cell versus temperature is analyzed as shown in Fig 2. It can be seen from the graph that, when the temperature increases from 25 °C to 50 °C, the efficiency of the cell has seen approximately linear decrement for all absorber layer a significantly and linearly decrease. CIGS absorber layer with thickness of 3 μm shows 8.14% degradation on efficiency from the temperature 25 °C to 50 °C while thickness of 2 μm and 1 μm shows degradation of 7.96% and 7.44% respectively from the gradient over the same range of temperature. Therefore, the average degradation for these layers can be calculated to be about 7.87% gradient. Temperature coefficient is a vital component role in order to determine the solar cell efficiency as well as to quantify the sensitivities of the performance of any solar cell. It is mainly determined by the absorber layer composition and doping concentration. Typically if the temperature of solar cell increases, the efficiency decreases almost linearly for all silicon and thin film solar cells [9]. In this study, the thickness of the absorber layer shows little effect to the change in the temperature coefficient. This suggests that the thickness of the absorber layer has small contribution to the change of efficiency due to the change of operating temperature. However, higher temperature coefficient will lead to loss mechanism of the CIGS solar cell. Furthermore, higher temperature would also cause higher recombination carriers effect in CIGS solar cell layers area that will drop the performance as well.

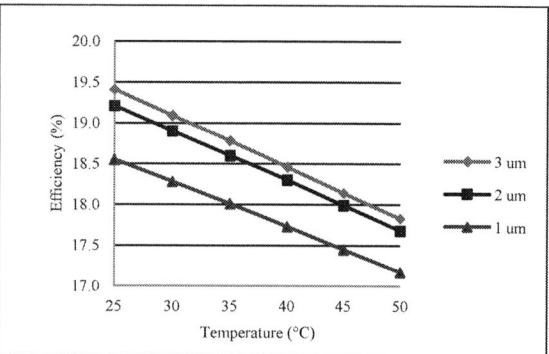

Fig 2. CIGS solar cell efficiency versus temperature

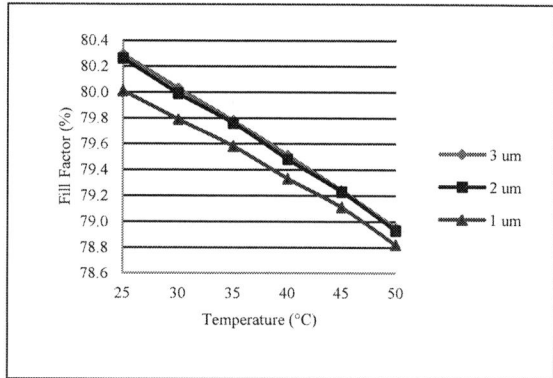

Fig 3. CIGS solar cell fill factor versus temperature

Fig 3 shows the effect of temperature on CIGS solar cell fill factor. From the graph it shows that the fill factor also decreases when the operating temperature rises. From the graph as well, the 3 μm thickness and 2 μm CIGS absorber layer solar cell show similar behavior of around 1.67% to 1.66% gradient degradations of fill factor. For thickness of 1 μm, slightly lower degradation of around 1.49% gradient was recorded. The degradations are similar for all thickness, where for efficiency 3 μm seems higher degradation than other thickness. Thus the average degradation of the fill factor from Fig 3 can be calculated to be 1.61% gradient. Again, these suggest that the thickness of the absorber layer has small contribution to the changes in fill factor due to the change of temperature. From the simulation results, it can be said that the performance of this CIGS solar cell has a direct correlation between efficiency and fill factor. When the efficiency decreases, the fill factor will decrease as well. This is because the efficiency in solar cell has a directly proportional to the fill factor [10]. As explained earlier, temperature coefficient exists instead of operating temperature of the solar cell. Besides, when the temperature coefficient increases, the fill factor will drop. This correlation can be determined in equation below that has been published in [11] as:

$$\beta_{FF_0} = \frac{1}{FF_0} \frac{dFF_0}{dT_c} \approx (1 - 1.02\ FF_o)\left(\frac{1}{V_{OC}} \frac{dV_{OC}}{dT_c} - \frac{1}{T_c}\right) \quad (1)$$

Where $\beta_{FF_0}, FF_0, T_c,$ and V_{oc} are temperature coefficient, fill factor, solar cell temperature and open circuit voltage respectively. From this equation, note that any temperature changes in solar cell, it is clearly will affect open circuit voltage as well apart from of fill factor. Furthermore major contribution that leads to detrimental of these solar cell performances is band gap energy in CIGS solar cell materials. This study only covers CIGS absorber layer material band gap which is already fixed to 1.20 eV into the simulation. CIGS material has direct band gap energy. Usually the band gap energy for CIGS absorber layer can be tuned from 1.0 eV to 1.7eV depending on compositions of the Indium and Gallium [12]. The band gap energy also has similar correlation between temperature effects. Taken together, these results suggest that higher temperature will decrease the band gap energy. When the band gap energy decreases in CIGS absorber layer, more photons will have energy higher than band gap energy to break free from the bond and create electron-hole pair which leads to an increase in photo-generated current. For this reason, the efficiency, fill factor and open circuit voltage tend to has detrimental issue [13], [14].

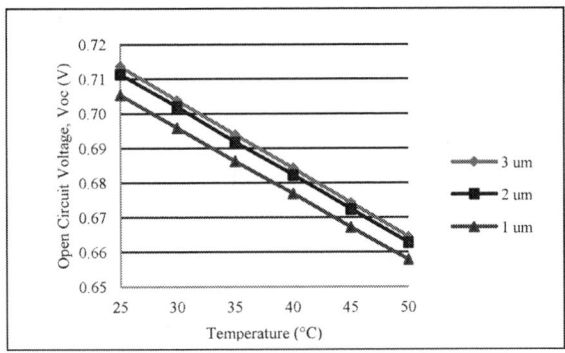

Fig 4. CIGS solar cell open circuit voltage versus temperature

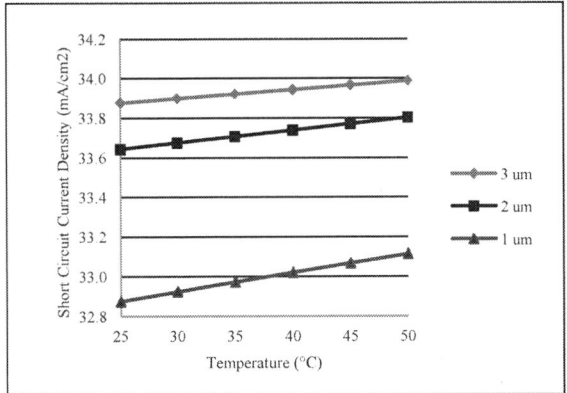

Fig 5. CIGS solar cell short circuit current vesus temperature

Another key point in this study is the open circuit voltage (V_{oc}) and the short circuit current density (J_{sc}). These two output characteristics would determine the efficiency of the solar cell as well. Fig 4 and Fig 5 have shown the opposite characteristics between them. When the temperature rise linearly, the V_{oc} of CIGS solar cell decrease for all thicknesses of the absorber layer. Surprisingly the J_{sc} of the CIGS solar cell slightly increase when simulating to higher temperature. From Fig 4, the 3 µm thickness CIGS absorber layer solar cell shows 7.00% gradient degradations of open circuit voltage due to temperature effect. Moreover, the 2 µm and 1 µm also show similar behavior as 6.75% and 6.67% gradient degradations respectively. That means total average degradation of open circuit voltage for three thicknesses of CIGS solar cells are 6.81% gradient. However in Fig 5, the J_{sc} shows slightly increase about 0.32%, 0.47% and 0.72% for absorber layer thickness 3 µm, 2 µm and 1 µm respectively as the average slope increment is 0.50%. This slope increment essentially is expected because most semiconductor band gaps decrease due to high temperature as the J_{sc} depends on the number of photons that able to to create an electron-hole pair. Hence it would create a loss mechanism factor such as parasitic absorption or surface recombination [15]. This finding confirms the association of efficiency, fill factor, V_{oc} and J_{sc} as output value of this solar cell. By all means the I-V characteristics of CIGS solar cell rely on temperature dependence. Besides, it is also has significant correlation between band gap energy reductions as defined earlier.

To understand the effect of temperature for CIGS solar cell, a few points need to be considered. Previous study, reported that by theoretically, most of parameter for common solar cell performance was temperature dependence [16]. From the study, it was found that the thickness of the CIGS absorber layer does not have significant contribution to the degradation of solar cell performance since all thickness resulted in similar performance trend. However, it is found that the 3 µm absorber layer produces better efficiency, fill factor, Voc and Jsc in comparison to thinner absorber layer as shown in Fig 2-5. This is because the absorber layer is a p type region area where the wavelengths of the illumination are absorbed. If the absorber layer is thick, more photons of the longer wavelength will be absorbed which in turn contributes to generation of electron–hole pair (EHP). The efficiency of the develop model was found to be 19.41% with the thickness of the absorber layer or 3um.

IV. CONCLUSION

In this paper, the analysis of CIGS solar cell on temperature effect has been simulated successfully using SCAPS software. The parameters were defined based on the previous study. From the simulations, it is clearly justify that the temperature significantly has effect on the CIGS solar cell performances. The efficiency, fill factor and open circuit voltage show a little percentage degradation when temperature increase. The thickness of CIGS absorber layer shows that the 3 µm CIGS absorber layer has better efficiency compare the two others which is 19.41%.

ACKNOWLEDGEMENT

The authors would like to thank Universiti Tenaga Nasional (UNITEN) and UNITEN R&D Sdn Bhd for the grant through this project.

REFERENCES

[1] P. Chopra, P. Paulson, V. Dutta, "Thin-film solar cells: An overview," Progress in Photovoltaics: Research and Applications, vol. 12, no. 2-3, pp. 69-92, 2004.

[2] T. Feurer, P. Reinhard, E. Avancini, B. Bissig, J. Löckinger, P. Fuchs, R. Carron,T. Weiss, J. Perrenoud, S. Stutterheim, S. Buecheler, and A.N. Tiwari, "Progress in thin film CIGS photovoltaics – Research and development, manufacturing, and applications.," Prog. Photovolt: Res. Appl., p. 645–667., 2016.

[3] Kato, Takuya, "Cu(In,Ga)(Se,S)2 solar cell research in Solar Frontier: Progress and current status," Japanese Journal of Applied Physics, vol. 56, no. 8S2, 2017.

[4] P. Jackson, D. Hariskos, E. Lotter, S. Paetel, R. Wuerz, R. Menner, W. Wischmann and M. Powalla, "New world record efficiency for Cu(In,Ga)Se2 thin-film solar cells beyond 20%.," Prog. Photovolt: Res. Appl., vol. 19, pp. 894-897, 2011.

[5] M. G. Deceglie, T. J. Silverman, B. Marion and S. R. Kurtz,, "Metastable changes to the temperature coefficients of thin-film photovoltaic modules," in IEEE 40th Photovoltaic Specialist Conference (PVSC), Denver, CO, 2014.

[6] M. Burgelman, P. Nollet and S. Degrave, "Modelling polycrystalline semiconductor solar cells," Thin Solid Films, Vols. 527-53, pp. 361-362, 2000.

[7] Nima Khoshsirat, Nurul Amziah Md Yunus, Mohd Nizar Hamidon, Suhaidi Shafie, Nowshad Amin,, "Analysis of absorber layer properties effect on CIGS solar cell performance using SCAPS,," Optik - International Journal for Light and Electron Optics,, vol. 126, no. 7-8, pp. 681-686, 2015.

[8] F.B. Dejene, "The structural and material properties of CuInSe2 and Cu(In,Ga)Se2 prepared by selenization of stacks of metal and compound precursors by Se vapor for solar cell applications," Solar Energy Materials and Solar Cells, vol. 93, no. 5, pp. 577-582, 2009.

[9] Olivier Dupré, Rodolphe Vaillon,Martin A. Green, Thermal Behaviour of Photovoltaic Devices, vol. 42, Cham, Switzerland: Springer Nature, 2017, pp. 9524-9532.

[10] Arno Smets, Klaus Jäger, Olindo Isabella, Rene Van Swaaij, Miro Zeman, Solar Energy: The physics and engineering of photovoltaic conversion, technologies and systems, Delft: UIT Cambridge Ltd, 2015.

[11] Jef Poortmans and Vladimir Arkhipov, Thin Film Solar Cells, Leuven, Belgium: John Wiley & Sons Ltd, 2007.

[12] Priyanka Singh, N.M. Ravindra, "Temperature dependence of solar cell performance - an analysis," Solar Energy Materials and Solar Cells, vol. 101, pp. 36-45, 2012.

[13] H. Heriche, Z. Rouabah, N. Bouarissa,, "New ultra thin CIGS structure solar cells using SCAPS simulation program,," International Journal of Hydrogen Energy,, vol. 42, no. 15, pp. 9524-9532, 2017.

[14] M. Theelen, A. Liakopoulou, V. Hans, F. Daume, H. Steijvers, N. Barreau, Z. Vroon, and M. Zeman, "Determination of the temperature dependency of the electrical parameters of CIGS solar cells," Journal of Renewable and Sustainable Energy, vol. 9, no. 2, 2017.

Comparative Study of the Temperature Effects on n-type and p-type Silicon Solar Cells by Numerical Simulation

[1] A.A. Zulkefle
Department of Diploma Studies
Faculty of Electrical
Engineering,
Universiti Teknikal Malaysia
Melaka
Melaka, Malaysia
aizan@utem.edu.my

[2] Z. Zakaria
Department of Electrical
Engineering Technology
Faculty of Engineering
Technology, Universiti Teknikal
Malaysia Melaka
Melaka, Malaysia
zaihasraf@utem.edu.my

[3] M. Zainon
Department of Electrical
Engineering Technology
Faculty of Engineering
Technology, Universiti Teknikal
Malaysia Melaka
Melaka, Malaysia
maslan@utem.edu.my

[4] A.I.A Rahman
Department of Electrical
Engineering Technology
Faculty of Engineering
Technology, Universiti Teknikal
Malaysia Melaka
Melaka, Malaysia
idil@utem.edu.my

[5] Z.A. Baharudin
Department of Electrical
Engineering Technology
Faculty of Engineering
Technology, Universiti Teknikal
Malaysia Melaka
Melaka, Malaysia
zikri@utem.edu.my

[6] M.A.M. Hanafiah
Department of Electrical
Engineering Technology
Faculty of Engineering
Technology, Universiti Teknikal
Malaysia Melaka
Melaka, Malaysia
ariff@utem.edu.my

[7] M.A. Sepee
ACIS Technology Sdn Bhd
Taman Paya Rumput Indah
Melaka, Malaysia
adam@acistechnology.com

Abstract—In this study, the influence of ambient temperature on the performance of p and n types of silicon solar cells have been investigated. The PC1D modeling software is used to simulate and analyze the photovoltaic properties of both types of silicon solar cells with the total thickness is restricted to 1µm and the ambient temperature is varied from 20 to 50°C. The simulation result exhibits the n type silicon solar cell give a better performance in term of short circuit current density compared to n type silicon solar cell. Apart from that, the conversion efficiency of silicon solar cells decrease linearly to ambient temperature due to higher recombination current. The efficiency of 5.58% is achieved for both types of silicon solar cells with ambient tempearture of 20°C.

Keywords—ambient temperature, silicon solar cell, p-type, n-type, PC1D

I. INTRODUCTION

Recently, the dominant photovoltaic market is silicon solar cells due to its abundance, maturity of manufacturing process, non-toxicity and high stability [1, 2]. The efficiency of average commercial crystalline silicon increased from 12 to 16% over the past decade [3, 4]. Since the thickness of commercial silicon solar cells is about 200–300 µm that leads to higher cost for the final product, the diminish of its thickness is required to cut down on material costs [5, 6]. Furthermore, reduction of solar cells thickness will deteriorate the performance of the photovoltaic cell due to weak absorption of indirect bandgap structure. The efficiency of 25.7%, a V_{OC} of 725 mV, a FF of 83.3% and a J_{SC} of 42.5 mA/cm^2 for 200 µm thick cell was the best results reported for n-type silicon solar cell [7, 8]. On the other hand, the best performance recorded for 250 µm thick of p-type silicon solar cell was 24.4% of the efficiency, 696 mV of the V_{OC}, 83.6% of the FF and 42.0 mA/cm^2 of the J_{SC} [7, 9].

In contrary, the performance of solar photovoltaic decreases with increasing operating temperature due to higher recombination current. Several researchers have intensively investigated the inevitable effect of temperature on the silicon solar cell characteristics in both experimental and theoretical work [10]-[16]. However, the performance of 1µm thick for phosphorus and boron doping of silicon solar cells through simulation has not been well documented. In our previous work [25], effects of thickness between silicon germanium and germanium solar cells were investigated using PC1D simulation. This issue serve motivation to study on effect of temperatures on n-type and p-type silicon solar cells using PC1D simulation.

In this study, we have investigated the temperature dependence of the 1µm thick silicon solar cell characteristics with different structure of pn junction using numerical simulation. PC1D software simulation has been chosen for this research due to its user-friendly system [17]-[21].

II. GENERAL FORMULATION

The equations such as the short circuit current density, open circuit voltage, fill factor and efficiency of solar cell are well known in the literature.

The short circuit current density, J_{sc} is given by

$$J_{sc} = q \int_{hv=E_g}^{\infty} \frac{dN_{ph}}{dhv} d(hv) \qquad (1)$$

Recombination losses, shadowing losses, reflection losses and ohmic losses are the amount of losses that need to be considered to determine the value of J_{sc}.

The open circuit voltage represents the maximum voltage that the solar cell can produced. The expression for V_{oc} as

$$V_{oc} = \frac{kT}{q} \ln\left(\frac{J_{sc}}{J_0} + 1\right) \qquad (2)$$

where the V_{oc} is related to J_{sc} and J_0.

Fill factor is the ratio of the maximum power output at the maximum power point to the product of the short circuit current density and open circuit voltage which can be written as

$$FF = \frac{P_{max}}{V_{oc}J_{sc}} \qquad (3)$$

The conversion efficiency of a solar cell is defined the as ratio of the power output to the power input and can be written as

$$\eta = \frac{V_{oc}J_{sc}FF}{P_{in}} \qquad (4)$$

where P_{in} is the intensity of the radiation incident.

III. NUMERICAL MODELING

By solving the equations of basic semiconductor device physics, the ideal characteristics for the photovoltaic cells can be defined [22].

A. Poisson's Equation

Poisson's equation relates the electric field produced by a set of stationary charges with the electric potential created.

$$\frac{d\xi}{dx} = \frac{\rho}{\epsilon} \qquad (5)$$

where, ξ is the material's permittivity.

B. Current Density Equations

Current density is essential equation consists of current density of electrons and holes in doped semiconductors as well as both the drift and diffusion currents. The following expressions for the current density of electrons and holes are given in equation (6) and (7).

$$J_e = q\mu_e n\xi + qD_e\frac{dn}{dx} \qquad (6)$$

$$J_h = q\mu_h n\xi + qD_h\frac{dp}{dx} \qquad (7)$$

C. Continuity Equations

The continuity equation states that the rate of change of electron/hole density due to the differences between direction of electron and hole flux plus the generation and minus the recombination.

$$\frac{1}{q}\frac{dJ_e}{dx} = U - G \qquad (8)$$

$$\frac{1}{q}\frac{dJ_h}{dx} = -(U - G) \qquad (9)$$

IV. METHODOLOGY

Fig. 1 exhibits the device schematic for both structures with an area of 100cm². Both solar cell parameters are tabulated in Table I and Table II, where the total thickness for both devices are 1µm. The surface recombination velocities of front and rear surfaces are 1x10⁶cm/s and 1x10⁵cm/s, respectively whilst the front surface reflectance is 10% across the solar spectrum. The ambient temperature is varied from 20°C to 50°C.

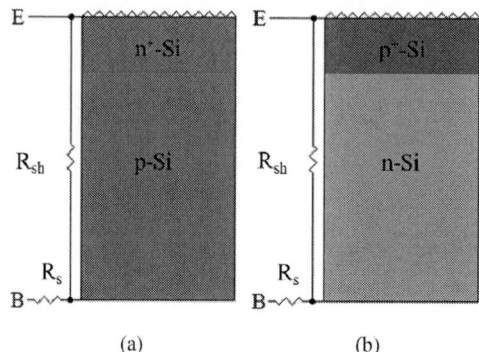

(a)   (b)

Fig. 1. Solar cell schematics used in PC1D program

(a) p-Si solar cell (b) n-Si solar cell

TABLE I.   PARAMETERS FOR P-SI SOLAR CELL

Device Structure	Thickness (µm)	Doping Concentration (cm⁻³)
n⁺-Si	0.1	10^{20}
p-Si	0.9	10^{16}

TABLE II.   PARAMETERS FOR N-SI SOLAR CELL

Device Structure	Thickness (µm)	Doping Concentration (cm⁻³)
p⁺-Si	0.1	10^{20}
n-Si	0.9	10^{16}

V. RESULTS & DISCUSSIONS

The performance variation as a function of temperature are shown in Fig. 2. For PC1D simulations, the temperature is varied from 20 to 50 degree celcius which is the ambient temperature. From Fig. 2, it can be clearly seen that the similar patterns of the curve for open circuit voltage (V_{OC}) and efficiency (η). The curves of open circuit voltage and efficiency are gradually decreased from the temperature of 20 to 50°C. Meanwhile, the curves of short circuit current density (J_{SC}) and fill factor (FF) are different with open circuit voltage and efficiency. On the other hand, Priyanka Singh et. al. [23] has presented justification that reverse saturation current increases with increasing temperature, and thus, V_{oc} decreases which reduces the fill factor and conversion efficiency as well. As the temperature increased, the short circuit current density (J_{SC}) for p and n types are slightly increased from 14.77mA/cm² and 15.04mA/cm² to 14.82mA/cm² and 15.08mA/cm², respectively. This is due to light harvesting that contribute additional photocurrent generation.

978-1-5386-5284-8/18 $31.00 © 2018 IEEE

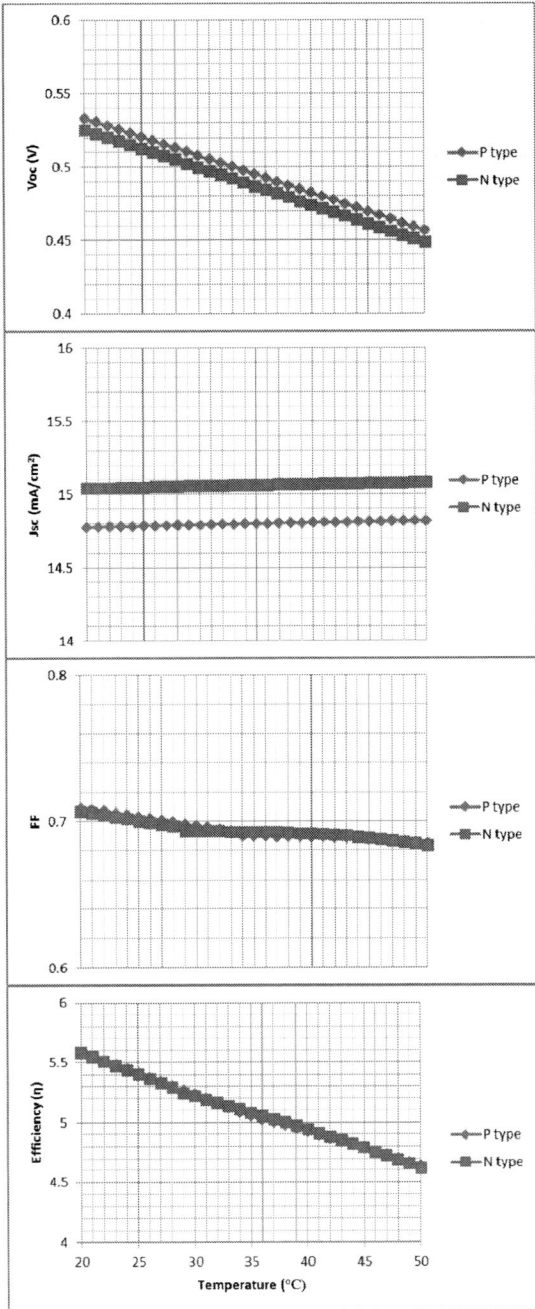

Fig. 2. Effect of temperature on both types of Si solar cell

However, the fill factor for both types almost constant at 0.70 that implies the FF parameter does not to be affected by ambient temperature. The efficiency for both types of silicon solar cell gradually decreased from 5.58% to 4.63% when the temperature is increased from 20 to 50°C. Thus, both solar cell's efficiency are inversely proportional to temperature where solar energy gets converted to heat [24]-

[26]. This simulation result is supported by Amelia et. al [27] and Khalis et. al [28] that the ambient temperatures could affect the performance parameters of solar cell.

VI. CONCLUSION

As a conclusion, both solar cells posses same characteristics where a highly efficient of 5.58% silicon solar cells at 20°C have been achieved from the numerical simulation. The performance for both types of solar cell are affected by ambient temperature where energy gap for both solar cells tends to decrease as the temperature is increased. As a result, the efficiency degradation of silicon solar cells can be observed. In addition, n type solar cell has higher short current density than p type solar cell due to higher carrier lifetime. Last but no least, a passive or active cooling systems can be proposed to improve the silicon solar cell performance.

Acknowledgment

The authors would like to acknowledge and appreciate the contribution of the Universiti Teknikal Malaysia Melaka (UTeM) through its research grant with code PJP/2017/FKE-CARE/S01565.

References

[1] F.Treble, Milestones in the development of crystalline silicon solar cells, Proc. World RenewableEnergy Congress, Florence, Italy, 1998, 473-478.

[2] Swapnil Dubey, Jatin Narotam Sarvaiya, Bharath Seshadri, "Temperature Dependent Photovoltaic (PV) Efficiency and Its Effect on PV Production in the World – A Review", Energy Procedia, Vol. 33 (2013), pp. 311 – 321

[3] T. Saga, Advances in Crystalline Silicon Solar Cell Technology for Industrial Mass Production, NPG Asia Materials, vol. 2, 2010, pp. 96-102.

[4] L.A. Dobrzański, M. Szczęsna, M. Szindler, A. Drygała, Electrical properties mono- and polycrystalline silicon solar cells, Journal of Achievements in Materials and Manufacturing Engineering, vol. 59, Issue 2, 2013, pp. 67-74

[5] K.R. Catchpole and A. Polman, Plasmonic solar cells, Optics Express, vol. 16, Issue 26, 2008, pp. 21793-21800.

[6] X. Zhang, Y. Yu, J. Xi, T. Liuand X. Sun, The plasmonic enhancement in siliconnanocone hole solar cells with back locatedmetal particles, Journal of Optics, vol. 17, no. 1, 2015, pp. 1-5.

[4] M. A. Green, Y. Hishikawa, E. D. Dunlop, D. H. Levi, J. Hohl - Ebinger, A. W.Y. Ho - Baillie, Solar cell efficiency tables (version 52), Progress in Photovoltaics Research and Applications, vol. 26, Issue 7, 2018, pp. 427-436.

[8] A. Richtera, J. Benicka, F. Feldmanna, A. Fella, M. Hermlea, S. W. Glunza, n-Type Si solar cells with passivating electron contact: Identifying sources for efficiency limitations by wafer thickness and resistivity variation, Solar Energy Materials and Solar Cells, vol. 173, 2017, pp. 96-105.

[9] J. Zhao, A. Wang, M. A. Green, F. Ferrazza, 19.8% efficient "honeycomb" textured multicrystalline and 24.4% monocrystalline silicon solar cells, Applied Physics Letters, vol. 73, no. 14, 1998, pp. 1991-1993.

[10] J.J. Wysocki and P. Rappaport, Effect of Temperature on Photovoltaic Energy Coversion, Journal of Applied Physics,vol. 31, 1960, pp. 571-578.

[11] W. Luft,Effects of Electron Irradiation on N on P Silicon Solar Cells,Advanced Energy Conversion, vol. 5, Issue 1, 1965, pp. 21-40.

[12] R. K. Yasui and L. W. Schmidt, Performance Characteristics of Ti-Ag Contact N/P and P/N Silicon Solar Cells, in 8th IEEE Photovoltaic Specialist Conference, 1970, pp. 110-122.

[13] D. J. Curtin and R. W. Cool, Qualification Testing of Laboratory Produced Violet Solar Cells, in 10th IEEE Photovoltaic Specialist Conference, 1973, pp. 139-152.

[14] L. J. Goldhammer and L. W. Slifer Jr.,ATS-6 Solar Cell Flight Experiment Through 2 Years in Orbit, in 12th IEEE Photovoltaic Specialist Conference, 1976, pp. 199-207.

[15] A. Agarwal, V. K. Tewary, S. K. Agarwal and S. C. Jain, Temperature Effects in Silicon Solar Cells,Solid-State Electronics, vol. 23, Issues10, 1980, pp. 1021-1028.

[16] J. R. Hauser and P. M. Dunbar, Performance Limitation of Silicon Solar Cells, IEEE Transactions Electron Devices, vol. 24, 1977, pp. 305-321.

[17] B.R. Losada, A. Moehlecke, J.M. Ruiz, A. Luque, Development of solar cells on microcrystalline alloys of Si1-xGex, 13th European PVSEC, Nice, 1995, pp. 925-928.

[18] E. Christoffel, L. Debarge, A. Slaoui, Modeling of multilayer SiGe based thin film solar cells, in 26th IEEE Photovoltaic Specialist Conference, 1997, pp. 783-786.

[19] M. Zainon, Z. Zakaria, S.A. Shahahmadi, M.A.M. Bhuiyan, M.M. Alam, K. Sopian, N. Amin, Effects of Germanium Layer on Silicon/Germanium Superlattice Solar Cells, in 39th IEEE Photovoltaic Specialist Conference, 2013, pp. 3484-3486.

[20] F.K. LeGoues, B.S. Meyerson, J.F. Morar, Anomalous strain relaxation in SiGe thin films and superlattices, Physical Review Letters, vol. 66, no. 22, 1991, pp. 2903-2906.

[21] D. A. Clugstonand P. A. Basore, Modelling free-carrier absorption in solar cells,Progressin Photovoltaics: Researchand Applications, vol. 5, 1997, pp. 229-236.

[22] M.A. Green, Solar Cells: Operating Principles, Technology and System Applications, vol.1. Kensington New South Wales: University of New South Wales, 1982.

[23] Priyanka Singh, N.M. Raindra, Temperature Dependence of Solar Cell Performance – An Analysis, Solar Energy Materials & Solar Cells, vol. 101, 2012, pp. 36-45.

[24] W. Shockley and H. J. Queisser, Detailed Balance Limit of Efficiency of p-n Junction Solar Cells, Journal of Applied Physics, vol. 32,1990, pp. 510-519.

[25] A. A. Zulkefle, M. Zainon, Z. Zakaria, M. A. Mat Hanafiah, N. H. Abdul Razak, S. A. Shahahmadi, M. Akhtaruzzaman, K. Sopian and N. Amin, A Comparative Study between Silicon Germanium and Germanium Solar Cells by Numerical Simulation, Applied Mechanics and Materials, vol. 761, 2015, pp. 341-346.

[26] R.T. Mohamad, M. Zainon, M. A. Mat Hanafiah, Effect of Temperature on 1 Micron Thick Silicon Solar Cell, International Journal of Applied Engineering Research, vol. 10, No. 11, 2015, pp. 29127-29133.

[27] A.R. Amelia, Y.M. Irwan, W.Z. Leow, M. Irwanto, M. Zhafarina, I. Safwati, Investigation of the effect temperature on photovoltaic (PV) panel output performance, International Journal on Advanced Science, Engineering and Information Technology, vol. 6, No. 5, 2016, pp. 682-688.

[28] M Khalis, R Masrour, G Khrypunov, M Kirichenko, D Kudiy and M Zazoui, Effects of Temperature and Concentration Mono and Polycrystalline Silicon Solar Cells: Extraction Parameters, Journal of Physics: Conference Series, vol. 758, No. 1, 2016, pp. 1-10.

Optimization of Baseline Parameters and Numerical Simulation for Cu(In,Ga)Se₂ Solar Cell

Fazliyana Za'abar
URND Sdn. Bhd.
Universiti Tenaga Nasional
Kajang, Malaysis
fazliyana@uniten.edu.my

Ahmad Wafi Mahmood Zuhdi
Dept. of Electronics and
Communication Engineering
Universiti Tenaga Nasional
Kajang, Malaysia
wafi@uniten.edu.my

Mohd. Shaparuddin Bahrudin
URND Sdn. Bhd.
Universiti Tenaga Nasional
Kajang, Malaysia
shaparuddin@yahoo.com

Siti Fazlili Abdullah
Dept. of Electronics and
Communication Engineering
Universiti Tenaga Nasional
Kajang, Malaysia
siti@uniten.edu.my

M. Najib Harif
Faculty of Applied Sciences
Universiti Teknologi MARA
Shah Alam, Malaysia
najib@ns.uitm.edu.my

Azri Husni Hasani
URND Sdn. Bhd.
Universiti Tenaga Nasional
Kajang, Malaysia
azri.husni@uniten.edu.my

Abstract—For the purpose of designing a highly efficient Cu(In,Ga)Se₂ (CIGS) solar cell, an understanding of the structural, optical and electronic properties of each constituent layers in the heterojunction cell is very crucial. Important parameters such as thickness, doping concentration, electron affinity and band gap energy are identified to govern the electrical characterization of the cell. In this paper, an extensive study on the effects of these parameters on the short circuit current density (J_{SC}) and open circuit voltage (V_{OC}) known as *J-V* characteristics is performed. Optimized values of each parameter obtained from different numerical simulations are summarized and presented. An optimal CIGS solar cell model is later identified and simulated using Silvaco ATLAS software. Performance analysis is carried out on the completed cell under standard irradiance with air mass 1.5 (AM1.5) spectrums. This proposed model provides simulated conversion efficiency of 23.58% and fill factor (FF) of 77.89% which is in agreement with experimental efficiencies found in literature.

Keywords—*CIGS solar cell, numerical simulation, parameters optimization, proposed baseline data, Silvaco ATLAS software, conversion efficiency*

I. INTRODUCTION

CIGS solar cell has emerged as a promising alternative to conventional silicon solar cells, leading other thin-film technologies with an efficiency reaching 22.6%, as certified in 2016 [1], [2]. Despite this high efficiency level, CIGS has not yet attained its full potential. The efficiency of current solar devices is limited by optical, collection, and recombination losses. According to Shockley-Queisser limit, maximum theoretical conversion efficiency of *p-n* junction-based solar cell is 32.8% at a band gap of 1.15 eV [3]. If all loss mechanisms were addressed at the same time, an efficiency approaching this level would be technically feasible.

Improvement of CIGS solar cell efficiency requires device optimization. Numerical simulations are advantageous for optimization because all the input parameters of the model in terms of device and material properties are well controlled. Parameters such as layer thickness, doping concentration, band gap, electron affinity, and many more can affect solar cell's characteristics and its electrical performance. Currently, a number of simulation packages such as AMPS-1D, SCAPS-1D, and Silvaco ATLAS are commonly used to perform an in-depth simulation study of CIGS solar cell for the purpose of optimization

This paper starts with an introduction to the structure of a typical CIGS solar cell and the importance of each layers' material properties. Afterwards, the results from different simulation studies on CIGS cell are discussed and optimized baseline parameters which can be used to develop a theoretical model of the cell are presented. Subsequently, numerical simulation was conducted using the obtained layer data. Starting with a conventional n-ZnO/n-CdS/p-CIGS structure, we simulated the parameters of current density-voltage (*J-V*) characteristics. The purpose of the study was to set optimum baseline parameters for CIGS solar cell and evaluate its characteristics with valid experimental results.

II. CIGS SOLAR CELL STRUCTURE AND SELECTION OF MATERIAL PARAMETERS

Generally, a typical CIGS solar cell structure consists of a top contact, followed by three layers of semiconductor material: doped zinc oxide (n-ZnO) which serves as the window layer, cadmium sulfide (CdS) as the n-type buffer layer, and CIGS as a p-type absorber layer. Molybdenum (Mo) is commonly used as a back contact in the stack, with soda-lime glass as the substrate. The standard structural layout of a CIGS cell is illustrated in Fig. 1.

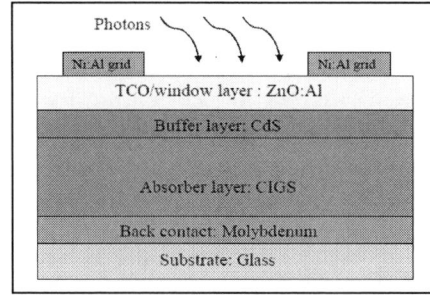

Fig. 1. Schematic representation of a typical thin-film CIGS solar cell

This basic structure could be modified in terms of layer thickness, cell surface area, and material properties. Our focus lies in particular on the influence of thickness and carrier concentration of the absorber-buffer-window region, as well as band gap energy of the absorber layer on the electrical parameters. Properties of these layers are evident to directly contribute in the performance of the cell [4]–[14].

A. Layers Thickness

ZnO is used both as a window or a transparent conducting oxide (TCO) layer and also as n-type junction. TCO serves as the first contact between photons and the cell thus this layer should allow the largest possible percentage of photons to reach the absorber layer while simultaneously allowing for good conduction of current generated by the cell [15]. ZnO is commonly doped with Aluminum to enhance the electrical conductivity [16]. Similarly, the role of CdS buffer layer other than optimizing the band alignment of the CIGS solar device [17] is to be as much transparent as possible, allowing a maximum sunlight absorption in the absorber layer [13]. Therefore, in the CIGS structure, ZnO and CdS layers are very thin. Contrary to ZnO and CdS, the CIGS layer acts as the absorber and is very important in determining the cell efficiency. It is reported that the performance of a CIGS cell increases with the increased thickness of the CIGS absorber layer.

B. Doping Concentration

Theoretically, donor (N_D) or acceptor (N_A) concentration of the component layers of CIGS solar cell can be determined from the capacitance voltage measurement [18], [19]. The effect induced by N_A and N_D on the V_{OC} and J_{SC} can be observed from the diode equations presented in [20].

C. Conduction band and valence band effective density of states

For direct band gap material such as ZnO, CdS, and CIGS, the recommended relations of effective masses for electrons and holes are discussed in [19]. Based on this, the effective densities of states in the conduction band N_C and the valence band N_V can be obtained via an equation presented in [20]. According to the equation, the N_C and N_V are temperature dependent. For CIGS solar cell, N_C and N_V at 25°C (300K) are determined to be 2.2×10^{18} cm^{-3} and 1.8×10^{19} cm^{-3} respectively. The same values are used as simulation parameters in [4], [6], [7], [9], [12], [18], [21]–[24].

D. Band gap energy

The band gaps of ZnO and CdS semiconductors are constant [20]. Conversely, in the chalcopyrite $Cu(In_{1-x},Ga_x)Se_2$ alloy system, the band gap can be tuned by varying the relative amount of Gallium (Ga) in the composition denoted by compositional ratio $x = (Ga)/(Ga+In)$. According to [19], the band gap can range from 1.04 eV for pure $CuInSe_2$ to 1.67 eV for pure $CuGaSe_2$. The band gap of CIGS layer is dependent on "x" following (1) as in below:

$$E_g(x) = 1.02 + 0.67x + 0.11x(x-1) \qquad (1)$$

From a number of simulation studies [8], [11], [19], [24]–[27], it is shown that V_{OC} increases linearly with an increase in band gap energy as expected, while J_{SC} decreases. The trade-off between V_{OC} and J_{SC} leads to an optimal value of absorber layer band gap energy. Currently, CIGS solar cell with $x = 0.3$ which corresponds to a band gap energy between 1.1 to 1.2 eV yields optimum efficiency both in simulation and

laboratory results [1], [28]. Electron affinity of the CIGS material is also dependent on the Ga composition of the material and the relation can be deducted from curve fitting as shown in [8]. The difference between electron affinity of CIGS absorber layer and CdS buffer layer plays an important role in the band alignment at the buffer/absorber interface [29].

III. DISCUSSION ON THE OPTIMIZATION STUDIES OF ZnO, CdS, AND CIGS MATERIAL PROPERTIES

For the purpose of cell performance optimization, the variation of photovoltaic-parameters as a function of thickness and doping concentration were simulated in a number of investigations. Optimum values for the two properties determined through simulations discussed in recent publications are presented in Table 1.

TABLE 1. PHYSICAL PARAMETERS PROPERTIES USED IN SIMULATIONS AND EXPERIMENTS FOR OPTIMAL CELL

Sim.[a] /Exp.[b]	Thickness (nm)			Doping Concentration (cm^{-3})			η %
	ZnO	CdS	CIGS	ZnO	CdS	CIGS	
[4][a]	20	50	4000	1×10^{16}	1×10^{16}	1×10^{16}	25.90
[9][a]	150	40	3000	1×10^{17}	2×10^{19}	1×10^{18}	23.20
[11][a]	200	50	3000	1×10^{18}	5×10^{18}	1×10^{19}	20.34
[14][a]	NA[c]	60	4000	NA[c]	1×10^{17}	1×10^{17}	21.4
[30][a]	800	50	3000	1×10^{18}	1×10^{17}	8×10^{16}	19.75
[15][a]	400	50	2000	NA[c]	NA[c]	NA[c]	17.2
[1][b]	150-200	30-50	2500-3000	NA[c]	NA[c]	NA[c]	22.6

[a.] Simulation [b] Experiment [c] Information not available in publication

A. Effect of ZnO, CdS, and CIGS thickness on V_{OC}, J_{SC}, and conversion efficiency

Simulation studies performed by [4], [5], [9], [12], [31] show that both the cell efficiency and J_{SC} are significantly influenced by the variation of the ZnO window layer thickness. As the ZnO film thickness increases, the optical transmission and the electrical resistivity of this layer decrease. Reduction in J_{SC} in case of increased thickness is attributed to the low optical transmission of light that allows less amount of light to be absorbed to the CIGS layer thus affecting the creation of electron-hole pairs [4], [5], [9], [12]. Conversely, if the thickness of this layer is to be too thin, the cell performance degrades due to the increase of loss because of reflection [4], [12]. It is suggested that the best efficiency should be achieved using ZnO window layer with an optimum thickness that is from 100 to 200 nm.

Analysis of the results from simulation [4], [9], [12], [22] show that except for V_{OC} which remains constant, all remaining photovoltaic parameters (J_{SC}, η, and FF) are decreasing with the increase of CdS buffer layer thickness. When CdS thickness increases, a large number of short wavelength photons are absorbed in this layer before reaching the absorber layer, leading consequently to degraded cell performance [24]. This is because the absorbed photons in CdS do not contribute significantly in the collected photo-current. Hence, it is preferable to minimize its thickness in order to reduce optical absorption loses. Though CIGS cell with thinner buffer layer shows higher performance, thicknesses less than 40nm currently is not reachable due to limitation in fabrication techniques and instruments. Based on these considerations, the range of 40 nm to 60 nm is determined to be the optimized thickness of buffer layer in CIGS solar cell [7], [24], [32], which is in agreement with CdS thickness (50 nm) commonly adopted in the lab [1], [28].

The general performance of a CIGS cell increases with the increased thickness of the CIGS absorber layer but with a much slower rate over 3000 nm [6], [8], [14]. By increasing the absorber layer thickness, more photons with longer wavelengths can be collected in this layer which will contribute to more generation of electron-hole pairs (EHP). This will results in higher J_{SC} and V_{OC} which will collectively increase the conversion efficiency of the cell [4]–[6], [9], [10], [25]. Though a thicker absorber layer benefits for complete photons absorption, too thick of the layer will lead to material waste and will eventually promote recombination. Moreover, simulations in [8], [18], [24], [31] show that for film thickness exceeding 3000 nm, the values of J_{SC} and V_{OC} are almost unchanged which will not result in major efficiency gains. On the contrary, if the absorber layer thickness is reduced, the back contact will be very close to the depletion region, thus promoting more recombination events at the back contact. Due to the recombination, less amount of electrons will contribute to the quantum efficiency, affecting both V_{OC} and J_{SC} [4], [6], [10], [24], [25].

B. Effect of ZnO, CdS, and CIGS doping on V_{OC}, J_{SC}, and conversion efficiency

Based on the diode equations [20] discussed earlier, it can be observed that as the values of N_A and N_D increases, the saturation current I_S decreases which leads to a logarithmic increase in the open-circuit voltage, V_{OC}. On the other hand, with increasing doping level, the output current I of the cell is controlled by two limiting factors: higher drift velocity of the majority carriers which increases the output current and increased minority carrier recombination rate that will reduce the output current. Therefore, the output current depends on which of the two factors dominates the other at a particular doping concentration [11], [30].

According to [4], [9], [30], as N_D of ZnO window layer increases, the cell efficiency increases as well and becomes almost constant when N_D reaches a value of 1×10^{18} cm^{-3}. The increase in output conversion efficiency is attributed to the improve collection of photo-generated carriers as the doping concentration increases. The same trend can be observed in the efficiency variation versus CdS buffer layer carrier concentration (N_D) [4], [9], [13], [18]. Above 10^{17} cm^{-3}, the efficiency increases only slightly then starts to decrease as N_D becomes higher. Therefore, the optimized doping level of the CdS buffer layer is restricted to 1×10^{17} cm^{-3} [18], [30]. For CIGS layer, as N_A increases further than 10^{16} cm^{-3}, the efficiency of the solar cell reduces due to the recombination process at the back contact, suggesting 10^{16} cm^{-3} to be in optimum range [4], [18], [30], [33], [34]. These results are in good agreement with default values presented in [19].

IV. Proposed Baseline Data for Numerical Simulation

Based on recent publications and trends, the recommended values for different parameters required for numerical simulation of CIGS solar cell is presented in this section. The values are displayed in Table 2.

TABLE 2. BASELINE PARAMETERS FOR OPTIMAL CELL

Parameters	Layers		
	ZnO	CdS	CIGS
Thickness (nm)	150	50	3000
Band gap, E_g (eV)	3.3	2.4	1.27
Donor concentration, N_D(cm^{-3})	1×10^{18}	1×10^{17}	0
Acceptor concentration, N_A(cm^{-3})	0	0	2×10^{16}
Conduction band effective density of states, N_C (cm^{-3})	2.2×10^{18}	2.2×10^{18}	2.2×10^{18}
Valence band effective density of states, N_V (cm^{-3})	1.8×10^{19}	1.8×10^{19}	1.8×10^{19}

V. Simulation Approach Using Silvaco Atlas Software

Properties of the device layers defined in Table 2 were used as input parameters to design a theoretical model of CIGS solar cell. Then, the output characteristics of the optimized cell were numerically simulated.

The schematic device structure of CIGS solar cell developed using Silvaco simulation tool is shown in Fig. 2. The stack is composed of Molybdenum (Mo) back contact, a p-type wide-band gap absorber layer (CIGS), followed by n-type buffer layer made of CdS and a window/TCO layer made of doped ZnO of thickness as per Table 2 respectively.

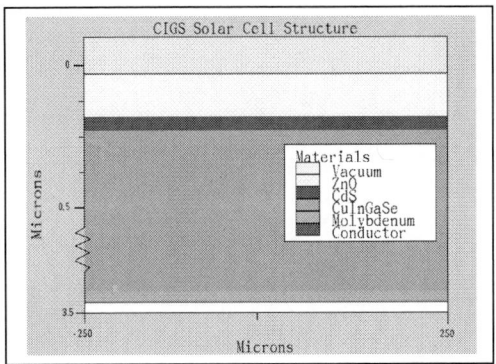

Fig. 2. Schematic CIGS solar cell structure used for simulation.

Fig 3. shows current density-voltage *(J-V)* characteristics curve for CIGS solar cell simulated under standard test condition of AM 1.5 illumination spectra, 1000 W/m^2 irradiance, and 25°C (300K) temperature.

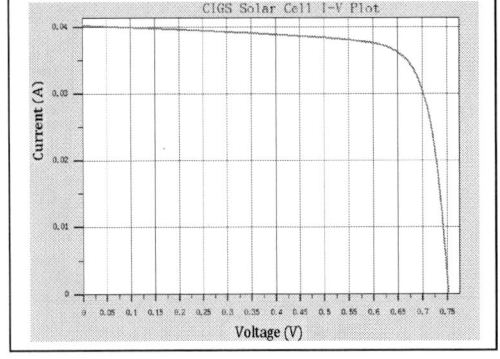

Fig. 3. J-V characteristic curve for optimized CIGS solar cell

The resulting performance parameters of the V_{OC}, J_{SC}, FF, and η determined using *J-V* characteristics are displayed in Table 3, and compared with gathered experimental values [1],

[35]. It can be observed that the efficiency is slightly higher than the highest recorded efficiency of 22.6% [1], [2] observed experimentally. Based on the simulation's results and analysis given in previous sections, the simulated cell's V_{OC} is improved by setting the absorber layer band gap at optimum value which is around 1.2 eV. The cell's J_{SC} on the other hand is enhanced by setting the absorber, buffer and window layers thickness at optimum value which leads to the increase in absorber layer's absorption rate. However, the simulated cell is not an exact comparison to the experimental cell reported in [1], [35]. This is because the information on some of the parameters used such as doping concentration, band gap, and exact thickness of each layer are not reported for the experimental cell in the reference. In addition, the simulation of CIGS cell in this work is performed in an ideal environmental condition hence moderate resistive loss is expected out of the cell. The changes in CIGS solar cell efficiency introduced by the presence of grain boundary (GB) are also not considered in our baseline case. According to [36] GBs that are presence within the space-charge region (SCR) will lower the V_{OC}, whereas GBs in the bulk material are observed to reduce the J_{SC} of the cell. Nonetheless, the close agreement between experimental and simulation values from this work validates our set of parameters as a baseline data for numerical simulation and modelling of CIGS solar cell.

TABLE 3. PERFORMANCE PARAMETERS - SIMULATION (FROM THIS WORK) VS EXPERIMENT

Performance Parameter	This work	Experiment [1][a]	Experiment [35][b]
Open Circuit Voltage, Voc (V)	0.75	0.74	0.72
Short Circuit Current Density, Jsc (mA/cm²)	40.27	37.80	39.40
Fill Factor, FF (%)	77.89	80.60	78.20
Efficiency, η (%)	23.58	22.60	22.30

[a] P. Jackson et al. 2016   [b] R. Kamada et al. 2016

Conclusion

In this paper, based on gathered simulation and experimental studies, we provide a guideline that should be considered when assigning input parameters for numerical simulation of CIGS thin film solar cell. Numerical simulation is proven to play a vital role to understand the basic factors influencing and limiting the electrical parameters of the cell and to optimize the performance before experimental procedure can be conducted. The combination of optimum parameters gathered from recent simulation studies are subsequently used in our simulation. In optimum condition, we achieved conversion efficiency of 23.58% and fill factor of around 78%. The simulated results are in close agreement with experimental values though the efficiency is 0.98% higher than the champion laboratory CIGS cell efficiency. The analysis and examination of the simulation results suggest that the optimization of material properties has an advantage in improving the cell efficiency.

References

[1] P. Jackson, R. Wuerz, D. Hariskos, E. Lotter, W. Witte, and M. Powalla, "Effects of heavy alkali elements in Cu(In,Ga)Se2 solar cells with efficiencies up to 22.6%," *Phys. Status Solidi - Rapid Res. Lett.*, vol. 10, no. 8, pp. 583–586, 2016.

[2] M. A. Green *et al.*, "Solar cell efficiency tables (version 50)," *Prog. Photovoltaics Res. Appl.*, vol. 25, no. 7, pp. 668–676, 2017.

[3] S. Siebentritt, "What limits the efficiency of chalcopyrite solar cells?," *Sol. Energy Mater. Sol. Cells*, 2011.

[4] H. Heriche, Z. Rouabah, and N. Bouarissa, "High-efficiency CIGS solar cells with optimization of layers thickness and doping," *Optik*

(Stuttg)., vol. 127, no. 24, pp. 11751–11757, 2016.

[5] H. Movla, E. Abdi, and D. Salami, "Simulation analysis of the CIGS based thin film solar cells.," *Opt. - Int. J. Light Electron Opt.*, vol. 124, no. 22, pp. 5871–5873, 2013.

[6] P. Chelvanathan, M. I. Hossain, and N. Amin, "Performance analysis of copper-indium-gallium-diselenide (CIGS) solar cells with various buffer layers by SCAPS," *Curr. Appl. Phys.*, vol. 10, no. SUPPL. 3, pp. 387–391, 2010.

[7] M. Mostefaoui, H. Mazari, S. Khelifi, A. Bouraiou, and R. Dabou, "Simulation of High Efficiency CIGS Solar Cells with SCAPS-1D Software," *Energy Procedia*, vol. 74, pp. 736–744, 2015.

[8] N. Khoshsirat, N. A. Md Yunus, M. N. Hamidon, S. Shafie, and N. Amin, "Analysis of absorber layer properties effect on CIGS solar cell performance using SCAPS," *Optik (Stuttg).*, vol. 126, no. 7–8, pp. 681–686, 2015.

[9] S. Dabbabi, T. Ben Nasr, and N. Kamoun-Turki, "Parameters optimization of CIGS solar cell using 2D physical modeling," *Results Phys.*, vol. 7, pp. 4020–4024, 2017.

[10] S. Tobbeche and H. Amar, "Two-dimensional Modelling and Simulation of CIGS thin-film solar cell," *J. New Technol. Mater.*, vol. 4, no. 1, pp. 89–93, 2014.

[11] M. Asaduzzaman, M. Hasan, and A. N. Bahar, "An investigation into the effects of band gap and doping concentration on Cu(In,Ga)Se2 solar cell efficiency," *Springerplus*, vol. 5, no. 1, 2016.

[12] H. Firoozi and M. Imanieh, "Improvement the Efficiency CIGS Thin Film Solar Cells by Changing the Thickness Layers," *Int. J. Eng. Comput. Sci.*, vol. 6, no. 7, pp. 2319–7242, 2017.

[13] A. Sylla, S. Touré, and J. P. Vilcot, "Numerical modeling and simulation of CIGS-based solar cells with ZnS buffer layer," *ARPN J. Eng. Appl. Sci.*, vol. 13, no. 1, pp. 64–74, 2018.

[14] S. Banerjee, Y. K. Ojha, K. Vikas, and A. Kumar, "High Efficient CIGS based Thin Film Solar Cell Performance Optimization using PC1D," *Int. Res. J. Eng. Technol.*, vol. 3, no. 6, pp. 385–388, 2016.

[15] M. M. Islam *et al.*, "Thickness study of Al:ZnO film for application as a window layer in Cu(In1-xGax)Se2thin film solar cell," *Appl. Surf. Sci.*, vol. 257, no. 9, pp. 4026–4030, 2011.

[16] M. Ajili, N. Jebbari, N. Kamoun Turki, and M. Castagné, "Effect of Al-doped on physical properties of ZnO Thin films grown by spray pyrolysis on SnO 2 : F/glass," *EPJ Web Conf.*, vol. 29, p. 2, 2012.

[17] S. Ullah, "Thin film solar cells based on copper-indium-galium selenide (CIGS) materials deposited by electrochemical techniques," 2017.

[18] M. A. M. Bhuiyan, M. S. Islam, and A. J. Datta, "Modeling , Simulation and Optimization of High Performance CIGS Solar Cell," *Int. J. Comput. Appl.*, vol. 57, no. 16, pp. 26–30, 2012.

[19] M. Gloeckler, A. L. Fahrenbruch, and J. R. Sites, "Numerical modeling of CIGS and CdTe solar cells: setting the baseline," *3rd World Conf. onPhotovoltaic Energy Conversion, 2003. Proc.*, vol. 1, pp. 491–494, 2003.

[20] S. M. Sze and K. K. Ng, *Physics of Semiconductor Devices*. 2007.

[21] M. Elbar and S. Tobbeche, "Numerical Simulation of CGS/CIGS Single and Tandem Thin-film Solar Cells using the Silvaco-Atlas Software," *Energy Procedia*, vol. 74, pp. 1220–1227, 2015.

[22] A. Benmir and M. S. Aida, "Analytical modeling and simulation of CIGS solar cells," *Energy Procedia*, vol. 36, pp. 618–627, 2013.

[23] M. B. Hosen, A. N. Bahar, M. K. Ali, and M. Asaduzzaman, "Modeling and performance analysis dataset of a CIGS solar cell with ZnS buffer layer," *Data Br.*, vol. 14, pp. 246–250, 2017.

[24] X. Shang *et al.*, "A numerical simulation study of CuInS2solar cells," *Thin Solid Films*, vol. 550, pp. 649–653, 2014.

[25] N. Amin, P. Chelvanathan, M. I. Hossain, and K. Sopian, "Numerical modelling of ultra thin Cu(In,Ga)Se2solar cells," *Energy Procedia*, vol. 15, no. 2011, pp. 291–298, 2012.

[26] M. A. Rifat, M. Imtiaz, M. Khan, and K. Alam, "Simulation Study on the Effects of Changing Band Gap on Solar Cell Parameters," *2016 9th Int. Conf. Electr. Comput. Eng. Dhaka, Bangladesh*, pp. 86–89, 2016.

[27] S. Oladapo, B. M. SOUCASE, and B. AKA, "Numerical Simulation And Performance Optimization Of Cu(In,Ga)Se2 Solar Cells," *IOSR J. Appl. Phys.*, vol. 8, no. 4, pp. 01–11, 2016.

[28] P. Jackson *et al.*, "Properties of Cu(In,Ga)Se2solar cells with new record efficiencies up to 21.7%," *Phys. Status Solidi - Rapid Res. Lett.*, vol. 9, no. 1, pp. 28–31, 2015.

[29] N. Khoshsirat and N. A. A. M. Yunus, "Copper-Indium-Gallium-diSelenide (CIGS) Nanocrystalline Bulk Semiconductor as the Absorber Layer and Its Current Technological Trend and Optimization," in *Nanoelectronics and Materials Development*, 2016.

[30] S. M. Shamim, A. Sarker, M. R. Ahmed, and M. F. Huq, "Performance Analysis on the Effect of Doping Concentration in Copper Indium Gallium Selenide (CIGS) Thin-film Solar Cell," *Int. J. Comput. Appl.*, vol. 113, no. 14, pp. 975–8887, 2015.

[31] D. A. Columbus, "Design and Optimization of Copper Indium Gallium Selenide Solar Cells For Ligthweight Battlefield Application," no. June, pp. 33–34, 2014.

[32] N. Khoshsirat and N. A. Md Yunus, "Numerical simulation of CIGS thin film solar cells using SCAPS-1D," *2013 IEEE Conf. Sustain. Util. Dev. Eng. Technol.*, pp. 63–67, 2013.

[33] M. F. M. Fathil *et al.*, "The impact of minority carrier lifetime and carrier concentration on the efficiency of CIGS solar cell," *Semicond.*

Electron. (ICSE), 2014 IEEE Int. Conf., pp. 24–27, 2014.

[34] F. T. Zohora, M. A. M. Bhuiyan, and S. Saimoom, "Simulation and Optimization of High Performance CIGS Solar Cells," *Int. Conf. Mech. Eng. Renew. Energy 2015*, 2015.

[35] K. Rui *et al.*, "New World Record Cu(In,Ga)(Se,S)2 Thin Film Solar Cell Efficiency Beyond 22%," *2016 IEEE 43th Photovolt. Spec. Conf. PVSC 2016 Portland(OR) USA, June 2016.*, pp. 3–7, 2016.

[36] M. Gloeckler, W. K. Metzger, and J. R. Sites, "Simulation of Polycrystalline Cu(In,Ga)Se2 Solar Cells in Two Dimensions," in *Mater. Res. Soc. Symp. Proc. Vol.*, 2005.

Modelling and Simulation of Photovoltaic Solar Cell using Silvaco TCAD and Matlab Software

Azri Husni Hasani
URND Sdn. Bhd.
Universiti Tenaga Nasional
Kajang, Malaysia
azri.husni@uniten.edu.my

Siti Fazlili Abdullah
Dept. of Electronics and
Communication Engineering
Universiti Tenaga Nasional
Kajang, Malaysia
siti@uniten.edu.my

Ahmad Wafi Mahmood Zuhdi
Dept. of Electronics and
Communication Engineering
Universiti Tenaga Nasional
Kajang, Malaysia
wafi@uniten.edu.my

Mohd. Shaparuddin Bahrudin
URND Sdn. Bhd.
Universiti Tenaga Nasional
Kajang, Malaysia
shaparuddin@yahoo.com

Fazliyana Za'abar
URND Sdn. Bhd.
Universiti Tenaga Nasional
Kajang, Malaysis
fazliyana@uniten.edu.my

M. Najib Harif
Faculty of Applied Sciences
Universiti Teknologi MARA
Shah Alam, Malaysia
najib@ns.uitm.edu.my

Abstract—In this paper, a modelling approach for a photovoltaic solar cell has been proposed which begins with the development of a solar cell up to enabling the solar cell to be implemented at circuit level simulations. This modelling approach is useful in the photovoltaic field to have an initial or overall observation on the effects toward the photovoltaic system. The modelling approach begins with modelling a thin film Cu(In,Ga)Se$_2$ (CIGS) solar cell using Silvaco TCAD (Technology Computer-Aided Design) software with a predefined baseline parameters. The electrical parameters as well as the I-V curve of the TCAD model are obtained and the data is exported to be post-processed in Matlab software. Key parameters of the TCAD model are used to develop an equivalent electrical model. The Single-diode model topology is implemented for simplicity. In order to test the validity of the single-diode model, the I-V curve is compared to the I-V curve of the TCAD model. As an extension, the I-V curves are also presented across different temperatures in order to test the accuracy of the single-diode model

Keywords—modelling, photovoltaic, solar cell, circuit level, TCAD, Matlab, I-V curve, single-diode

I. INTRODUCTION

Solar energy is a popular and emerging renewable energy source due to the theoretically infinite resource from the sun. Besides, there are no moving parts involved in the conversion of solar energy to electrical energy which makes it almost maintenance free. Solar energy is associated with the term photovoltaic (PV) which covers the conversion of solar energy into electrical energy using semiconductor materials known as the solar cell. Typically, an isolated, stand-alone, or off-grid PV system consists of three main components such as the solar module, the charge controller module, and the battery system. The solar module is a group of solar cells connected together to achieve a higher solar energy output. The charge controller module is where Maximum Power Point Tracking (MPPT) algorithm is implemented, and the battery system is to store the converted solar energy for later usage.

It can be observed that there are several areas available for research, such as the solar cell, the MPPT algorithm, and the power management of the battery system. The solar cell is the first component of the PV system where the sun's

irradiance is taken as the input to be converted into electrical energy. The maximum output power is directly proportional to the efficiency of the solar cell. Thus, there are abundance of researches done to develop solar cells that are capable of converting solar energy to electrical energy at a higher efficiency. This includes developing solar cells from different kinds of materials and processes [1]. Besides the solar cell, there are also significant researches done on the MPPT algorithm to be implemented in the PV system. The objective is to ensure the PV system operates at the Maximum Power Point (MPP) of the solar cell by continuously tracking the MPP. There are continuous researches done to improve the conventional algorithm such as the Perturb and Observe (P&O) algorithm [1]. Besides, there are also new algorithms with higher complexity being introduced such as the fuzzy logic approach and the grey wolf algorithm [3], [4]. The battery system is another section in the PV system where till date, there are researches done to implement new or improve currently available power management systems [5], [6].

A modelling approach for PV solar cell which is a combination of Silvaco TCAD and Matlab software is proposed. The modelling approach begins with defining the baseline parameters of the solar cell. The resulting electrical characteristics and the non-linear I-V curve of the solar cell will be used to develop an equivalent electrical model of the solar cell. The validity of the model is tested at different temperatures such as 280K, 300K, and 320K.

The electrical model is used to provide an initial insight on the performance of the solar cell when implemented in a PV system. This includes the impact towards the PV system when changes are made to the baseline parameters of the solar cells. This is significant in the PV field as the performance of a solar cell can be observed as a single cell or as a whole PV system.

II. DESCRIPTION OF THE MODELLING APPROACH

Solar cell device development and solar PV system are usually developed separately. During the solar cell device development, the baseline parameters of the solar cell are manipulated to improve the efficiency and electrical characteristics of the solar cell. For example, to improve the efficiency of a thin film CIGS solar cell, the thickness,

978-1-5386-5284-8/18 $31.00 © 2018 IEEE

doping concentration, electron affinity, and band gap energy are manipulated [7]. The results of the changes are discussed and concluded at the solar cell level.

For the latter, solar PV system includes development of the MPPT algorithms and the power management system. MPPT algorithms are usually developed and tested with a solar cell or solar module with commercially available data [2]. This is also similar on the development of PV power management systems where commercially available data of the solar cell or solar module are implemented [5].

In the proposed modeling approach, a new step is introduced to link between the solar cell device development and the solar PV system. It is the implementation of the solar cell as a part of a complete PV system. At this point, the solar cell can be implemented in the development of the MPPT algorithms as well as the power management system. With this approach, the flexibility of the PV simulation flow will improve in terms of the capability to manipulate the baseline parameters of the subjected solar cell.

III. SIMULATION OF SOLAR CELL USING SILVACO TCAD SOFTWARE

For the modelling approach, a CIGS solar cell is designed using Silvaco TCAD based on a predefined baseline parameters. These parameters such as shown in Table 1 are chosen to model an optimal thin-film CIGS solar cell based on results of prior researchers [8], [9], and [10].

TABLE 1. PREDEFINED BASELINE PARAMETERS FOR CIGS SOLAR CELL

Parameters	Layers		
	ZnO	CdS	CIGS
Thickness (nm)	150	50	3000
Band gap, E_g (eV)	3.3	2.4	1.27
Donor concentration, N_D (cm^{-3})	1x10^{18}	1x10^{17}	0
Acceptor concentration, N_A (cm^{-3})	0	0	2x10^{16}
Conduction band effective density of states, N_C (cm^{-3})	2.2x10^{18}	2.2x10^{18}	2.2x10^{18}
Valence band effective density of states, N_V (cm^{-3})	1.8x10^{19}	1.8x10^{19}	1.8x10^{19}

Fig. 1 shows the designed TCAD solar cell model with the baseline parameters described in Table 1. The structure of the TCAD model is composed of Molybdenum (Mo) back contact, a p-type wide-band gap absorber layer (CIGS), followed by n-type buffer layer made of cadmium sulphide (CdS) and a window layer made of doped zinc oxide (ZnO).

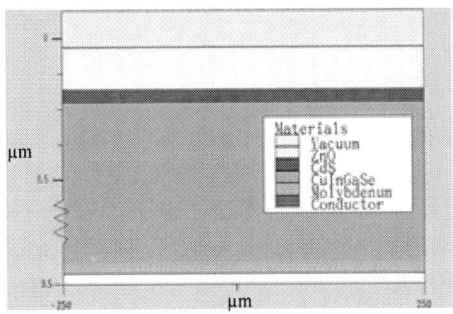

Fig. 1. TCAD model structure designed with Silvaco TCAD

The I-V curve of the designed solar cell at standard test condition (STC) usually at 300 K and 1000 W/m^2 and air mass 1.5 (AM 1.5) is depicted in Fig. 2. From the I-V curve, short-circuit current (I_{sc}), open-circuit voltage (V_{oc}), current at MPP (I_{mp}), and voltage at MPP (V_{mp}) is 40.3 mA, 0.752 V, 36.3 mA, and 0.649 V respectively.

These parameters will be used to develop an equivalent electrical model of the TCAD model in the next step of this modelling approach using Matlab software. Besides the I-V curve, a logfile of the TCAD model is available as the output of Silvaco TCAD. The logfile containing electrical parameters of the TCAD model can be exported into comma separated values (csv) file by utilizing TonyPlot (Silvaco's interactive visualization tool) for post-processing with other software, which is described in the next section of this paper.

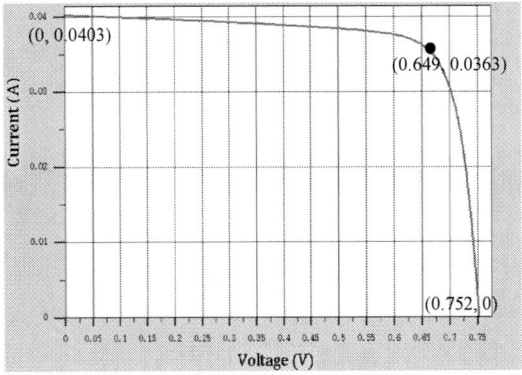

Fig. 2. I-V curve of the TCAD model designed with Silvaco TCAD

IV. MODELLING OF THE SOLAR CELL USING MATLAB SOFTWARE

The main purpose of using Matlab software is to construct an equivalent electrical model of the TCAD model, which is to reproduce the non-linear I-V curve. In a standard non-linear I-V curve, three marked points are highlighted such as the short-circuit (0, I_{sc}), MPP (V_{mp}, I_{mp}), and open-circuit (V_{oc}, 0) as shown in Fig. 3.

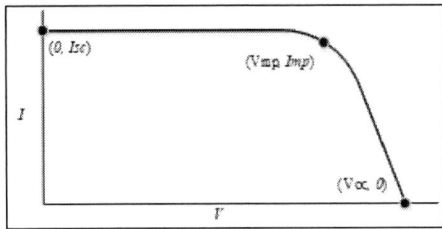

Fig. 3. Standard non-linear I-V curve with the three marked points

Previous PV system studies have utilized different circuit topologies to represent solar cells such as the single-diode model, two-diode model, and three-diode model [11]-[13]. In this work, the single-diode model circuit topology is chosen because the model offers good compromise between simplicity and accuracy [11]. Fig. 4 shows the electrical circuit of a practical single-diode solar cell with the equivalent series and parallel resistance.

978-1-5386-5284-8/18 $31.00 © 2018 IEEE

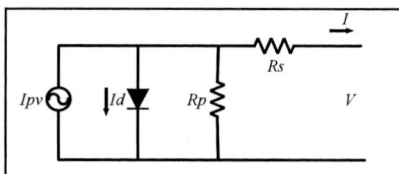

Fig. 4. Electrical circuit representation of a practical single-diode solar cell

The I-V characteristics of a practical single-diode solar cell is mathematically described as below [14].

$$I = I_{pv} - I_o \left[e^{\left(\frac{V + R_s I}{V_t \alpha} \right)} - 1 \right] - \frac{V + R_s I}{R_p} \quad (1)$$

where: I_{pv} = PV current
I_o = Diode reverse saturation current
V_t = Thermal voltage
α = Diode ideality constant
R_s = Equivalent series resistance
R_p = Equivalent parallel resistance

From equation (1), the thermal voltage, $V_t = kT/q$ where k is the Boltzmann constant ($1.3806503 \times 10^{-23}$ J/K), T is the temperature in Kelvin, and q is the electron charge ($1.60217646 \times 10^{-19}$ C). The PV current, I_{pv} depends on the solar irradiation and the temperature of the solar cell where it is described as shown below [15].

$$I_{pv} = \left(I_{pv,n} + K_I \Delta_T \right) \frac{G}{G_n} \quad (2)$$

where: $I_{pv,n}$ = PV current at STC
K_I = Short-circuit current temperature coefficient
Δ_T = Difference between actual and nominal temperature
G = Irradiation
G_n = Irradiation at STC

I_o on the other hand, is further elaborated as in equation (3) [15].

$$I_o = \frac{I_{sc,n} + K_I \Delta_T}{e^{([V_{oc,n} + K_V \Delta_T]/\alpha V_t)} - 1} \quad (3)$$

where: $I_{sc,n}$ = Short-circuit current at STC
$V_{oc,n}$ = Open-circuit voltage at STC
K_V = Open-circuit voltage temperature coefficient

Equations (1) – (3) are used to develop the single-diode model. The key parameters of the TCAD model are implemented such as $I_{sc,n}$, $V_{oc,n}$, I_{mp}, and V_{mp}. The temperature coefficients, K_I and K_V are taken from the average of commercially available CIGS solar cells datasheet [16]-[20]. The values of the equivalent resistances R_s and R_p are obtained through Newton-Raphson iteration method [11]. Fig. 5 shows the reconstructed I-V curve of the single-diode model based on the I-V curve of the TCAD model at STC.

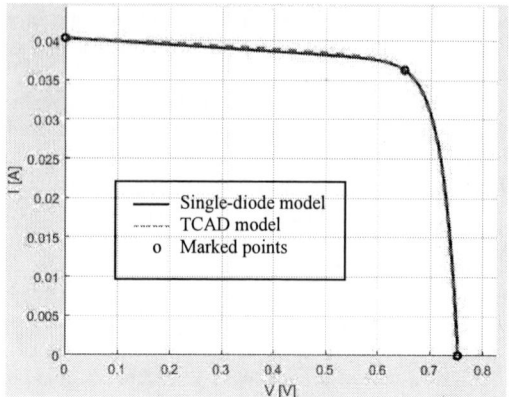

Fig. 5. I-V curve of single-diode model and I-V curve of TCAD model

From Fig. 5, it can be observed that the single-diode model can reconstruct the I-V curve of the TCAD model with great accuracy. This observation validates the I-V curve of the single-diode model to that of the TCAD model. The curves are exactly matched at the three marked points denoted by the dots since these points are used as the basis in developing the single-diode model. Slight error gaps are observed at other points of the I-V curve. This is the limitation of the single-diode model, although the error gaps can be reduced by increasing the number of iterations in finding the values of R_s and R_p.

V. VALIDATING THE SINGLE-DIODE MODEL AT DIFFERENT TEMPERATURES

To further validate the I-V curve of the single-diode model, a set of I-V curves of the single-diode model and TCAD model are plotted at different temperatures. This is to ensure the single-diode model is able to accurately represent the TCAD model for further usage in the PV system. Fig. 6 shows the I-V curves of the single-diode model and the I-V curves of the TCAD model at three different temperatures such as 280K, 300K, and 320K.

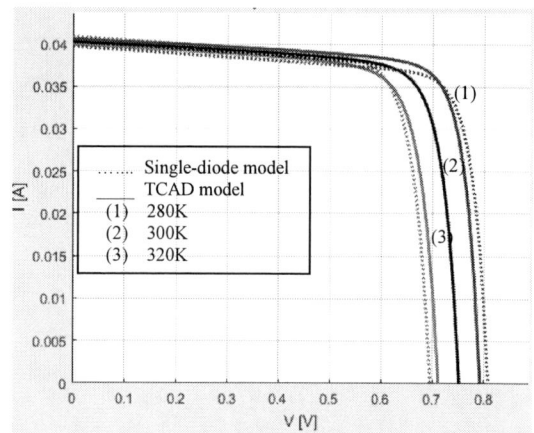

Fig. 6. I-V curves of the single-diode model and TCAD model at different temperatures

As shown in Fig. 6, the I-V curves of the single-diode model and TCAD model are denoted by the dotted line and solid line respectively. The different temperatures are denoted by the colors and numbers where blue (1) is for

280K, black (2) is for 300K, and red (3) is for 320K. The absolute error between the I-V curves are as described in Table 2.

TABLE 2. ABSOLUTE ERROR BETWEEN SINGLE-DIODE MODEL AND TCAD MODEL AT DIFFERENT TEMPERATURES

Temperature	280K		320K	
	V_{oc}	I_{sc}	V_{oc}	I_{sc}
Single-Diode Model (a)	0.807V	38 mA	0.696 V	41 mA
TCAD Model (b)	0.790V	41 mA	0.710 V	40 mA
Absolute Error gap ($\|[a\text{-}b]/b \times 100\|$)	2.15%	7.32%	1.97 %	2.50 %

The I-V curves are almost accurate at temperature 300K. However, when the temperature is increased or decreased to 320K and 280K respectively, the absolute error between the I-V curves increases. From Table 2, it can be observed that the single-diode model is able to maintain at most an absolute error of 7.32% which is at 280K. This result indicates that the single-diode model is sufficiently accurate to the TCAD model at temperature ranging from 280K to 320K. This is significant in implying the validity of the single-diode model to replace the TCAD model for circuit level simulations.

VI. CONCLUSION & DISCUSSION

In this paper, a modelling approach has been proposed to be implemented in PV systems. The approach is an extension of the conventional modelling approach for solar cells by introducing another step which enables circuit implementation of the solar cell. This is significant in the PV field since the approach is able to provide initial or overall insight on how the baseline parameters of a solar cell affects the performance of the PV system.

The modelling approach is a combination of Silvaco TCAD and Matlab software where Silvaco TCAD is used to develop a TCAD model of a thin-film CIGS solar cell from a predefined baseline parameters and Matlab software is used to post-process the output file. Matlab software is also used to develop an equivalent single-diode model to represent the TCAD model. The I-V curve of the single-diode model is validated against the I-V curve of the TCAD model at three different temperatures such as 280K, 300K, and 320K. At most, the absolute error between the curves is at 7.32%.

This modelling approach is not only limited to model and simulate thin-film solar cells. As long as there are information on the electrical characteristics and the I-V curve of a solar cell, this modelling approach can be utilized to develop an equivalent electrical model for circuit level simulations.

REFERENCES

[1] M. A. Green, Y. Hishikawa, E. D. Dunlop, D. H. Levi, J. Hohl-Ebinger and A. W. Y. Ho-Baillie, "Solar Cell Efficiency Tables (Version 51)," *Prog Photovolt Res Appl*, no. 26, pp. 3-12, 2018.

[2] J. Ahmed and Z. Salam, "An Enhanced Adaptive P&O MPPT for Fast and Efficient Tracking under arying Environmental Conditions," *IEEE Transactions on Sustainable Energy*, 2018.

[3] U. Yilmaz, A. Kircay and S. Borekci, "PV system fuzzy login MPPT method and PI control as a charge controller," *Renewable and Sustainable Energy Reviews*, vol. 81, no. 1, pp. 994-1001, 2018.

[4] S. Mohanty, B. Subudhi and P. K. Ray, "A New MPPT Design using Grey Wolf Optimization Technique for Photovoltaic System under Partial Shading Conditions," *IEEE Transactions on Sustainable Energy*, vol. 7, no. 1, pp. 181-188, 2016.

[5] Y. E. Abu Eldahab, N. H. Saad and A. Zekry, "Enhancing the Design of Battery Charging Controllers for Photovoltaic Systems," *Renewable and Sustainable Energy Reviews*, vol. 58, pp. 646-655, 2016.

[6] A. Mirzaei, M. Forooghi, A. A. Ghadimi, A. H. Abolmasoumi and M. R. Riahi, "Design and Construction of a Charge Controler for Stand-Alone PV/Battery Hybrid System by using a New Control Strategy and Power Management," *Solar Energy*, vol. 149, pp. 132-144, 2017.

[7] J. Ramanujam and U. P. Singh, "Copper Indium Gallium Selenide based Solar Cells - A Review," *Energy & Environmental Science*, vol. 10, no. 6, pp. 1306-1319, 2017.

[8] H. Heriche, Z. Rouabah and N. Bouarissa, "High-Efficiency CIGS Solar Cells with Optimization of Layers Thickness and Doping," *Optik - International Journal for Light and Electron Optics*, vol. 127, no. 24, pp. 11751-11757, 2016.

[9] M. Mostefaoui, H. Mazari, S. Khelifi, A. Bouraiou and R. Dabou, "Simulation of High Efficiency CIGS Solar Cells with SCAPS-1D Software," *Energy Procedia*, vol. 74, pp. 736-744, 2015.

[10] N. Khoshsirat, N. A. Md Yunus, M. N. Hamidon, S. Shafie and N. Amin, "Analysis of Absorber Layer Properties Effect on CIGS Solar Performance using SCAPS," *Optik - International Journal for Light and Electron Optics*, vol. 126, no. 7-8, pp. 681-686, 2015.

[11] M. G. Villalva, J. R. Gazoli and E. R. Filho, "Comprehensive Approach to Modeling and Simulation of Photovoltaic Arrays," *IEEE Transactions on Power Electronics*, vol. 24, no. 5, pp. 1198-1208, 2009.

[12] K. Ishaque, Z. Salam and H. Taheri, "Simple, Fast, and Accurate Two-Diode Model for Photovoltaic Modules," *Solar Energy Materials and Solar Cells*, vol. 95, no. 2, pp. 586-594, 2011.

[13] K. Nishioka, N. Sakitani, Y. Uraoka and T. Fuyuki, "Analsis of Multicrystalline Silicon Solar Cells by Modified 3-Diode Equivalent Circuit Model Taking Leakage Current through Periphery into consideration," *Solar Energy Materials and Solar Cells*, vol. 91, no. 13, pp. 1222-1227, 2007.

[14] H. S. Rauschenbach, Solar Cell Array Design Handbook: The Principles and Technology of Photovoltaic Energy Conversion, New York: Van Nostrand Reinhold Company, 1980.

[15] D. Sera, R. Teodorescu and P. Rodriguez, "PV Panel Model based on Datasheet Values," *IEEE International Symposium on Industrial Electronics (ISIE)*, pp. 2392-2396, 2007.

[16] "CIGS Thin-Film Solar Modules," STION, 2012. [Online]. Available: http://www.solardesigntool.com/components/module-panel-solar/Stion/2376/STN-150/specification-data-sheet.html.

[17] "TS CIGS Series High-Efficiency CIGS Solar Module," TSMC Solar, [Online]. Available: www.tsmc-solar.com/Assets/.../TS_CIGS_Series_C2_Datasheet_EU-EN_01-2015.pdf.

[18] "CIGS Solar Cell," MiaSole, 2015. [Online]. Available: miasole.com/wp-content/uploads/2014/09/SolarCell_Datasheet_5.pdf.

[19] "High Performance CIGS Thin-Film Solar Modules," STION, 2013. [Online]. Available: http://www.solargy.com.sg/downloads/STO%20135%20to%20150W.pdf. [Accessed 7 March 2018].

[20] "SoloPanel Model SP1," SOLOPOWER, 2013. [Online]. Available: solopower.com/wp-content/uploads/.../solopower_solopanel_sp1_product_specs.pdf. [Accessed 7 March 2018].

978-1-5386-5284-8/18 $31.00 © 2018 IEEE

Design and broadband verification of RF amplifier for 4G/ 5G front-end modules

Basuki Rachmatul Alam

Electronics Research Group, School of Electrical Engineering and Informatics,
Institute of Technology Bandung
Bandung 40132, Indonesia
E-mail : br_alam@yahoo.com; basuki.rachmatul@stei.itb.ac.id

Abstract— The RF amplifier for broadband front end module of 4G or 5G terminals should be capable of handling transmission rate over 100Mbps to over 1Gbps with low BER and low non-linearity spectrum products. This paper reports broadband verification of a front-end RF amplifier designed for broadband 4G and 5G transmission. The RF amplifier employs an Infineon SiGe Bipolar transistor matched to 50Ω load and source impedance through 3-section and 2-section L-C network, respectively. Broadband S-parameter characteristic was verified showing encircling S_{21} behavior around $|s_{21}| \cong 7$ over hundreds of MHz centering at 2.4GHz. The S_{11} and S_{22} show encircling curves centering the origin with relatively small radius. The broadband performance of LTE FDD 64QAM signal transmission thru the RF Amplifier was verified on the signal spectrum and modulation quality by utilizing Keysight M9381A Vector Signal Generator (VSG), M9391A Vector Signal Analyzer (VSA) and VSA 89600. Broadband signal of LTE FDD 64QAM was generated by Keysight M9381A Vector Signal Generator (VSG) by setting 64QAM modulation and LTE signaling parameters resulting broadband LTE FDD signal with bandwidth of 20MHz at frequency of 1.8GHz and 2.4GHz. The LTE FDD 64QAM signal spectrum and modulation constellation measured at the output of front-end RF Amplifier by Keysight M9391A VSA and VSA 89600 subsequently. Broadband modulation constellation, spectrum as well as EVM summary of the LTE FDD signal were observed by VSA 89600 at frequency 1.8 and 2.4GHz. Out-of band noise spectrum and non-linearity products of the 20MHz LTE FDD 64QAM signal increase at the output of front-end RF amplifier at frequency 1.8GHz, though less visible at 2.4GHz. The 64QAM vector-point constellation of the LTE FDD 64QAM signal in 2.4GHz is relatively still intact, eventhough the EVM slightly degrades to 11.478 % from 10.772% after amplified by front-end RF amplifier.

Keywords—broadband, front-end RF amplifier, LTE FDD, 64QAM, modulation constellation, EVM

I. INTRODUCTION

Broadband RF telecommunication has been evolving lately into the present state of fully migration to 4G LTE technology, while the development of higher density and faster 5G technology is underway and heading toward few pilot units of provider in field trial. Several 5G systems from well known telecom providers such as Qualcomm and Sony-Ericsson have just been tested. 5G technology promises higher density and data throughput up to tenths of Gbps transmission. Higher efficiency and density of 5G

system require high precision of digital vector modulation and high multichannel data transmission of OFDM dan OFDMA with tight spacing between signal channel to neighboring channels. Tight spacing of multichannel of OFDMA of 78.125 KHz spacing 12.8 μs OFDM Symbol 0.8/ 1.6 /3.2 μs Cyclic of the new WLAN IEEE 802.11ax standard requires high linearity of RF front-end amplifier of 5G access terminals. While the LTE Release 8 has 180kHz Resource Block (RB) with 12 subcarriers in each RB where each subcarrier has spacing of 15kHz to next subcarrier. This tight spacing needs high spectrum fidelity of signal delivered from linear front end RF amplifier, besides low phase noise of the carrier oscillator. Large order of vector modulation of 5G transmission up to 4096-QAM necessitates meticulous baseband signal processing and signal transmission handling. Very low EVM (error vector magnitude) besides low floor noise and low phase noise must be mintained to preserve the magnitude and phase of high order modulation of the 5G broadband signal. Near-40dB or less than 0.01% EVM must be attained for 4096-QAM modulation of 5G transmission. For a semi-5G transmission of new WLAN 802.11ax EVM must maintained to less than -35dB, this means less than 0.03% of EVM. This high quality of digital modulation performance might be delivered from a SDR-FPGA module or a FPGA development board to a fast DAC or IIR Filter and converted to high microwave frequency by Balanced I-Q Up Converter before being amplified by front-end RF amplifier. Therefore, front-end RF amplifier must be designed according to broadband and linear amplifier design approaches to suppress out-of band noise and spurious as well as non-linearity spectrum products [4,5,6,7,8]. Multitone characterization can be applied to observe non-linear behavior of RF front-end amplifier : RF amplifier for transmitting and LNA (low noise amplifier) for receiving the broadband signal. Up-conversion and down conversion I-Q balanced mixer must be employed to achieve high image and IMD (intermodulation distortion) rejection [9,10]. Among those non-linear products, the 3rd order intermodulation (IM3) is the closest to fundamental signal carrier compared to 2nd order intermodulation (IM2) and higher order IMD, therefore is more detrimental than other non-linear products to broadband 4G and 5G multichannel and high density transmission. To achieve high linearity broadband transmission, the 3rd order intermodulation (IM3) must be suppressed by implementing several approaches. Broadband impedance matching and load pulling methods are the most practical approaches compared to implementation of linearization circuit topologies such as predistortion, feedback or feedforward topology. Load-pull method may yield optimum load to deliver the optimum output and the maximum power added efficiency. Since the optimum load obtained from load-pull process is the load at most linear position within the I-V characteristics, the non-linear products can be expected at the most supressed level.

This paper will confer broadband design approach and verification of an RF amplifier for front end module for 1.8GHz and 2.4GHz broadband transmission utilizing real 4G LTE FDD 64-QAM signal. The initial design work has been done in RF Microelectronics Laboratory of Electronics group at School of Electrical Engineering and Informatics of Institute of Technology Bandung, while the implementation as well as measurement has been done in RF and Smartphone test facility of Teaching Factory of Manufacturing Electronics (TFME) at State Polytechnics of Batam campus..

II. DESIGN AND IMPLEMENTATION

Broadband impedance matching has been implemented to design of RF front-end amplifier to match the optimum load impedance acquired from the load pulling simulation step to the RF transistor being used for the amplifier. The RF transistor for the RF front end amplified was an Infineon BJT BFP780, biased as class A amplifier [13]. The optimum load impedance was obtained, $Z_{Lopt} = (13.13 - j3.03)\Omega$. Impedance matching process was initiated by originating from Z_{Lopt} by tracing constant resistance until intersecting Q factor-curve on Smith Chart and continued on constant conductance arc several times with last constant conductance arc equal to G=1 and reaching the origin, $Y_O = 1$. This output impedance matching results in Γ_S should be matched to S_{11} of the BFP780. Since the Z_{in} set at $Zo = 50\Omega$, therefore the broadband impedance matching starts from $Z_1(S_{11})$ through several (three) L-C sections to origin, in similar fashion to the preceding output impedance matching approach. Figure 1, below shows the RF front-end amplifier module.

Figure 1. RF front-end module with broadband impedance matching at the output and input network.

III. MEASUREMENT AND ANALYSIS

Broadband signal transmission from the output of front-end RF amplifier was measured and characterized on its performance quality such as modulation constellation and EVM (Error Vector Magnitude) as the most crucial parameters. The S-parameter of RF front-end amplifier module was measured using Keysight PXIe VNA 9371M. While, the broadband signal performance was verified using Keysight PXie (Vector Signal Analyzer) M9391 VSA, M9381A VSG and VSA89600 Signal Studio.

A. Broadband Measurement

The broadband signal has been measured and characterized using Keysight PXie VSG (Vector Signal Generator) M9381A, VSA M9391A and VSA 89600. The VSG M9381A was set to generate LTE FDD 64-QAM signal with bandwidth of 20MHz at 1.8GHz and 2.4GHz. The VSG M9381A and VSA 89600 in PXie mainframe are shown in figure 2 below.

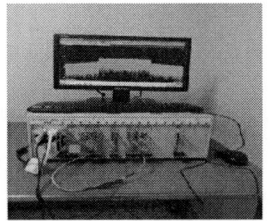

Figure 2 : LTE FDD 64QAM signal at the output of front-end RF amplifier measured by Keysight PXie M9381A VSG and M9391A VSA

The LTE FDD 64QAM signal measurement at 1.8GHz in figure 2 above shows some degree on non-linearity where the noise magnitude outside the 20MHz band of original LTE signal has also been increasing as well. The broadband LTE performance of front-end RF amplifier was verified by observing LTE FDD 64QAM signal from the M9381A generator at the output of front-end RF amplifier. The broadband performance of LTE FDD 64QAM signal at the input and output of the front end RF amplifier at 2.4GHz are shown in figure 3(a) and 3(b). The broadband parameters of the LTE FDD QAM signal encompass of spectrum bandwidth, modulation constellation of 64QAM in 64-vector points in 4 quadrant and EVM (error-vector- magnitude) summary as shown in figure 3(a) and 3(b) at the input and output of front-end RF amplifier, respectively. The spectrum of LTE FDD 64QAM output signal in figure 3(b) of front-end RF amplifier compared to input signal in figure 3(a) show 8dB gain within band of 20MHz, however the noise of out-of-band spectrum increase by around 8dB below 2.4GHz – 12MHz and 5-8dB above 2.4GHz + 12MHz. The modulation constellation of LTE FDD 64QAM signal at the output (trace A) in figure 3(b) exhibits more blurring vector points compared to constellation at the input of front-end RF amplifier trace A in figure 3(a). The quality degradation of vector modulation from graphical point of view is in line with the error summary (trace C upper right window) of figure 3(a) and 3(b) of the LTE FDD signal at the input and output of front-end RF amplifier. The EVM percentage levels of the LTE signal at the input and output of front-end RF amplifier are of

978-1-5386-5284-8/18 $31.00 © 2018 IEEE

10.772% and 11.478% as shown in trace C of figure 3(a) and 3(b), respectively.

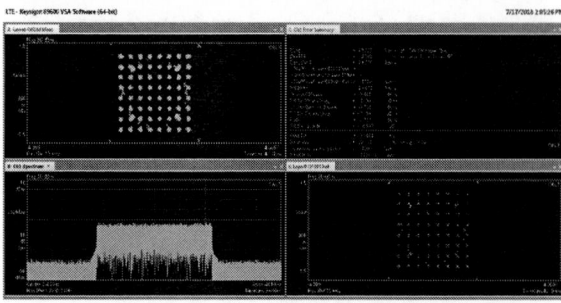

(at input of front-end RF amplifier)

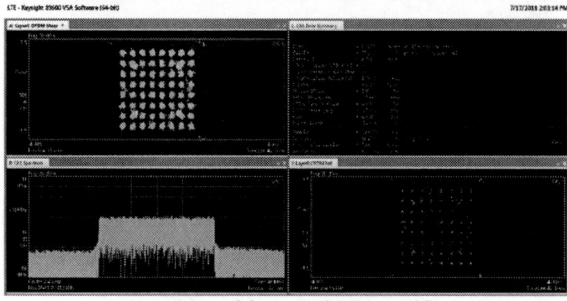

(at output of front-end RF amplifier)

Figure 3: Spectrum, modulation constellation and EVM in VSA 89600 window tiles of LTE FDD 64 QAM signal at the input (a) and output (b) of front-end RF amplifier.

B. Analysis on broadband measurement of LTE FDD 64QAM of front-end RF amplifier.

From broadband measurement of front-end RF amplifier employing LTE-FDD 64QAM signal at 1.8GHz and 2.4GHz, the signal spectrum show in band gain of around 8dB. However, out-of-band spectrum noise is also boosted to more than 7dB with maximum of 10dB,. Broadband performance of LTE FDD signal at the output of front-end RF amplifier is also degrading as shown graphically by broadening of all 64 vector points in 4 quadrants of 64QAM modulation. Summary of EVM in trace C of the output of front-end RF amplifier in figure 3(b) compared to EVM summary at the input in figure 3(a) shows degradation from 10.772 % to 11.478% of EVM and 2.677% to 4.795% of data EVM, in accordance to more blurring of vector points of 64QAM constellation at the output front-end RF amplifier. This evidence might support the broadband impedance matching is very crucial to sustain the broadband performance of front-end RF amplifier for 4G and 5G transmission.

IV. CONCLUSION AND REMARKS

Broadband performance verification of a front-end RF amplifier has been accomplished through measurement employing real LTE FDD 64QAM signal. Degradation of broadband performance quality have been observed at the output of front-end RF amplifier design as shown by degradation of modulation constellation of 64QAM vector points graphically and numerically, where the latter is evident as EVM increasing by one to several points. These findings verify the importance of broadband design approach of front-end RF amplifier thru broadband impedance matching implementation. Partial or incomplete broadband design execution has resulted in some degradation broadband signal performance as evidence found from the broadband measurement.

ACKNOWLEDGMENT (Heading 5)

Author would like to express appreciation to some of past students who has worked under his supervision for RF amplifier design and non-linearity effects either for his degree in undergraduate or in post-graduate, to share data portion of his co-publications, his final project or his thesis. Author would like to express sincere gratitude to all technical staff of TFME for their dedicated work in PCB fabrication and RF front-end amplifier assemblies.

REFERENCES

[1]. Agilent Technology: "3GPP Long Term Evolution: *System Overview, Product Development, and Test Challenges*": Application Note, 2009.

[2]. A. Khusaidi, B.R. Alam, "Characterization and Modeling of a GaAs HFET toward design of a Linear High Power Amplifier", APMC 2000, Sydney, Australia, December, 2000.

[3]. Saripudin, A, M.T. Hutabarat M.T, B.R. Alam, "Characterization of Non-linear Behavior of GaAs HFET Power Amplifier IC Based on Multitone Simulation", IEEE-APCCAS, Singapore, December 2002.

[4]. A.Saripudin, M.T. Hutabarat, B.R. Alam, "Nonlinear Characterization of RF LDMOS Power Amplifier Under Multitone Excitation" Asia Pacific Microwave Conference (APMC) 2003, Seoul, Korea, November 2003.

[5]. Kenington, Peter. B, *"High-Linearity RF Amplifier Design*. Artech House. Inc, 2000

[6]. A.Saripudin, I.B. Armanto, M.T. Hutabarat , B.R. Alam, *"The Effect of Bias and Load Impedance to Non-linearity Characteristics of RF LD-MOSFET"*, Proceedings of 2011 International Conference on Electrical Engineering and Informatics (ICEEI), July 2011, Bandung, Indonesia.

[7]. Xu,S.; Foo, P.; Wen, J.; Liu, Y.; Lin, F. & Ren, C,"RF LDMOS with extreme low parasitic feedback capacitance and highhot-carrier immunity",*Electron Devices Meeting*, 1999 IEDM Technical Digest. International (1999), 201-204).

[8]. Basuki R. Alam, *"Broadband Performance of multi-sections of step-line impedance matching of LD-MOSFET RF Amplifier at 900MHZ LTE Band"*, Proceedings of ICEEI 2015, Bali, 10-11 August 2015.

[9]. Ahmad Sidik, Maulana Yusuf Fathany, and Basuki Rahmatul Alam, *"Design of Broadband Low Noise Amplifier (LNA) 4G LTE TDD 2.3 GHz for Modem Application"*, 2015 International Symposium on Intelligent Signal Processing and Communication Systems (ISPACS) November 9-12, 2015.

[10]. Felix Gunawan, Basuki Rachmatul Alam, "Design and Modulation Analysis of cascade LNA for L-Band low magnitude signal", TSSA Proceedings, 2016 10th, October 6-7, 2016, Denpasar, Bali, Indonesia.

[11]. Aris Agung Pribadi, Vita Awalia Mardiana, Basuki Rachmatul Alam, *"Design Power Amplifier Using Load Pull Method in WLAN 802.11 ax Access Point Application"*, ISESD Proceedings, IEEE International Symposium on Electronics and Smart Devices, Jogjakarta October 17 – 19, 2017.

[12] Basuki R Alam, *"Non-linearity consideration in the design and implementation of RF Front-end medium power amplifier for broadband 4G and 5G telecommunication"*, Keynote Speech. ISESD Proceedings, IEEE International Symposium on Electronics and Smart Devices, Jogjakarta October 17 – 19, 2017.

978-1-5386-5284-8/18 $31.00 © 2018 IEEE

A Compact Bidirectional Pseudo Floating Gate Front-End for Resonating Sensors based on NAND and NOR logic Gates.

Luca Marchetti ,Yngvar Berg, Mehdi Azadmehr
Department of Microsystems (IMS)
University College of Southeast Norway
Email: Luca.Marchetti@usn.no

Abstract—In this paper we present an alternative method to realize a compact bidirectional Front-End for resonating sensors based on Ring Down Method (RDM) and pseudo floating gate amplifier (PFGA). This goal is achieved by using the voltage buffer inside the PFGA for both biasing of the amplifier and actuation of the sensor. The proposed analog front-end doesn't require any coupling capacitor, minimizing the area occupied by the circuit. Simulations in AMS-350nm CMOS technology with a power supply of 3.3V and measurements on a prototype implemented with discrete components have been used to verify the correct operation of the system. The Front-end was realized by using the integrated circuits (ICs) CD4007UBE and ALD1103, and tested with a real piezoelectric transducer (MURATA MA40S4R), characterized by a resonant frequency of 40kHz. Measurement and simulation results show that the proposed circuit is a good candidate to realize a compact and bidirectional pseudo floating gate front-end, which resembles the circuit of NAND and NOR logic gates.

Keywords—**Bidirectional, Front-End, Resonating Sensor, Pseudo Floating Gate Amplifier.**

I. INTRODUCTION

The pseudo floating gate amplifier (PFGA,Fig.1(a)) has been introduced in [1], in order to provide a simple solution to realize ultra low voltage (ULV) circuits. The main advantages of using this type of amplifier are low power consumption and compactness of the circuit. In the last 20 years, PFGAs have been improved and used to realize numerous analog and digital circuits, showing the high flexibility of this amplifier. Recently, techniques to realize resonating sensor front-ends based on bidirectional PFGA and Ring Down Method (RDM) [2] have been investigated [3]. The schematic of this front-end is shown in Fig.1(b).

The bidirectional behaviour of this circuit is achieved by swapping the power supply of the PFGA. The main drawbacks of this system are: first, the large decoupling capacitor values, which are difficult to implement in CMOS technology; second, the unwanted actuation of the sensor due to the swapping of the power supply, which should not affect the dynamic of the system but only change the directionality of the circuit and third, the need of extra external circuitry to generate the correct timing for the signals V_{IN} and Φ. In order to overcome these problems and provide a more

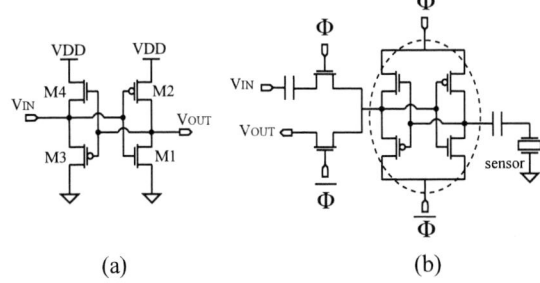

Fig. 1. Pseudo Floating Gate Amplifier (PFGA) and Bidirectional PFGA. (a) PFGA Schematic. (b) Schematic of the bidirectional front-end proposed in [3].

compact solution, we investigated an alternative way to use PFGA to realize bidirectional front-end for resonating sensors. The paper is organized as follows: Section II describes the working principle of the proposed circuit. Section III shows the simulation results of the proposed front-end implemented in AMS-350nm CMOS technology. Section IV describes the measurement setup utilized to test the fabricated prototype. Section V shows the measurement results and finally, Section VI provides the conclusions of this work.

II. NAND/NOR PFGA WORKING PRINCIPLE

The schematic of a PFGA is shown in Fig.1(a) and the working principle is well described in [1]. The PFGA was born as a unidirectional amplifier. However, by applying a wide amplitude signal at the output terminal, we can turn on M3 and M4, affecting the value of the PFGA input voltage. This situation can be achieved by satisfying one of these two conditions: $V_{OUT} > V_{DD}/2 + V_{thn}$ or $V_{OUT} < V_{DD}/2 - |V_{thp}|$ and will allow to realize a bidirectional device based on PFGA. Where: VDD is the power supply of the amplifier and V_{thn} and V_{thp} are the threshold voltages of the NMOS and PMOS respectively. A simple method to implement this PFGA consists in using a power supply so that: $|V_{thp}| < V_{DD}/2 < V_{DD} - V_{thn}$ and forcing the amplitude of the output signal to assume the value of one of the power supply rails.

978-1-5386-5284-8/18 $31.00 © 2018 IEEE

Unfortunately, by using this method, the power consumption of the PFGA increases significantly, and in extreme cases, can permanently damage the transistors in the circuit. For instance, when V_{OUT} is forced to VDD, then M4 turns on and the input voltage of the PFGA increases from $V_{DD/2}$ until $V_{IN} = V_{DD} - V_{thn}$ and M4 turns off again. At this point, M1 is subjected to high power consumption because its drain source voltage is equal to the power supply of the circuit, and the current absorbed by M1 is greater than its original value when V_{IN} was equal to $V_{DD}/2$. Similar phenomenon occurs when the output is forced to 0V, but in this case, M2 is the source of major power consumption of the amplifier. In conclusion, it is necessary to modify the original circuit of the PFGA in order to realize a bidirectional device and to avoid high power consumption. In this work, we propose two simple solutions shown in Fig.2(a,b).

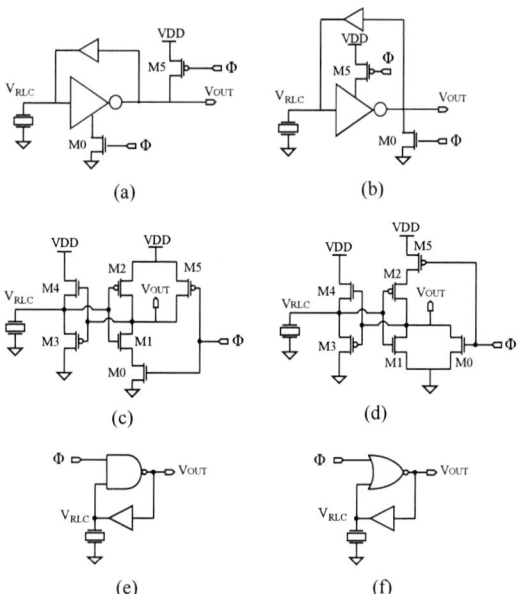

Fig. 2. NAND/NOR PFGA Front-End. (a) PFGA amplifier in NAND configuration. (b) PFGA amplifier in NOR configuration. (c) Transistor implementation of the NAND PFGA front-end (d) Transistor implementation of the NOR PFGA front-end. (e) NAND-PFGA logic view. (f) NOR-PFGA logic view.

Fig.2(c,d) represent the transistor implementation of the circuits in Fig.2(a,b) respectively. M0,M1,M2,M5 in Fig.2(c) resemble a NAND logic gate as shown in Fig.2(e), while the same transistors in Fig.2(d) resemble a NOR logic gate as shown in Fig.2(f). For this reason, we call the circuits in Fig.2(a,c,e) NAND-PFGA and the circuits in Fig.2(b,d,f) NOR-PFGA. The working principle of these two circuits are similar and in both cases the operation of the Front-End is divided in two phases: actuation mode ($\Phi_{NAND} = 0, \Phi_{NOR} = 1$) and read-out mode ($\Phi_{NAND} = 1, \Phi_{NOR} = 0$). During the actuation mode, the output voltage of the NAND-

PFGA is forced to VDD by M5, then the voltage buffer turns on, and the input voltage value is pulled up by M4. The input voltage variation actuates the sensor, which starts to vibrate at its own resonant frequency. At the same time, the inverter is cut-off because M0 turns off, which avoids high power dissipation on M1 and minimizes the power consumption of the whole amplifier. Next, the system passes in read-out mode, where M5 turns off, M0 turns on and the Front-End behaves like a normal PFGA. Finally, the electrical signal generated by the sensor can be read-out at the output terminal (V_{OUT}) of the PFGA. On the other hand, in NOR-PFGA front-end, the excitation of the sensor starts when the output terminal is forced to 0V by M0. This event turns on the PMOS M3, which pulls down the voltage across the sensor actuating it. In this case, M5 cuts off the inverter in the PFGA, avoiding high power consumption during the actuation mode. Once the sensor has been actuated, the system passes to read-out mode and the Front-End behaves like a normal PFGA. The key factor of this architecture is the voltage buffer design, which has to provide the biasing of the inverter during the read-out mode, and it has to provide the energy to actuate the sensor during the actuation mode. In order to satisfy the second requirement, the voltage buffer must be designed with large transistors dimensions, which also ensure to reduce the actuation time of the sensor. This design strategy will provide increasing static power consumption, if the system is realized in modern CMOS technology. However, it has no significant effects when this circuit is realized in low leakage CMOS technology. This is because the average current absorbed by the buffer is negligible, in respect to the average current absorbed by the inverter of the PFGA. In fact, during the actuation mode, the buffer absorbs high current for a very short time, while during the read-out mode, the leakage currents of the buffer are still in the order of a few nA, which are much smaller than the current absorbed by the inverter when it works as amplifier (tens of μA). In conclusion, the NAND and NOR PFGA take advantage of the large voltage buffer dimensions in the circuit, in order to ensure the correct biasing of the inverter during the read-out mode, and to provide energy to the sensor during the actuation mode.

III. SIMULATION RESULTS

The proposed front-ends have been simulated in AMS-350nm CMOS technology, with a power supply of 3.3V and loaded with the Butterworth Van Dike model of MURATA sensor ($L_m = 58mH, C_S = 300pF, R_b = 320\Omega, C_E = 2.2nF$) [4]. The aspect ratio of the MOSFET utilized for the implementation of NAND and NOR PFGA are reported in Table I.

The transistors of the inverter (M1,M2) have been designed in order to provide a static gain around -100V/V, while the large buffer transistors (M3,M4) dimensions optimizes the biasing of the PFGA, and they ensure enough energy to

TABLE I
FRONT-END TRANSISTOR DIMENSIONS

NAND-PFGA					
NOT		Buffer		Switches	
$(W/L)_1$	$(W/L)_2$	$(W/L)_3$	$(W/L)_4$	$(W/L)_5$	$(W/L)_0$
$\frac{2\mu m}{2\mu m}$	$\frac{7.5\mu m}{2\mu m}$	$\frac{10\mu m}{0.35\mu m}$	$\frac{32\mu m}{0.35\mu m}$	$\frac{0.4\mu m}{0.35\mu m}$	$\frac{1\mu m}{0.35\mu m}$
NOR-PFGA					
NOT		Buffer		Switches	
$(W/L)_1$	$(W/L)_2$	$(W/L)_3$	$(W/L)_4$	$(W/L)_5$	$(W/L)_0$
$\frac{2\mu m}{2\mu m}$	$\frac{7.5\mu m}{2\mu m}$	$\frac{10\mu m}{0.35\mu m}$	$\frac{32\mu m}{0.35\mu m}$	$\frac{5\mu m}{0.35\mu m}$	$\frac{0.4\mu m}{0.35\mu m}$

actuate the sensor as previously discussed. The dimensions of the switches M5 and M0 have been chosen to minimize their equivalent on resistance during the read-out mode.

First of all, static analysis of the PFGA made of M1, M2, M3, M4, NAND and NOR PFGA has been performed as shown in Fig.3.

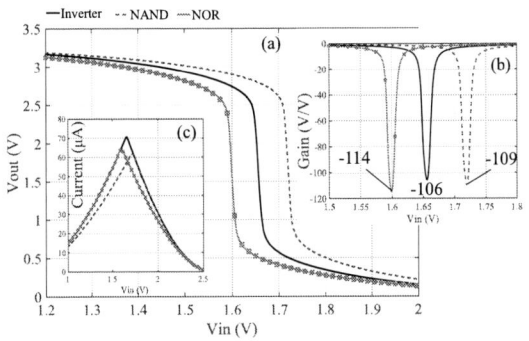

Fig. 3. Static analysis of the PFGA, NAND-PFGA and NOR-PFGA. (a) Voltage transfer characteristic curves (VTC). (b) Static gain (c) Static current absorbed.

This analysis is performed by varying the input voltage with an external source and monitoring the output voltage of the amplifiers. Therefore, in this analysis, the transistors M3,M4 cover no role. The static gain of the PFGA,NAND and NOR PFGA are -106V/V, -109V/V and -114V/V respectively. The current absorbed by PFGA, NAND and NOR PFGA are $70\mu A, 60\mu A, 65\mu A$ respectively. These amplifiers show low pass filter behaviour, which bandwidth has been extracted by performing AC analysis with a load capacitance of 1pF. Simulation results show that PFGA, NAND and NOR PFGA have a bandwidth of 377kHz, 316kHz, 301kHz respectively. Next, the dynamic performance of NAND-PFGA are represented in Fig.4.

On the other hand, the dynamic performance of the NOR-PFGA are shown in Fig.5.

For both NAND and NOR PFGA, the output signal is initially distorted after that the system passes in read-out mode. This is due to the "saturation" phenomenon characteristic of this amplifier, which is described in [5]. After some

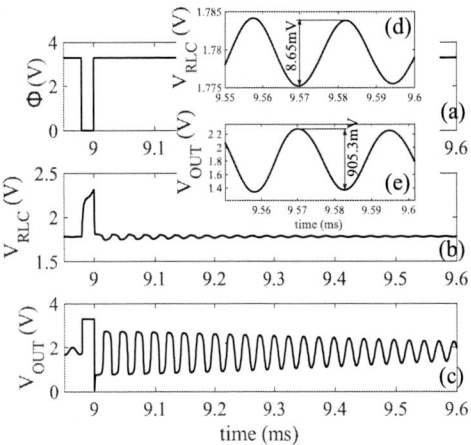

Fig. 4. NAND-PFGA Front End dynamic response. Simulated with a load capacitance of $C_L = 1pF$, Φ duty cycle = 98%, $f_\Phi = 1kHz$. (a) Control signal Φ. (b) Voltage across the sensor (c) Front-End output voltage. (d) Sensor voltage when the PFGA works in linear region. (e) Output voltage when the PFGA works in linear region.

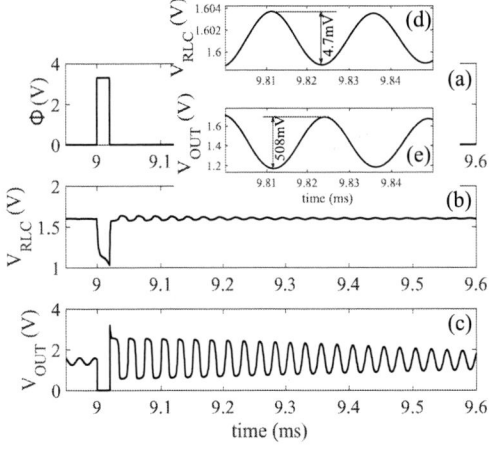

Fig. 5. NOR-PFGA Front End dynamic response. Simulated with a load capacitance of $C_L = 1pF$, Φ duty cycle = 2%, $f_\Phi = 1kHz$. (a) Control signal Φ. (b) Voltage across the sensor (c) Front-End output voltage. (d) Sensor voltage when the PFGA works in linear region. (e) Output voltage when the PFGA works in linear region.

time, when the oscillation generated by the sensor has been sufficiently damped, the input/output relation of the amplifier becomes linear. In this range, input and output amplitude have been compared in order to estimate the dynamic gain of the amplifier. The NAND-PFGA and the NOR-PFGA show a gain of -105V/V (905.3mV/8.65mV), -108V/V (508mV/4.7mV) respectively, which are very close to the values obtained during the static analysis. This means that the voltage buffers in NAND and NOR PFGA do not significantly affect the performance of the amplifier, and they provide just the biasing.

978-1-5386-5284-8/18 $31.00 © 2018 IEEE

Next, the current absorbed by the voltage buffer during the read-out mode has been measured to be a few nA in both circuits, which is a negligible value in respect to the current absorbed by the inverter (M1,M2). Finally, the average current of a whole read-out cycle (1 ms) has been measured for NAND and NOR PFGA and used to estimate the static power consumption of the circuit, which is $214\mu W$ for both circuits.

IV. MEASUREMENT SET-UP

A prototype of the proposed circuits has been fabricated in discrete components, in order to verify the correct operation of the system. The transistors have been implemented by using ICs CD4007UBE for the NAND and NOR transistors, and ALD1103 for the voltage buffer transistors, fed by a power supply of 3.3V. These ICs ensure that the transistor dimensions of the voltage buffer (M3,M4) are greater than the dimensions of the inverter transistors (M1,M2), which is required for achieving good performance. The integrated circuits have been connected as shown in Fig.6.

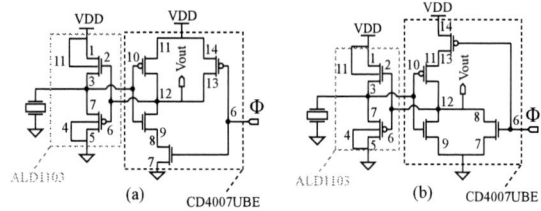

Fig. 6. Pin Mapping of ICs CD4007UBE and ALD1103 to build a NAND and NOR PFGA. (a) NAND-PFGA (b) NOR -PFGA.

Furthermore, the circuit has been tested with a real resonating transducer: MURATA MA40S4R, which model was previously used to perform the simulations of the circuit. Finally, the control signal Φ is characterized by a frequency of 1kHz.

V. MEASUREMENT RESULTS

Measurement Results for the NAND-PFGA front-end are shown in Fig.7.

Measurement results show that the front-end is able to excite and read-out the sensor, which oscillates at 40kHz. However, it is possible to observe other secondary phenomena, such as a nonlinear behaviour of the sensor, which produces a quasi-sinusoidal signal and the presence of a second order mode at 338kHz as shown in Fig.7(c). The circuit absorbs an average current of $44\mu A$, thus it is characterized by a static power consumption of $145\mu W$. NOR PFGA has also been implemented and measurement results are shown in Fig.8

Fig.8 shows that the NOR-PFGA presents the same performance as the NAND-PFGA, with the only difference is that the control signal Φ must be inverted. The static power consumption measured is $145\mu W$ also for NOR-PFGA.

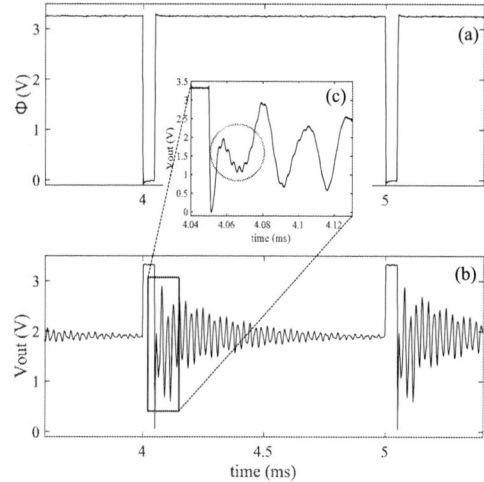

Fig. 7. NAND-PFGA response. Output terminal loaded with oscilloscope probe $C_L = 14pF$, $R_L = 10M\Omega$. $f_\Phi = 1kHz$ and duty cycle of 96%. (a) Control signal Φ, (b) Output signal of the Front-End (c) Second order mode oscillation after the transition from actuation to read-out mode

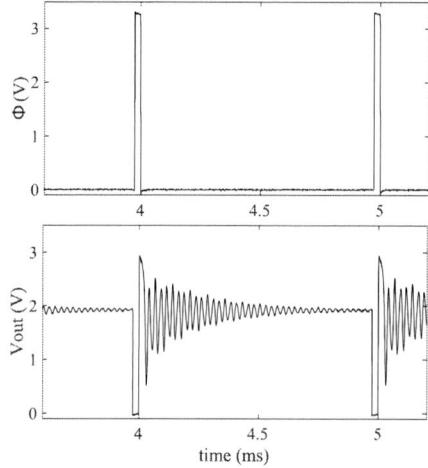

Fig. 8. NOR-PFGA response. Output terminal loaded with oscilloscope probe $C_L = 14pF$, $R_L = 10M\Omega$. $f_\Phi = 1kHz$ and duty cycle of 4%. (a) Control signal Φ, (b) Output Signal of the Front-End.

VI. CONCLUSION

In conclusion, we realized and verified the possibility of using a PFGA as a bidirectional front-end for resonating sensors. Simulations have been used to analyze the behaviour of the front-end, and measurements have been performed to confirm the operation of this circuit. The front-end is very compact because it doesn't need any coupling capacitor and it utilizes the voltage buffer for both biasing and sensor actuation purposes. The proposed circuit shows power consumption performance comparable to the other circuits described in previous works. Finally, a comparison between NAND and

NOR PFGA shows that the NAND PFGA is more compact. Future works will continue to investigate on the possibility to improve the PFGA for resonating sensor front-end, and will look forward to implement the proposed circuit in ASIC by using low and high leakage technologies, performing noise analysis and process variation analysis.

REFERENCES

[1] Y. Berg, O. Mirmotahari, and S. Aunet, "Pseudo floating-gate inverter with feedback control," in *2006 IFIP International Conference on Very Large Scale Integration*, Oct 2006, pp. 272–277.

[2] Y. Yan, Z. Zeng, C. Chen, H. Jiang, Z. y. Chang, D. M. Karabacak, and M. A. P. Pertijs, "An energy-efficient reconfigurable readout circuit for resonant sensors based on ring-down measurement," in *IEEE SENSORS 2014 Proceedings*, Nov 2014, pp. 221–224.

[3] M. Azadmehr, B. K. Khajeh, and Y. Berg, "A bidirectional circuit for actuation and read-out of resonating sensors," in *2014 IEEE Faible Tension Faible Consommation*, May 2014, pp. 1–4.

[4] L. Marchetti, Y. Berg, O. Mirmotahari, and M. Azadmehr, "Bidirectional front-end for piezoelectric resonator," in *2016 IEEE 13th International Conference on Networking, Sensing, and Control (ICNSC)*, April 2016, pp. 1–4.

[5] L. Marchetti, Y. Berg, and M. Azadmehr, "Design and modelling of a bidirectional front-end for resonating sensors based on pseudo floating gate amplifier," *Electronics*, vol. 6, no. 3, 2017.

2018 IEEE International Conference on Semiconductor Electronics (ICSE)

Design of Ultra Low Voltage Low Power DXCCII for Analog Signal Processing

Mohammad Faseehuddin[1], Jahariah Sampe[1], Sawal Hamid Md Ali[2]

[1]Institute of Microengineering and Nanoelectronics (IMEN), University Kebangsaan Malaysia (UKM),
43600 Bangi, Selangor, Malaysia
[2]Department of Electrical, Electronic and Systems Engineering, Universiti Kebangsaan Malaysia (UKM),
43600 Bangi, Selangor, Malaysia

faseehuddin03@siswa.ukm.edu.my, jahariah@ukm.edu.my

Abstract—**This paper presents the design of a versatile current mode active block namely, dual x current conveyor (DXCCII). The proposed DXCCII is capable of working under ultra low voltage low power (ULVLP) environment. The design uses floating gate (FG) technique and modified flip voltage follower (FVF) based differential pair to achieve operation under extremely low supply voltage and near rail to rail dynamic range. The ULVLP DXCCII is designed using 0.18μm technology parameters and simulated in spice software to verify its performance. The circuit works at a supply voltage of ±0.4V and exhibit a power dissipation of only 67μW. The proposed active block is suitable for biomedical signal processing.**

Keywords— *current mode; floating gate; flip voltage follower; filter; low power*

I. Introduction

The rapid growth in the research and development of portable devices, biomedical systems, wireless sensor nodes etc. has put serious constraints on operating supply voltage and power dissipation of the circuits [1-4]. The miniaturization of CMOS technology is making chips denser and faster but the technology scaling requires the supply voltage to be scaled proportionally to keep the electric fields within the semiconductor to safe limits and reduce power consumption. This trend is proving beneficial for digital circuits but it has posed serious challenges for the analog designers [1, 4] because the threshold voltage of the metal oxide semiconductor (MOS) transistor is not scaled in equal proportion to the supply voltage to minimize leakage in digital circuits. The threshold voltage has become a substantial fraction of the supply voltage especially in deep submicron technologies [1, 4-5]. For analog circuits to function properly the input signal must overcome the threshold voltage of the MOS transistor this eats up the a lot of the signal headroom and the output signal swing is greatly reduced causing severe performance degradation in analog circuits [1, 4-5]. The analog designers are adopting many innovative techniques to solve this drawback and design high performance ULVLP circuits. The design techniques for ULVLP circuits can be categorized into three groups. Operation level design [1], circuit level design [1] and device level design [1]. The first technique is based on operating the MOS transistor is sub threshold region of operation. In this region low supply voltage is required and the power dissipation is greatly reduced. The main drawback of this approach is that the dynamic range and operational frequency are drastically reduced. The second approach includes using innovative circuit design techniques like low

voltage cascode [5], level shifting [5] and flip voltage follower (FVF) [7] etc. The third approach works by modifying the conventional MOS transistor in such a way so as to make it work under ULVLP regime. The techniques include bulk driven technique [1, 4-6], bulk driven quasi floating gate technique [4, 8] and floating gate technique [9-10, 13]. All these techniques reduce the threshold voltage of the MOS transistor thereby increasing the dynamic range of the designed circuits.

The stringent requirements of ULVLP operation has shifted the focus of the researchers to current mode circuits. The operational amplifier (Op-amp) is the undisputed voltage mode analog building block for analog circuit design but it suffer from constant gain bandwidth, low slew rate and reduced dynamic range especially under LVLP operation these drawbacks render it unsuitable for ULVLP circuits. The current mode circuits on the other hand have many advantages over their voltage mode counter parts [11-12] like wide bandwidth, high slew rate, simple circuit, they are less effected by supply voltage scaling etc. The most versatile current mode active block is the second generation current conveyor (CCII) [11-12]. In past two decades it has been utilized in the design of multitude of application that include filters [11-12], oscillators [11-12], function generators [12], immittance simulators [11-12] etc. Numerous new topologies of current conveyor have evolved with time like the dual output current conveyor (DOCCII) [12], inverting current conveyor (ICCII) [12], current conveyor transconductance amplifier (CCTA) [12], dual x current conveyor (DXCCII) [12, 14] etc. The DXCCII is a versatile active block and it is found suitable for design of minimum components filters and oscillator structures [11, 13-14].

In this research the ULVLP implementation of DXCCII is proposed. The floating gate MOS technique is employed to design the DXCCII working at extremely low supply voltage of ±0.4V. The section II outlines the design procedure of ULVLP DXCCII. The section III gives the simulation results and discussion followed by conclusion.

II. Design of DXCCII

The DXCCII is a highly versatile current mode active block. It is functionally a combination of second generation current conveyor and Inverting current conveyor. It carries features of CCII and ICCII is one single integrated circuit leading to greater flexibility for analog designers. The block diagram of DXCCII is shown in Fig. 1 and the voltage current characteristics are presented in Equation1.

This research was funded by UKM internal grant (GUP-2015-021)

978-1-5386-5284-8/18 $31.00 © 2018 IEEE

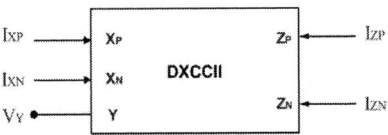

Fig 1. Block diagram of DXCCII

$$\begin{bmatrix} I_Y \\ V_{XP} \\ V_{XN} \\ I_{ZP} \\ I_{ZN} \end{bmatrix} = \begin{bmatrix} 0 & 0 & 0 \\ 1 & 0 & 0 \\ -1 & 0 & 0 \\ 0 & 1 & 0 \\ 0 & 0 & 1 \end{bmatrix} \begin{bmatrix} V_Y \\ I_{XP} \\ I_{XN} \end{bmatrix} \qquad (1)$$

A. Floating Gate Technique

Initially the floating gate transistors were primarily used in the design of memories but latter they are successfully employed in the design of low voltage low power (LVLP) analog circuits [9-10, 13]. In this technique the gate of the conventional MOS transistor is not directly connected to the input signals rather it is left floating [13]. The inputs are capacitively coupled to the gate of the MOS transistor using small value input capacitors. The multi input floating gate MOS (FG-MOS) transistor is shown in Fig. 2 (a). The FG-MOS can be fabricated in all CMOS technologies but the double poly technology is preferred one [4, 13]. During the fabrication process a residual charge Q_{FG} can get trapped in FG-MOS transistor resulting in voltage offset. This charge needs to me removed for precise function of the FG-MOS circuits. The unwanted charge can be removed by using Fowler-Nordheim (FN) tunneling, hot electron injection and ultraviolet light [13].

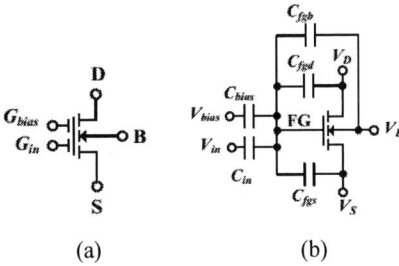

(a)   (b)

Fig. 2. (a) The symbol of FG-MOS (b) Equivalent circuit of FG-MOS

The voltage at the floating node as shown in Fig. 2(b) can be calculated using the law of conservation of charge which yields

$$V_{FG} = \frac{C_{in}V_{in} + C_{bias}V_{bias} + C_{GD}V_D + C_{GS}V_S + C_{GB}V_B + Q_{FG}}{C_{Total}} \qquad (2)$$

Where Q_{FG} is the residual charge trapped at the FGMOS during the fabrication process. The total capacitance $C_{Total} = C_{in} + C_{bias} + C_{GD} + C_{GS} + C_{GB}$, where C_{GD}, C_{GS}, C_{GB} are the parasitic capacitances associated with the source, drain and bulk region respectively. If the input capacitance is selected such that their sum is much greater than the parasitic capacitances then:

$$V_{FG} \approx K_1 V_{in} + K_2 V_{bias} \qquad (3)$$

Where $K_1 = \frac{C_{in}}{C_{Total}}$ and $K_2 = \frac{C_{bias}}{C_{Total}}$ are the equivalent weights. The equivalent threshold voltage of FG-MOS is given by Equation 4.

$$V_{Threshold-FG} = \frac{V_T - V_{bias}K_2}{K_1} \qquad (4)$$

Where V_T is the threshold voltage of a conventional gate driven MOS (GD-MOS) transistor. It can be inferred from the equation that by suitable choice of K_1, K_2 and V_{bias} the threshold voltage of the FG-MOS can be reduced or even made zero.

The effective transconductance of the FG-MOS is given by Equation 5.

$$g_{m,eff} = K_1 g_m = \frac{C_{in}}{C_{Total}} g_m \qquad (5)$$

The g_m is the transconductance of the conventional GD-MOS transistor. It can be clearly observed from the equation that the transconductance of FG-MOS is lower than the GD-MOS transistor. The transconductance of FG-MOS can be increased by increasing the value of the input capacitors at the expense of increased chip area.

B. CMOS Implementation of ULVLP DXCCII

The core of the circuit is the floating gate flipped voltage follower (FG-FVF) based differential pair [6, 8] as shown in Fig. 3. The transistor M03 of the differential pair provides the tail current. The transistor M03, input transistor M02 and transistor M05 forms the flipped voltage follower. The gate voltage of the M03 ensures that equal current flows in both the input devices. The transistors M04 and M05 form the active load of the differential pair. The use of FVF reduce the minimum supply required for operation down to $V_{DD} = |V_{GSTail}| + V_{DSSat.M05}$, where $V_{DSSat.M05}$ is the minimum drain to source voltage required to keep M05 saturated and V_{GSTail} is the gate to source voltage of the tail current source.

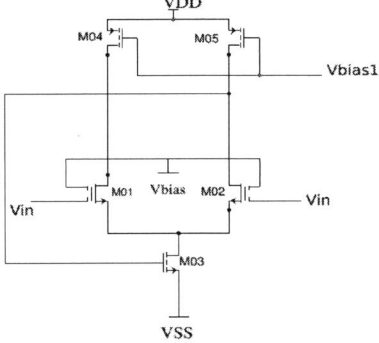

Fig. 3. The modified FG-FVF based differential pair

The multi input FG-MOS transistors are employed as input transistors of the differential pair to reduce the threshold voltage. The threshold voltage is lowered by properly selecting the input capacitors and connecting one input of the FG-MOS transistors to a suitable bias voltage there by facilitating rail to rail dynamic range and common voltage range.

The complete CMOS implementation of the ULVLP DXCCII is shown in Fig. 4. Two multi-input FG-FVF differential pairs are utilized for the design. The transistors M01-M05 and M06-M10 form the low voltage differential pairs. The transistors M11-M18 constitute the X_P and X_N terminals while transistors M19-M26 makeup the current output Z_P and Z_N terminals. The cascode transistors are used to increase the output impedance of current output Z terminals. The negative feedback is employed to reduce the impedance at the X_P and X_N terminals. The transistor M27 sets the bias current of the DXCCII. The voltage applied at the Y terminal appears at the X_P terminal and in inverted form at the X_N terminal. The current input at the X_P and X_N terminals appears at the current output Z_P and Z_N terminals.

The current following in Z_P and Z_N nodes are independent of each other. The Y, Z_P and Z_N terminals are high impedance and the current input X_P and X_N terminals are low impedance.

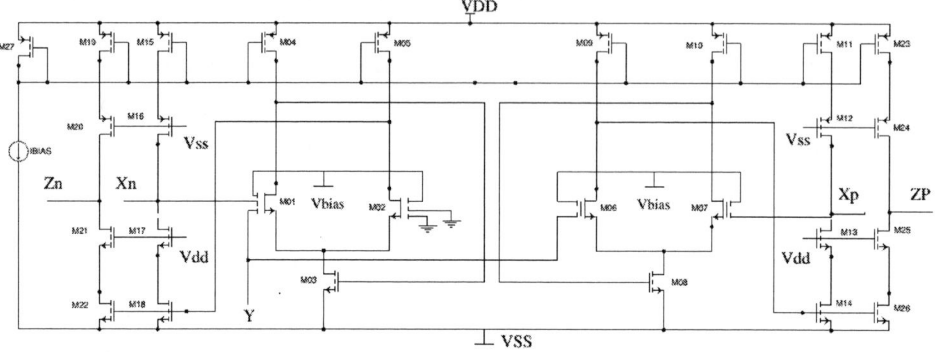

Fig. 4. The CMOS implementation of ULVLP DXCCII

III. SIMULATION RESULTS

The proposed DXCCII is simulated in spice software to confirm the theoretical findings. The 0.18μm parameters from TSMC are used to validate the design. The supply voltages are set at ±0.4V and the bias current is fixed at $I_B = 2\mu A$. One of the inputs of the FG-FVF differential pairs are tied to a bias voltage of 0.4V to reduce the threshold voltage. All the input capacitors are select as 0.1pF. The procedure outlined in [15] is used to obtain the various design metrics of the designed DXCCII. First, the dc voltage following characteristic is measured. A dc sweep of ±0.4V is applied at the Y terminal and the voltage at the X_P and X_N terminals is observed as can be deduced form Fig. 5 the DXCCII exhibits near rail to rail voltage swing.

TABLE I: THE ASPECT RATIOS OF THE TRANSISTORS

Transistors	Width (μm)	Length (μm)
M01-M02, M06-M07	4	0.72
M03-M08	8	0.72
M04-M05, M09-M10	5	0.72
M11, M15, M19, M23	100	0.72
M14, M18, M22, M26	100	2
M12, M16, M20, M24	50	0.72
M13, M17, M21, M25	50	2
M27	10	0.72

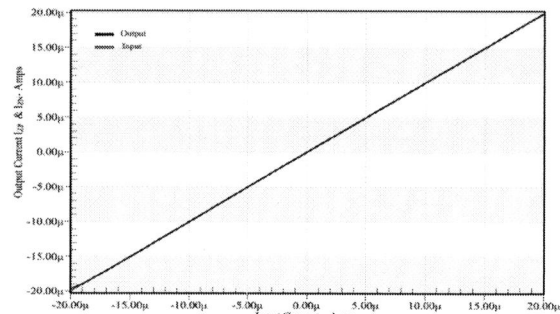

Fig. 6. The current transfer characteristics of the DXCCII

The transient analysis of voltage and current followers is also performed to check the input to output phase error and the signal processing capability of the DXCCII. A sinusoidal voltage signal of 100mV p-p at 400kHz is applied at the Y terminal and the voltages at the X_P and X_N terminals are noted, as can be seen from Fig. 7 the input and output signals follow each other with negligible phase error.

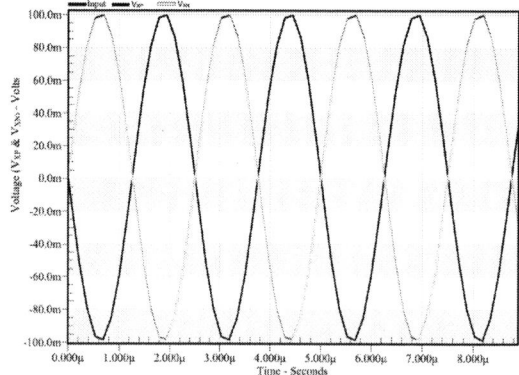

Fig. 7. Transient analysis results of the voltage followers

The AC analysis is carried out to obtain the voltage and current transfer bandwidths of the DXCCII. The voltage transfer bandwidth is measured with 100k Ω resistors attached to the X terminals. The AC responses are presented in Fig. 8-10 and the measured bandwidths are

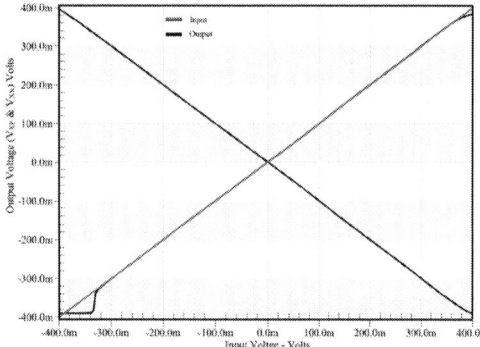

Fig. 5. The dc voltage transfer characteristic of the DXCCII

Second, the dc current transfer between the X_P, Z_P and X_N, Z_N terminals is studied. The current followers works till ±18μA as shown in Fig. 6.

(V_{XP}/V_Y)=23.82MHz, (V_{XN}/V_Y)=19.23MHz, (I_{ZP}/I_{XP})=26.977MHz and (I_{ZN}/I_{XN})=22.18MHz.

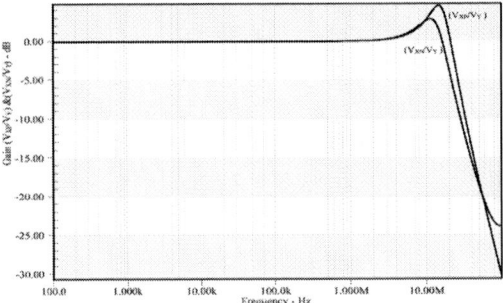

Fig. 8. The ac analysis of the voltage followers

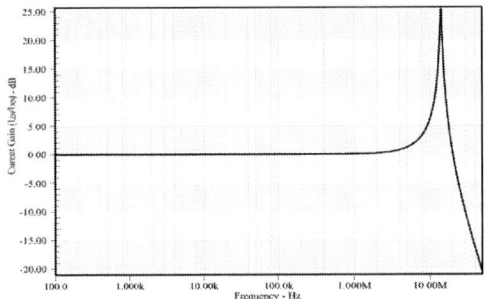

Fig. 9. The ac analysis of the current follower I_{XN} and I_{ZN} terminals

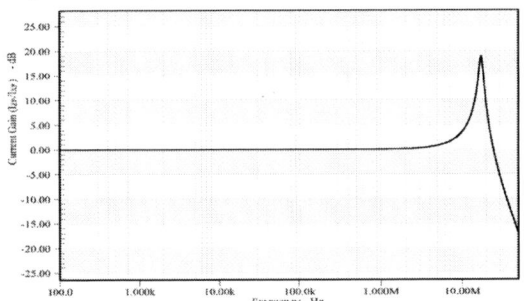

Fig. 10. The ac analysis of the current follower I_{XP} and I_{ZP} terminals

The important design parameters of the ULVLP DXCCII are summarized in Table 2 and compared with the low voltage DXCCII topology available in the literature. It is seen that the proposed active block requires low operating voltage and features low power dissipation. The proposed ULVLP DXCCII is suitable for low to moderate frequency operations especially in bio medical signal processing and speech processing.

TABLE II. THE PERFORMANCE PARAMETERS OF DXCCII

Parameters	Proposed	[14]
Supply Voltage (V)	±0.4	±0.75
Power Dissipation (μW)	67	200
Voltage transfer range (mV)	±340	-400 to 380
Current transfer range (μA)	±18	±60
Voltage transfer bandwidth (V_{XP}/V_Y) MHz	23.82	265
Voltage transfer bandwidth (V_{XN}/V_Y) MHz	19.23	350
Current transfer frequency (I_{ZP}/I_{XP}), (I_{ZN}/I_{XN}), MHz	26.977, 22.18	290, 908
Input Impedance $(X_P)/(X_N)$ Ω	200	6.6k
Technology	0.18μm	0.18μm

CONCLUSION

In this research the DXCCII capable of working under ULVLP regime is developed. The design utilize floating gate technique to overcome the threshold voltage limitation and provided near rail to rail dynamic range. The modified floating gate flipped voltage follower (FG-FVF) based differential pair is employed in the design to facilitated low supply voltage operation. The proposed DXCCII is designed in 0.18μm parameters from TSMC and simulated in spice. The proposed design works at an extremely low supply voltage of ±0.4V while consuming only 67μW of power. The important design parameters are measured and compared with the existing low voltage DXCCII to bring forth the merits of the proposed design. The ULVLP DXCCII is suitable for use in bio medical systems where signals in few kHz range need to be processed.

REFERENCES

[1] M. Faseehuddin, J. Sampe, M. S. Islam, "Designing ultra low voltage low power active analog blocks for filter applications utilizing the body terminal of MOSFET: A review," Asian Journal of Scientific Research, vol .9, pp. 106-121, 2016.

[2] J. Sampe, F. F. Zulkifli, N. A. A. Semsudin, M. S. Islam, B. Y. Majlis "Ultra low power hybrid micro energy harvester using RF, thermal and vibration for biomedical devices", International Journal of Pharmacy and Pharmaceutical Sciences, vol. 8, pp. 18-21, 2016.

[3] J. Sampe, T. N. T. Mohamad, D. D. Berhanuddin, "Higher Sensitivity RF-DC Rectifier for Ultra-Low Power Semi-Active RFID Tags", In Proceedings of the International Conference on High Performance Compilation, Computing and Communications ACM, pp. 131-135, 2017.

[4] F. Khateb, S. B. A. Dabbous, S. Vlassis, "A survey of non-conventional techniques for low-voltage low-power analog circuit design," Radioengineering, vol. 22, pp. 415-427, 2013.

[5] S. S. Rajput, S. S. Jamuar, "Low voltage analog circuit design techniques," IEEE Circuits and Systems Magazine, vol. 2, pp. 24-42, 2002.

[6] G. Raikos, S. Vlassis, C. Psychalinos, "0.5 V bulk-driven analog building blocks," AEU-International Journal of Electronics and Communications, vol. 66, pp. 920-927, 2012.

[7] R. G. Carvajal, J. Ramírez-Angulo, A. J. López-Martín, A. Torralba, J. A. G. Galán, A. Carlosena, F. M. Chavero,"The flipped voltage follower: A useful cell for low-voltage low-power circuit design," IEEE Transactions on Circuits and Systems I: Regular Papers, vol. 52, pp. 1276-1291, 2005.

[8] F. Khateb, "Bulk-driven floating-gate and bulk-driven quasi-floating-gate techniques for low-voltage low-power analog circuits design," AEU-International Journal of Electronics and Communications, vol. 68, pp. 64-72, 2014.

[9] Y. Berg, T. S. Lande, O. Naess, H. Gundersen, "Ultra-low-voltage floating-gate transconductance amplifiers,"IEEE Transactions on circuits and systems II: Analog and digital signal processing, vol. 48, pp. 37-44, 2001.

[10] J. Ramirez-Angulo, S. C. Choi, G. Gonzalez-Altamirano, "Low-voltage circuits building blocks using multiple-input floating-gate transistors," IEEE Transactions on Circuits and Systems I: Fundamental Theory and Applications, vol. 42, pp. 971-974, 1995.

[11] G. Ferri, N. C. Guerrini, "Low-voltage low-power CMOS current conveyors," Springer Science & Business Media, 2003.

[12] R. Senani, D. R. Bhaskar, A. K. Singh, "Current conveyors: variants, applications and hardware implementations," Springer, 2014.

[13] E. Rodriguez-Villegas, "Low power and low voltage circuit design with the FGMOS transistor, " IET, Vol. 20, 2006.

[14] S. A. Tekin, H. Ercan, M. Alçi, "A Versatile Active Block: DXCCCII and Tunable Applications," Radioengineering, vol. 23(4), pp. 1131, 2014.

[15] A. Fabre, O. Saaid, H. Barthelemy, "On the frequency limitations of the circuits based on second generation current conveyors," Analog Integrated Circuits and Signal Processing, vol. 7, pp. 113-129, 1995.

2018 IEEE International Conference on Semiconductor Electronics (ICSE)

Corner Mismatch Model for Fast Non-Monte Carlo Best and Worst Cases Simulation

Philip Beow Yew Tan, Chiew Ching Tan and Mohamad Marzuki Bin Mohd Fauzi

Silterra Malaysia Sdn. Bhd.

Kulim Hi-Tech Park 09000 Kulim,

Kedah, Malaysia.

philip_tan@silterra.com, chiewching_tan@silterra.com and mohamad_marzuki@silterra.com

Abstract— **In this paper, we propose a non-Monte Carlo corner mismatch model, that enables circuit designers to simulate the best-case and the worst-case mismatch variation of their circuit performance fast. Standard mismatch model needed to be run in statistical Monte Carlo simulation which is time-consuming. Corner mismatch model can be run fast in non-Monte Carlo simulation. The concept of corner mismatch model is similar to the process variation corner models such as Fast-Fast (FF) or Slow-Slow (SS) model. We have demonstrated that corner mismatch model is useful in designing mismatch circuit such as current mirror. Besides the fast non-Monte Carlo simulation, corner mismatch model also allows designers to determine how many sigma of mismatch variation to be applied to the transistors in the matching pair. This enables the designers to analyze the amount of mismatch variation that can cause a critical impact to the entire circuit performance.**

Keywords— non-Monte Carlo; corner mismatch model; device modeling; MOSFET mismatch.

I. INTRODUCTION

Mismatch variation is the variation caused by intra die or local differences between two identical transistors which are drawn side-by-side in an integrated circuit. While process variation is the die-to-die, wafer-to-wafer or lot-to-lot global manufacturing variation.

Throughout the years, mismatch variation has been explored and studied in [1] - [5] due to its well-known consequence on analog circuit especially circuits involved matching pairs. Transistors with a larger size have better matching compared to small-size transistors. Mismatch coefficient is the figure of merit for mismatch. Smaller the mismatch coefficient number better the matching. Mismatch coefficient is determined from the slope of the Pelgrom mismatch plots. Mismatch variation is insignificant for large devices. For small devices, mismatch variation is important and needs to be modeled accurately.

Usually, standard SPICE model only captures the process variation but not the mismatch variation. This means that a pair of same size transistors always has the exact matching linear threshold voltage, V_{tlin} and saturation drain current, I_{dsat}. Since mismatch has become significantly important for analog circuit design especially in technology nodes such as 0.18 μm and 0.13 μm, foundry provides additional mismatch models to fulfill the analog design requirement.

In 2016 [6], we have introduced a methodology to capture the MOSFET subthreshold swing mismatch with

BSIM4 model. Subthreshold swing mismatch is important for low power application, especially when the targeted region of operation is below threshold voltage. The main challenge in model subthreshold swing mismatch is it does not obey the Pelgrom rule. Instead of having a single mismatch coefficient (slope of the mismatch plot), it has two mismatch coefficients. Large devices have a smaller value of subthreshold swing mismatch coefficient compared to small devices. This means the subthreshold swing mismatch variation becomes much significant when device size is smaller than the Critical Area, A_c. We believe this is due to the small-size devices are more sensitive to subthreshold swing fluctuation compared to large-size devices.

Although mismatch models are available for circuit simulation including subthreshold region, mismatch model still has one major concern. The simulation time is long. This is because the mismatch model needs to be run using statistical Monte Carlo simulation. Monte Carlo simulation is used to investigate how the individual device mismatches of a circuit may accumulate and affect the entire circuit [7]. It is done by analyzing a large set of circuit instantiations with randomly varied devices.

A method to capture DC random offsets as equivalent AC pseudo-noises is proposed [8]. It uses the linear periodically time-varying (LPTV) noise analysis from RF circuit simulators to estimate the effects of the device mismatch on the circuit transient performance, such as logic path delay and oscillator frequency. This method is considered as an extension to the DCMatch analysis which can speed up the simulation time without performing the traditional Monte Carlo simulations.

Recently, SRAM operational mismatch (SOMM) corner model is introduced to model the SRAM mismatch variation [9]. The authors estimate the combination of transistor mismatch variation in a given SRAM bitcell, which represents the worst read or write performance. This worst mismatch combination is used to implement the SRAM bitcell corner model. The bitcell corner model is then used to evaluate the worst case condition of read or write operation in SRAM design.

In this paper, we propose a new non-Monte Carlo corner mismatch model that can be easily implemented on top of the standard foundry mismatch model for matching circuit design use. The concept is similar to the process variation corner models such as Fast-Fast (FF) case or Slow-Slow (SS) case. Corner mismatch model allows designers to inject mismatch into either one or both of the transistors in the

978-1-5386-5284-8/18 $31.00 © 2018 IEEE   230

matching pair. Designers are also allowed to define the number of mismatch sigma used in the simulation.

II. EXPERIMENT

Current mirror is a fundamental compositional element which is frequently used in the analog and digital circuit designs. Current mirror is popular among the circuit designers due to its inherent simplicity of operation and low cost associated with CMOS fabrication [10]. However, the simplicity of current mirror makes it particularly susceptible to device mismatch. Hence, we use current mirror circuits to verify the corner mismatch model.

Three NMOS current mirrors and three PMOS current mirrors circuits are designed and fabricated using 1.8 V Low Voltage 0.18 μm CMOS technology. All the current mirrors have a matching pair of transistors M_1 and M_2 with W/L = 3 μm / 1 μm. R_{ref} is connected to both the Drain and Gate of M_1 and R_{load} is connected to the Drain of M_2. Fixed R_{ref} of 106 kΩ is used. Three different R_{load} with values of 33 kΩ, 70 kΩ and 106 kΩ are used in the NMOS and PMOS current mirrors. Fig. 1 and Fig. 2 show the NMOS and PMOS current mirrors with R_{load} = 33 kΩ respectively.

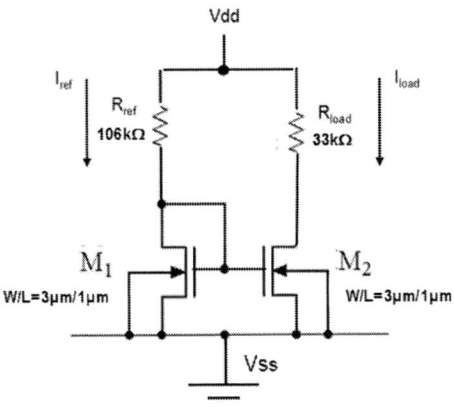

Fig. 1. NMOS Current mirror circuit with a matching pair of transistors M_1 and M_2 with W/L = 3μm/1μm and R_{ref} and R_{load} of 106 kΩ and 33 kΩ respectively.

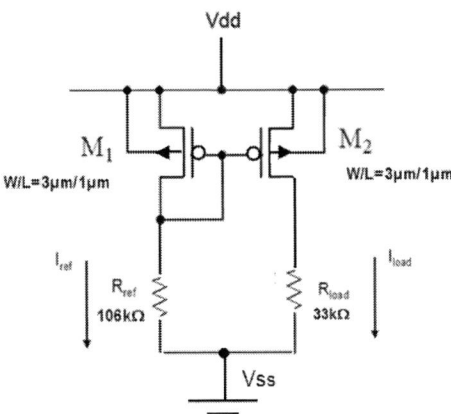

Fig. 2. PMOS Current mirror circuit with a matching pair of transistors M_1 and M_2 with W/L = 3μm/1μm and R_{ref} and R_{load} of 106 kΩ and 33 kΩ respectively.

III. CORNER MISMATCH MODEL

In this section, we will discuss the architecture of the corner mismatch model. We introduce *vth0n_mis* and *u0n_mis* mismatch skew parameters into core model, *nlvi.0* as shown in Fig. 3. Parameters *vth0n_mis* and *u0n_mis* are added into *vth0* and *u0* of the SPICE parameters respectively. The purpose is to add mismatch variation into V_{tlin} and I_{dsat} of the transistors. These two parameters are introduced the same way as we add process variation parameters (for example *skew_lv_vtn*) into the core model. This method allows the values of *vth0n_mis* and *u0n_mis* to be generated either randomly from Monte Carlo Mismatch simulation or as two fixed values from non-Monte Carlo corner mismatch simulation.

```
$ Core Model of Corner Mismatch Model
$
.model nlvi.0 nmos (
+ level = 54
+ vth0 = ... + skew_lv_vtn + vth0n_mis
+ u0 = ... + skew_lv_u0n+ u0n_mis
+... )
.endl
```

Fig. 3. Core model, nlvi.0 with corner mismatch parameters, vth0n_mis and u0n_mis to introduce mismatch variation to vth0 and u0 SPICE parameters.

Corner mismatch model is a subckt model that calls the core model. The corner mismatch model, *mnlv* is illustrated in Fig. 4. The mismatch parameters, *vth0n_mis* and *u0n_mis* are defined as instant parameters of the internal transistor, *xmint*. Mismatch coefficient for V_{tlin} and I_{dsat} are *avtn_lv* and *aidn_lv*. When run non-Monte Carlo simulation, *dvtn_lv* and *didn_lv* Gaussian parameters values are zero. Hence, defining *misvar = 3* represents three sigma of mismatch variation for a non-Monte Carlo simulation. Parameter *mult* represents the multiplier, *m*. Hence, it has the same value as the *m*.

```
$ Corner Mismatch Model
$
.param
+dvtn_lv = agauss(0,1,1)
+didn_lv = agauss(0,1,1)
$
.subckt mnlv d g s b
.param
+misvar = 0
+mult = 1
$
xmint d g s b nlv w='w' l='l' sa='sa' sb='sb' \
    vth0n_mis = '(dvtn_lv + misvar) * (avtn_lv/sqrt(w * 1 * mult))' \
    u0n_mis = '(didn_lv + misvar) * (aidn_lv/sqrt(w * 1 * mult))'
.ends
```

Fig. 4. Corner mismatch model implementation using subckt method.

Fig. 5 shows the current mirror netlist that uses the corner mismatch model for mismatch analysis. Transistor *xm0* and *xm1* are set to three sigma higher and lower mismatch by using parameter *misvar = 3* and *misvar = -3* respectively. This means total of six sigma mismatch difference between them. The simulation results of the netlist yield the maximum I_{load} corner case (the upper line in Fig. 7).

978-1-5386-5284-8/18 $31.00 © 2018 IEEE

```
$ Current Mirror Netlist for Mismatch + Process Corner Simulation
$
.temp 25
.lib 'bsim4d18v_lv_rev1_4p.modlib' d18v_lv_ff
$
v1 VDD 0 1.8
v2 VSS 0 0
$
.subckt NMOS_CM VDD VREF VSS VLOAD
R0 VDD VREF r=106k
R1 VDD VLOAD r=33k
XM0 VREF VREF VSS VSS mnlv w=3e-6 l=1e-6 sa=1e-6 \
  sb=1e-6 misvar=3
XM1 VLOAD VREF VSS VSS mnlv w=3e-6 l=1e-6 sa=1e-6 \
  sb=1e-6 misvar=-3
.ends
$
x1 VDD VREF VSS VLOAD NMOS_CM
$
.dc v1 0 1.8 1.8
.print dc i(x1.R1)
$
.end
```

Fig. 5. Netlist of current mirror that uses Corner Mismatch model for non-Monte Carlo mismatch analysis of maximum I_{load} corner case.

IV. RESULTS AND DISCUSSION

Silicon measurement results of 27 dice for I_{load} is shown in Fig. 6. Simulation results from Monte Carlo process variation only cannot match to the actual measured I_{load} variation. When include mismatch variation together with process variation to the Monte Carlo simulation, the results can match to the measured I_{load} variation. Since it is a random statistical Monte Carlo simulation, analog designers not able to accurately determine the maximum I_{load} variation due to mismatch.

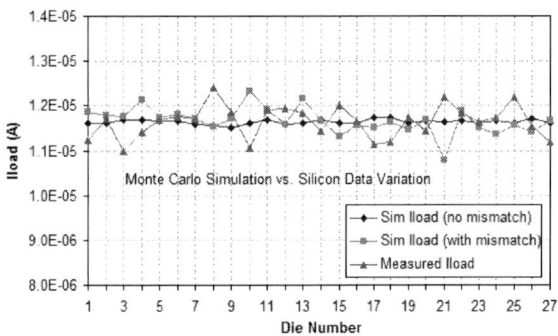

Fig. 6. NMOS Current Mirror I_{load} simulation results of Monte Carlo mismatch model versus measured data. R_{ref} = 106 kΩ and R_{load} = 33 kΩ.

By using the newly introduced corner mismatch model, analog designers able to perform non-Monte Carlo simulations to determine the minimum (min) and the maximum (max) I_{load} variation of the current mirror due to mismatch variation. Fig. 7 shows that the 27 dice measured I_{load} variation cannot be covered by the FF and SS process corners from simulation. When using the corner mismatch model together with the FF and SS process corner models, the simulation results show that the I_{load} variation can be covered by the simulation corners range. The variation range from FF and SS process corners is much smaller compared to the variation range from corner mismatch model. This indicates that mismatch variation has much significant impact compared to process variation for current mirror circuit.

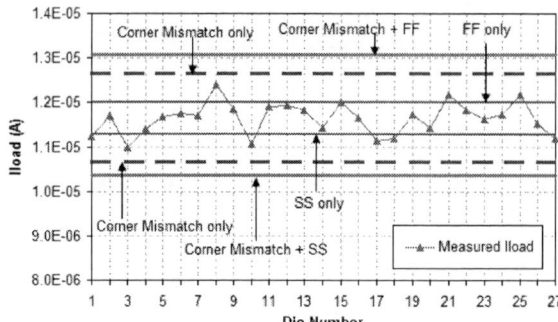

Fig. 7. NMOS Current Mirror I_{load} simulation results of different types of corner models versus measured data. R_{ref} = 106 kΩ and R_{load} = 33 kΩ.

Measured silicon data of all the current mirror circuits versus simulation data for I_{load} are plotted in Fig. 8 to Fig. 10 for NMOS current mirrors and Fig. 11 to Fig. 13 for PMOS current mirrors. The simulation results show that without mismatch model, the Monte Carlo simulation variation of the 27 dice (diamond) is far from the actual measured variation (triangle).

When includes the mismatch model, the Monte Carlo simulation variation (square) matches to the measured variation but the Monte Carlo simulation is time-consuming especially for complex circuitry. With corner mismatch model, the realistic min and max of I_{load} variation (lines) can be determined fast without the need to run Monte Carlo simulation.

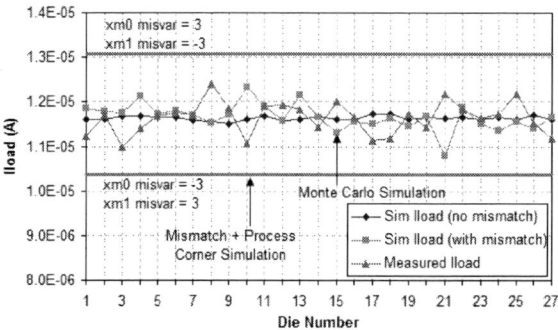

Fig. 8. NMOS Current Mirror I_{load} simulation results of no mismatch and with mismatch versus measured data for R_{ref} = 106 kΩ and R_{load} = 33 kΩ.

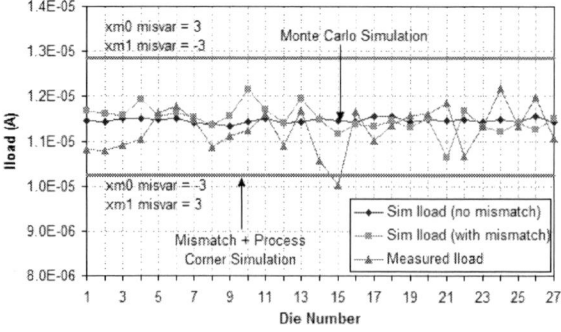

Fig. 9. NMOS Current Mirror I_{load} simulation results of no mismatch and with mismatch versus measured data for R_{ref} = 106 kΩ and R_{load} = 70 kΩ.

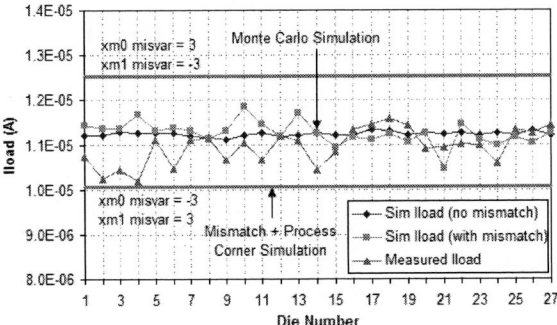

Fig. 10. NMOS Current Mirror I_{load} simulation results of no mismatch and with mismatch versus measured data for R_{ref} = 106 kΩ and R_{load} = 106 kΩ.

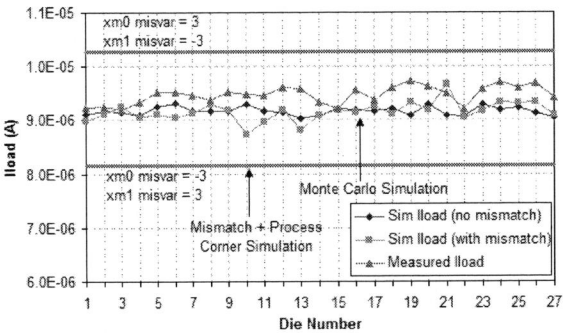

Fig. 11. PMOS Current Mirror I_{load} simulation results of no mismatch and with mismatch versus measured data for R_{ref} = 106 kΩ and R_{load} = 33 kΩ.

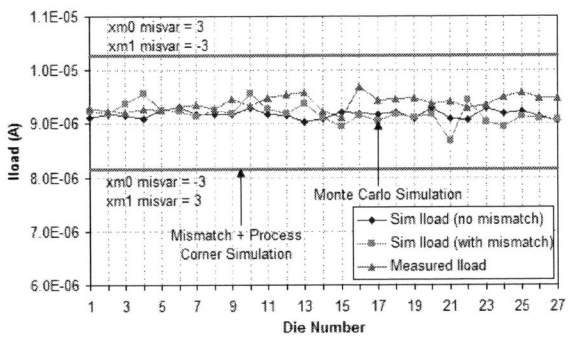

Fig. 12. PMOS Current Mirror I_{load} simulation results of no mismatch and with mismatch versus measured data for R_{ref} = 106 kΩ and R_{load} = 70 kΩ.

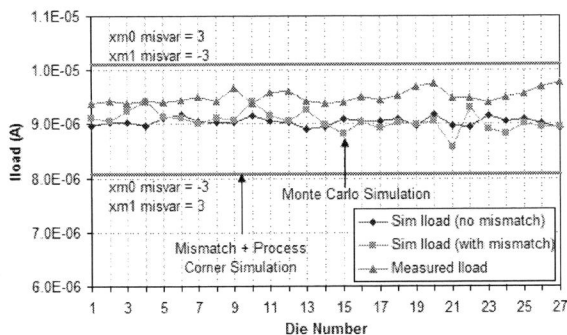

Fig. 13. PMOS Current Mirror I_{load} simulation results of no mismatch and with mismatch versus measured data for R_{ref} = 106 kΩ and R_{load} = 106 kΩ.

V. CONCLUSION

We have proposed a new corner mismatch model that allows designers to obtain the corner cases of mismatch characteristics fast without the needs to run the time consuming statistical Monte Carlo simulation. This is because the corner mismatch model can be run in non-Monte Carlo simulation. In this paper, we have used a simple current mirror to demonstrate how the corner mismatch model enables us to determine the I_{load} min and max variations due to mismatch. The corner mismatch model concept is similar to the process corner variation. Since process variation model covers both statistical Monte Carlo simulation and corner non-Monte Carlo simulation, mismatch variation model should also cover both of these simulations as well.

REFERENCES

[1] P. G. Drennan, C. C. McAndrew, "Understanding MOSFET mismatch for analog design", IEEE Journal of Solid-State Circuits, vol. 38, no. 3, pp. 450-456, 2003.

[2] X. Yuan, T. Shimizu, U. Mahalingam, et al., "Transistor mismatch properties in deep-submicrometer CMOS technologies", IEEE Transactions on Electron Devices, vol. 58, no. 2, pp. 335-342, 2011.

[3] N. Z. Butt and J. B. Johnson, "Modeling and analysis of transistor mismatch due to variability in short-channel effect induced by random dopant fluctuation", IEEE Electron Device Letters, vol. 33, no.8, pp. 1099-1101, 2012.

[4] Q. Zhang, C. Wang, H. Wang, et al., "Experimental study of gate-first FinFet threshold-voltage mismatch", IEEE Transactions on Electron Devices, vol. 61, no.2, pp. 643-646, 2014.

[5] C. C. McAndrew, M. Zunino and B. Braswell, "A Self-Amplifying Four-Transistor MOSFET Mismatch Test Structure", IEEE Transactions on Semiconductor Manufacturing, vol. 26, no. 3, pp. 273-280, 2013.

[6] X. E. Bee, M. Mohd Fauzi and P. B. Y. Tan, "Modeling of MOSFET Subthreshold Swing Mismatch with BSIM4 Model", IEEE International Conference on Semiconductor Electronics (ICSE), pp. 86-88, 2016.

[7] H. Hung and V. Adzic, "Monte Carlo SImulation of Device Variations and Mismatch in Analog Integrated Circuits", Proceedings of the National Conference on Undergraduate Research (NCUR), pp. 1-8, 2006.

[8] J. Kim, K. D. Jones and M. A. Horowitz, "Fast, Non-Monte-Carlo Estimation of Transient Performance Variation Due to Device Mismatch", IEEE Transaction nn Circuits and Systems, vol. 57, no. 7, pp. 1746-1755, 2010.

[9] T. H. Choi, H. Jeong, Y. Yang, J. Park and S. O. Jung, "SRAM Operational Mismatch Corner Model for Efficient Circuit Design and Yield Analysis", IEEE Transactions on Circuits and Systems, vol. 64, no. 8, pp. 2063-2072, 2017.

[10] T. Datta and P. Abshire, , "Mismatch Compensation of CMOS Current Mirrors using Floating-Gate Transistors ", IEEE International Symposium on Circuits and Systems, pp. 1823-1826, 2009.

978-1-5386-5284-8/18 $31.00 © 2018 IEEE

Characterization of SOI Film Thickness, Oxide Thickness and Charges with C-V Measurement

Ke Kian Seng Joseph
Technology Development
Infineon Technologies (Kulim)
Sdn Bhd,
Kulim, Malaysia
Joseph.ke@infineon.com

Tan Chan Lik
Technology Development
Infineon Technologies(Kulim)
Sdn Bhd
Kulim, Malaysia
ChanLik.Tan@infineon.com

Niew Soon Huat
Technology Development
Infineon Technologies(Kulim)
Sdn Bhd
Kulim, Malaysia
SoonHuat.Niew@infineon.com

Abstract— Capacitance-Voltage measurement is a crucial method to characterize and study the behavior of the device. In this work, the capacitance voltage characteristics of a partially depleted Silicon-On-Insulator MOSFET were analyzed and discussed. The important parameters like the gate oxide thickness, buried oxide thickness, silicon film thickness, fixed oxide charges and interface trapped charges were extracted from the capacitance voltage between the front gate and drain/source at different back gate voltage. The measured results were in good agreement with inline and XSEM result. The frequency dependency on the result is also observed and discussed in this paper.

Keywords—C-V Characteristics, Silicon-On-Insulator, PD MOSFET, Accumulation, Depletion, Inversion, Charges.

I. INTRODUCTION

Silicon on Insulator (SOI) has demonstrated many advantages over the conventional bulk MOSFET like lower parasitic junction capacitance, reduction of short channel effect, lower leakage current, elimination of latch up and immunity against radiation with improved device isolation, smaller layout, Improvement of subthreshold swing etc. [1] The manufacturing processes are also compatible with most conventional fabrication processes. But production costs for SOI substrate are much more expensive than bulk silicon, hence a proper characterization of the processes are crucial to ensure good device performance and to prevent low yielding SOI wafer.

Capacitance-Voltage (C-V) measurement is one of the crucial characterization method. In this paper, the C-V characteristics of a Partially Depleted (PD) SOI MOSFET were studied. For PD SOI, space charge region of the front and back gate do not interact and are separated by quasi-neutral region. The channel body is thus floating and will exhibit the floating body effect. The doping concentrations in the channel determine the depletion depth, leaving a neutral silicon region isolated from the grounded substrate by the Buried Oxide (BOX) beneath. The SOI film thickness in this study is thicker than the maximum front gate depletion width, the SOI will exhibits a floating body effect and it is regarded as a PD MOSFET.

C-V characteristic can reveal many important device parameters like gate oxide and buried oxide thickness, silicon film thickness, oxide charges, mobile ions, interface trapped density, doping profile etc. These are the important parameters which influence the electrical performance of the device substantially, for instance the SOI film thickness are one of the key parameter related to SOI MOSFET Breakdown voltage [2].

II. CHARACTERIZATION SET UP & MEASUREMENT STRUCTURE

The measurement structure of this study is shown in Fig. 1. The SOI thin film is made of N- doped with phosphorus and the drain/source is implanted with N+ dopant. It is a depletion mode N-MOSFET which is normally on. The front gate is made of N type in-situ phosphorus doped polysilicon. The capacitor area is 52152um2, the SOI film thickness is 200nm, Gate Oxide (GOX) thickness is 40nm and the BOX thickness is 400nm. The P-type well (Pwell) is made with high energy implantation of Boron through the BOX above the N+ substrate. The Pwell serves as the back gate to the SOI MOSFET. [3]

Capacitance is measured between gate and drain/source. The high terminal (CMH) of Capacitance Measurement Unit (CMU) is connected at Polysilicon gate, and the low terminal (CML) is connected at the drain/source which are tied together. The measurements were performed with Agilent 4284A Precision LCR meter. DC sweep voltage from negative voltage to positive voltage was applied at the front gate, causing the device to pass through accumulation, depletion, and inversion regions.

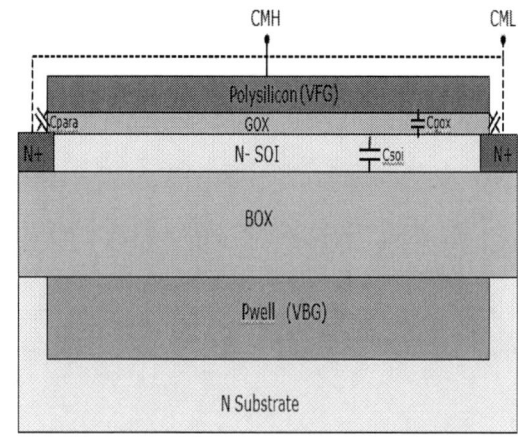

Fig. 1. Schematic Diagram of CV measurement structure of SOI MOSFET

Capacitance is the magnitude of the differential change in charge dQ on one plate as a function of differential change in voltage dV across the capacitor.

$$C = \frac{dQ}{dV} \qquad (1)$$

CMU measure the AC current induced by the capacitance of the MOSFET as in (2) and it is proportional to the measurement frequency [4].

$$I_{AC} = C \, dV_{AC}/dt \qquad (2)$$

The capacitance in AC signal frequency (f) domain is shown in (3).

$$C = I_{AC}/ 2\pi f V_{AC} \qquad (3)$$

III. RESULT AND DISCUSSION

SOI Film and GOX Thickness

Capacitance was measured between gate and drain/source with different back gate voltage (Vbg) at -75V, 0V and 75V. The front gate (Vfg) was swept from -6V to 6V, covering the accumulation, depletion and inversion region at front channel. The test frequency was set at 100KHz. AC source voltage applied was 30mV. The C-V curve obtained is shown in Fig. 2 below.

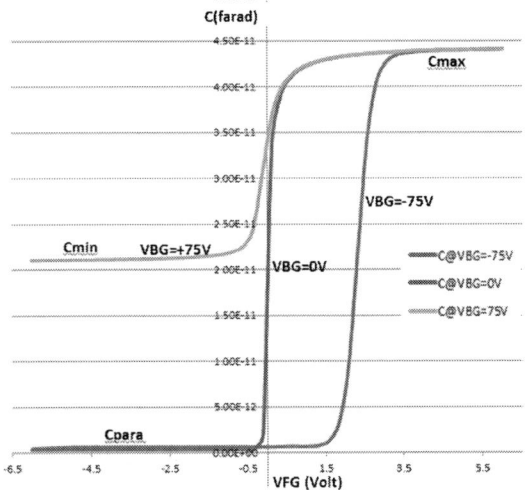

Fig. 2. CV measurement of SOI MOSFET with 3 different back gate voltage with Vfg sweep from -6V to +6V.

A strong positive DC bias voltage applied at front gate causes the majority carriers in the SOI film to acumulate near the gate oxide interface, since the carrier cannot get through the insulating layer, capacitance is at maximum (Cmax) in the accumulation region. Cgox is the gate oxide capacitance and Cpara represents the parasitic capacitances generated by overlapping of n+ area with the polysilicon gate area, it is the gate to drain/source overlap capcitance, extracted at channel off condition when Vfg was biased smaller than front channel threshold voltage at negative voltage. Cmax consists of Cgox and Cpara connected in parallel.

$$Cmax = Cgox + Cpara \qquad (4)$$

The Cgox and the Thickness of gate oxide (Tgox) can then be calculated with equation (5) & (6). εox is the permittivity of the oxide and it is 3.9 ε0 where ε0 is the permittivity of free space.

$$Cgox = Cmax - Cpara \qquad (5)$$

$$Tgox = \frac{\varepsilon ox}{Cgox} \qquad (6)$$

When Vbg was biased at high voltage (+75V), and Vfg swept at negative voltage, the majority carrier (electron) in the N- SOI got pushed away from the oxide interface and depletion region formed. The SOI film at this condition behaves like an dielectric, it is depleted of mobile carriers but filled with ionized charges, and the depletion width is at maximum. The C-V meaasured across the Gate and drain/source is now consist of the Cgox in series with the SOI capacitance (Csoi) as in (7). Noted that the polysilcion depletion effect was not considered in this study.

$$\frac{1}{Cmin - Cpara} = \frac{1}{Cgox} + \frac{1}{Csoi} \qquad (7)$$

The thickness of SOI (Tsoi) is then derived as in (8). W and L are the width and length of the drawn capitance area. εsi is the permittivity of silicon.

$$Tsoi = \varepsilon si * WL * \frac{Cmax - Cmin}{(Cmax - Cpara)(Cmin - Cpara)} \qquad (8)$$

When Vbg was biased at -75V, the flat band voltage shifted to more positive value as the threshold voltage of the front channel was increased by the back gate bias. The back gate coupling effect is seen. The calculated GOX and SOI thickness out of the C-V measurements were verifed with the values obtained through physical Cross Sectional MicroScopy (XSEM), and they were in good agreement.

Frequency Dependency on Measurement Result

The dependence of AC signal frequency on the C-V behaviour was studied for test frequency ranging from low frequency to high frequency at 1KHz, 10KHz, 100KHz and 100MHz. Sweeping voltage for Vfg was from -6V to 6V, else Vbg was set at -75V, 0V and +75V respectively. Fig 3. indicates that for Vbg=0V, the C-V curve do not shift with respond to the frequency variation. The 4 curves are overlap together. No frequency dependency seen on the result.

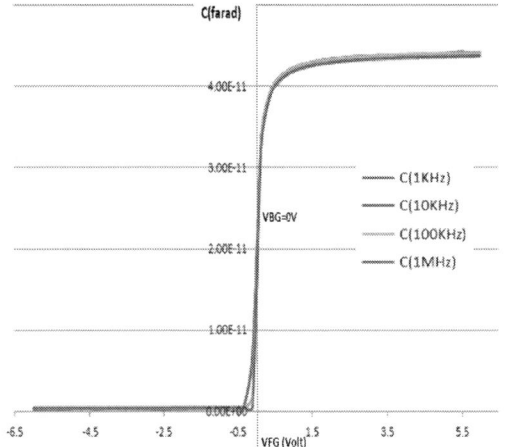

Fig. 3 C-V curve with Vbg=0. No frequency dependency seen on the measurement result.

978-1-5386-5284-8/18 $31.00 © 2018 IEEE

Similar observation can be seen with Vbg=+75V. The C-V curve do not shift with respond to the AC signal frequency variation as shown in Fig. 4 below.

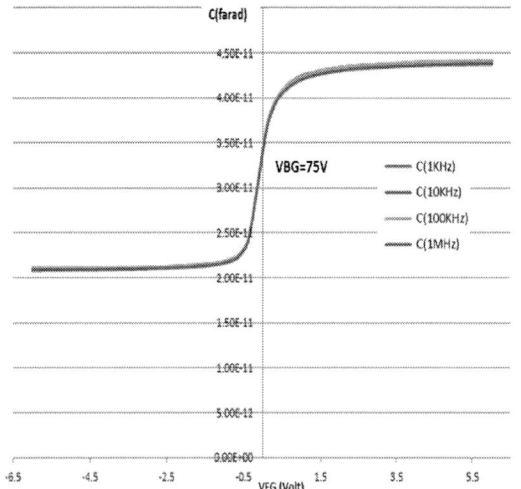

Fig. 4. C-V curve with Vbg = 75V. No frequency dependency seen on the measurement result.

But for Vbg=-75V, a strong frequency dependance is seen as in Fig. 5. The flatband voltage reduced as frequency decreased. Higher frequency shifted the flatband voltage to higher value.

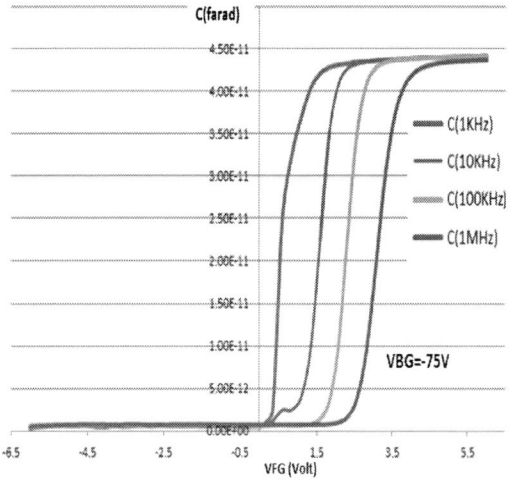

Fig. 5. C-V curve with Vbg=-75V. Strong frequency dependency seen on the measurement result.

At Vbg=-75V, the n- SOI body was in inversion. This frequency dependence occurs primarily in inversion since a certain time is needed to generate the minority carriers in the inversion layer [5]. The inversion charge is contributed by the minority carriers which have a definite generation and recombination rate. Now when the frequency applied

becomes greater then that rate, the inversion charge cannot follow the change. The majority and minority carrier response times to ac gate voltage are very different, the minority carrier response is typically as long as 0.01-1s and much slower than the frequency of bias at high frequency [6]. At high frequency AC signal, the minority carriers cannot be generated or be removed fast enough to adequately respond to the AC signal, but they do follow changes in gate bias.

Fixed Oxide Charges and Interface Trapped Charges

The C-V characteristics of 2 samples (Sample A and sample B) were collected from 2 different wafer fabrication plants to compare the charges level. The measurement was taken at Vbg=-75V and the result is shown as in Fig. 6.

The high frequency (100KHz) capacitance was used to evaluate the fixed oxide charge densisty. Sample B has high fixed oxide charge created during the gate oxidation process and the curve is shifted along the voltage axis with flatband shift to the left compared to sample A. The positive fixed charges shifted the flatband voltage by amount which equals the charge divided by the oxide capacitance. The flatband voltage shift due to fixed oxide charges is distinct at high frequency measurement.

The C-V curve at low frequency (1KHz) on the 2 samples shown the stretch out pattern, indicating the possibility of different level of interface trapped charges. The stretch out is more pronounced at low frequency as the traps and minority carrier inversion charges are responding to the ac signal. The interface trap charges respond is immediate when frequency is low [7]. Majority carriers can follow very high frequencies, but the minority carriers and a part of the interface states cannot follow high frequencies. The interface trapped charge density (Dit) estimation is better at low frequency for interface trap charges.

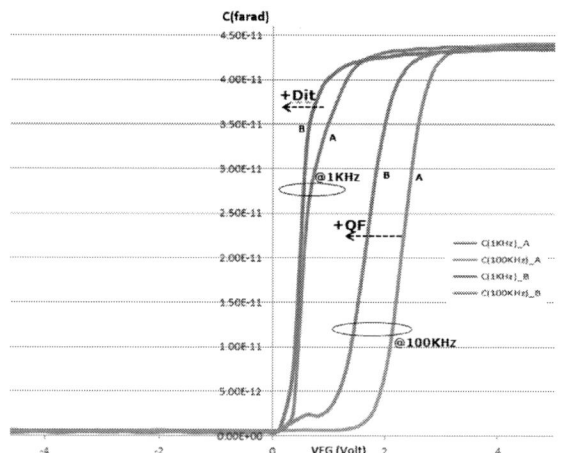

Fig. 6. C-V characteristic respond to fixed oxide charges and interface trapped charges at low and high frequency for samples from 2 different wafer fabrication plant. Vbg set at -75V.

BOX Thickness

The BOX thickness of SOI is an important parameter to model the device performance. The set up to characterize the BOX thickness is illustrated in Fig. 7.

978-1-5386-5284-8/18 $31.00 © 2018 IEEE   236

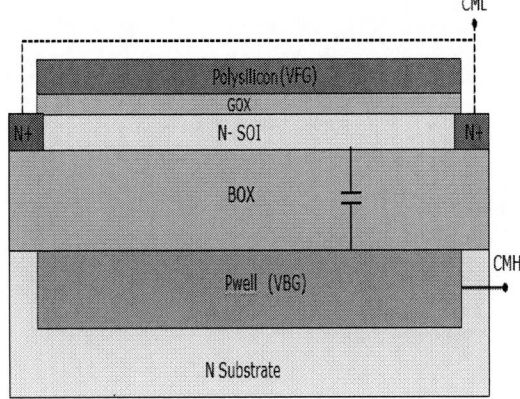

Fig. 7. Schematic diagram illustrates BOX thickness measurement set up

The CMH was connected at pwell, else the CML connected at drain/dource at SOI film. The back gate Vbg was swept from -40V to 40V. The capacitance between back gate and drain/source were measured at Vfg of -5V, 0V and +5V respectively. Frequency of 100KHz was used in this measurement. The measured C-V curve is shown in Fig. 8.

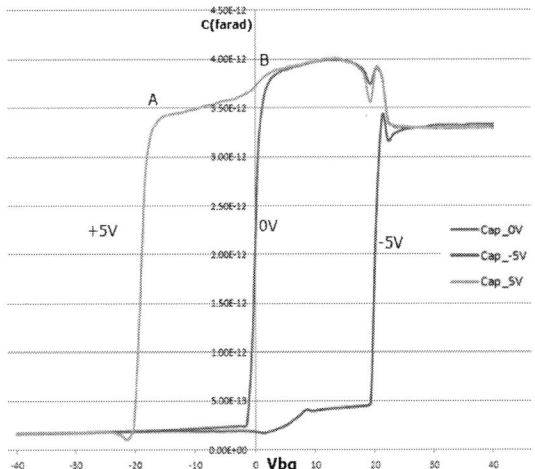

Fig. 8. C-V curve with Vbg sweep from -40V to 40V. The Vfg set at +5V, 0V and -5V.

Due to the coupling effect, when Vfg at -5V, the back gate flatband voltage shifted to the right meaning that the required back channel threshold voltage become higher. But when Vfg ws biased at +5V, the back gate required more negative voltage (~-20V) to invert the channel and turn off the normally on depletion MOSFET.

Capacitance of BOX (Cbox) was obtained when the channel was in accumulation, measured with high positive

Vbg, with Vfg=0V. The BOX thickness can then be derived by dividing the permittivity of oxide by Cbox.

$$\frac{1}{C} = \frac{1}{Cbox} + \frac{1}{Cdeppwell} \qquad (9)$$

When Vbg swept below 0V at negative voltage, the pwell which is p-type become depleted and the depletion capacitance Cdeppwell was in series with Cbox, causing the total capacitance to reduce as can be seen from Point B to Point A in Fig. 8.

There is a hump in capacitance seen when Vbg at 0V to 20V, the effect is still under investigation and remain out of the scope of this study.

IV. CONCLUSION

The C-V characteristics of PD SOI MOSFET have been presented. Important parameters like SOI film thickness, GOX thickness, BOX thickness, fixed oxide charges and interface trapped charges were derived from the C-V curve and discussed together with the frequency dependency effect. The C-V characteristics provide a wealth of information to evaluate the performance of MOSFET fabrication process and to model the device performance.

ACKNOWLEDGMENT

The authors would like to thanks Dirk Priefert from Infineon Germany for his valuable advice on the measurement methodology, and to Ray Ang Chung Keow from Infineon Kulim Characterization lab for the support on C-V measurement.

REFERENCES

[1] B. Vandana, "Study of Floating Body Effect in SOI Technology", International Journal of Modren Engineering Research. Vol. 3. Issue 3, June 2013.

[2] Y.Hu, H.Liu, Q. Xu, L. Wang J., Wang, S. Chen, P. Zhao, Y. Wang, G. Wang., " Dimension Effect on Breakdown Voltage of Partial SOI LDMOS ", Journal of the Electron Devices Society. Vol 1, No 5, May 2017.

[3] D. Priefert, " Test prohram Kerf C5SOI", Test Porgram Kerf C5SOI Document, pp 17, 2017.

[4] L. Stauffer, "Fundamentals of Semiconductor C-V Measurements" Keithley Instruments Inc. Article in EE, Evaluation Engineering 31(1):18-21, Jan 2009

[5] B.V.Zeghbroeck, "Principles of Semiconductors Devices", eBook, 2014.

[6] D. Tomaszewski, L. Lukasiak, A. Zareba and A. Jakubowski, "An impact of Frequency on Capacitances of Partially-Depleted SOI MOSFETs" Journal of Telecomnunications and Information Technology. 3-4/2000.

[7] N.P. Maity, R. R. Thakur, R. Maity, R.K. Thapa, S. Baishya. "Analysis of Interface Charge using Capacitance-Voltage Method for Ultra Thin HfO2 Gate Dielectric Based MOS Devices" Procedia Computer Science Vol 57, pg 757-760, 2015.

Comparative Study of Si Based Micromachined Patch Antenna Operating at 5 GHz for RF Energy Harvester

Noor Hidayah Mohd Yunus[1,2], Jahariah Sampe[1] and Jumril Yunas[1]
[1]Institute of Microengineering and Nanoelectronics
Universiti Kebangsaan Malaysia
43600 Bangi, Selangor, Malaysia
noorhidayahm@unikl.edu.my, jahariah@ukm.edu.my, jumrilyunas@ukm.edu.my

Alipah Pawi[2]
[2]Communication Technology Section
Universiti Kuala Lumpur British Malaysian Institute
53100 Gombak, Selangor, Malaysia
alipah@unikl.edu.my

Abstract— **This paper presents a comparative study of Si based micromachined and Si based regular rectangular microstrip patch antenna operating at 5 GHz (unlicensed Industrial, Scientific and Medical (ISM) band) frequency for radio frequency (RF) energy harvester. The work is aimed to find alternative substrate modes for integrating the antenna with the harvester circuitry design. The micromachined antenna is based on a modification of silicon (Si) wafer using micromachining technology. In this design, the Si material is etched horizontally underneath the printed patch to produce a cavity consisting of an air-filled and the Si structure in a defined thickness ratio. The air cavity synthesizes a low effective dielectric constant around the radiating patch and improving the radiation gain. The design has been simulated and analysis in CST Microwave Studio software (CST MWS). The results validate superior parameter of the micromachined over regular antenna where the gain has increased by 166.2% while the -10 dB bandwidth performed. Thus the idea of micromachining is relatively useful for designing an efficient radiator and broad bandwidth antenna for integrated RF energy harvester applications.**

Keywords— *RF energy harvester, micromachining, CST MWS, integrated antenna, radiation gain, microstrip patch antenna*

I. INTRODUCTION

Harvesting of RF energy is an attractive way for supplying energy while energy resources are limited and costly. Transmission of electromagnetic waves continuously from base station, communication tower, television and gadgets such mobile phone and radio, results in abundance of scattered electromagnetic waves in surrounding. These waves can be harvested by antenna system that convert the electromagnetic waves into useful electrical power [1-6]. In this study, the microstrip patch antenna is preferred for the intended RF energy harvester application.

Microstrip patch antennas have been widely used in communication applications such as navigation, remote sensing and telemetry due to their planar, low manufacturing cost, simplicity, integratable characteristics and enormous CAD software tools of design and analysis. Extensive rigorous research of microstrip patch antenna is incorporated to improve radiation characteristics and bandwidth widening. A microstrip patch antenna printed on high resistivity or high effective dielectric constant ($\varepsilon_r \geq 10$) substrate shows performance degradation on surface waves excitation. Thus, reduction in bandwidth, lower radiation efficiency and degraded radiation patterns of the antenna [7-8]. Generally such antennas print on low effective dielectric constant ($\varepsilon_r \leq 10$) substrates and characterized by higher radiation efficiency and maximum bandwidth as reported by numerous of theory and experimental researches. Only several experimental methodologies have been set forth to resolve on the excitation of high dielectric constant substrate in microstrip patch antenna. Physical high dielectric constant substrate alteration such reduction of the substrate thickness yields very small amounts of radiation resistance.

Micromachined antenna is a conception idea of the radiation characteristics which is achieved by a micromachining technique to produce an air cavity of the Si substrate in a defined thickness ratio. More specifically, the radiating patch is printed on a dielectric section comprised of air and Si substrate. Micro-electromechanical systems (MEMS) fabrication technology has contributed to the micromachining development [9]. The advantages of MEMS micromachined structure are it provides efficient radiation, wider bandwidth and the complete antenna system and other systems can be integrated and fabricated on the same Si substrate [7-9]. This study carries out on a comparative parameter of a Si based micromachined and regular rectangular microstrip patch antenna without cavity on Si substrate resonating at 5 GHz microwave band.

This work is supported by HiCOE project AKU95 and internal grant GUP-2015-21 by Universiti Kebangsaan Malaysia.

II. Si BASED MICROSTRIP PATCH ANTENNA

Micromachined antenna structures consist of a patch over the cavity and microstrip line. The microstrip line fed into the patch with the input impedance antenna of 50 Ω. Micromachining technique is used to produce an air cavity of the Si dielectric region that has thickness ≤ 50% of the original Si substrate thickness. In this micromachined antenna model, the hollowed air cavity (ε_r = 1) underneath the patch of 255 μm thick from the original Si substrate of 525 μm thickness (ε_r = 11.9, electric conductivity = 0.00025) is designed. The hollowed walls are slanted owing to the anisotropic nature when the chemical etching is performed as shown in Fig.1. The cavity sized and structure are designed to reduce the effective ε_r suitable for RF antenna application. The effective dielectric constant ε_{reff} formula by a cavity model is estimated by the following expression [7,10]:

$$\varepsilon_{reff} = \varepsilon_{cavity} \left(\frac{L + 2\Delta L \dfrac{\varepsilon_{fringe}}{\varepsilon_{cavity}}}{L + 2\Delta L} \right) \quad (1)$$

$$\frac{\varepsilon_{fringe}}{\varepsilon_{cavity}} = \frac{\varepsilon_{air} + (\varepsilon_{sub} - \varepsilon_{air})x_{air}}{\varepsilon_{air} + (\varepsilon_{sub} - \varepsilon_{air})x_{fringe}} \quad (2)$$

where

$$\varepsilon_{cavity} = \frac{\varepsilon_{air}\varepsilon_{sub}}{\varepsilon_{air} + (\varepsilon_{sub} - \varepsilon_{air})x_{air}} \quad (3)$$

In the expression, ε_{cavity} is the relative dielectric constant of the mixed substrate. ε_{fringe} is the relative dielectric constant in fringing fields. ΔL is the extension length of the antenna. The concept of micromachined antenna should reduce the ε_{reff}.

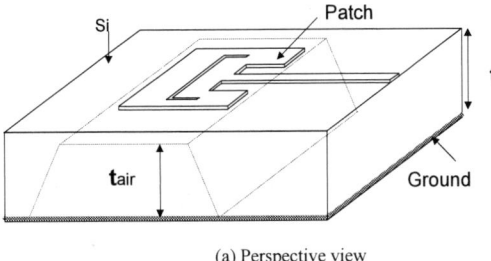

(a) Perspective view

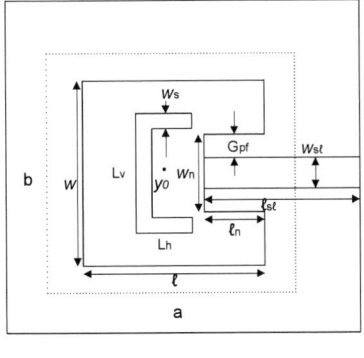

(b) Top view

Fig. 1. Geometry of micromachined microstrip patch antenna with air-Si region.

The antennas are designed so that the device should be able to resonate at 5 GHz. The optimum dimensions of a Si based micromachined antenna and Si based regular patch that resonance at 5 GHz is shown in Table I. A 1.0 μm thick metal layer of the patch and ground sides are mounted on Si substrate for both antenna structures. Here in case the micromachined antenna structure resonance at 5 GHz based on the Si substrate in which has no cavity (a, b and t_{air} = 0), the resonant frequency initially is shifted from 5 GHz to 7.039 GHz as shown in Table II. By physically altering the patch and Si substrate dimension, the 5 GHz resonant frequency of the regular antenna can be obtained. Property optimization by adjusting the antenna dimensions using CST MWS is implemented.

TABLE I.   DIMENSIONS OF ANTENNA FOR 5 GHZ RESONANT FREQUENCY ON SI (ε_R = 11.9) SUBSTRATE

Dimension (mm)	L	W	a	b	t_{air}	t	l	w
Regular	48	48	0	0	0	0.525	39.1	38
Micro-machined	30	28.8	22	22	0.255	0.525	17.5	20.6

The effects of the dimensional variation are further studied by comparing the antenna dimension of the regular antenna in case the cavity exists. From the desired 5 GHz resonant frequency, it is observed that the regular antenna size should be enlarged by an average of 44.63% when compared to the micromachined antenna size. This means that the antenna size is larger for material with higher dielectric constant. The antenna parameters such as return loss parameter, far-field pattern, bandwidth and 3 dB beam widths will be discussed in Section III.

III. CAD SIMULATION RESULT AND DISCUSSION

CAD model of the antenna is simulated by CST MWS software. The simulated return loss results are plotted in Fig. 2. It is observed that the antennas is resonating at 5 GHz with a return loss of -20.6547 dB, and -4.4834 dB for Si micromachined antenna and Si based regular antenna, respectively. The Si based regular antenna with a return loss of above -10 dB, accordingly no valuation which related to the -10 dB bandwidth. The result of -10 dB bandwidth for the Si micromachined antenna is 56.5 MHz. It reveals that, higher dielectric constant based on Si substrate micromachining technique such micromachined antenna improves bandwidth parameter.

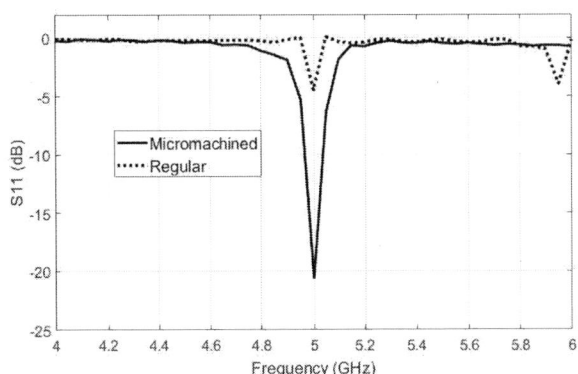

Fig. 2. Return loss comparison resonating at 5 GHz.

The variation of return loss versus resonant frequency by varying t_{air} depth on Si based micromachined antenna is shown in Fig. 3. Here by adjusting the t_{air} depth, the parametric resonant frequencies also can be changed. The return loss is lower than -15 dB at the frequencies for the different values of the t_{air} depth. As expected, the -10 dB bandwidth parameter is performed in all resonant frequencies. The variance between the parameter patterns can be attributed by the different effective dielectric constant ε_{reff} of the cavity from the different t_{air} depth. Compared to the case of no t_{air} depth ($t_{air} = 0$), the ε_{reff} of the substrate is evidently high and the bandwidth will decrease partly owing to the increase of the Si substrate height.

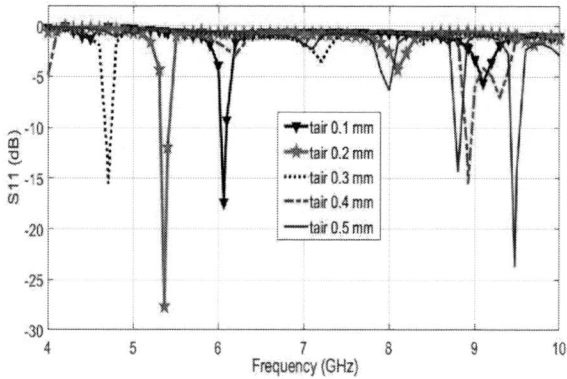

Fig. 3. Variation of return loss for different values of t_{air} depth on Si substrate.

The simulated E-plane (y-z plane) and H-plane (x-z plane) radiation patterns of the Si based antenna is shown in Fig. 4. The E-plane pattern is symmetric toward z-axis and the H-plane is asymmetric toward the x-z axis. These phenomena are caused by the patch and the phase of the slot, respectively. The radiation patterns are omnidirectional beam and good realized gain of 5.493 dBi with much smoother antenna radiation pattern as shown in Fig. 4(a). The E-plane is wider than the H-plane for the linearly polarized radiation wave that more than 5 dBi gain. In contrast, minus realized gain of 3.638 dBi for the radiation pattern shown in Fig. 4(b). It means that the radiation mechanism of Si based micromachined antenna is better than the Si based regular antenna. As soon as the Si based regular antenna is replaced by making an air cavity, the antenna gain increases satisfactory.

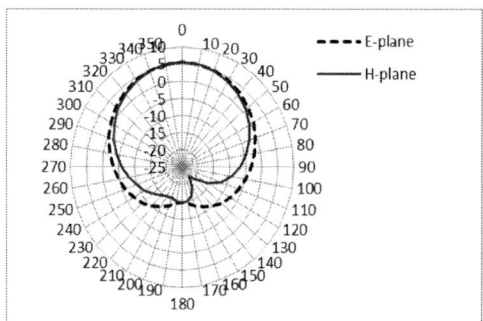

(a) Micromachined antenna

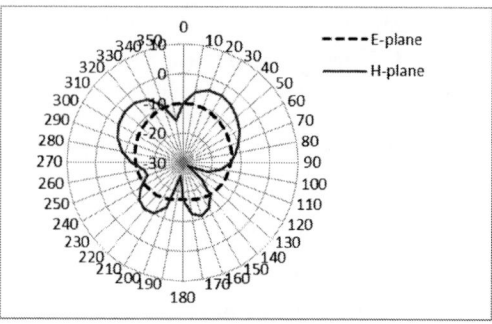

(b) Regular antenna

Fig. 4. E-plane and H-plane at 5 GHz on Si substrate

Detailed CAD analysis of Si based micromachined antenna is carried out by varying air-cavity depth. An estimate depth of t_{air} is assumed vertical by considering also the anisotropic nature of the hollowed cavity walls. The t_{air} is varied from no air cavity (0 mm) until full air cavity (0.525 mm), maintaining all the other dimensions constant. The resonant frequency shift, -10 dB bandwidth, radiation gain and 3 dB plane beam widths are observed as shown in Table II. Since the centre frequency is fixed at 5 GHz, the Si based micromachined antenna having increased bandwidth performance and fair wider 3 dB angular beam widths in both E-plane and H-plane. It shows that, 3 dB beam width for E-plane is wider than H-plane for different t_{air} depth on Si substrate. The Si based micromachined antenna at 5 GHz also gives an antenna gain better than 5 dBi over the beam width planes. However, shifted to the higher resonant frequency increases the radiation gain that effect from the variation of t_{air} depth.

TABLE II.   CAD ANALYSIS OF SI BASED MICROMACHINED ANTENNA BY VARYING TAIR DEPTH ON SI SUBSTRATE

t_{air} (mm)	Resonant frequency (GHz)	-10 dB bandwidth (%)	Gain (dBi) at Resonant frequency	3 dB E-plane beam width	3 dB H-plane beam width
0	7.039	0.56	7.605	74.7°	66.7°
0.200	5.365	1.09	5.978	88.5°	79.0°
0.255	5.000	1.27	5.493	93.3°	82.3°
0.300	4.708	1.05	4.960	96.9°	85.0°
0.400	8.920	0.75	5.562	58.1°	49.9°
0.500	9.478	0.55	7.168	85.6°	51.7°
0.525	7.462	0.28	5.048	84.5°	57.3°

IV. CONCLUSION

A comparative study of a Si based micromachined antenna and Si based regular rectangular microstrip patch antenna at operating 5 GHz microwave-band has been performed. Trade off consideration in choosing the dielectric constant ε_r of the substrate material has a significant role in the antenna performance and the physical size dimension. The results obtained point out the advantage of the Si based micromachined antenna over the Si based regular antenna in improving the radiation gain and bandwidth. The air cavity within the Si substrate is significantly indicated that S11 parameter better than 20 dB, with reduced antenna size. An effective results of more than 5 dBi radiation gain and 56.5

MHz frequency bandwidth, respectively meet the desired requirement of integration between the Si based micromachined antenna operating at 5 GHz with the RF energy harvester circuit.

ACKNOWLEDGMENT

The authors acknowledge the gratitude of support from Universiti Kebangsaan Malaysia under HiCOE Project AKU95 and internal grant GUP-2015-21.

REFERENCES

[1] Yunus, N. H. M., Sampe, J., Yunas, J., & Pawi, A. (2017, October). Parameter Design of Microstrip Patch Antenna Operating at Dual Microwave-band for RF Energy Harvester Application. In *IEEE Regional Symposium on Micro and Nanoelectronics, RSM 2017*, pp. 92-95.

[2] Salah-Eddine Adami, Plamen Proynov, Geoffrey S. Hilton, Guang Yang, Chunhong Zhang, Dibin Zhu, Yi Li, Steve P. Beeby, Ian J. Craddock, and Bernard H. Stark. (2018). A Flexible 2.45-GHz Power Harvesting Wristband With Net System Output From− 24.3 dBm of RF Power. *IEEE Transactions on Microwave Theory and Techniques*, vol. *66, no. 1, pp. 380-395.*

[3] Sampe, J., Yunus, N. H. M., Yunas, J., & Pawi, A. (2017). Ultra-low Power RF Energy Harvesting of 1.9 GHz & 2.45 GHz Narrow-band Rectenna for Battery-less Remote Control. *Int. J. Information and Electronics Engineering*, vol. *7, no. 3, pp. 118-122.*

[4] Yunus, N. H. M., Sampe, J., Yunas, J., & Pawi, A. (2017). MEMS Based RF Energy Harvester for Battery-less Remote Control: A Review. *American Journal of Applied Sciences, vol. 14. no. 2, pp. 316-324.*

[5] Zulkifli, F. F., Sampe, J., Islam, M. S., & Mohamed, M. A. (2015). Architecture of Ultra Low Power Micro Energy Harvester using RF Signal for Health Care Monitoring System: A Review. *American Journal of Applied Sciences, vol. 12, no. 5, pp. 335-344.*

[6] J. Sampe, N. A. A. Semsudin, F. F. Zulkifli, M. S. Islam, and M. Z. A. Razak. (2017). Hybrid Energy Harvester Based on Radio Frequency, Thermal and Vibration Inputs for Biomedical Devices. *Asian Journal of Scientific Research, vol. 10, no. 2, pp. 79–87.*

[7] Papapolymerou, I., Drayton, R. F., & Katehi, L. P. (1998). Micromachined Patch Antennas. *IEEE Transactions on Antennas and Propagation, vol. 46, no. 2, pp. 275-283.*

[8] Singh, V. K. (2010). Ka-band Micromachined Microstrip Patch Antenna. *IET microwaves, antennas & propagation, vol. 4, no. 3, pp. 316-323.*

[9] Liu, P., Chang, L., Li, Y., Zhang, Z., Wang, S., & Feng, Z. (2017). A Millimeter-Wave Micromachined Air-Filled Slot Antenna Fed by Patch. *IEEE Transactions on Components, Packaging and Manufacturing Technology, vol. 7, no. 10, pp. 1683-1690.*

[10] Bahal, I. J. and P. Bhartia, Microstrip Antennas, Artech House, Boston, MA, 1985.

2018 IEEE International Conference on Semiconductor Electronics (ICSE)

Design of Phase Frequency Detector (PFD),Charge Pump (CP) and Programmable Frequency Divider for PLL in 0.18um CMOS Technology

Anim Arifah Ahmad, Sawal Hamid Md Ali, Noorfazila Kamal,
Siti Raudzah Abdul Rahman

Dept. of Electrical, Electronic & System
National University of Malaysia,
43600 Bangi, Selangor.
animad_90@yahoo.com, sawal@ukm.edu.my,
fazila@ukm.edu.my, raudzah88@gmail.com

Masuri Othman,
Institute of Microengineering and Nanoelectronics
National University of Malaysia,
43600 Bangi, Selangor.
masuri57@gmail.com

Abstract— **In this paper, a Phase Frequency Detector (PFD) Charge Pump (CP) and programmable frequency divider Phase Locked Loop (PLL) for Bluetooth Low Energy (BLE) are presented. It is implemented using 180nm CMOS technology. The programmable frequency divider consists of Dual Modulus Prescaler divide by 15/16, 7-Bit Programmable Counter (P), and 6-Bit Swallow Counter (S). The divider will operate between 2.4 – 2.48 GHz frequency with 40 channels frequency hopping. The design has been simulated with 1.8 Vdd and consumed 3.51mW power.**

Keywords—PLL;phase frequency detector; charge pump; programmable frequency divider; dual modulus prescaler; main counter;swallow counter

I. Introduction

Nowadays, the main challenges for getting high speed wireless communication system and the demand for the high operating frequency in low voltage, small area and low power components that merged performances with the capability to be constructed economically in large quantities [1]. PLL as frequency synthesizers plays an important role in a radio frequency (RF) system. A phase locked loop is one of the major parts for transceiver. Its role is to provide a very stable clock and in many situations the PLL is required to be programmable. The ability to produce programmable frequency clock generator has allowed the transceivers to operate at different frequencies and hence allows the implementation of frequency hoping that makes the information transmission safer.

There are different types of PLL depending on their applications. Generally PLL consists of Phase Frequency Detector (PFD), Charge Pump (CP), Loop Filter (LF), Voltage Controlled Oscillator (VCO) and Frequency Divider as a feedback. The PLL architecture used is shown in Fig. 1. Most of the research work of the PLL is described in many publications [2]–[5].

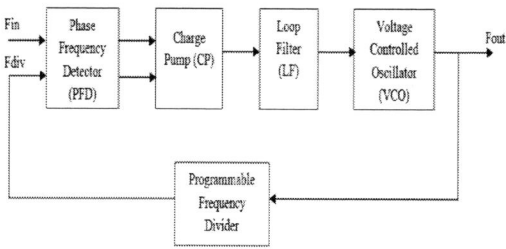

Fig. 1. PLL Basic Block Diagram

II. The Specification of PLL

The PLL was designed for the new standard Bluetooth Low Energy 4.0 (BLE) that operates in the ISM band at 2.4 GHz. The PLL has ability to be programmable from 2.4 GHz to 2.48 GHz. The BLE working group has set the bandwidth (BW) of 2MHz bandwidth per channel, with adaptive frequency hopping as many as 40 channels. All the 40 channels represent each frequency. For example,channel zero (0) is set at frequency 2400 MHz, channel one (1) at 2402 MHz until channel 40 at frequency 2480 MHz.

III. The Design of Phase Frequency Detector and Charge Pump

A conventional architecture of the PFD is shown in the Fig. 2. It is based on two D flip-flops (DFF) and delayed RESET AND-gate [6]. The output of the PFD depends on both the phase and frequency of the inputs SET. Phase Frequency Detector is a digital circuit detecting phase or frequency difference between reference signal (F_{ref}) and feedback signal (F_{div}) by comparing them and generates two output signals which are UP and DOWN. If reference signal is leading than feedback signal, UP signal will produce pulse. Likewise the feedback signal, if it is leading the DOWN signal will high. If there is no phase difference UP and DN signals will be zero. The generated output signals will go to the next block which is Charge Pump (CP).

978-1-5386-5284-8/18 $31.00 © 2018 IEEE   242

The charge pump circuit is shown in Fig. 3. The charge pump will take the output signal from PFD and translates them into currents and will go through a loop filter (LF) before transmitted to Voltage Controlled Oscillator (VCO). The output voltage will control the output frequency of the VCO. Basically, the charge pump consists of two (2) current sources and (2) switches [7]. There are three states in the charge pump correspond to the output to the loop filter which are charging state, discharging state and zero current state.

PFD will generate pulse width signal of the UP and DOWN signal. The charge pump will produce zero current state which means the charging and discharging is equal. When the feedback output frequency is leading the reference frequency, the PFD will activate the DOWN signal and deactivate the UP signal. Hence, switch S1 will be opened and switch S2 will be closed. The current ICP will flow out from the filter and reduce the V_{out}.

The lock condition of the PLL is established when the divider output frequency is the same as the reference frequency. The PFD will also deactivate and the both charge pump switches will be opened until the divider output frequency changes. Since switches are open, there is no current path formation, hence no current will flow into or out from the filter.

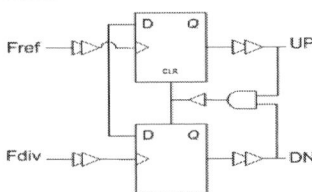

Fig. 2. PFD Architecture

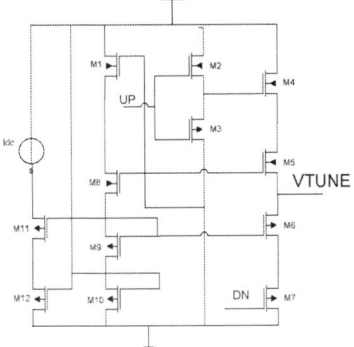

Fig. 3. Charge Pump Circuit

A. Simulation result of the PFD and CP

The simulation was done for the Phase Frequency Detector (PFD) and Charge Pump separately. Fig. 4 shows that the PFD produces UP pulse when frequency reference is leading while Fig. 5 shows the down pulse when reference signal (F_{ref}) is lagging. Fig. 6 showed the charge pump current (Icp) between UP and DOWN which is about 200uA. To avoid any mismatch of charge pump in PLL, the tuning voltage of VCO should be between 0.4V to 1.4V.

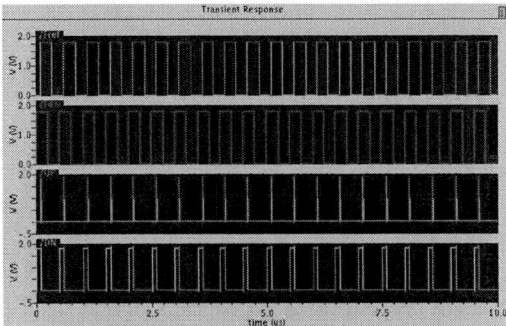

Fig. 4. PFD - When the frequency reference (F_{ref}) is leading.

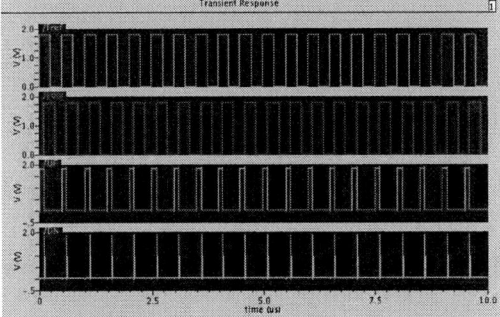

Fig. 5. PFD - When the frequency reference (F_{ref}) is lagging.

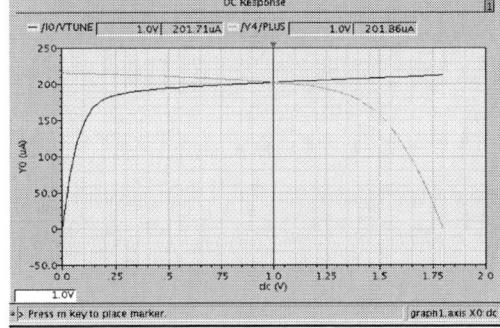

Fig. 6. Charge Pump Current between UP and DOWN

IV. THE DESIGN OF PROGRAMMABLE DIVIDER

The role of the programmable frequency divider is produce 40 channels for frequency hopping. The block diagram of the dual modulus 15/16 divider is shown in Fig. 7. One of the approaches to produce these frequency divider is using four (4) sets of TSPC divide by 2 with three (3) input OR gate and two set of multiplexers as shown in Fig. 8.

The will divide high frequency with a fixed value prescaler. The role of prescaler to reduce the PLL from high frequency(GHz) to more manageable frequencies (MHz). The prescaler operates in two (2) modes (MC) ie divide by N which is by 15 and divide by N+1 which is 16, depending on the control input to the prescaler.

Fig. 9, shows the complete programmable frequency divider that includes another two programmable counters which are 7-bit Main Counter (P) and the 6-bit Swallow Counter (S) to control the division ratio of the frequency. The division relation between the division (N, N+1), main counter (P) and swallow counter (S) is given by :

$$F_{div} = S(N+1) + (P-S)N = NP + S \qquad (1)$$

Based on equation (1), channel zero (0), will produce 2400MHz and go forth until channel 40 that will produce 2480MHz.

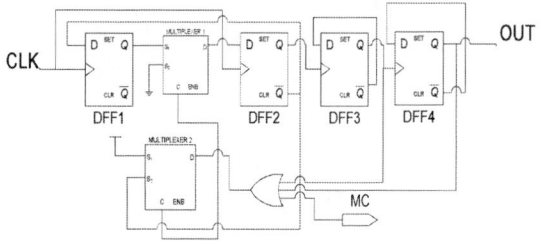

Fig. 7. Block Diagram of Divide 15/16

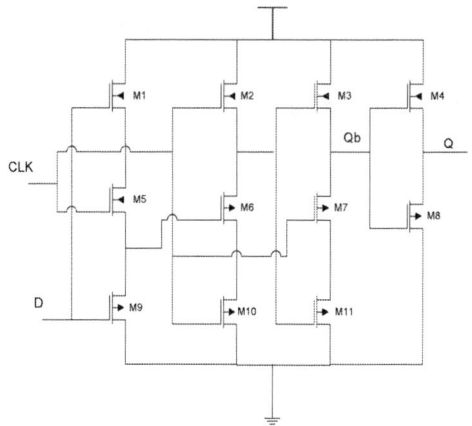

Fig. 8. TSPC Divide By 2 [8]

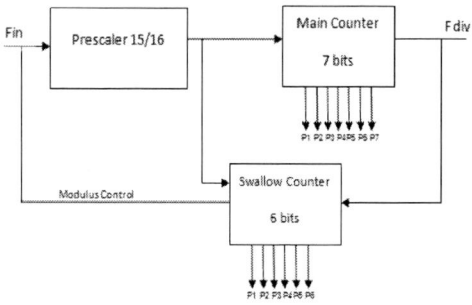

Fig. 9. Programmable Frequency Divider Block Diagram

A. The designed of 15/16 Prescaler

A 15/16 prescaler is used for the programmable divider as shown in Fig. 7. The prescaler consist of control logic signal

know as MC. It will decide either to divide by 15 or 16 depending on the signal given. If the signal is high, the prescaler will divide by 15. If the signal is low, the prescaler will divide by 16. To provide 2 MHz output (F_{div}), the value of S-counter should be less than P counter value (S <P) to satisfy the topology of pulse-swallow [9].

B. The design of the Main Counter (P) and Swallow Counter

The prescaler will receive the input from the VCO output and P and S counter work in the frequency range 150 - 160 MHz in order to obtain 2 MHz frequency output.

The 7-bit Main Counter (P) value is set to 80. The counter used in this design is 7-bit asynchronous down counter. It is designed with seven (7) reloadable TSPC D flip-flops (DFF) and end of count (EOC) detector.

The Swallow Counter used in this paper is consists of 6-bit asynchronous down counter. It is designed almost the same with the Main Counter but the reloadable D flip-flops for the Swallow Counter needs an extra logic function (MC) to be incorporated. It is added to reduce the switching activities in the reloadable D flip-flop.

From the Table I, shows the exact values for the Main Counter and the Swallow Counter to achieve a different frequency division ratio for frequency synthesizer.

TABLE I. VALUES FOR PROGRAMMABLE FREQUENCY DIVIDER

Frequency Division (FD)	Prescaler (N/N+1)	Main Counter (P)	Swallow Counter (S)
2400 - 2480	N=15	P =80	S=0-40

C. Simulation results of the Programmable Frequncy Divider

The simulation results of the Programmable Frequency Divider's block including prescaler, Main Counter and Swallow Counter are shown in Fig. 10, Fig. 11 and Fig. 12 showed the result of the input frequency which is 2.4 GHz. Fig.10 and Fig. 11 show the operation of the divider from 2.4GHz input frequency for two different control signal (MC = 0) and (MC =1) respectively. Fig.12 shows the output of the programmable frequency divider which will produce 2MHz feedback frequency.

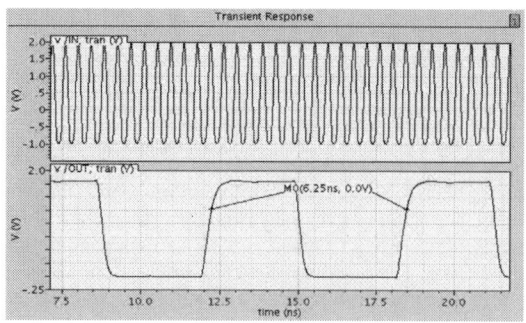

Fig. 10. Divide by 15 (MC = 0)

978-1-5386-5284-8/18 $31.00 © 2018 IEEE   244

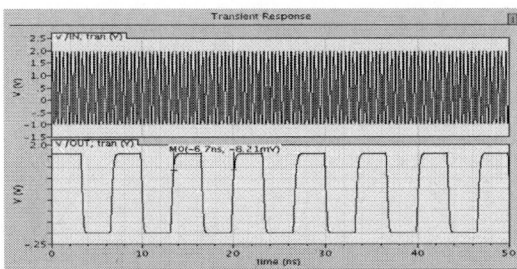

Fig. 11. Divide by 16 (MC = 1)

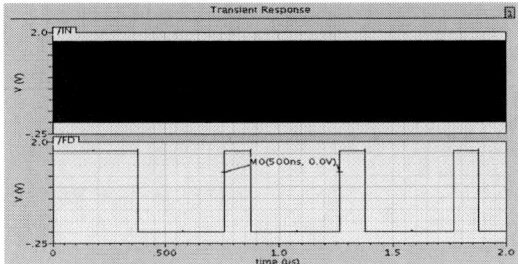

Fig. 12. Final result for the Programmable Frequency Divider (F_{in} = 2.4 GHz , F_{div} = 2 MHz at channel)

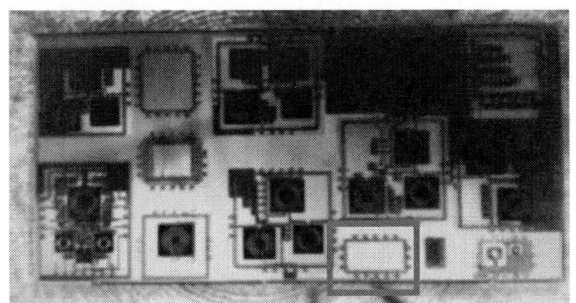

Fig. 13. Programmable Frequency Divider fabricated chip

TABLE II. PERFORMANCES COMPARISON OF PROGRAMMABLE FREQUENCY DIVIDERS IN 0.18UM TECHNOLOGY

Performances	[10]	[11]	[12]	This work
Power Supply, V	1.8	1.8	1.8	1.8
Operating freq, (GHz)	Up to 4	Up to 3.4	Up to 5	Up to 3
Division Ratio	15/16	481-496	32/33	1200-1240
Power Consumption, mW	3.625	6.25	4.3	3.51

V. CONCLUSION

This paper presents the design of Phase Frequency Detector (PFD) Charge Pump (CP) and Programmable Frequency Divider for PLL based on Silterra 0.18um CMOS for Bluetooth operation frequency

Technology with division ratio from 2400 - 2480 MHz with 40 channels frequency hopping . The simulation results showed the current consumption for the PFD and CP is 200 µA while the divider is 1.95mA at supply voltage of 1.8 V. Fig. 13 shows the fabricated chip, the measurement on this chip is ongoing. Table II shows the comparison between the proposed and the previous works.

ACKNOWLEDGEMENT

The author would like to thank the lecturer and staff of National University of Malaysia, Collaborative Microelectronic Design Excellence Centre (CEDEC), USM for providing Cadence Virtuoso tools and also to Silterra for facilitating the chip fabrication.

REFERENCES

[1] N. Deepak, D. Seshchalam, and K. Raghavendra, "A 12GHz Programmable Frequency Divider," pp. 199–203, 2016.

[2] N. M. H. Ismail and M. Othman, "Low Power Phase Locked Loop Frequency Synthesizer for 2 . 4 GHz Band Zigbee," *Appl. Sci.*, vol. 2, no. 2, pp. 337–343, 2009.

[3] N. Ickes, A. P. Chandrakasan, A. Pll, and U. P. Rf, "A 0.68V 0.68mW 2.4GHz PLL for ultra-low power RF systems," pp. 2–6, 2016.

[4] C. C. Boon, M. A. Do, K. S. Yeo, and J. G. Ma, "Multiple Modulus Fractional-N Frequency Divider Using N + 1 / 2 Division," *Control*.

[5] A. P. Loop, K. O. Kenneth, S. Member, and A. Abstract, "A 50-GHz Phase-Locked Loop in 0.13- um CMOS," vol. 42, no. 8, pp. 1649–1656, 2007.

[6] O. Abdelfattah, I. Shih, and G. Roberts, "A 0 . 55-V 1-GHz Frequency Synthesizer PLL for Ultra-Low-Voltage Ultra-Low-Power Applications," 2015.

[7] K. K. Patel and N. D. Patel, "Phase Frequency Detector and Charge Pump For DPLL Using 0.18µm CMOS Technology," *Int. J. Emerg. Technol. Adv. Eng.*, vol. 3, no. January, pp. 55–58, 2013.

[8] P. S. M. Ranjan, "A 1 . 8 Ghz-2 . 4 Ghz Fully Programmable Frequency Divider And A Dual-Modulus Prescaler For High Speed Frequency Operation In PLL System Using 250nm Cmos Technology," vol. 2, no. August, pp. 1510–1517, 2012.

[9] R. A. Dabhi and M. E. E. C. Student, "A Low Power 1MHz Fully Programmable Frequency Divider in 45nm CMOS Technology," vol. 1, no. 1, pp. 39–46, 2014.

[10] H. Liu, X. Zhang, Y. Dai, and Y. Lv, "Low power consumption high speed CMOS dual-modulus 15/16 prescaler for optical and wireless communications," *Optoelectron. Lett.*, vol. 7, no. 5, pp. 341–345, 2011.

[11] N. M. H. Ismail and M. Othman, "CMOS programmable divider for zigbee frequency synthesizer," *3rd Int. Conf. Signals, Circuits Syst. SCS 2009*, vol. 16, no. 2, pp. 1–3, 2009.

[12] L. Yan, H. Siliang, W. Donghui, and H. Chaohuan, "A 5-GHz programmable frequency divider in 0.18-um CMOS technology," *J. Semicond.*, vol. 30, no. 5, p. 055004, 2009.

2018 IEEE International Conference on Semiconductor Electronics (ICSE)

Design of 2.4 GHz CMOS LC Tank Voltage Controlled Oscillator (VCO) for PLL using 0.18 μm CMOS Technology

Siti Raudzah Abdul Rahman, Sawal Hamid Md Ali,
Noorfazila Kamal, Anim Arifah Ahmad
Dept. of Electrical, Electronic & System
National University of Malaysia
43600 Bangi, Selangor
raudzah88@gmail.com, sawal@ukm.edu.my, fazila@ukm.edu.my,
animad_90@yahoo.com

Masuri Othman
Institute of Microengineering and Nanoelectronics
National University of Malaysia
43600 Bangi, Selangor
masuri57othman@gmail.com

Abstract—**This paper presents the design of a 2.4 GHz LC Voltage Controlled Oscillator implemented using 0.18 μm CMOS technology. The LC VCO achieves a simulated phase noise of -97.76 dBc/Hz at 1MHz of offset frequency. The output frequency of VCO can be tuned from 2.4 GHz to 2.48 GHz frequency range which correspond to an 80 MHz tuning range Bluetooth device.**

Keywords— Voltage Controlled Oscillator, PLL, Phase Noise

I. INTRODUCTION

Phase locked loop (PLL) is one of the important components in a transceiver. Its role is to provide a very stable clock and in many situations, the PLL is required to be programmable. The ability to produce programmable frequency clock generator has allowed the transceivers to operate at different frequencies and hence allows the implementation of frequency hopping that makes the information transmission safer.

Although there are different types of PLL depending on their applications, generally PLL consist of Phase Frequency Detector (PFD), Charge Pump (CP), Loop filter (LF), Voltage Controlled Oscillator (VCO) and programmable divider (PDiv) [1], [2], [3], [4]. The PLL architecture is shown in Fig.1.

There are some issues for CMOS-based RFIC: CMOS model is ideal for digital IC design and thus RF CMOS requires new development in the CMOS model. Secondly, its highest resistivity is about 20 Ohms-cm (depending on the dopant level). As a result, a strong interaction between parts of circuits on the same substrate can take place. The passive devices tends to have low Q due to high losses in the substrate [5].

The other shortcomings of the CMOS are its low power rating and high noise level. Thus, extra careful is required when designing mm wave RF integrated circuits (FRIC).

The objective of this paper is to design and optimize the Voltage Controlled Oscillator (VCO) for 2.4 GHz Bluetooth frequency.

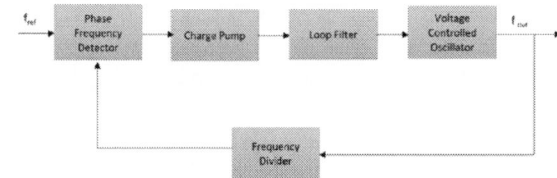

Fig. 1. Block in a PLL

II. VOLTAGE CONTROLLED OSCILLATOR

A. RFIC CMOS circuits for VCO

In the literature, two commonly used circuit techniques for VCO are ring oscillator and LC Tank [6]. Ring oscillator is known to produce relatively low oscillation frequency (< 1 GHz) and suffers from poor phase noise [7]. Thus, most of the high frequency applications and mm-wave RF makes use of LC tank oscillator [8]. The most popular topology is the cross-coupled as shown in Fig. 2 [9].

The circuit consists of three parts: the LC Tank which is the frequency selective part and the active circuit made of cross-coupled transistor pair and the current source to provide current for stable operation. The cross-coupled transistor pair and the current source transistor could be either NMOS or PMOS. As a result, four different topologies could be built as shown in Fig. 3.

Generally, the NMOS transistor pair with the PMOS current source configuration is used due to the lower power consumption and better frequency tuning range.

Further, a reduction in the power consumption by is obtained by using CMOS cross couple pair with the current source removed to avoid the problem with voltage headroom.

978-1-5386-5284-8/18 $31.00 © 2018 IEEE

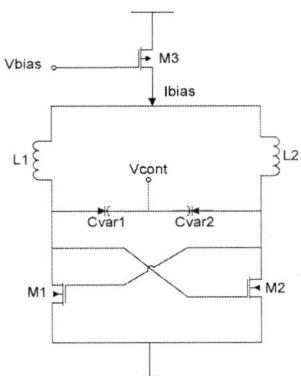

Fig. 2. Cross-coupled VCO

MOS varactors or PN-junction varactors are used in the LC Tank to achieve frequency tuning capability. The cross coupled transistor pair represents the one-port oscillator to provide a negative resistor to compensate for the energy loss in the LC tank in order to sustain the oscillation with the frequency ω given by:

$\omega = \frac{1}{\sqrt{LC}}$; Where L is the inductor and C is the total capacitance. The total capacitance includes the varactor capacitance and the parasitic capacitances in the circuit. Focusing on the LC frequency selective part, its equivalent circuit could be represented as a parallel combination of Lp, Cp, and Rp where Lp, Cp, and Rp are the equivalent inductor, capacitor and resistor/ loss of the whole LC tank respectively. Lp, Cp, and Loss Rp are given as [10]:

$$Lp = L(1 + \frac{1}{Q_L^2}) \qquad (1)$$

$$Cp = C(\frac{1}{Q_C^2}) \qquad (2)$$

$$Rp = [RL \cdot (1 + Q_L^2)] \; // \; [RC \cdot (1 + Q_C^2)] \qquad (3)$$

Where the $Q_L = \omega_L / RL$ and $Q_C = 1/\omega c Rc$ are the Q-factor of L and C respectively. If the Q_L and Q_C is big, then $L_P = L$ and $C_P = C$. The total Q of the tank Q-tank is

$$\frac{1}{Qtank} = +\frac{1}{Q_L} + \frac{1}{Q_C} \qquad (4)$$

For low frequency (several GHz), the values of Q_L is between 5 to 10 while for Q_C could be as high as 100 [11]. However, in the mm wave frequency, the Q_L is expected to increase. Thus, in the mm-wave, Q_C is expected to be dominant in term of loss and phase noise reduction. However, these values do not take into consideration of the output buffer and parasitic element in the circuit.

B. Condition for starting the oscillations

The equivalent resistance at the input of the cross coupled is given by the equation (5) with the transconduction of the two cross-coupled transistors are the same [12].

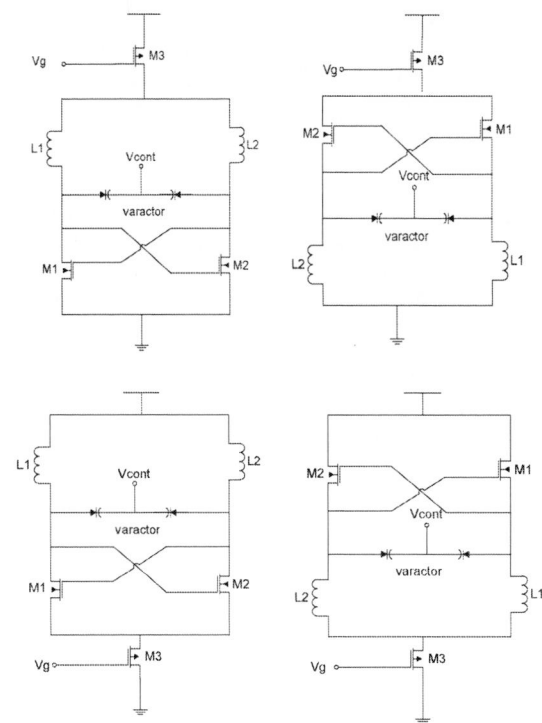

Fig. 3. Different topologies of cross-coupled VCO

$$Rin = \frac{Vin}{Iin} = -\frac{2}{gm} \qquad (5)$$

It is assumed that there is no channel modulation effect.

The absolute value of R_{in} should be as small as possible (so that gm is as large as possible) to produce oscillation. However, as the oscillation frequency increase, the parasitic components in the circuit also increases and thus may prohibit the circuit form oscillating.

Based on the model of the one-port oscillator described earlier, the oscillation will start if the condition shown by equation (6) is met [13]:

$$2R_P - 2/gm > 0 \qquad (6)$$

Where R_P is the equivalent resistor/loss in the half circuit. In term of R_P, the condition to start the oscillation is gm. $R_P \geq \alpha$ where α has a value between 2 to 3.

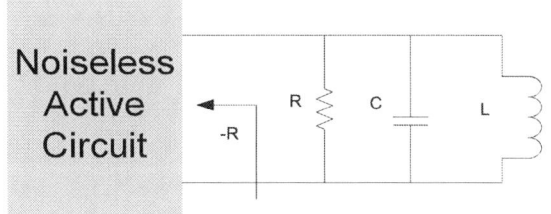

Fig. 4. Ideal one-port oscillator

C. The Phase Noise of the LC Tank VCO

The phase noise is another important criteria for the design of VCO [14]. In ideal situation and in application such as Local Oscillator, the VCO must produce pure sine wave whose spectrum has the shape of an impulse. Unfortunately, the nonlinear time-variant events that take place in system give rise to multiple sine waves of different frequencies and amplitude. The phase noise is said to spread the noise skirt around the carrier frequency. One important source of noise comes from R_P where its current noise spectral density is given by equation (7). The phase noise is defined as is the ratio of single-sideband noise power in a unit bandwidth at offset $\Delta\omega$ over the carrier power [15] as given in equation (8)

$$i_n^2/\Delta f = 4KT/R_P \tag{7}$$

$$L\{\Delta\omega\} = 10\log[\frac{2kT}{Pcarrier}(\omega o/2Q\Delta\omega)^2] \tag{8}$$

Equation (8) shows that the phase noise is proportional to $1/(\Delta\omega)^2$ or increase by $-20\ dBc/dec$. From the equation, the phase noise $L\{\Delta\omega\}$ can be reduce by:

1. Increasing the carrier power
2. By increasing the Q as $L\{\Delta\omega\}$ is inversely proportional to the square of the Q-factor.

Unfortunately, there are many other noise sources that contribute to the phase generated from the loss resistor RP. Thus, the value calculated from the $L\{\Delta\omega\}$ is the minimum value of phase noise. As observed by D. B Leeson [16], there is also $(1/\Delta\omega)^3$ as he proposed in equation (9)

$$L\{\Delta\omega\} = 10\log\{\frac{2FKT}{P_{carrier}} \cdot [1 + \left(\frac{\omega_o}{2Q \cdot \Delta\omega}\right)^2] \cdot (1 + \frac{\Delta\omega_{1/f^3}}{|\Delta\omega|})\} \tag{9}$$

Where F is an empirical parameter called device excess noise number, $2FKT/Pcarrier$ is the noise floor and $\Delta\omega_{1/f^3}$ refer to the corner frequency between $(1/\Delta\omega)^3$ and $(1/\Delta\omega)^2$ region. A number of research projects have been carried out to determine the value of F [17]. If the noise sources are from the LC, the white noise from the transistor cross-coupled pair and the white noise from the NMOS current source, the F parameter can be calculated based on equation (10)

$$F = 1 + \gamma_{ccp} + (4/9) + \gamma_{bias}g_{mbias} \cdot R_P \tag{10}$$

Where γ_{ccp} is the channel thermal noise coefficient of the MOS transistor, $+ \gamma_{bias}$ is the channel thermal noise coefficient of the current source and g_{mbias} is the transconductance of the current source. Fig. 5. shown phase noise.

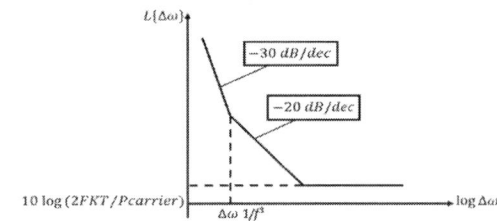

Fig. 5.   Phase Noise

III.   SIMULATION RESULT

The design has been simulated using 0.18 µm technology with a supply voltage of 1.8 V. There are four parameters have been simulated to get the performance of Voltage Controlled Oscillator namely output swing, phase noise, tuning range and gain. Fig. 6. shown the proposed schematic of the LC-VCO.

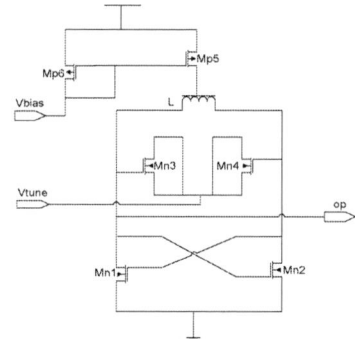

Fig. 6.   The Schematic of the LC-VCO

A.   Output Swing

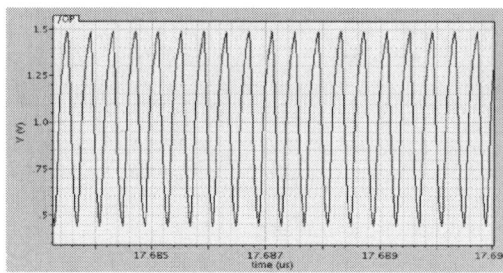

Fig. 7.   Transient response of the LC-VCO

Transient analysis of the VCO is shown in Fig. 7. The value of the center frequency is 2.4 GHz. The amplitude of this LC- Tank is about 1 V. Then the stabilization time of the output is around 20 µs.

B.   Phase Noise

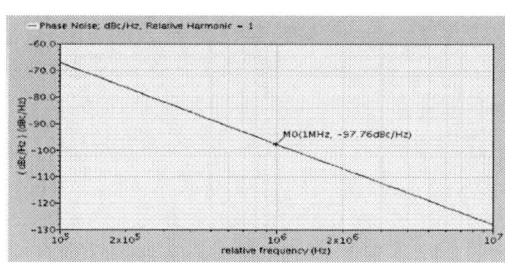

Fig. 8.   Phase noise of the LC-VCO at f_o = 2.4 GHz

The phase noise of the LC-VCO is shown in Fig. 8. The phase noise at 1 MHz offset frequency is -97.76 dBc/Hz from the center frequency 2.4 GHz. Table I. shows the phase noise for different offset frequency.

Table I. Phase Noise at various offset frequency for LC- VCO

Offset Frequency	Phase Noise
10 kHz	-36.41 dBc/Hz
100 kHz	-67.1 dBc/Hz
1 MHz	-97.76 dBc/Hz
10 MHz	-128.0 dBc/Hz
100 MHz	-156.3dBc/Hz

C. Tuning Range and Gain

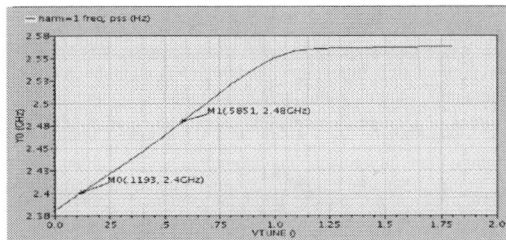

Fig. 9.   Tuning range of the LC-VCO

The tuning range of the LC-VCO is shown in Fig. 9. The tuning range of the LC-VCO is 80 MHz from 2.4 GHz – 2.48 GHz. The linearity of LC-VCO form 0.1193 V to 0.5851 V. Hence, the gain is 171.75 MHz/V.

D. Perfomance Comparison with other literature

Table II. Performance comparison of LC-VCO

Specification	This work	[2]	[9]
Technology (μm)	0.18	0.18	0.18
Center Frequency (GHz)	2.4	2.4	2.4
Supply Voltage (V)	1.8	3.3	1.8
Phase Noise (dBc/Hz @ 1 MHz offset)	-97.76	-135.6	-91
Tuning Range	80 MHz	1.04 GHz	300 MHz
Power Consumption (mW)	3.78	6.17	0.9

Table II summaries the performance comparisons between the proposed work and previous work for LC-VCO. The proposed work achieved better phase noise compared to other works with comparable power consumption.

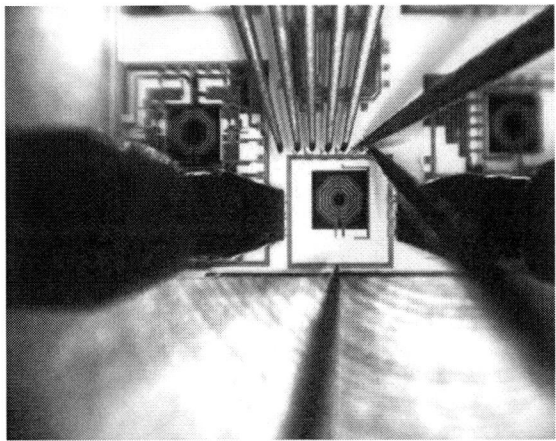

Fig. 10.   The fabrication chip of the LC-VCO

IV.   CONCLUSIONS

A 2.4 GHz – 2.48 GHz VCO for Bluetooth application has been successfully designed. The design is implemented using Silterra 0.18 μm CMOS technology process. Simulation results show that the tuning range is from 2.381 GHz to 2.562 GHz with the phase noise of -97.76 dBc/Hz at 1 MHz offset and the power consumption of 3.78 m. Fig. 10. shown the fabrication chip of the LC-VCO, the measurement still on going at CEDEC, USM.

ACKNOWLEDGMENT

The authors acknowledge with gratitude to lecturer at UKM. The research was partially sponsored by the TalentCorp Malaysia and collaborate with CEDEC, USM. Also, the chip fabrication was donated by Silterra Malaysia. The simulation was performed with Cadence Virtuoso Software.

REFERENCES

[1] V. Prasad and C Sharma, " A Review of Phase Locked Loop," International Journal of Emerging Technology and Advanvce Engineering, vol. 2, pp. 98-104, June 2012.

[2] M. V. Bhat, S. Jain, M. P. Srivatsa, M. Nithing and H. M. Kittur, " Design of Low Phase Noise Voltage Controlled Oscillator for Phase Locked Loop," 2017 International conference on Microelectronic Devices, Circuits System (ICMDCS).

[3] M. A. Joy and G. Manoj, "Analysis of Phase Locked Loop using 180nm technology for Bluetooth application," International Journal of Advanced Research in Computer and Communication Engineering, vol. 2, no. 3, pp. 1481- 1484, March 2013.

[4] C. P. Charjan. and A. S. Joshi, "Implementation of 2.4GHz Phase Locked Loop using Sigma Delta Modulator," International Journal of Application or Innovation in Engineering of Management, vol. 3, pp. 528-533, March 2014.

[5] D. Solanki, R. Chandel, T. Alam and A. Nishad," Design and Analysis of LC-VCO using MEMS Spiral Inductor," International Journal of Micro and Nano System, vol. 1, pp. 47-51, 2011.

[6] T. Miyazaki, M. Hashimoto, and H.Onodera, "A performance comparison of PLLs for clock generation using ring oscillator VCO and LC Oscillator in a Digital CMOS process," Proceeding of the 2004 Asia and South Pacific Automation Conference, IEEE press, 2004

[7] R. K. Patil and V. G. Nasre, " A performance comparison of current starved VCO and source Coupled VCO for PLL in 0.18μm CMOS Process," International Journal of Engineering and Innovative Technology, vol. 1, pp. 48-52, Feb. 2012.

[8] A. Nasri and M. Yargholi, " Design of a LC-Tank CMOS Voltage Controlled Oscillator," International Journal of Research in Electrical and Electronic Technology, vol. 1, pp. 33-37, June. 2014

[9] V. Anusha and K.C. Narasimhamurthy, " Design a Voltage Controlled Oscillator for 2.4GHz Wireless Applications," International Journal of Computer Application, vol. 1, pp. 31-35 September 2015.

[10] H. M. P. C. Jayaweer and A. Muhtaroglu, "Model Based Optiomization of Integrated Low Voltage DC-DC Converter for Energy Harvesting Apllications," Journal of Physics, 2016.

[11] A. P. V. D. Wel, S. L. J. Gierink, R. C. Frye, V. Boccuzzi and B. Nauta, "A Robust 43 GHz VCO in Standard CMOS for OC-768 SONET Apllications," IEEE Journal of Solid-State Circuits, pp. 345-348, August 2004.

[12] G. Haramkar, R.P. Patil," Design of a 2.4 GHz Low Phase Noise CMOS LC VCO for Wireless Applications," International Journal of Innovative Research in Electrical, Electronics, Instrumentation and Control Engineering, vol. 4, no. 6, June, 2016

[13] Z. Safarian and H. Hashemi,"Wideband Multi-Mode CMOS VCO Design using Coupled Inductors," IEEE Transactions on Circuits and Systems, vol. 56, no.8, August 2009.

[14] B. Razavi, "A study of Phase Noise in CMOS Oscillator," IEEE. Journal of Solid-State Circuits, vol. 31, no.3, March 1996

[15] C. Y. Yang and M. T, Tsai," High-Frequency Low Noise Voltage-Controlled LC- Tank Oscillator using a Tunable Inductor Technique," IEICE Trans. Electron, vol. E89-C, November 2006.

[16] D.B. Leeson, "A Simple model of feedback oscillator noise spectrum," in Proceeding of the IEEE 54(2), pp. 329-330, 1996.

[17] P. Andreani and X. Wang,"On the Phase-Noise and Phase-Error Performances of Multiphase LC CMOS VCOs," IEEE Journal of Solid-State Circuits, vol. 39, no. 11, November 2004.

First Principle Study of Graphene-Carbon Nanotubes Hybrid (GCH) Structure for Advanced Nanoelectronics Devices

Lee Li Theng[1], Iskandar Yahya[1,2], Mohd Ambri Mohamed[1], Mahamad Fariz Mohamad Taib[3]

[1]*Institute of Microengineering & Nanoelectronics,*
Universiti Kebangsaan Malaysia,
43600 UKM, Bangi, Selangor, Malaysia.
[2]*Faculty of Engineering and Built Environment,*
Universiti Kebangsaan Malaysia,
43600 Bangi, Selangor, Malaysia.
[3]*Faculty of Applied Science,*
Universiti Teknologi Mara,
40450 Shah Alam, Selangor, Malaysia.
Email : P89043@siswa.ukm.edu.my

Abstract—Although graphene and carbon nanotubes (CNT) exhibit remarkable electronic and mechanical properties, these two carbon allotropes face limitations due to their nanoscale size. The combination of graphene and CNT are suggested to overcome the problem of bundling and stacking in CNT and graphene which limit their performance. First principle study is carried out on the electronic properties of graphene-carbon nanotubes hybrid (GCH) structure using BIOVIA Material Studio software. Density Functional Theory (DFT) is used to calculate its band structure and density of states. The results show that when semiconducting CNT combine with graphene, its band gap is reduced compare to the pristine CNT. However, when metallic CNT is combine with graphene, the changes in band gap is depend on CNT's chirality. Chiral CNT will decrease the band gap of GCH, zigzag CNT causes not much changes in its GCH, while armchair CNT will increase the band gap of its GCH. Therefore, the result shows that it is possible to tune the band gap of GCH by combining graphene sheet with different type of CNT depending on our needs.

Keywords—graphene, carbon nanotubes (CNT), hybrid, first principle study, density functional theory (DFT)

I. Introduction

Carbon nanotubes (CNT) were first discovered by Iijima in 1991 [1]. There are two types of CNT, single-walled CNT (SWCNT) and multi-walled CNT (MWCNT). SWCNTs have tunable electronics characteristics which can be either metallic or semiconducting depend on its chirality. Graphite that is commonly found in pencil is made from million layers of graphene that stack together. Andre and Kostya are two scientists who were the first to isolate a monolayer graphene in 2004 [2]. Since then, they start to investigate the potential and properties of graphene and published their paper in Science which was then sparked a global explosion in graphene research. In 2010, their work in graphene had earned them a Nobel prize.

Then, there is an idea come out to make a composition of these two carbon allotropes. Matsumoto and Saito were the first to study this kind of structures [3]. They published a paper in 2002 which focusing on geometry and electronics states of graphene sheets covalently linked with carbon nanotubes. This earliest paper about graphene-carbon nanotubes hybrids (GCH) structures is a simulation work.

GCH structures can be categorized into two main types. One is without covalent bond between graphene and CNT while the other one is CNT covalently bonded with graphene. There are a lot of experimental work reported in synthesis of GCH structures. Without covalent bond, CNTs are just simply lie on graphene's surface. The main benefit of this kind of architecture is that it offers a three-dimensional conductive carbon network with open channels for electron transfer [4-6]. Direct growth method can provide the possibility of covalent bond between graphene and CNTs. There are one step and two steps fabrication technique reported using chemical vapor deposition process [7-9]. GCH with covalent bond can offer better characteristics; such as high transportation rate of electrolyte ions, electrons, and highly conductive due to synergetic effect between CNT and graphene [10].

From earlier research done, we know Graphene-carbon nanotubes hybrid (GCH) structures shows promising properties. Besides, it can be implemented in a wide area of application. Thus, if we can have better understanding about electronics properties of GCH, it can be a great advantage to the society in return. Moreover, most study on GCH structures consists of experimental work only. There is lack of simulation study on GCH structures..

II. Methodology

A. BIOVIA Material Studio

BIOVIA Materials Studio software is used in this work. It is a complete modeling and simulation software that is designed for researchers especially in the field of materials science and chemistry. It can be used to predict and understand the relationship between a material's atomic and molecular structure with its properties and behavior.

There are a few choices of quantum tools available in Materials Studio that can be used for calculations. For example, there are CASTEP, DMol³, DFTB+, and VAMP. In this project, we choose CASTEP for our calculations. CASTEP is a leading code for calculating the properties of materials from first materials. The reasons of choosing CASTEP is that it is specialize in simulates the properties of solids, interfaces, and surfaces of a wide range of materials including ceramics, semiconductors, and metals using a plane-wave density functional method. This meets our requirements to study the interface between carbon nanotubes and graphene.

B. Modeling of GCH structure

Graphene and CNT structure are directly import from the library file and used for the calculation. First, we import the graphene structure and made a supercell of 6x6x1. We must ensure that the size of graphene is larger than the diameter of CNT which attached to it. Next, preferable CNT's chirality are chosen and import from the library. After that, graphene and carbon nanotube are combined using build layer option. Graphene and CNT are selected as layer 1 and layer 2 respectively, and build as crystal. Hydrogen atom is added to the upper side of CNT to terminate the reaction on the upper side. Resulting structure is as shown in Fig 1 below.

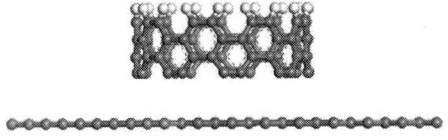

Fig. 1. *Modeled GCH structure without covalent bond.*

To build a GCH structure with a covalent bond, it is more complicated and challenging. Both graphene and CNT structure are imported from library first. Next, a whole which match the diameter of CNT is made by deleting carbon atom from graphene layer. After that, cavalent bond is made by connecting the carbon atom from CNT to carbon atom in graphene layer. The resulting structure is then tidy up using the clean tool. Finally, Hydrogen atom is also added to the upper part of CNT to terminate it.

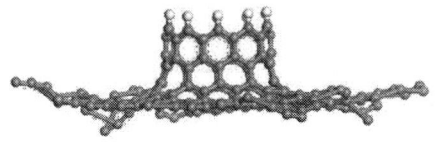

Fig. 2. Modeled GCH structure with covalent bond.

C. Geometry Optimization & Electronics Properties Calculation

CASTEP is used for the calculation of band structure and Density of states. Prior to the calculation on electronic properties, geometry optimization is performed to ensure the modeled structure reach the minimum energy states before further calculation is performed. Perdew, Burke and Ernzerhof (PBE) functional which belongs to the class of generalized gradient approximation (GGA) is used in this work. K-points set of fine quality was used and the maximum Self Consistent Field (SCF) cycles is set to 9999 to ensure the structure reach the minimum energy states before the calculations ends. All the parameter are the same when running calculation on electronic properties. Band structure and Density of states are chosen for the next calculation after the structure is geometrically optimized.

III. RESULTS & DISCUSSIONS

To verify the result from Material Studio, we had run the calculation on pristine monolayer graphene, metallic CNT and semiconducting CNT. Fig. 3 to Fig. 5 below shows the band structure of monolayer graphene, metallic CNT and Semiconducting CNT. The simulated band gap values are tabulated in Table 1. Comparison is made between the result from Material Studio software and a published result [11]. The results show that our result has not much different from the published work.

TABLE I.   COMPARISON BETWEEN THEORETICAL VALUE WITH THIS WORK.

Structure	Energy Band Gap (eV)	
	Theoretical	*This work*
Monolayer graphene	0	0.053
CNT (7,7)	0	0.122
CNT (10,0)	1.073	0.874
CNT (10,5)	0.993	0.849

Next, energy band gap calculation on different type of CNT with graphene sheet is performed. We choose 4 semiconducting CNTs and 4 metallic CNT with diameter range from 6 Å to 12.5 Å. From Figure 1, it is shown that all GCH structures with semiconducting CNTs has a lower energy band gap compared to its pristine CNT. For metallic CNT GCH structures, GCH (12,3) shows lower band gap compared to pristine CNT while all other GCH with metallic CNT shows larger band gap compared to pristine CNT. However, the increase in band gap for GCH (9,0) is very small and almost negligible. GCH (6,6) and (9,9) shows an obviously larger band gap after combine with graphene sheet. Both of it are armchair CNTs. To verify this, calculations is performed on another armchair CNT with chirality (7,7). The result shows that the band gap of its GCH had increased too. Thus, it can be observed that the change in band gap is regarding to the chirality of CNT attached to the graphene and GCH structure with armchair CNT will increase the band gap compared to its pristine CNT.

The GCH modeled in Figure 1 above is without covalent bond in between the interface of CNT and graphene. Next, we also performed calculation on GCH structure with covalent bond at the interface of CNT and graphene. However, not all CNT is able to form a stable GCH structure with covalent bond. Among the GCH structures without covalent that we had performed calculations, CNT (10,5), (12,3) and (9,9) failed to form a stable GCH structure with covalent bond. It fails to reach a minimum energy state which is stable to run for further calculations. It is believed that CNT (10,5) and (12,3) are chiral CNT which have a more complicated arrangement of carbon atoms compared to armchair and zigzag CNTs while CNT (9,9) has a large diameter than common CNT which is usually less than 10 Å. All of the GCH structures with covalent bond has smaller band gap compared to GCH structure without covalent bond and pristine CNT as shown in Figure 3 below. Except for GCH (9,0), its GCH structure with covalent bond has a slightly larger band gap compared to GCH without covalent bond. However, the increase in band gap is so small. The presence of covalent bond provides a path for electron to pass through easily from CNT to graphene. Thus, GCH structure has smaller band gap which makes electron to pass through from valence band to conduction band easily.

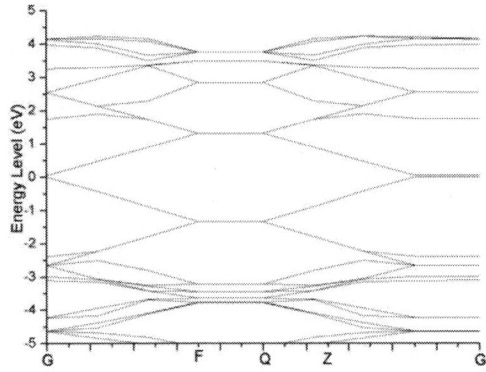

Fig. 3.   *Band structure of a monolayer graphene.*

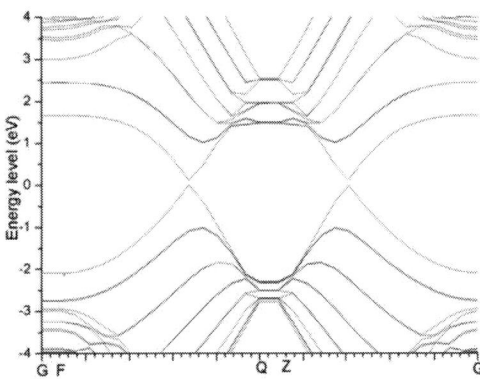

Fig. 4.   *Band Structure of Metallic CNT (7,7)*

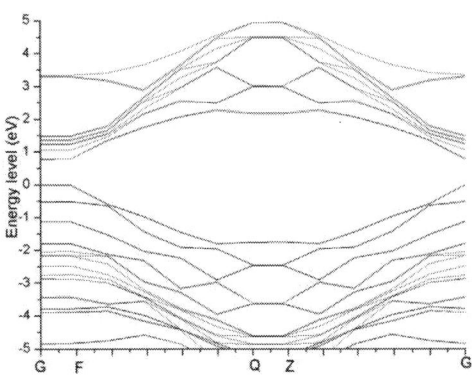

Fig. 5.   *Band structure of semiconducting CNT (10,0)*

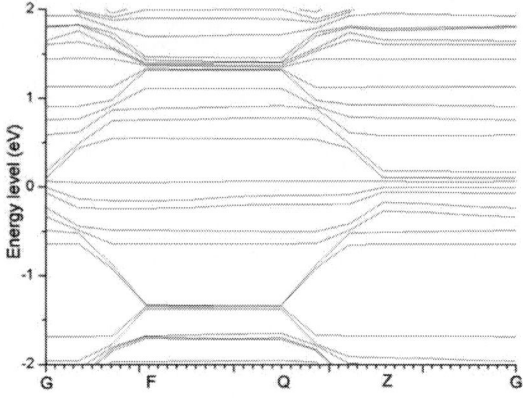

Fig. 6.   *Band structure of GCH (10,0) structure without covalent bond.*

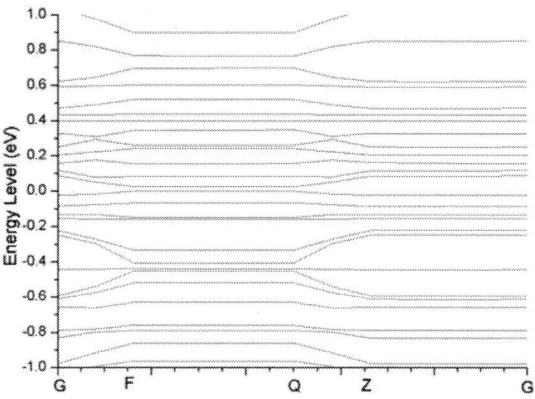

Fig. 7.   *Band structure of GCH (8,0) structure with covalent bond*

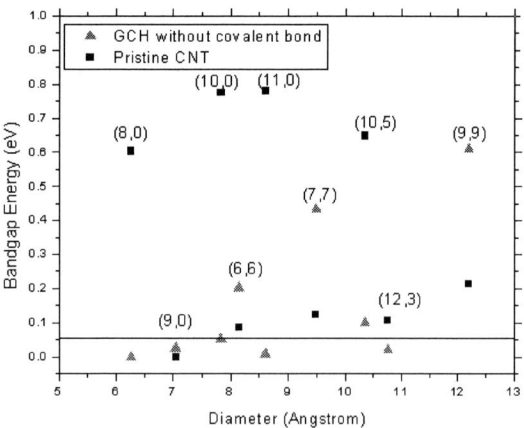

Fig. 8.   Band gap comparison between pristine CNT & GCH without covalent bond.

978-1-5386-5284-8/18 $31.00 © 2018 IEEE

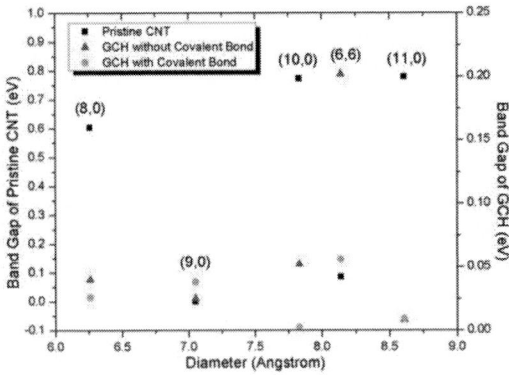

Fig. 9. Band gap comparison between pristine CNT, GCH without covalent bond and GCH with covalent bond.

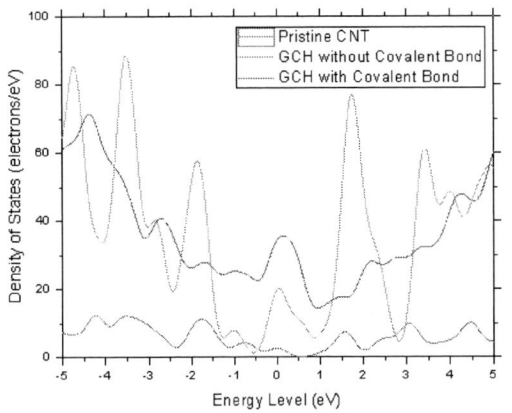

Fig. 10. *Density of ststes comparison between pristine semiconducting CNT (8,0), its corresponding GCH without covalent bond & with covalent bond.*

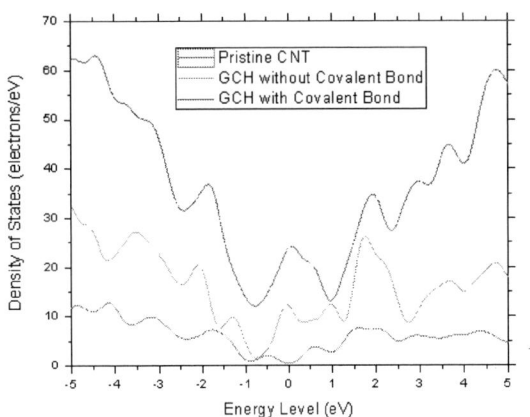

Fig. 11. *Density of ststes comparison between pristine metallic CNT (9,0), its corresponding GCH without covalent bond & with covalent bond.*

Next, comparison between the DOS of pristine CNT and GCH structure with and without covalent bond are made. Fig. 10 and Fig. 11 shows the DOS of semiconducting CNT with its corresponding GCH structure and metallic CNT with its corresponding GCH structure

respectively. From Fig. 10 & 11, we can observe that GCH structure has a higher number of electrons at each energy level as compared to pristine CNT. This is due to the decrease in band gap of GCH structure compared to pristine CNT. GCH with covalent bond does not shows a vanishing of DOS. This is due to the presence of covalent bond make the structure to be always filled with electron in every energy level.

IV. CONCLUSION

The band gap of GCH structure can be altered by controlling the type of CNT that is attached to the graphene. The possibility of altering GCH band gap can only be done on GCH without covalent bond. All GCH structure with covalent bond shows a very low band gap that is almost 0 eV. Therefore, depending on our needs, we can decide the type of GCH that is suitable to use in applications.

ACKNOWLEDGMENT

This work was supported in-part by research grants FRGS/1/2015/TK04/UKM/02/2 and LRGS/2015/UKM-UKM/NANOMITE/04/01 from the Ministry of Higher Education Malaysia.

REFERENCES

[1] S. Iijima, "Helical microtubules of graphitic carbon", *Nature*, vol. 354, no. 6348, pp. 56-58, 1991.

[2] A. Geim and K. Novoselov, "The rise of graphene", *Nature Materials*, vol. 6, no. 3, pp. 183-191, 2007

[3] T. Matsumoto and S. Saito, "Geometric and Electronic Structure of New Carbon-Network Materials: Nanotube Array on Graphite Sheet", Journal of the Physical Society of Japan, vol. 71, no. 11, pp. 2765-2770, 2002.

[4] J. Shao, W. Lv, Q. Guo, C. Zhang, Q. Xu, Q. Yang and F. Kang, "Hybridization of graphene oxide and carbon nanotubes at the liquid/air interface", Chem. Commun., vol. 48, no. 31, pp. 3706-3708, 2012.

[5] J. Liu, L. Zhang, H. Wu, J. Lin, Z. Shen and X. Lou, "High-performance flexible asymmetric supercapacitors based on a new graphene foam/carbon nanotube hybrid film", Energy Environ. Sci., vol. 7, no. 11, pp. 3709-3719, 2014.

[6] D. Yu and L. Dai, "Self-Assembled Graphene/Carbon Nanotube Hybrid Films for Supercapacitors", The Journal of Physical Chemistry Letters, vol. 1, no. 2, pp. 467-470, 2009.

[7] S. Singal, A. Srivastava and Rajesh, "Electrochemical Impedance Analysis of Biofunctionalized Conducting Polymer-Modified Graphene-CNTs Nanocomposite for Protein Detection", Nano-Micro Letters, vol. 9, no. 1, 2016.

[8] T. Chen, Q. Zhang, M. Zhao, J. Huang, C. Tang and F. Wei, "Rational recipe for bulk growth of graphene/carbon nanotube hybrids: New insights from in-situ characterization on working catalysts", Carbon, vol. 95, pp. 292-301, 2015.

[9] Y. Kim, W. Song, S. Lee, C. Jeon, W. Jung, M. Kim and C. Park, "Low-temperature synthesis of graphene on nickel foil by microwave plasma chemical vapor deposition", Applied Physics Letters, vol. 98, no. 26, p. 263106, 2011.

[10] Z. Fan, J. Yan, L. Zhi, Q. Zhang, T. Wei, J. Feng, M. Zhang, W. Qian and F. Wei, "A Three-Dimensional Carbon Nanotube/Graphene Sandwich and Its Application as Electrode in Supercapacitors", Advanced Materials, vol. 22, no. 33, pp. 3723-3728, 2010.

[11] R. Weisman and S. Bachilo, "Dependence of Optical Transition Energies on Structure for Single-Walled Carbon Nanotubes in Aqueous Suspension: An Empirical Kataura Plot", Nano Letters, vol. 3, no. 9, pp. 1235-1238, 20

Electrokinetic Behavior and Stability of Solder Powders in Aqueous Media

Terence Lucero F. Menor
Department of Mining, Metallurgical and
Materials Engineering
University of the Philippines Diliman
Quezon City, Philippines
tfmenor@up.edu.ph

Manolo G. Mena
Department of Mining, Metallurgical and
Materials Engineering
University of the Philippines Diliman
Quezon City, Philippines
manolo.mena@coe.upd.edu.ph

Herman D. Mendoza
Department of Mining, Metallurgical and
Materials Engineering
University of the Philippines Diliman
Quezon City, Philippines
judge.mendoza@gmail.com

Abstract— Solder pastes are widely used in creating mechanical, thermal and electrical connection between electronic components. Continued miniaturization of electronic assemblies faces manufacturers with the difficult challenge of dispensing solder pastes in extremely precise and repeatable manner. The most common problem in solder paste dispensing is the clogging of dispensers which is caused by the agglomeration and settling of solder powder suspensions. Degree of agglomeration in a suspension highly depends on the electrokinetic behavior of the particles. In this work, microelectrophoresis was employed to investigate the effect of pH and electrolyte concentration on the electrokinetic behavior and stability of Sn, SAC305, and PbSn5Ag2.5 suspensions in aqueous media. Results revealed that the electrokinetic behavior of the three solder powder suspensions are similar, and this similarity can be attributed to the surface composition of the powders. Electrokinetic measurements showed that the zeta potentials of the solder powders are highly dependent on pH and electrolyte concentration with isoelectric points between pH 3.5 and 5.5. Results were verified through settling tests.

Keywords— *solder powder, electrokinetic behavior, isoelectric point, stability, surface analysis*

I. INTRODUCTION

Soldering is a well-known metallurgical joining method in which filler metals are melted below 425°C. In the electronics industry, soldering is necessary in creating mechanical, thermal and electrical connection between electronic components [1].

Solder paste is the most important material used in electronic assemblies. Solder paste is a homogeneous, stable suspension of solder powder in a flux binder. The binder may either be aqueous or non-aqueous and contains flux and other substances that determine the flow properties of the paste. Typically, solder pastes are composed of 30-70% solder powder by volume or about 88-91% solder powder by weight [2]. Most defects in soldering such as poor release, slumping and bridging originate from poor understanding of the flow properties and processing of solder pastes [3].

Due to the increasing global customer demand for miniaturized, hand-held, and pocket electronics product, electronics manufacturing industries have utilized advanced Integrated Circuit packages to achieve the manufacture of smaller, lighter, faster, and cheaper products [4], [5]. Manufacturers face the difficult challenge of depositing

---

Department of Science and Technology – Engineering Research and Development for Technology

Heraeus Materials Singapore Pte. Ltd.

---

solder paste in extremely precise and repeatable patterns. Microdispensers and stencils with finer apertures are used for this purpose. However, the reduction in the size of the dispenser and stencil apertures will usually result to increased clogging and incomplete transfer of paste to the printed circuit board (PCB) pads [3], [4], [6].

Agglomeration may occur in solder powder suspensions due to the attraction between particles. This leads to uneven distribution of solder particles in solder pastes and affects the flow properties of the paste resulting in poor solder finish. In some cases, high degree of agglomeration results in clogging of dispensers.

When a solid particle is immersed in a liquid medium, surrounding ions tend to distribute themselves around the surface of the particle resulting in a distribution of charges known as the electrical double layer (EDL). EDL is caused by spontaneous charging of particle surfaces in a liquid medium [7]. Surface charging of suspensions may be obtained from adsorption or orientation of dipolar molecules on the particle surface, electron transfer between the solid and the liquid, selective adsorption of ions onto the surface, and dissociation of ions from the solid phase into the liquid.

Agglomeration of particles occurs when the attractive van der Waals force exceeds the repulsive electrostatic force. For a stable suspension, wherein there is little or no agglomeration, the electrostatic force between particles must be sufficiently high. Since the electrostatic force is dependent on the magnitude of surface charge (and zeta potential ζ) of the particles, knowledge and understanding of the factors affecting the zeta potential of the particles will result to the control of the stability of the particles in the suspension.

The zeta potential is a good indicator of the stability of a suspension. The higher the magnitude of the zeta potential is, meaning the more the positive or the more negative the zeta potential is, the higher the repulsive forces between the particles and the more stable the particles are. When the particles are close to the isoelectric point or the point of no charge ($\zeta = 0$), the particles tend to agglomerate. The common dividing line between stable and unstable suspension is usually taken as +30mV and -30mV. Increasing the magnitude beyond these values will result in an increase on the stability of the suspensions. However, these values may vary depending on the type of the suspension (polymer, metal oxides, or metal sols) [8].

Despite the importance of electrokinetics on the stability of solder powder suspensions, there is no known literature on the effect of solution parameters on the electrokinetic behavior of solder powder suspensions. The aim of this study is to investigate effect of pH and electrolyte concentration on the electrokinetic behavior and stability of solder powder suspensions in aqueous media. Solder powders under investigation were Sn, SAC305 and PbSn5Ag2.5. The surface composition was then analyzed and related to the electrokinetic behavior of the solder powder suspensions.

II. METHODOLOGY

A. Materials

Solder powders studied were Sn (Sigma-Aldrich, USA), Type 7 SAC305 (Heraeus, Singapore) and Type 8 PbSn5Ag2.5 (Heraeus, Singapore). Powders were used without further purification. Since electrokinetic behavior of suspensions is highly affected by the composition of the solid particles, the composition of the powders was determined through X-ray fluorescence (XRF) and X-ray diffraction (XRD) analyses.

B. Suspension Preparation

Aqueous solutions of known pH and electrolyte concentration were prepared and placed in a glass beaker. The pH of the solutions was adjusted by using either HNO_3 or KOH. KNO_3 was used as electrolyte and studied concentrations were 10^{-2}, 10^{-3} and 10^{-4} M. Solder powders were then added to the solutions with solid loadings of 1 g/L. The suspensions were sonicated for five minutes before electrophoretic mobility measurements were conducted.

Samples were analyzed using optical microscopy in order to determine the average particle size of the suspended powders. Samples for microscopy were obtained by taking portions of the suspensions immediately after sonication. The portions taken were then placed on glass slides and air dried for microscopy.

C. Zeta Potential Determination

Electrophoretic mobilities of the suspensions were obtained using a Zeta Meter 4.0 (USA) with Zeiss DR Microscope module and GT-2 type quartz cell. The equipment uses microelectrophoresis in obtaining the electrophoretic mobility of the particles. Obtained electrophoretic mobilities were then converted to zeta potential.

Zeta potential was calculated using the steps discussed by Delgado et al. [9] and Olson [10]. The steps involve obtaining the radius of the particles, and calculating κ^{-1}, called the Debye length or the EDL thickness, given by

$$\kappa^{-1} = [\varepsilon_{rs}\varepsilon_0 k_b T/(\Sigma z_i^2 e^2 n_i)]^{1/2} \tag{1}$$

in which ε_{rs} is the relative permittivity of the solution, ε_0 is the electric permittivity in vacuum, k_b is the Boltzmann's constant, T is temperature, z_i is the ionic valence, e is the elementary charge, and n_i is the charge number and concentration of ion. Based on the values of the particle radius and Debye length, the appropriate model to calculate zeta potential was chosen.

TABLE I.   COMPOSITION OF SOLDER POWDERS FROM XRF

Powder	Weight Percent (%)			
	Sn	Pb	Ag	Cu
Sn	99.70	0.004	0.04	0.02
SAC305	96.21	0	3.13	0.44
PbSn5Ag2.5	4.85	93.13	1.94	0

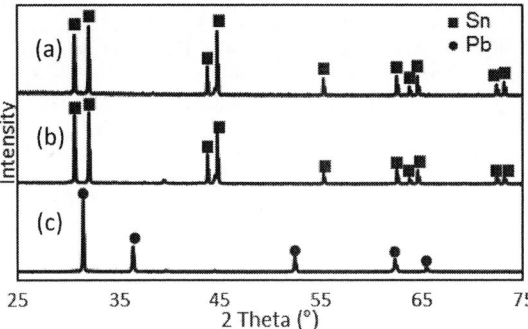

Fig. 1.   XRD spectra of (a) Sn, (b) SAC305 and (c) PbSn5Ag2.5 powders

D. Surface Analysis

Surface composition of each type of solder powder was determined through Auger electron spectroscopy (AES) using JAMP-9500F Auger Microprobe (Jeol, USA) to detect the presence of surface impurities which can affect the electrophoretic mobility of the suspensions.

E. Settling Test

Settling tests were performed in test tubes. 10 mL suspensions with solid loadings of 1 g/L were prepared and sonicated for 5 minutes before the test. The suspensions were illuminated to clearly show the suspended particles through Tyndall effect. Images of the suspensions were taken at different time intervals.

III. RESULTS AND DISCUSSION

A. Material Characterization

Table II shows the results obtained from the XRF analysis. Results obtained is in agreement with the data obtained from the manufacturers. Trace elements detected include Ni, Cd, and In. Fig. 1 shows the XRD results obtained. As shown, only the major components (i.e. Sn and Pb) of the solder powders were detected and these components are in their metallic form.

B. Zeta Potential of Solder Powder Suspensions

Average particle radii of Sn, SAC305, and PbSnAg2.5 obtained from microscopy are 2.13, 2.00, and 2.87 μm, respectively. The appropriate model in calculating the zeta potential of the solder powder suspensions based from the obtained particle radius and calculated Debye length from (1) is the Helmholtz-Smoluchowski equation. The equation relates the electrophoretic mobility (u_e) to zeta potential by

$$u_e = (\varepsilon_{rs}\varepsilon_0\zeta)/\eta \tag{2}$$

where η is the dynamic viscosity of the liquid.

978-1-5386-5284-8/18 $31.00 © 2018 IEEE

As mentioned earlier, zeta potential is a good indicator of stability and the higher the magnitude of the zeta potential of the suspension is, the more stable the particles are. Fig. 2 shows the zeta potential of Sn, SAC305 and PbSn5Ag2.5 suspensions at varying pH and KNO_3 concentration calculated from (2). At neutral pH, zeta potentials of the three solder powders are negative, indicating that the surfaces of the particles are negatively charged. Decreasing the pH towards the acidic region resulted in charge reversal and isoelectric point of the suspensions is between pH 3.5 to 5.5. High magnitudes of zeta potential were obtained at basic conditions with values of -38.8, -55.4, and -37.7 mV for Sn, SAC305 and PbSn5Ag2.5, respectively. These results indicate that the repulsive force between suspended particles of the three solder powders is strongest in basic media.

The zeta potential of the three solder powders decreased as KNO_3 concentration increased. This observation was expected since addition of inorganic salts results in the reduction of the double layer thickness thus decreasing the zeta potential of the suspensions [11]. As shown, maximum zeta potential values obtained at 10^{-2} M KNO_3 concentration in the basic region are -19.3, -34.6 and -19.8 mV for Sn, SAC305 and PbSn5Ag2.5, respectively, which are much lower than the values obtained at 10^{-4} M KNO_3 concentration.

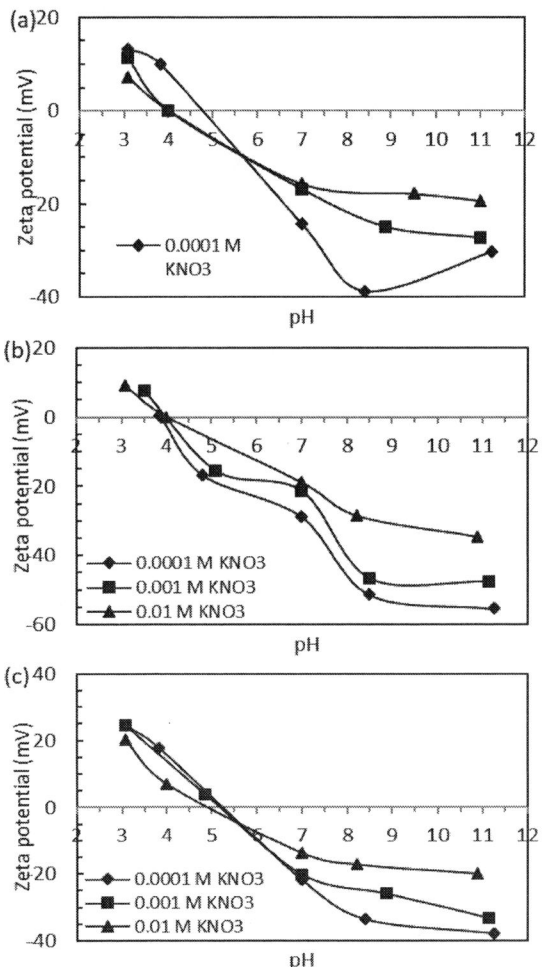

Fig. 2. Effect of pH and KNO_3 concentration on zeta potential of a) Sn, (b) SAC305, and (c) PbSn5Ag2.5 suspension

C. Surface Analysis

From the zeta potential curves presented in Fig. 2, it can be observed that the three solder powders showed similarity in their electrokinetic behavior despite the differences in their bulk composition. Surface analysis was done to explain this similarity. Fig. 3 shows the AES spectra obtained from the surface of the three solder powders.

The AES spectra indicate that the surfaces of the solder powders contain C, Sn and O, with presence of Pb in PbSn5Ag2.5. Measured high carbon intensities on the surfaces of the powders are attributed to carbon contamination due to air exposure. Anjard [12] noted that a clean metal surface is tightly polar in nature and will prefer to absorb organic compounds. Hence, carbonaceous material is formed on the surface.

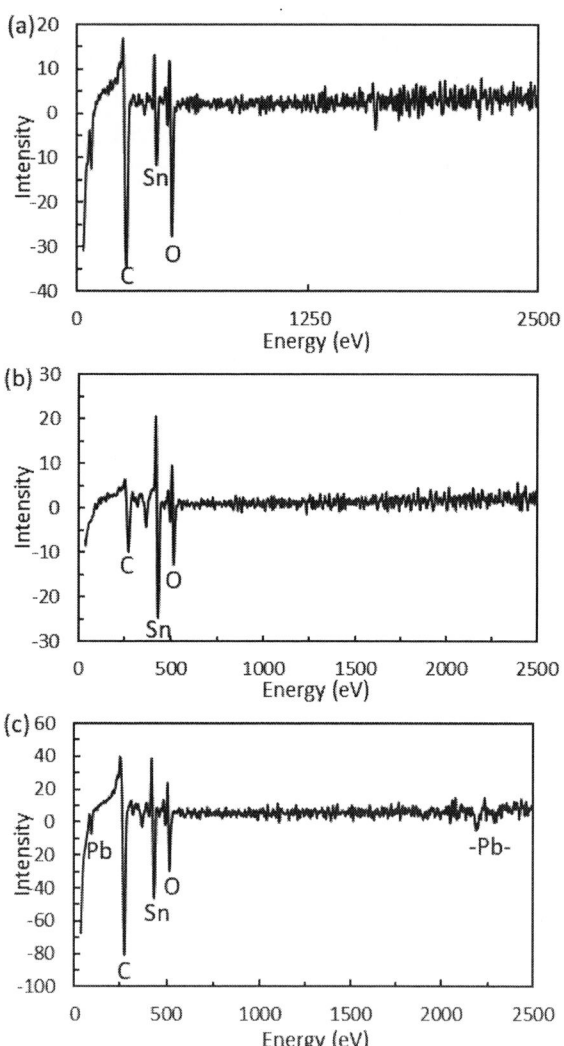

Fig. 3. AES spectra taken from the surface of (a) Sn, (b) SAC305 and (c) PbSn5Ag2.5

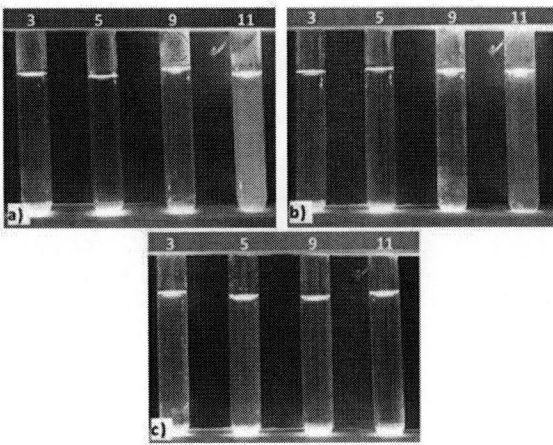

Fig. 4. Suspensions with varying pH and KNO_3 concentration of (a) 10^{-4} M, (b) 10^{-3} M, and (c) 10^{-2} M after 24 hrs

High Sn content was expected in Sn and SAC305. However, it is notable that PbSn5Ag2.5 also showed high Sn intensity on its surface despite having only a total of 4.85% Sn based on XRF analysis shown in Table 1. There is also high oxygen content on the surface of the powders despite the absence of oxides in the XRD spectra presented in Fig. 1. High Sn content on the surface of PbSn5Ag2.5 is mainly due to the surface segregation of Sn. Studies [12]–[15] have observed the surface segregation of Sn in Sn-containing solder powders. Surface segregation occurs in Pb-Sn powders with Sn content ranging from 0.5% to 10%. Since tin is preferentially oxidized than the lead present, tin segregation is enhanced by exposing clean surface to oxygen indicating that presence of oxygen is the driving force of tin diffusion to the surface along the grain boundaries [14]. This results in high amount of Sn, mostly in oxide form, on the surface of the Sn-containing solder powders. This thin film of oxide is inevitably formed upon exposure to oxygen and serves as a protective layer and prevents further oxidation of the solder powders [16], [17]. The native oxide formed by tin is SnO and SnO_2 [18]. This implies that majority of the surface of the powders used in this study are also covered with tin oxides.

D. Stability of Solder powder suspensions

Fig. 4 shows the effect of pH and electrolyte concentration on the stability of SAC305 suspensions after 24 hrs. As expected, suspension with 10^{-4} M KNO_3 and pH 11 exhibited good stability since this suspension have high zeta potential as shown in Fig. 2. Increasing KNO_3 concentration to 10^{-3} M resulted in a slight decrease on the stability of the suspension at pH 11 as exhibited by the decrease on the turbidity of the solution. Poor stability was exhibited at 10^{-2} M KNO_3.

All Sn and PbSn5Ag2.5 suspensions exhibited poor stability after 24 hrs. This can be explained by the low zeta potential obtained from both suspensions. As shown in Fig. 2, maximum magnitudes of zeta potential of Sn and PbSn5Ag2.5 suspensions are -38.8, and -37.7 mV respectively, whereas maximum magnitude of zeta potential of SAC305 suspension is -55.4 mV. This indicates that the dividing line between stable and unstable solder powder suspensions in aqueous solutions is around -40mV or -50

mV. Although Sn composition is more similar with SAC305 than PbSn5Ag2.5, obtained zeta potential values from Sn is closer to the values obtained from PbSn5Ag2.5. This observation can also be attributed to the high carbon and oxygen intensities on the surface of Sn and PbSn5Ag2.5 powders. The contaminants present on the surface of the powders might have affected the zeta potential obtained. This can also be the reason for the low suspension stability observed from the two powders.

IV. CONCLUSION

Sn, SAC305 and PbSn5Ag2.5 are negatively charged in neutral aqueous suspensions as shown by the electrophoretic measurements. Solder powder suspensions are more stable in the basic region and charge reversal occurs in the acidic region. Particle agglomeration is expected at high ionic strengths as evident from the decrease in the magnitude of zeta potential measured at high salt concentrations. High degree of agglomeration is also expected near the isoelectric point between pH 3.5 and 5.5. Electrokinetic behavior of the different types of solder powder suspensions are comparable which can be attributed to the similarities in their surface composition as verified by AES. Only SAC305 suspensions with high magnitude of zeta potential are stable after 24 hrs.

ACKNOWLEDGMENT

T.L.F. Menor would like to thank April Joy Garete for assisting during experimentation and Jill Manapat and John Kenneth Cruz for assisting in electrophoretic mobility measurements.

REFERENCES

[1] M. Abtew and G. Selvaduray, "Lead-free Solders in Microelectronics," Mater. Sci. Eng., vol. 27, pp. 95–141, 2000.

[2] S. Mallik, N. N. Ekere, A. E. Marks, A. Seman, and R. Durairaj, "Modeling the Structural Breakdown of Solder Paste Using the Structural Kinetic Model," JMEPEG, vol. 19, pp. 40–45, 2010.

[3] C. Billotte and P. J. Carreau, "Rheological characterization of a solder paste for surface mount applications," Rheol Acta, vol. 45, pp. 374–386, 2006.

[4] E. H. Amalu et al., "A study of SnAgCu solder paste transfer efficiency and effects of optimal reflow profile on solder deposits," Microelectron. Eng., vol. 88, no. 7, pp. 1610–1617, 2011.

[5] H. Amalu, N. N. Ekere, and S. Mallik, "Evaluation of rheological properties of lead- free solder pastes and their relationship with transfer efficiency during stencil printing process," Mater. Des., vol. 32, pp. 3189–3197, 2011.

[6] D. Ashley and S. J. Adamson, "Advancements in Solder Paste Dispensing," PennWell Corporation, 2008.

[7] K. Schießl, F. Babick, and M. Stintz, "Calculation of double layer interaction between colloidal aggregates," Adv. Powder Technol., vol. 23, no. 2, pp. 139–147, 2012.

[8] S. Vallar, D. Houivet, J. El Fallah, D. Kervadec, and J. Haussonne, "Oxide Slurries Stability and Powders Dispersion : Optimization with Zeta Potential and Rheological Measurements," J. Eur. Ceram. Soc., vol. 19, pp. 1017–1021, 1999.

[9] A. V. Delgado, F. Gonzalez-Caballero, R. J. Hunter, L. K. Koopal, and J. Lyklema, "Measurement and interpretation of electrokinetic phenomena," J. Colloid Interface Sci., vol. 309, no. 2, pp. 194–224, 2007.

[10] E. Olson, "Zeta Potential and Colloid Chemistry.," J. GXP Compliance, vol. 16, no. 1, pp. 81–96, 2012.

[11] R. Hac, A. Genç, B. Bak, A. Preparation, and M. W. F. Emulsions, "Evaluation of Droplet Sizes from Video Images for Metal Working Fluids," Int. J. Environ. Chem. Ecocological, Geol. Geophys. Eng., vol. 7, no. 11, pp. 757–761, 2013.

[12] R. P. Anjard, "Factors influencing the quality of solder powder used in solder pastes for thick film circuits, printed circuitboards and the electronics industry in general," Powder Technol., vol. 36, no. 2, pp. 189–202, 1983.

[13] S. Cho, J. Yu, S. K. Kang, and D.-Y. Shih, "Oxidation study of pure tin and its alloys via electrochemical reduction analysis," J. Electron. Mater., vol. 34, no. 5, pp. 635–642, 2005.

[14] R. J. Bird, "Corrosion-Resistant Lead-Indium and Lead-Tin Alloys: Surface Studies by Photo-Electron Spectroscopy (ESCA)," Met. Sci. J., vol. 7, no. 1, pp. 109–113, Jan. 1973.

[15] D. D. Hillman and L. S. Chumbley, "Characterization of tin oxidation products using sequential electrochemical reduction analysis (SERA)," Solder. Surf. Mt. Technol., vol. 18, no. 3, pp. 31–41, Jul. 2006.

[16] S. Zhang, Y. Zhang, and H. Wang, "Effect of oxide thickness of solder powders on the coalescence of SnAgCu lead-free solder pastes," J. Alloys Compd., vol. 487, no. 1–2, pp. 682–686, 2009.

[17] J. F. Kuhmann et al., "Oxidation and Reduction Kinetics of Eutectic SnPb , InSn , and AuSn : A Knowledge Base for Fluxless Solder Bonding Applications," IEEE Trans. Compon. Packaging. Manuf. Technol., vol. 21, no. 2, pp. 134–141, 1998.

[18] A. J. Bevolo, J. D. Verhoeven, and M. Noack, "A Leels and Auger Study of the Oxidation of Liquid and Solid Tin," vol. 134. North-Holland Publishing Company, pp. 499–528, 1983.

978-1-5386-5284-8/18 $31.00 © 2018 IEEE

Challenges in Developing Thin Profile, Smaller Flip Chip Bump Pitch FCBGA Packaging

Shaw Fa Lim
Package Engineering
Advanced Micro Devices, Inc. (AMD)
Bayan Lepas, Malaysia
shaw-fa.lim@amd.com

Keith Newman
Package Reliability
Advanced Micro Devices, Inc. (AMD)
Santa Clara, US
Keith.Newman@amd.com

Abstract- **The influence of substrate copper density distribution, substrate bump coplanarity, stiffener attach process, and substrate clamping by magnetic boat during die attach were evaluated. The substrate warpage behavior throughout the package assembly process was characterized using shadow moiré. Balanced substrate copper density distribution, pre-stiffener substrate before flip chip bump reflow, and substrate clamping during reflow reduced flip chip solder bridging fall-out. The decrease in solder bridging was due to the lower substrate warpage seen during die attach. In particular, solder bridging fall-out was well-correlated to die attach area warpage. Substrate with and without clamping during reflow has met the package reliability requirement.**

Keywords—Flip Chip Bump Solder Bridging, FCBGA, Smaller Pitch, Substrate Warpage.

I. INTRODUCTION

FCBGA packaging in portable CPU products increasingly requires a lower profile and higher die I/O density. These requirements have driven thinner package substrates and smaller flip chip bump pitches. These changes, in turn, pose challenges in developing high-yield, cost effective, and reliable FCBGA packages.

High I/O density requires multiple electrical metal routing layers in the package substrate, increasing substrate thickness. To keep overall package height low, the laminate core thicknesses of build-up style substrates are typically reduced. Reducing core thickness decreases mechanical stiffness of a substrate, potentially resulting in increased substrate warpage.

In the application described in this paper, a rectangular substrate, 25mm x 35mm x 0.54mm (thickness) with a thin core thickness of 210um, was selected to meet overall package height requirements. A flip chip solder bump pitch of 150um was needed to support the high I/O density. Unfortunately, flipchip solder bumps at this fine pitch increase the sensitivity of substrate warpage during the die attach process, sometimes creating flip chip bump solder bridging. A lower substrate warpage during reflow is preferred so that solder bridging does not occur with fine pitch flipchip solder bumps. The use of thinner substrate laminate cores affects solder bridging likelihood in first-level interconnect, as well as second-level BGA package coplanarity. In this study, we investigated numerous design, materials, and assembly processes that relate to substrate warpage, and subsequent assembly yield, package reliability, and BGA package coplanarity. The substrate warpage behavior during the reflow process was characterized using shadow moiré.

II. THIN PACKAGE CHALLENGES DURING DEVELOPMENT

A die size 10.93 x 19.20 mm, 0.38 thickness and substrate size 25x35, 0.54mm thickness with 3-2-3 metal build up was used in the package development. The package was attached with stiffener to ensure BGA coplanarity met the manufacturing requirement.

Figure 1: Package Layout.

The package assembly process started from mass-reflow die attach, de-flux, underfill dispense and cure, stiffener attach and ended with BGA solder ball mount. A range of copper (Cu) density distributions between top and bottom layers of the substrate were evaluated. The impact of stiffener attach (SA) to the substrate assembly process before and after flip chip die bump reflow was evaluated. Mass reflow process was used in die bump reflow. A die bump pitch of 150um and varying flip chip bump coplanarity were evaluated. No substrate clamping during die attach was used.

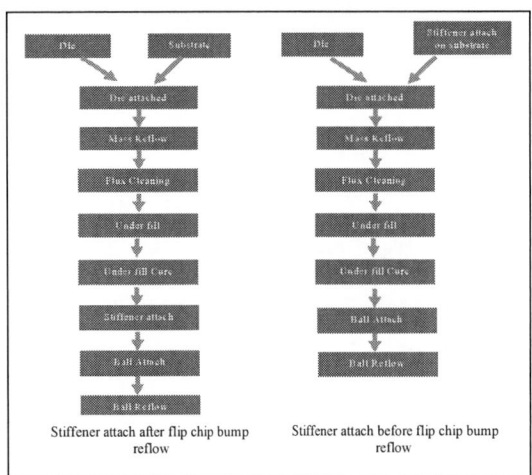

Stiffener attach after flip chip bump reflow   Stiffener attach before flip chip bump reflow

Figure 2: Assembly process for SA before and after flip chip bump reflow.

978-1-5386-5284-8/18 $31.00 © 2018 IEEE

The initial design of experiment (DOE) matrix consist of different substrate condition, substrate suppliers and stiffener attach before/after die reflow were evaluated. All legs were 100% open/short tested and contact non-wet and solder bridging were confirmed using x-ray and cross section techniques.

Solder bridging (SB) from DOE Leg 1 are reported in Figure 3; no contact non- wet (CNW) issue was observed.

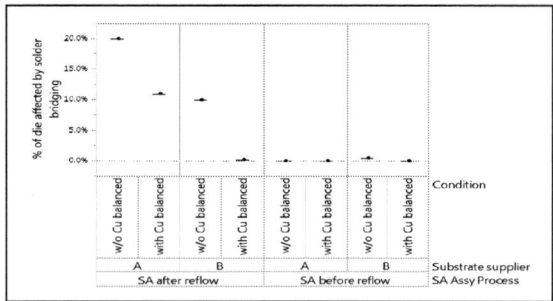

Figure 3: Solder bridging fall out versus different DOE conditions.

A. Substrate copper density top and bottom balance

It was found that top/bottom copper density imbalance contributed to solder bridging (see Figure 4). In this evaluation, substrate supplier A had a greater copper imbalance than supplier B, resulting in a larger package warpage during flip chip bump reflow. The observation was confirmed with shadow moiré; the warpage of substrate supplier A was 89um at 225°C (cooling) compared to 62um for supplier B.

During peak temperature of flip chip bump reflow, high temperature caused the substrate to warp upwards in a concave shape. In contrast, the die remains relatively flat during the die attach process. Consequently, adjacent flip chip bumps in the die corner are more likely to bridge during reflow. The substrate warpage effect can be seen at Figure 5, located at corner A and C. The decreased gap between the die and substrate pushed adjacent, molten solder bumps to join, resulting in solder bridging. The gap height between the die and substrate surface are shown in Table 1 for an extreme case.

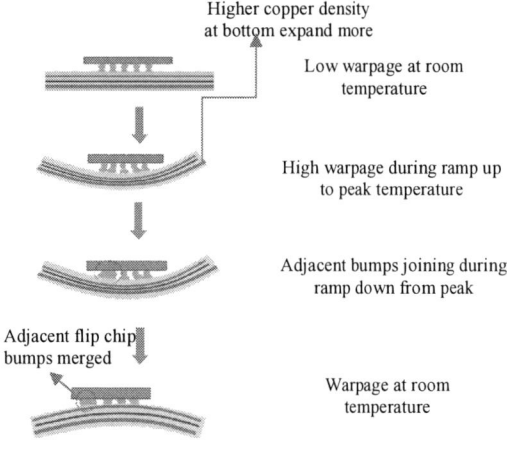

Figure 4: Package warpage model during reflow

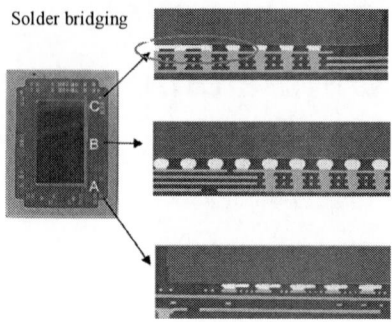

Figure 5: Cross section of SB fallout unit (extreme case).

Table 1: Gap height between die and substrate surface (extreme case)

Zone	A	B	C
Gap between die surface and SR surface (in micrometer)	25.8	61.5	23.1

B. Stiffener Attach before and after flip chip bump reflow

From the design of experiment (DOE), attachment of the stiffener before flip chip bump reflow showed lower solder-bridging fall-out compared to stiffener attach after flip chip bump reflow. Stiffener 2mm width x 0.3mm thickness was used.

An advantage of stiffener attach before die bump reflow is that the substrate is warped in an upwards direction (convex shape) during reflow due to the increased expansion of the stiffener (see Figure 6). The observation was confirmed from a measured shadow moiré data value of +52um. This change in warpage decreases the tendency of molten solder to be squeezed during reflow. However, final package BGA solder ball coplanarity was found higher when stiffener attach before die bump reflow was used.

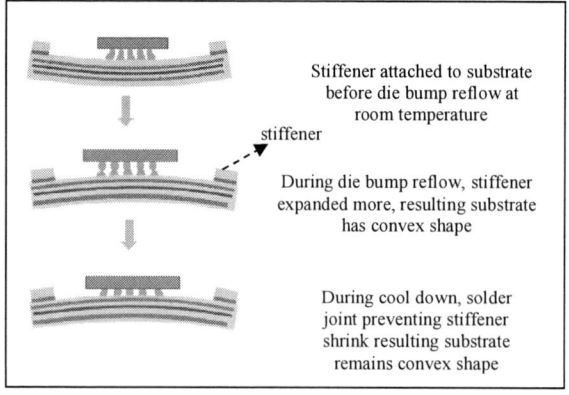

Figure 6: Pre-Stiffener attached to substrate before flip chip package warpage model.

C. Reducing flip chip Bump Coplanarity

Flip chip bump coplanarity was evaluated to understand the correlation to the SB fall-out. Figure 7 shows that higher flip chip die bump coplanarity will result in higher SB fallout. Reduction in flip chip bump coplanarity comes at an increased substrate cost, so it is a trade-off.

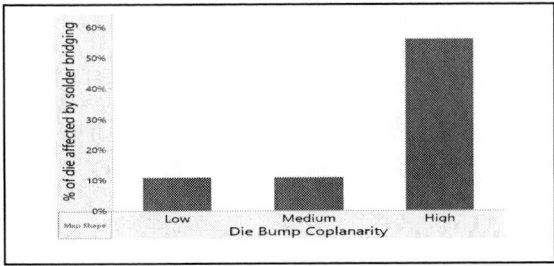

Figure 7: Flip chip Coplanarity versus SB fall-out.

D. Clamping process during flip chip bump reflow

Substrate clamping during die bump reflow was evaluated to assess if the SB fallout could be improved. Custom-made assembly boats with magnets embedded at four side of each pocket were fabricated. The top metal cover is pulled down to the boat by magnet force and flattens the substrate during reflow temperature as shown in Figure 8. SB fall-out and shadow moiré on substrate with a magnetic boat process were collected.

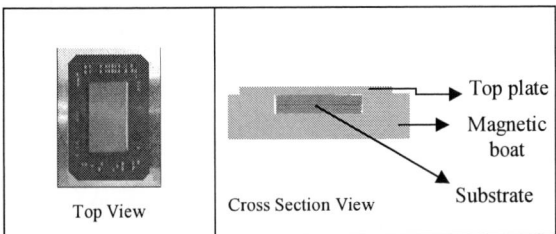

Figure 8: Magnetic Boat Structure.

From the experiment data shown in Figure 9, it was found that magnetic boats minimize SB fall-out significantly compared to normal boats without top clamping. The improvement is seen across all substrates (with and without copper balanced) and suppliers (A and B).

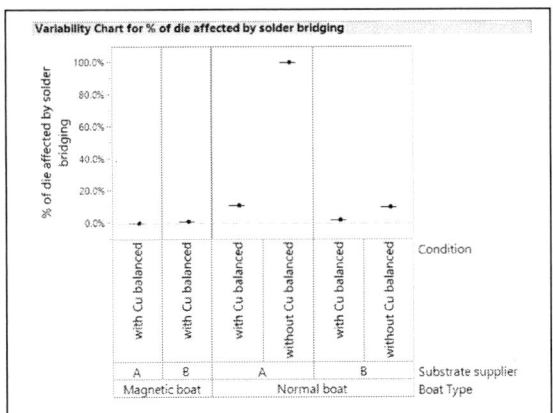

Figure 9: SB Fallout versus Boat Type.

The improvement was attributed to the lower substrate warpage when the top plate is constrained by the magnetic boat during reflow temperature. At a liquidus temperature of 225°C, substrate warpage is less than -50um when a magnetic boat is used, compared to greater than -50um with a normal assembly boat. The substrate warpage across reflow temperature is illustrated in Figure 10.

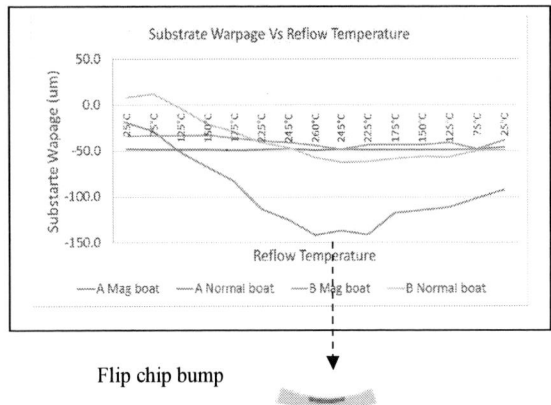

Figure 10: Substrate warpage across reflow temperature.

From the contour plot at 225°C (cooling) in Figure 11, the substrate warpage for the unit with top clamping magnetic boat showed lower warpage ~40um versus ~62um for the unit without top clamping.

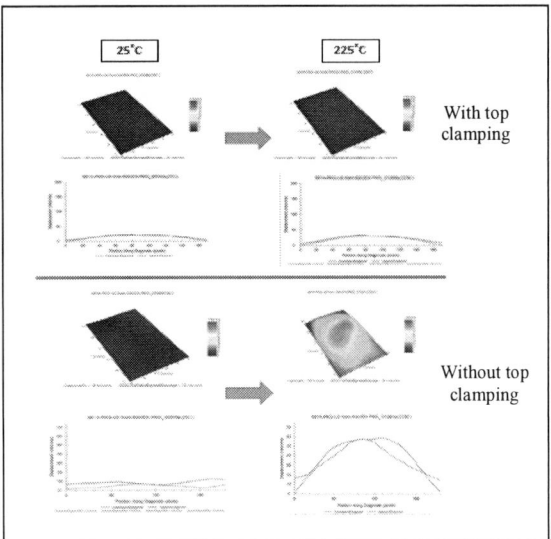

Figure 11: Clamping Effects on Substrate Warpage.

From the understanding of warpage behavior through shadow moiré on a bare substrate without die, a model of a die attached to substrate during flip chip bump reflow was proposed as illustrated in Figure 12.

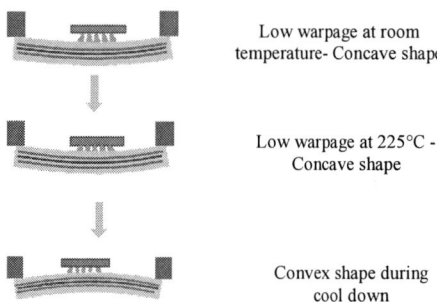

Low warpage at room temperature- Concave shape

Low warpage at 225°C - Concave shape

Convex shape during cool down

Figure 12: Die attach to substrate formation model.

Die attach area warpage was more relevant to die bump solder bridging compared to substrate warpage since die to substrate bump joining occurred at this area. As such, a relationship between die attach area warpage at reflow temperature 225°C (cooling) and solder bridging yield was established from the experiment. Figure 13 illustrated that the higher die area substrate warpage, the higher the SB fall-out.

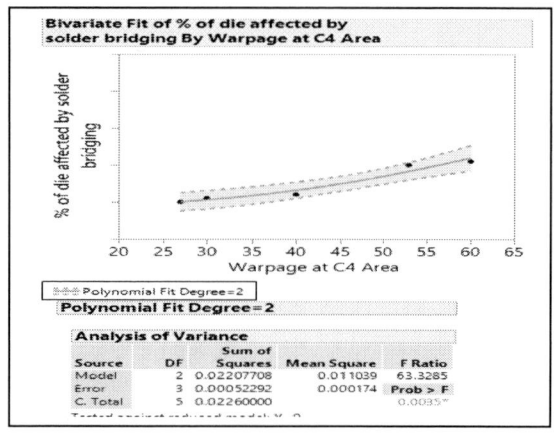

Figure 13: SB fallout versus Warpage at flip chip area.

Package warpage comparison between with and without substrate clamping during die bump reflow is illustrated in Figure 14. Both package warpages are within 100um which are meeting JEDEC requirement between -140um and 220um during reflow [1].

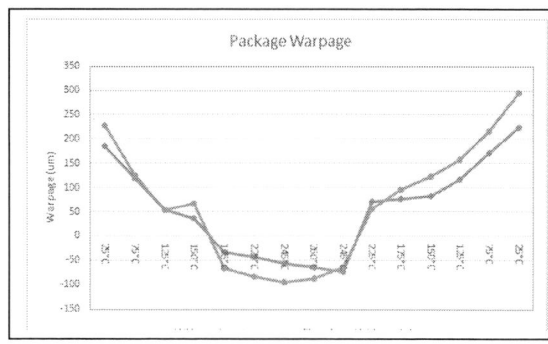

Figure 14: Package warpage comparison with/without substrate clamping during die reflow.

Reliability data was collected for assembly processes with/without substrate clamping. The reliability results passed in both cases as illustrated in Table 2.

Table 2: Reliability Data for both package with/without substrate clamping

Reliability Test Plan		Package without substrate clamping (normal boat)	Package with substrate clamping (magnetic boat)
1.	Preconditioning J-Std- 020 level 3 (192 hrs soak)	Pass	Pass
2.	TCG1200 (-40°C to 125°C)	Pass	Pass
3.	UHAST (130°C/85%RH) 96 hrs	Pass	Pass
4.	HTS (150°C) 1000 hrs	Pass	Pass

III. CONCLUSION

An optimized metal balance substrate and assembly process condition were achieved to meet desirable product yield, reliability, and package requirements. The substrate warpage model established through the evaluations has allowed us to better understand solder bridging performance. Careful consideration during package development should be put on substrate metal balance, warpage behavior during die bump reflow characterization, and stiffener attach process. Substrate clamping using a magnetic boat during die bump reflow significantly reduced substrate warpage and solder bridging. The quantified relationship between solder bridging performance and die attach area warpage provides a foundation for continued FCBGA thin package success.

ACKNOWLEDGMENT

The authors would like to thank Ho Soo Fong (TF-AMD) for conducting experiments and data collection with us in developing a thin profile, smaller Flip Chip Bump Pitch FCBGA package.

REFERENCES

[1] Reflow Flatness Requirements for Ball Grid Array Packages, SPP-024, Issue A, Jedec Publication 95, pp. 3.24-3.

[2] Sean S. Too, Mohammad Khan, Kevin Lim, Mike Loo, WC Lau, Azlina Nayan, SF Ng, BL Peh and Edwin Goh, "Thin-core MCM assembly development for high-performance server microprocessor", IEEE, pp. 517-522, 2011.

[3] Mamaru K, Daisuke M, Masateru K, Manabau W, Kenji F, Nobutaka I and Hitoshi S, "Low Warpage Coreless Substrate for IC Packages", Transactions of The Japan Institute of Electronics Packaging, pp.55-62, Vol.5, No.1, 2012.

2018 IEEE International Conference on Semiconductor Electronics (ICSE)

Ultrasonic sensor system with a 94 Mrad total-ionizing-dose tolerance

Shinya Fujisaki and Minoru Watanabe
Electrical and Electronic Engineering
Shizuoka University
3-5-1 Johoku, Hamamatsu, Shizuoka 432-8561, Japan
Email: tmwatan@ipc.shizuoka.ac.jp

Abstract—The Fukushima Daiichi nuclear power plant experienced multiple meltdown events after losing all power after an earthquake and its related tsunami disaster in 2011. Since the Fukushima Daiichi nuclear power plant still has regions of intense radiation, robots with a high radiation tolerance are necessary to complete work in decommissioning the reactors. Such robots frequently require proximity sensors. We have chosen an ultrasound sensor as a radiation-hardened sensor because ultrasound sensors consist of piezoelectric ceramic, which is very robust against radiation. However, although the ultrasound sensor itself is robust against radiation, since a semiconductor receiver circuit must be constructed to amplify the signal received from the ultrasound sensor, we have designed a radiation-hardened receiver circuit and have constructed an ultrasonic sensor system. This paper reports evaluation results demonstrating that the radiation-hardened ultrasonic sensor system has at least a 94 Mrad total-ionizing-dose tolerance.

I. Introduction

The Fukushima Daiichi nuclear power plant sustained several meltdown events after losing all power after an earthquake and its related tsunami disaster in March 2011 [1]–[4]. To date, although the Fukushima Daiichi nuclear power plant site has been cleaned up to some extent, regions with intense radiation of over 100 Sv/h have been confirmed around the damaged reactors and melted nuclear fuel underneath the reactors. Tokyo Electric Power Co. Ltd., the owner of the Fukushima Daiichi nuclear power plant, has planned to decommission the nuclear power plant reactors. Currently, we presume that robots must be able to function and perform tasks in 1000 Sv/h intense radiation conditions as the worst case during the decommissioning process [5],[6].

During the decommissioning process, autonomous robots must perform tasks instead of humans. Since autonomous robots frequently require certain proximity sensors, we have chosen ultrasound sensors as radiation-hardened proximity sensors because ultrasound sensors consist of piezoelectric ceramic, which is not a semiconductor device. For that reason, they are extremely robust against radiation. Nevertheless, although the ultrasound sensor itself is robust against radiation, we must design a radiation-hardened receiver circuit since a receiver circuit using semiconductor devices must be used for

978-1-5386-5283-1/18/$31.00 ©2018 IEEE

Fig. 1. Commercially available ultrasonic sensor unit (HC-SR04).

amplification of the weak signal generated by the ultrasound sensor.

Currently, various radiation-hardened VLSIs with total-ionizing-dose tolerances of 300 krad to 1 Mrad are available [7]–[11]. Under intense radiation conditions of 1000 Sv/h, the radiation-hardened VLSIs can function only for around 3–10 hours. Although the working lifetimes of the current radiation-hardened VLSIs are insufficient, new radiation-hardened optically reconfigurable gate arrays have been under development. The radiation tolerances of optically reconfigurable gate arrays are greater than 200 Mrad [12]–[18]. The radiation-hardened receiver circuit is intended for use along with optically reconfigurable gate arrays.

This paper presents a perfect radiation-hardened ultrasound sensor system including transfer and receiver ultrasound sensors and a radiation-hardened receiver circuit. The total-ionizing-dose tolerance has been confirmed experimentally using a Cobalt 60 gamma radiation source as at least a 94 Mrad total-ionizing-dose.

II. Commercially available ultrasonic sensor

First, a commercially available ultrasonic sensor unit has been tested under an approximately 30 TBq Cobalt 60 gamma radiation source. The ultrasonic sensor unit specifications are presented in Table I. A photograph is presented in Fig. 1. We

Fig. 2. Co 60 Gamma radiation experiment of the commercially available ultrasonic sensor unit (HC-SR04).

TABLE I
SPECIFICATIONS OF THE COMMERCIALLY AVAILABLE ULTRASONIC SENSOR UNIT
(HC-SR04).

Measurement range	2 - 400 cm
Resolution	0.3 cm
Voltage	5.0 V
Current	15 mA
Operating frequency	40 KHz
Size	45 × 20 × 15 mm

have prepared three identical HC-SR04 modules. All three modules were exposed to the Cobalt 60 gamma radiation source, as shown in Fig. 2. From that exposure, all three modules were broken by radiation at 50 krad. The results demonstrate that the total-ionizing-dose tolerances of the three commercially available ultrasonic sensors were less than 50 krad. In the radiation condition of 1000 Sv/h of the Fukushima Daiichi nuclear power plant, the useful life of the ultrasonic sensor unit is limited to 30 min. That radiation tolerance is insufficient for decommissioning work.

However, the ultrasonic sensor devices have never been damaged by radiation. Only, semiconductor devices that control the ultrasonic sensor device have been damaged by radiation.

III. ULTRASONIC SENSOR DEVICE

A photograph of an ultrasonic sensor device extracted from the HC-SR04 module is shown in Fig. 3. Although semiconductor devices are vulnerable to radiation [7]–[11],

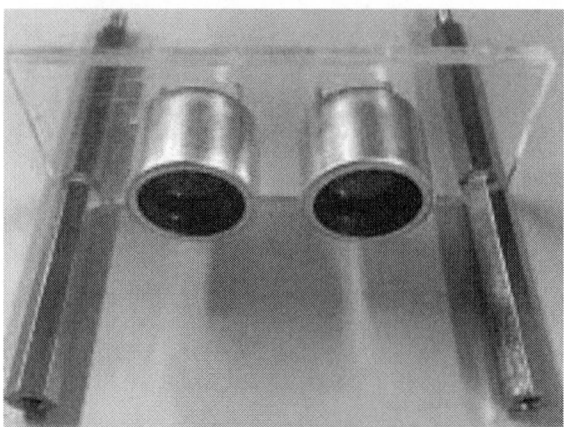

Fig. 3. Ultrasonic sensor device.

ultrasonic sensor devices never have any semiconductor element. Therefore, the ultrasonic sensor devices themselves are very robust against radiation. The total-ionizing-dose tolerance of the ultrasonic sensor device was measured using the 30 TBq Cobalt 60 gamma radiation source. To date, we have confirmed that the total-ionizing-dose tolerance of the ultrasonic sensor device itself is at least 300 Mrad as shown in Fig. 4. No degradation has been confirmed even after 300 Mrad irradiation. Therefore, the total-ionizing-dose tolerance of the ultrasonic sensor unit could also be increased if the total-ionizing-dose tolerance of semiconductor parts needed to control the ultrasonic sensor device could be increased.

IV. ULTRASONIC TRANSMITTER CIRCUIT

The transmitter circuit of the ultrasonic sensor device is simple. The ultrasonic sensor device can be driven directly by a field programmable gate array (FPGA). For this experiment, we used a Cyclone V FPGA on a DE1SOC FPGA board. A block diagram of the transmitter circuit is portrayed in Fig. 5. The Cyclone V FPGA generates a 40 kHz 3.3 V rectangular wave that is sent to an ultrasonic sensor device as shown in Fig. 5. Since the signal current is only 1.14 mA, which is sufficiently lower than the drive current of 10 mA of an I/O on the Cyclone V FPGA, the ultrasonic sensor device can be driven directly by an I/O on Cyclone V FPGA. At that time, the power consumption of the ultrasonic sensor is 1.88 mW. The sending ultrasonic signal is received on an ultrasonic receiver circuit. The time difference between sending of an ultrasonic signal and receiving the ultrasonic signal reflected by an obstacle are measured to calculate the distance from the ultrasonic sensor to the obstacle. The maximum measurement distance is assumed as 30 cm.

V. ULTRASONIC RECEIVER CIRCUIT

The circuit diagram of the ultrasonic receiver circuit is shown in Fig. 6. Since the output voltage from an ultrasonic sensor device is low when receiving an ultrasonic signal reflected by an obstacle, the signal must be amplified. Therefore,

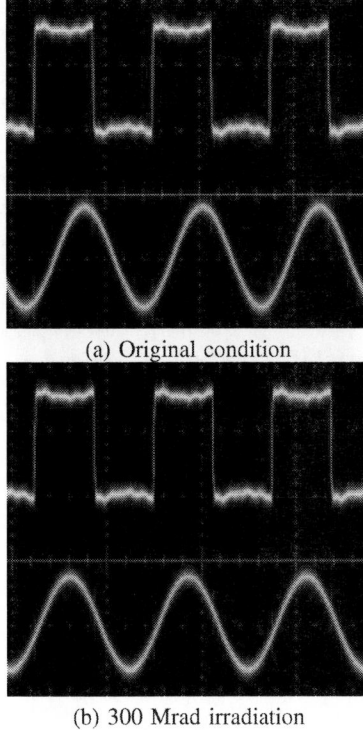

(a) Original condition

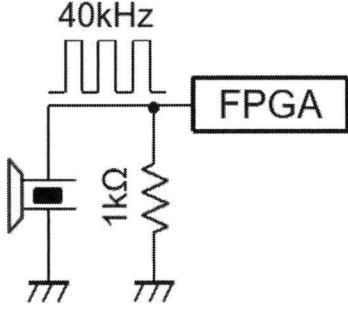

(b) 300 Mrad irradiation

Fig. 4. Waveforms of transmitter and receiver signals of the ultrasonic sensor device as shown in the upper part and the lower part, respectively. No degradation has been confirmed even after 300 Mrad irradiation.

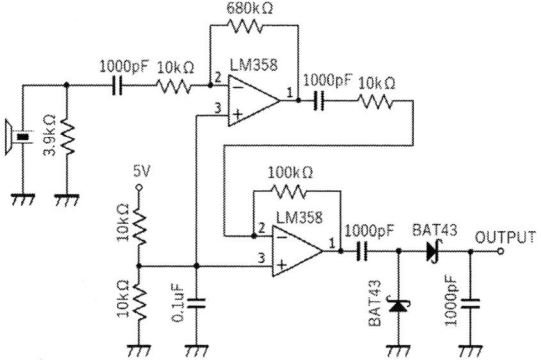

Fig. 6. Circuit diagram of the ultrasonic receiver circuit.

Fig. 7. Radiation experiment to assess the ultrasonic sensor system using an approximately 30 TBq Cobalt 60 gamma radiation source.

VI. RADIATION EXPERIMENT RESULTS

The developed ultrasonic sensor system was exposed to an approximately 30 TBq Cobalt 60 gamma radiation source, as shown in Fig. 7. We confirmed that the total-ionizing-dose tolerance of the radiation-hardened ultrasound sensor system is at least a 94 Mrad. Even if the ultrasonic sensor system is exposed to a 94 Mrad total-ionizing-dose, the ultrasonic sensor system is still working correctly. Distance measurement tests using the ultrasonic sensor system have been conducted as shown in Fig. 8. Any distance from 0 cm to 30 cm in a 5 cm interval can be measured correctly with high accuracy by the ultrasonic sensor system.

The radiation tolerance of the proposed system is 1,880 times higher than that of a commercially available ultrasound sensor system. In the radiation condition of 1000 Sv/h of the Fukushima Daiichi nuclear power plant, the useful lifetime of the proposed ultrasonic sensor system is longer than 940 hours.

Fig. 5. Circuit diagram of the ultrasonic transmitter circuit.

the ultrasonic receiver circuit requires a certain semiconductor amplifier. For this study, we have chosen commercially available components or non-radiation-hardened devices. However, a radiation-hardened bipolar transistor based amplifier or LM358 was selected. The signal of the ultrasonic device was amplified on two stages. Finally, the amplified signal is sent to the Cyclone V FPGA.

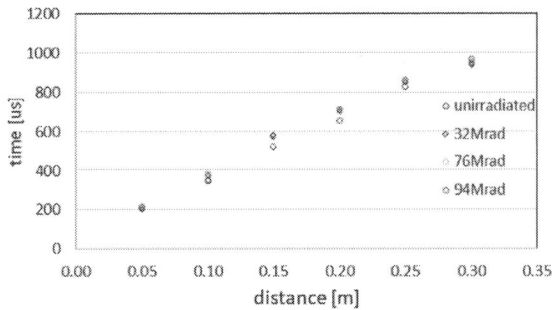

Fig. 8. Irradiation result.

The radiation tolerance of the proposed ultrasonic sensor system is sufficient for use in robots used for decommissioning operations.

Currently, a Cyclone V FPGA was used for the ultrasonic sensor system. The Cyclone V FPGA device is especially vulnerable to radiation. However, radiation-hardened optically reconfigurable gate arrays are undergoing continued development [12],[13]. The total-ionizing-dose tolerance of radiation-hardened optically reconfigurable gate arrays is over 200 Mrad, which is 200 times higher radiation tolerance than that of current radiation-hardened VLSIs. By replacing the Cyclone FPGA with an optically reconfigurable gate array, 94 Mrad total-ionizing-dose tolerance can be realized perfectly.

VII. Conclusion

This paper has presented a proposal for a new radiation-hardened ultrasound sensor system. We have confirmed that the total-ionizing-dose tolerance of the radiation-hardened ultrasound sensor system is at least 94 Mrad. That radiation tolerance is 1,880 times higher than the tolerance of commercially available ultrasound sensor systems. In intense radiation conditions of 1000 Sv/h at the Fukushima Daiichi nuclear power plant, the lifetime of the proposed ultrasonic sensor system can be estimated as longer than 940 hours. The radiation tolerance of the proposed ultrasonic sensor system is sufficient for application in robots that are designed for decommissioning operations.

Acknowledgment

This research was partly supported by the Initiatives for Atomic Energy Basic and Generic Strategic Research No. 283101.

References

[1] I. Gunawan, A. Gorod, L. Hallo, T. Nguyen, "Developing a system of systems management framework for the Fukushima Daiichi Nuclear disaster recovery," International Conference on System Science and Engineering, pp. 563-568, 2017.

[2] L. Schmidt, A. Horta, S. Pereira, A. Delicado, "The Fukushima nuclear disaster and its effects on media framing of fission and fusion energy technologies," International Conference on Advancements in Nuclear Instrumentation Measurement Methods and their Applications, pp. 1-11, 2015.

[3] R.C. Baumann, "Determining the Impact of Alpha-Particle-Emitting Contamination From the Fukushima-Daiichi Disaster on Japanese Semiconductor Manufacturing Sites," IEEE Transactions on Nuclear Science, Vol. 59, Issue 4, pp. 1186-1196, 2012.

[4] Ohno, K. et al. Robotic control vehicle for measuring radiation in Fukushima Daiichi Nuclear Power Plant. IEEE International Symposium on Safety, Security, and Rescue Robotics, pp. 38-43, 2011.

[5] T. Sakaue, S. Yoshino, K. Nishizawa, K. Takeda, "Survey in Fukushima Daiichi NPS by combination of human and remotely-controlled robot," IEEE International Symposium on Safety, Security and Rescue Robotics, pp. 7-12, 2017.

[6] A.R. Jones, A. Griffiths, M.J. Joyce, B. Lennox, S. Watson, J. Katakura, K. Okumura, K. Kim, M. Katoh, K. Nishimura, K. Sawada, "On the design of a remotely-deployed detection system for reactor assessment at Fukushima Daiichi," IEEE Nuclear Science Symposium, Medical Imaging Conference and Room-Temperature Semiconductor Detector Workshop, pp. 1-4, 2016.

[7] Nadim F. Haddad; Ronald D. Brown; Richard Ferguson; Andrew T. Kelly; Reed K. Lawrence; Daniel M. Pirkl; John C. Rodgers, "Second generation (200MHz) RAD750 microprocessor radiation evaluation," European Conference on Radiation and Its Effects on Components and Systems, pp. 877-880, 2011.

[8] Xi Qin; Changqing Feng; Deliang Zhang; Bin Miao; Lei Zhao; Xinjun Hao; Shubin Liu; Qi An , "Development of a High Resolution TDC for Implementation in Flash-Based and Anti-Fuse FPGAs for Aerospace Application," IEEE Transactions on Nuclear Science, Vol. 60, Issue 5, pp. 3550-3556, 2013.

[9] R. Glein, F. Rittner, A. Heuberger, "Detection of solar particle events inside FPGAs," European Conference on Radiation and Its Effects on Components and Systems, pp. 1-5, 2016.

[10] J.J. Wang, D. Dsilva, N. Rezzak, S. Varela, S. Cui, "Single Event Effects Testing on the SERDES, Fabric Flip-Flops and PLL in a Radiation-Hardened Flash-Based FPGA-RT4G150," IEEE Radiation Effects Data Workshop, pp. 1-6, 2016.

[11] R. Monreal, C. Carmichael, G. Swift, "Single-Event Characterization of Multi-Gigabit Transceivers (MGT) in Space-Grade Virtex-5QV Field Programmable Gate Arrays (FPGA)," IEEE Radiation Effects Data Workshop, pp. 1-8, 2011.

[12] T. Fujimori, M. Watanabe, "Parallel light configuration that increases the radiation tolerance of integrated circuits," Optics Express, Vol. 25, Issue 23, pp. 28136-28145, 2017.

[13] T. Fujimori, M. Watanabe, "High-speed scrubbing demonstration using an optically reconfigurable gate array," Optics Express, Vol. 25, Issue 7, pp. 7807-7817, 2017.

[14] R. Moriwaki, M. Watanabe, "Optical configuration acceleration on a new optically reconfigurable gate array VLSI using a negative logic implementation," Applied Optics, vol. 52, no. 9, pp. 1939-1946, 2013.

[15] Y. Ueno, M. Watanabe, "Fiber remote configuration for an optically reconfigurable gate array with four configuration contexts," Optics Communications, vol. 283, issue 23, pp. 4614-4618, 2010.

[16] T. Mabuchi, M. Watanabe, "A superimposing acceleration and optimization method of optical reconfiguration speed without any increase of laser power," Applied Optics, vol. 49, no. 22, pp. 4120?4126, 2010.

[17] M. Watanabe, T. Shiki, F. Kobayashi, "Scaling prospect of optically differential reconfigurable gate array VLSIs," Analog Integrated Circuits and Signal Processing, vol. 60, issue 1-2, pp. 137-143, 2009.

[18] S. Kubota, M. Watanabe, "Programmable Optically Reconfigurable Gate Array Architecture and its writer," Applied Optics, vol. 48, issue 2, pp. 302-308, 2009.

978-1-5386-5284-8/18 $31.00 © 2018 IEEE

Effect of DRIE on the Structure of Si based Filtration Pore Arrays Fabricated with Double Side Aluminium Coating Layer

Kamarul 'Asyikin Mustafa[1,2], Jumril Yunas[1], Azrul Azlan Hamzah[1], Wan Ammar Fikri Wan Ali[1], Burhanuddin Yeop Majlis[1]

[1]Institute of Microengineering and Nanoelectronics, Universiti Kebangsaan Malaysia, Bangi, Malaysia
[2]Department of Electrical and Electronics Engineering, Faculty of Engineering,
Universiti Pertahanan Nasional Malaysia, Sungai Besi, Malaysia
burhan@ukm.edu.my, kamarul.asyikin@siswa.ukm.edu.my

Abstract—This paper reports the effect of DRIE process performed to define silicon filtration passageway underneath the aluminium based uniform pore filtration membrane and the membrane cavity on the back end of the silicon substrate. The aluminium based pore arrays for filtration is fabricated on top of the bare silicon substrate with uniform pore size and pore-to-pore gap at cryogenic temperature DRIE. Initially, the aluminium layer is sputtered on both end surface as a foundation of uniform pore arrays structure fabrication and as a DRIE mask for filtration membrane cavity on the back. Both aluminium based uniform pore filtration membrane with underneath silicon etch depth and its filtration cavity structure on the back are observed after 10 minutes and 1 h 45 minutes of DRIE process respectively. The silicon etch rate selectivity over aluminium is evaluated by verifying it's etch depth after the DRIE process. The etched area of silicon substrate underneath the fabricated aluminium based pore filtration membrane is then performs as filtration passageway when it unites with etched silicon area conducted for filtration cavity structure which started from the back end of the silicon substrate. All results have been verified by optical microscope and scanning electron microscopy. The result of the study should be beneficiary for the development of a reproducible filtration membrane with uniform pore that able to separate waste from required nutrients based on molecule size.

Keywords— aluminium, uniform pore arrays, silicon, DRIE, filtration

I. INTRODUCTION

People receiving dialysis treatment still acquires nutrient supplement after the treatment due to the filtration membrane used in the dialysis machine still could not fully mimic the role plays by our native kidneys in separating wastes from required solutes in blood. This is due to perforated pores in filtration membrane used in dialysis treatment are not uniform in size and thus could not selectively prevent wastes from passing through the filtration membrane together with required nutrients in blood based on molecule size [1], [2]. Hence, to accommodate this issue, we work on fabrication of reproducible uniform pore arrays in filtration membrane using silicon as the substrate [3], [4], offering alternative as one of possible dialysis membrane that could enhance filtration capability based on molecule size selectivity. Negative charge dielectrophoresis may be manipulated to prevent clogging of

the filtration membrane and thus it could be used repeatedly and no further cleaning is required [5].

In this work, we report deep reactive ion etching (DRIE) process focused on the fabrication of silicon cavity using aluminium mask. We utilize DRIE process to obtain both silicon passageway underneath aluminium based uniform pore arrays [6] and also filtration cavity on the substrate back end that will then form as a filtration membrane when through-substrate holes are fabricated on the substrate [7]. The final structure will be implemented for filtration system. The DRIE process is conducted at cryogenic temperature in an environment of sulfur hexafluoride (SF_6) and oxygen (O_2) mixture [8]. Low temperature is superior in improving the silicon etching anisotropy and reducing the etch rate of the aluminium coating layer [9]. This DRIE method is an alternative method to the one reported in our previous work using KOH + IPA wet etching method [10] to obtain silicon filtration cavity. Initially, aluminium coating is sputtered as the uniform pore arrays pattern foundation on one of the substrate surface, and as DRIE mask for filtration cavity on the back end to protect unexposed area from dry etch strike.

II. FABRICATION PROCESS

A <100>-oriented 500 µm thick double side polished bare silicon substrate is used as the started sample with a dimension of 20 mm x 20 mm. After standard cleaning procedure, the bare silicon substrate is ready for subsequent fabrication steps. The cleaning procedure is repeated when the DRIE process for uniform pore arrays is completed and ready for the back end silicon membrane structure fabrication. The uniform circle pore arrays is designed to have a diameter of 1.8 µm with pore-to-pore distance set to 5 µm. Meanwhile, the membrane square frame is designed to have 2.2 mm side lengths.

Both surface of the bare silicon substrate is initially sputtered with aluminium of 120 nm thickness and afterward designed uniform pore arrays pattern is transferred onto the aluminium layer on one of the substrate's surface by exposing the substrate coated with positive photoresist under UV light exposure. Then, aluminium etching process is conducted to etch away all aluminium thickness from aluminium exposed area to form uniform pore structure on the aluminium sputtered coating layer. Then, the substrate is subjected to DRIE process to etch silicon beneath the aluminium based uniform pore arrays structure to form a passageway for filtration purpose. The pore structure etched on the aluminium

---

K.A.M. thanks Universiti Pertahanan Nasional Malaysia for the study leave given, and Ministry of Higher Education of Malaysia for SLAI scholarship granted.

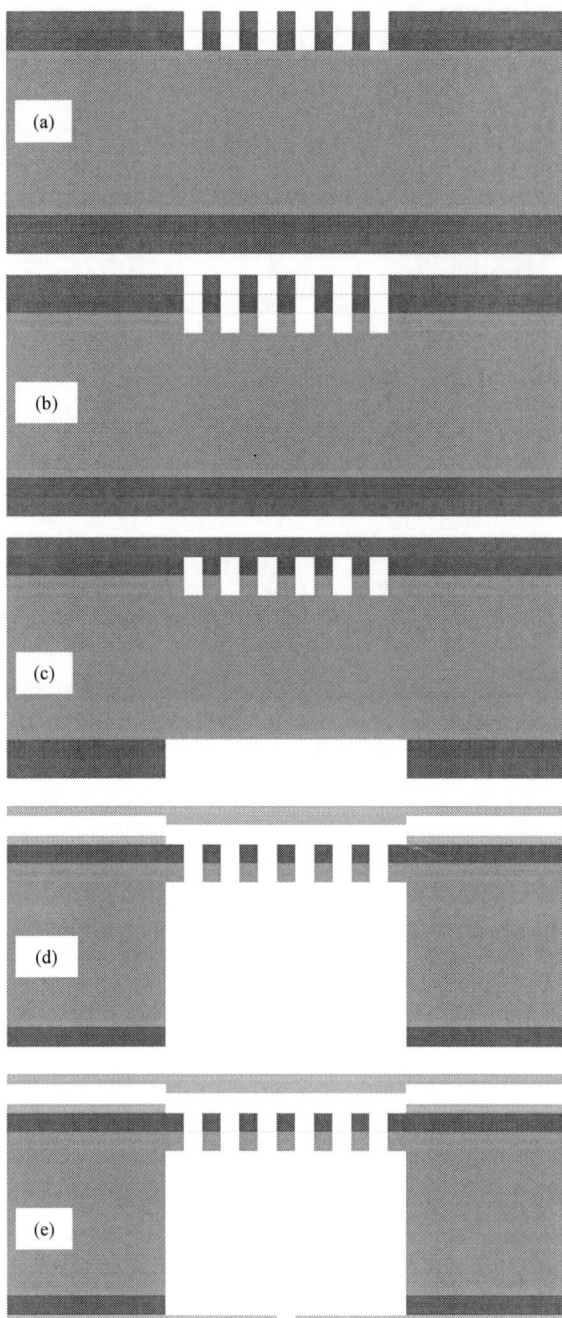

Fig. 1. Fabrication process for silicon membrane with uniform pore arrays (a) pore photoresist spin coat, pattern development and aluminum etching for pore pattern, (b) silicon DRIE for pore pattern, (c) back surface photoresist spin coat, filtration cavity pattern development, filtration cavity aluminium etching, (d) Sealing the top side aluminium based membrane using PDMS, silicon DRIE for back side of the substrate, (e) Sealing the back side filtration cavity using PDMS.

coating layer will later on function as membrane with its thickness will be the thickness of final filtration membrane.

Afterwards, after pore photoresist removal and standard cleaning procedure, the back end of the substrate is ready for membrane structure patterning where the process steps are the same as detailed in our previous work [10], with process differences are only in soft-bake parameters which set at 120 °C for two minutes, and the usage of bare silicon with aluminium coating layer instead of silicon nitride substrate utilized in [10]. Post aluminium etching process and prior to DRIE process for silicon filtration cavity structure etching, a PDMS filtration sealing layer is bonded on front end of the substrate treated by corona discharge treatment to improve the adhesion to silicon surface [11]. This PDMS filtration sealing layer is prepared beforehand using Perspex mould to give a filtration shape to poured PDMS solution. This PDMS filtration sealing layer also functions to protect the surface from the grease usually used to affix substrate to DRIE sample's carrier. To form a filtration cavity on the silicon substrate, it is required to resume the DRIE process for filtration cavity structure until a thorough filtration passageway is obtained when the passageway beneath aluminium based uniform pore arrays united with the etched silicon filtration cavity structure. Fig. 1 demonstrated fabrication steps for both uniform pore arrays and filtration cavity fabrication.

The filtration membrane structure etching on silicon substrate is conducted using Oxford Instrument Plasmalab 100 System, utilizing SF_6 as the etchant. RF and ICP forward power are set at 80 W and 600 W respectively. The process is set up at cryogenic temperature of -115±5 °C, SF_6 flow rate of 60 sccm and O_2 flow rate of 13.5 sccm. The duration of DRIE process is controlled so that anticipated aluminium based uniform pore arrays silicon membrane is achievable.

The sample is then analysed under scanning electron microscopy (SEM) to verify silicon etched depth that forms the aluminium based uniform pore arrays filtration membrane.

III. RESULT AND DISCUSSION

Fig. 2 shows the microphotograph of uniform pore arrays pattern developed on photoresist layer. The aluminium layer is then etched for 10 minutes using mixture of phosphoric acid, nitric acid, and deionized water in the ratio of 16: 1: 4 at room temperature, respectively while Fig. 3 shows the structure of the etched aluminium layer after photoresist removal post aluminium etching process.

From both Fig. 2 and Fig. 3, it is observed that the pore size widening and pore-to-pore gap reduction are achieved which estimated due to long duration of aluminium etching process in ensuring all exposed aluminium thickness are etched away. The pore size has increased to 2.29 µm, which is about 27% larger from the initial size. Duration of the wet etching process should be optimized later on to obtain identical pore size and pore-to-pore distance as those obtained post pattern development process. Fig. 3 also shows that some of the pores are not etched after aluminium etching process is conducted. This is due to those no etched pore pattern is not properly developed during pattern development process. To verify either there is remaining photoresist in the circle pore area, a cross-section view should be conducted since the status could not be confirmed by optical microscope only due to it was designed to have 1.8 µm diameter only. However, this verification process was omitted so that it is possible to progress to following processes to form a filtration membrane. Thus, a future work of verification process for optimization of pattern development and aluminium etching process using different sample set should also be conducted. After completed aluminium etching process, the substrate is ready for DRIE process.

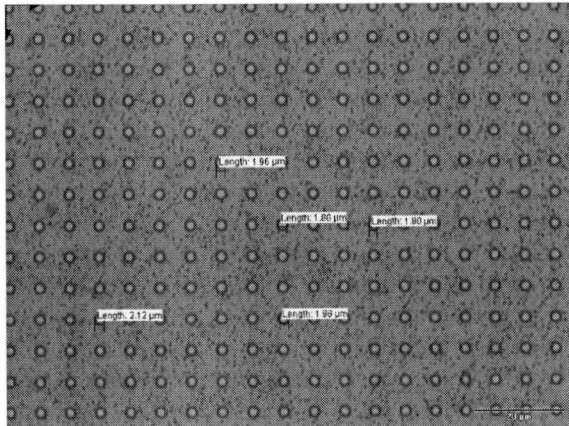

Fig. 2.   Post pattern development process of uniform pore arrays.

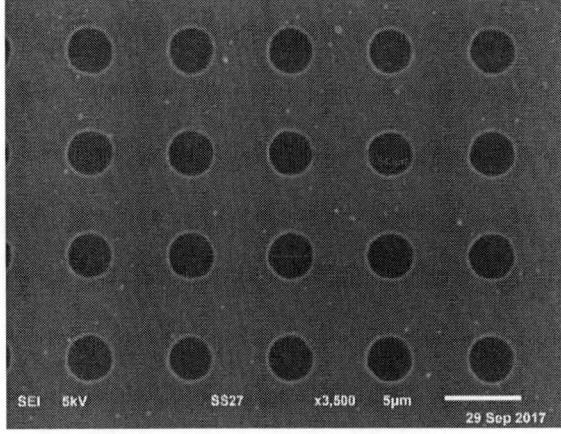

Fig. 4.   SEM of aluminium based uniform pore arrays structure.

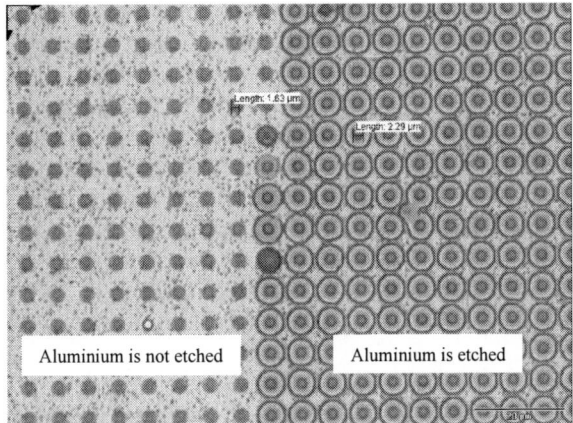

Fig. 3.   Post aluminium etching process of uniform pore arrays.

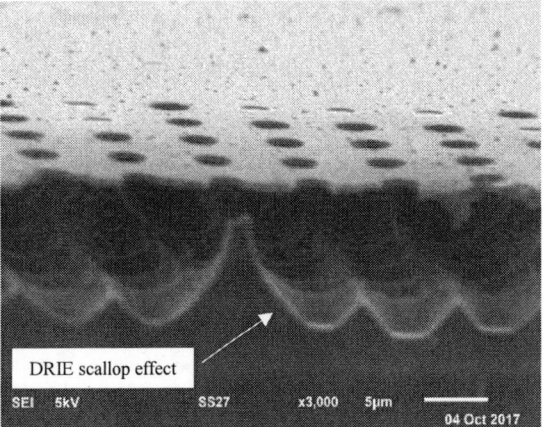

Fig. 5.   SEM cross section of aluminium based uniform pore filtration membrane with underneath silicon etched depth.

A. Aluminium Coating Layer as Uniform Pore Filtration Membrane

SEM observation of post DRIE substrate structure after 10 minutes DRIE process is demonstrated in Fig. 4. The lateral etch caused an aluminium based pore widening of about 25% compared to aluminium pattern. For vertical etching, an approximately 10 μm depth of silicon cavity beneath the aluminium based uniform pore filtration membrane is etched as shown in cross section view in Fig. 5. This equivalents to an etch rate of 1 μm min⁻¹ at the cryogenic temperature of -115±5 °C, SF_6 flow rate of 60 sccm and O_2 flow rate of 13.5 sccm. The obtained DRIE profile is similar to those reported in [12], where scalloping effect on the etched silicon wall resulted from the DRIE process is observed. The aluminium layer which acts as the filtration membrane with uniform perforated pore is still in good condition even after subjected to DRIE process.

B. Aluminium Coating Layer as DRIE Mask

Prior to DRIE process, PDMS filtration sealing layer is bonded on front end of the substrate treated by corona discharge treatment working as protection from remaining grease on DRIE sample carrier and also as the filtration sealing layer itself. Using aluminium coating layer as DRIE mask for filtration cavity fabrication on the back end of the silicon substrate, an approximately 227.514 μm of silicon etch depth is observed as shown in Fig. 6. A vertical etch rate of about 2.2 μm min-1 was achieved at same DRIE parameter setting as mentioned earlier in section III.A. A straight side walls of the filtration cavity structure is obtained with the setting. Thus, to ensure that the cavity at the back end of the substrate meets the aluminium based uniform pore filtration membrane on the front end, about 223 minutes or 3 hours and 43 minutes of DRIE duration with the same parameter setting are required.

Fig. 6. SEM of back end silicon membrane with aluminium as the etch mask.

IV. CONCLUSION

In conclusion, we presented a straightforward method for fabrication of aluminium based uniform pore arrays for filtration using double side aluminium coating layer. The uniform pore arrays is patterned beforehand on selected optimized photoresist layer on top of aluminium coating before it is subjected to aluminium etching process to transfer the pattern into the aluminium coating layer and followed by subsequent DRIE process to create passageway underneath the aluminium based uniform pore arrays for filtration purpose. The filtration function will be realized once this created passageway joins with filtration membrane cavity etched area conducted starting from the back end of the silicon substrate. Different etch rate is observed for fabrication of passageway beneath the aluminium based uniform pore membrane and filtration cavity, indicates that the size of structure that is etched influences the etching rate. On the substrate back end, aluminium coating acts as the filtration cavity pattern mask while aluminium coating on the front end itself acts as the aluminium based uniform pore membrane. Aluminium based filtration membrane can be used to simplify pore fabrication of silicon based filtration. Final structure of the aluminium based uniform pore filtration membrane will afterwards assembled with PDMS sealing layer and tubing to work as a complete set of filtration device.

ACKNOWLEDGMENT

This work is partly supported by Universiti Kebangsaan Malaysia under DPP-2018-006 Dana Pembangunan Penyelidikan PTJ, Ministry of Higher Education of Malaysia under PRGS/1/2017/TK05/UKM/02/1, and also the French RENATECH network and its FEMTO-ST technological facility.

REFERENCES

[1] K. A. Mustafa, J. Yunas, A. A. Hamzah, and B. Yeop Majlis, "Finite Element Analysis on Mechanical Characteristic of Nanoslit Filtration Membrane for Artificial Kidney," in International Conference in Semiconductor Electronics, 2016, pp. 21–24.

[2] A. A. Hamzah, H. E. Zainal Abidin, B. Yeop Majlis, M. Mohd Nor, A. Ismardi, G. Sugandi, T. Y. Tiong, C. F. Dee, and J. Yunas, "Electrochemically deposited and etched membranes with precisely sized micropores for biological fluids microfiltration," J. Micromechanics Microengineering, vol. 23, no. 7, p. 74007, 2013.

[3] N. Marsi, B. Yeop Majlis, F. Mohd-Yasin, and A. A. Hamzah, "The fabrication of back etching 3C-SiC-on-Si diaphragm employing KOH + IPA in MEMS capacitive pressure sensor," Microsyst. Technol., vol. 21, no. 8, pp. 1651–1661, 2014.

[4] A. A. Hamzah, J. Yunas, B. Yeop Majlis, and I. Ahmad, "Sputtered encapsulation as wafer level packaging for isolatable MEMS devices: A technique demonstrated on a capacitive accelerometer," Sensors, vol. 8, no. 11, pp. 7438–7452, 2008.

[5] F. Wan Yunus, A. A. Hamzah, M. R. Buyong, J. Yunas, and B. Yeop Majlis, "Negative charge dielectrophoresis by using different radius of electrodes for biological particles," in Proceedings of the 2017 IEEE Regional Symposium on Micro and Nanoelectronics (RSM 2017), 2017, pp. 84–87.

[6] "Metal Membranes, Innovative Membrane Technology," Metalmembranes, 2018. [Online]. Available: http://www.metalmembranes.com/. [Accessed: 15-Apr-2018].

[7] C. Duan, W. Wang, and Q. Xie, "Review article: Fabrication of nanofluidic devices," Biomicrofluidics, vol. 7, no. 2, pp. 026501-1-026501-41, 2013.

[8] K. R. Williams, K. Gupta, and M. Wasilik, "Etch Rates for Micromachining Processing-Part II," J. Microelectromechanical Syst., vol. 12, no. 6, pp. 761–778, 2003.

[9] K. Grigoras, L. Sainiemi, J. Tiilikainen, A. Säynätjoki, V.-M. Airaksinen, and S. Franssila, "Application of ultra-thin aluminum oxide etch mask made by atomic layer deposition technique," J. Phys. Conf. Ser., vol. 61, pp. 369–373, 2007.

[10] K. A. Mustafa, J. Yunas, A. A. Hamzah, and B. Yeop Majlis, "Application of BOE and KOH + IPA for Fabrication of Smooth Nanopore Membrane Surface for Artificial Kidney," in Proceedings of the 2017 IEEE Regional Symposium on Micro and Nanoelectronics (RSM 2017), 2017, pp. 18–21.

[11] J. Alvankarian and B. Yeop Majlis, "A new UV-curing elastomeric substrate for rapid prototyping of microfluidic devices," J. Micromechanics Microengineering, vol. 22, no. 3, p. 035006, 2012.

[12] Z. A. Syed Mohammed, M. A. S. Olimpo, D. P. Poenar, and S. Aditya, "Smoothening of scalloped DRIE trench walls," Mater. Sci. Semicond. Process., vol. 63, no. December 2016, pp. 83–89, 2017.

2018 IEEE International Conference on Semiconductor Electronics (ICSE)

Shallow Trench Isolation Stress Effect on 45 Degree Rotated MOSFET Layout

Chiew Ching Tan[1,2], Philip Beow Yew Tan[1] and M.K. Md Arshad[2,3]

[1]Silterra Malaysia Sdn. Bhd., Kulim Hi-Tech Park 09000 Kulim, Kedah, Malaysia.
[2]School of Microelectronic Engineering, Universiti Malaysia Perlis (UniMAP), 02600 Pauh, Perlis, Malaysia.
[3]Institute of Nano Electronic Engineering, Universiti Malaysia Perlis, 01000 Kangar, Perlis, Malaysia
chiewching_tan@silterra.com, philip_tan@silterra.com and mohd.khairuddin@unimap.edu.my

Abstract— **This paper describes the Shallow Trench Isolation (STI) stress effect on 45 degree rotated MOSFET layouts. 45 degree rotated PMOS layout with large S_a (active space) shows 15% increase in saturation drain current (I_{dsat}) as compared to the non-rotated ones. On the other hand, PMOS layout with minimum S_a and same rotation show 3.8% I_{dsat} enhancement only. This is because when rotated 45 degree, the channel orientation changes from <110> to <100> axis. This enhances the PMOS I_{dsat} with large S_a. At minimum S_a, the I_{dsat} enhancement is small because the PMOS is at high STI stress. For <110> or non-rotated layouts, the STI stress is observed to increase the I_{dsat} by 14%. The impact of STI stress effect on PMOS with minimum S_a is negligible at <100> or rotated. For NMOS, no Idsat enhancement or degradation is observed on 45 degree rotated layouts. Finally, we have studied the effect of using 45 degree rotated PMOS in ring oscillator circuit.**

Keywords—MOSFET; stress; notch orientation; channel orientation; mobility

I. INTRODUCTION

As the CMOS transistors scale down to deep submicron region, the mechanical stress effect induced by Shallow Trench Isolation (STI) on carrier transport in the channel becomes more and more important. According to Scott et al. [1], stress from the trench isolation edge is the main factor that affects layout sensitivity as well as the carrier mobility, junction leakage and the hot-electron lifetime of the device.

Traditionally, CMOS transistors have been fabricated on (100) silicon wafers with a notch that defines the <110> channel orientation in the wafer. (100) surface orientation was chosen in the early days of CMOS technology mainly due to its high electron mobility and it also yielded the low interface trap density [2]. But, the drawback of this substrate is the PMOS transistors suffer from low hole mobility.

As the drive current enhancement becomes important, the selection of the traditional configuration of transistor in (100) surface orientation and <110> current flow direction become less than ideal. For NMOS transistor, electrons have the best electron mobility in the conventional (100) silicon wafer compared to (111) or (110) wafers, and both <100> and <110> directions of the current flow are almost equivalent. But for PMOS, the hole mobility can be increased by 16% when the channel in the (100) wafer is aligned with the <100> current flow direction [2].

The question then arises is what is the best way to enhance both the drive current for NMOS and PMOS

transistors. One method is to use the conventional wafers and varying the device architecture, while another method involves wafers that are engineered to provide access to both crystalline orientations [3].

In [4, 5], we have studied the stress effect of STI on the two wafers with different channel orientation. In this paper, both <110> and <100> channel orientation transistors are simultaneously fabricated on the same (100) wafer with notch aligned to <110> current flow using layout modification. As shown in Fig. 1, <100> channel orientation transistor is by rotating the transistor layout by a 45° with respect to the <110> notch orientation. The objective of investigations presented here is to study the mobility enhancement and the STI x-stress effect on the 45° rotated MOSFET layouts.

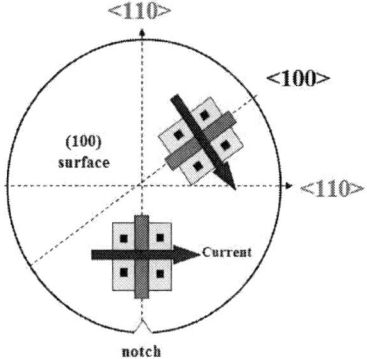

Fig. 1. (100) silicon substrate with <110> and <100> channel orientation transistors on a same wafer.

II. EXPERIMENT

In this work, both <110> and <100> channel orientation transistors are fabricated on same wafer. The voltage of the power supply, V_{dd} was 1.8V. On the same wafer, we have two sets of CMOS transistor, i.e. <110> and <100> channel orientations transistors. We randomly choose the 15 dice for each set of the transistor. All the transistors have the symmetrical source and drain, $S_a = S_b$. S_a and S_b represent the active space in source and drain sides respectively. For all the measurements in this experiment, I_{dsat} was measured at $V_d = V_g = 1.8V$ and $V_s = V_b = 0V$.

978-1-5386-5284-8/18 $31.00 © 2018 IEEE

In the first experiment, we studied the effect of x-stress on the transistor with <110> and <100> channel orientation. The test structures with fixed Width/Length = 10 μm/0.18μm and five different S_a (0.52 μm, 0.6 μm, 0.8 μm, 1.2 μm, and 5 μm) are used in this experiment. In the second experiment, we have the test structures with different degree of rotation. NMOS and PMOS transistors with fixed W/L = 10 μm/0.18 μm and two fixed S_a are used. S_a = 5μm has five degrees of rotation: 0°, 10°, 20°, 30° and 45°. While S_a = 0.52 μm has two degrees of rotation: 0° and 45°.

Two ring oscillator (RO) circuits are designed and fabricated using rotated versus non-rotated PMOS transistors to investigate their impact on the circuit performance. NMOS with W/L = 12 μm/0.5 μm and PMOS with W/L = 22 μm/0.5 μm are used in the ROs. Minimum S_a of 0.48 μm is used in both the ring oscillators.

III. RESULTS AND DISCUSSION

Firstly, we studied the effect of channel orientation on the x-stress effect and its impact on the drive strength of the device. Fig. 2 shows the NMOS I_{dsat} plot of transistor with W/L = 10μm/0.18μm for <110> and <100> channel orientations. In both of the case, NMOS I_{dsat} decreases as the S_a decreases, and the rate of decrease are the same. This indicates that the STI x-stress has the same impact on NMOS transistor in both <110> and <100> channel orientations.

Fig. 3 shows the PMOS I_{dsat} normalized to S_a = 5 μm versus S_a for <110> and <100> channel orientations respectively. Smaller S_a means higher STI stress effect. STI stress effect increases the PMOS I_{dsat} in <110> channel orientation but has less impact in <100> channel orientation. This shows that the STI x-stress effect on the PMOS transistor is negligible in <100> channel orientation.

The results from Fig. 2 and Fig. 3 are in agreement with the previously studied results [4, 5]. STI x-stress effect has same impact, 6% Idsat reduction on NMOS transistors in <110> and <100> channel orientations. While STI x-stress effect has different impact on PMOS transistors in <110>, and <100> channel orientations with 13% and 2% Idsat enhancements respectively.

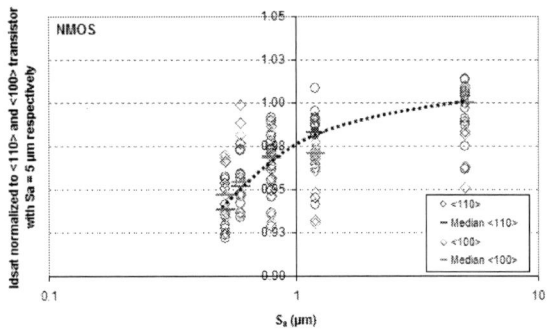

Fig. 2. NMOS normalized I_{dsat} versus S_a with W/L = 10 μm/0.18 μm for <110> and <100> channel orientation

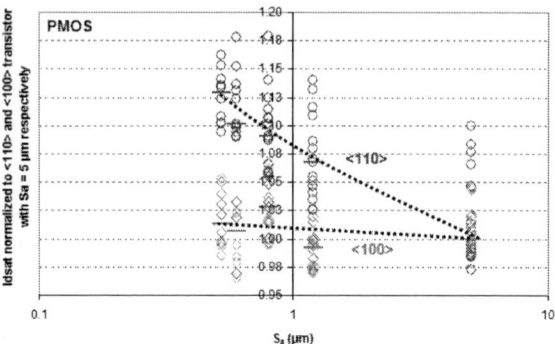

Fig. 3. PMOS normalized I_{dsat} versus S_a with W/L = 10 μm/0.18 μm for <110> and <100> channel orientation

Fig. 4 shows the plot of NMOS I_{dsat} versus the degree of rotation of transistor with W/L = 10 μm/0.18 μm. A non-rotated transistor in this experiment is the same as a transistor in <110> channel orientation. While a 45° rotated transistor is the same as a transistor in <100> channel orientation. From the plot, the NMOS I_{dsat} value range from 0° to 45° rotation are almost the same. This means no significant I_{dsat} change is observed when the NMOS transistor is rotated to 45°. Fig. 4 also indicates that STI x-stress has the same impact on NMOS transistor in both non-rotated and 45° rotated layout.

Fig. 5 shows the PMOS I_{dsat} versus the degree of orientation of transistor with W/L = 10 μm/0.18 μm and Sa = 5 μm. The results show that the I_{dsat} of the PMOS transistor with S_a = 5 μm is increased by 14% when the PMOS transistor is rotated from 0° to 45°. For S_a = 0.52 μm, PMOS I_{dsat} increase by 3.8% only when the transistor rotated from 0° to 45°. This means a significant I_{dsat} enhancement is observed only for PMOS with S_a = 5 μm but not for PMOS with S_a = 0.52 μm when the PMOS transistor is rotated to 45°.

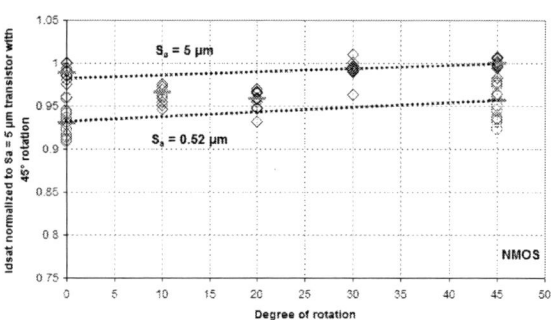

Fig. 4. NMOS normalized I_{dsat} versus degree of rotation of transistor with W/L = 10μm/0.18μm.

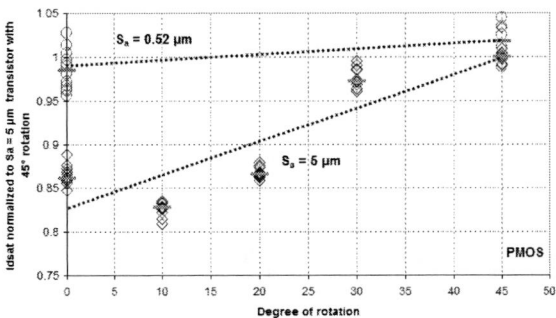

Fig. 5. PMOS normalized I_{dsat} versus degree of rotation of transistor with W/L = 10μm/0.18μm.

A ring oscillator circuit comprises of CMOS inverter stages and each of the inverters consists of NMOS and PMOS transistors. Table 1 shows the gate delay results of ring oscillators (ROs) with non-rotated NMOS and PMOS in both non-rotated and rotated layouts. Only non-rotated NMOS transistor is used in this experiment because the I_{dsat} of non-rotated and rotated NMOS are almost equivalent. The result shows that the gate delays for both the ROs are in the range of 190 ps. This proves that when rotate PMOS with minimum S_a the hole mobility enhancement is less significant.

TABLE 1: GATE DELAY RESULTS FOR RING OSCILLATOR WITH FIXED NON-ROTATED NMOS AND PMOS IN NON-ROTATED AND 45° ROTATED LAYOUTS.

Die Number	Gate Delay (ps)	
	Non-rotated NMOS and PMOS	*Non-rotated NMOS and 45° Rotated PMOS*
Die 1	196.6	199.0
Die 2	189.9	192.8
Die 3	185.9	189.1
Die 4	191.0	194.4
Die 5	196.6	199.0
Die 6	195.0	197.0
Die 7	187.5	191.4
Die 8	181.7	185.3
Die 9	187.5	191.8
Die 10	195.2	197.7
Average	190.7	193.8

IV. THEORETICAL EXPLANATION

In this section, we used the constant energy valley of conduction and valence band to explain how the STI x-stress affect the performance of non-rotated and rotated NMOS and PMOS transistors. Saturation drain current, I_{dsat} of the transistors is used to evaluate the performance of the transistor. The I_{dsat} varies at different STI stress effects (different S_a values) and at different channel orientations (different transistor rotation). As shown in the equation (1), I_{dsat} is calculated based on the number of carriers and their mobility [6]:

$$J = qn_c\mu E \qquad (1)$$

J is the current density, q is the electron charge, n_c is the number of carriers and E is the electric field. The carriers for

NMOS transistor are electron while the carrier for PMOS transistor are holes. The effect of CMOS transistor channel orientation on the electron and hole mobilities has been studied by many authors for several decades [7-10]. The mobility of the carriers is determined by their directional effective mass in the current flow direction which is same as the channel orientation.

The constant energy valleys of conduction band for electrons in <110> and <100> channel orientations transistors are shown in Fig. 6. In this paper, <110> and <100> channel orientations are equivalent to non-rotated and rotated transistors respectively. Compressive x-stress in either <110> or <100> channel orientations cause higher energy state in (5, 6) valleys with high electron effective mass and lower energy state in (1, 2, 3, 4) valleys with low electron effective mass. More electrons are occupying the (1, 2, 3, 4) valleys with low electron effective mass as the electrons prefer to occupy the lower energy state. Hence, the overall electron mobility is reduced and results in a lower I_{dsat}.

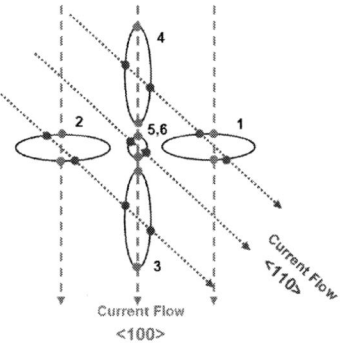

Fig. 6. Constant energy valley of conduction band for electrons in <110> and <100> current flow direction.

The increase of hole mobility in <100> channel orientation compared to <110> direction can be understood from the valence band structure of silicon. In the valence band of silicon, there are light hole bands and heavy hole bands. According to Sayama et al. [7], the effective mass of the heavy hole in the <100> channel orientation is lighter compared to <110> channel orientation.

Fig. 7 shows the constant energy valleys of valence band for holes in <110> and <100> channel orientations transistors. When the current is flowing in <110> direction and compressive stress acts on the same direction, more holes will occupy the [$\bar{1}$10] lobes. The effective mass of [$\bar{1}$10] lobes is lighter compared to the [110] lobes in <110> current flow direction. Therefore, the overall hole mobility is increased and resulted a higher PMOS I_{dsat}. For PMOS with <100> current flow direction, the x-stress compresses 45° to both the [$\bar{1}$10] and [110] lobes on the same direction as the current flow. The distribution of the holes in both the lobes are equivalent, only insignificant hole effective mass reduction in <100> current flow direction. Hence, the PMOS I_{dsat} enhancement is less significant.

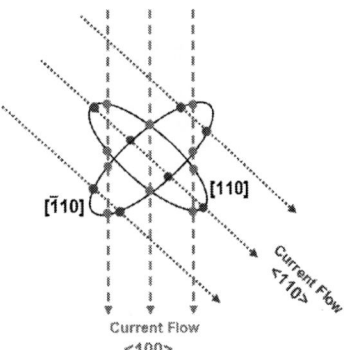

Fig. 7. Constant energy valley of valence band for holes in <110> and <100> current flow direction.

V. CONCLUSION

We have studied the effect of mobility enhancement and STI stress effect on 45° rotated MOSFET layout. PMOS layout with 45° rotated and large S_a shows 15% increase in I_{dsat} compare to a non-rotated. In contrast, at similar rotation angle, the PMOS layout with minimum S_a only shows 3.8% I_{dsat} enhancement. This is because when rotated 45 degree, the channel orientation changes from <110> to <100> that enhances the PMOS I_{dsat} with large S_a. At minimum S_a, the I_{dsat} enhancement is less because the PMOS is already at high STI stress that has increased the I_{dsat} by 13% when non-rotated. Also, STI stress effect on PMOS with minimum S_a is negligible when rotated. No I_{dsat} enhancement or degradation is observed on 45° rotated NMOS. Finally, we also have demonstrated the effect of using 45° rotated PMOS in ring oscillator circuit.

REFERENCES

[1] G. Scott, J. Lutze, F. Nouri and M. Manley, " NMOS Drive Current Reduction Caused by Transistor Layout and Trench Isolation Induced Stress". IEEE International Electron Devices Meeting (IEDM). pp. 827-830, 1999.

[2] L. Chang, M. Ieong and M. Yang, "CMOS Circuit Performance Enhancement by Surface Orientation Optimization" in IEEE Transactions on Electron Devices, vol. 51, no. 10, pp. 1621-1627, Oct. 2004.

[3] Y. Nishi and R. Doering, "Handbook of Semiconductor Manufacturing Technology", 2nd Edition, CRC Press, 2007.

[4] C. C. Tan and P. B. Y. Tan, "Shallow Trench Isolation Stress Effect on CMOS Transistors with Different Channel Orientations", IEEE International Conference on Semiconductor Electronics (ICSE)., 2016.

[5] C. C. Tan and P. B. Y. Tan, "Experimental Investigation and Physical Explanation of Shallow Trench Isolation Stress Effect in MOSFETs", IEEE Regional Symposium on Micro and Nanoelectronics (RSM)., 2017.

[6] S. M. Sze, Physics of Semiconductor Devices, John Wiley and Sons, Inc., 1981.

[7] H. Sayama, Y. Nishida, H. Oda, T. Oishi, S. Shimizu, T. Kunikiyo, K. Sonada, Y. Inoue and M. Inuishi, "Effect of <100> Channel Direction for High Performance SCE Immune pMOSFET with Less Than 0.15μm Gate Length", IEDM 1999.

[8] T. Komoda, A. Oishi, T. Sanuki, K. Kasai, H. Yoshimura, K. Ohno, M. Iwai, M. Saito, F. Matsuoka, N. Nagashima and T. Noguchi, "Mobility Improvement for 45nm Node by Combination of Optimized Stress Control and Channel Orientation Design", IEDM 2004.

[9] P. Yang, W. S. Lau, S. W. Lai, V. L. Lo, S. Y. Siah and L. Chan, "Effects of Swicthing from <110> to <100> Channel Orientation and Tensile Stress on n-Channel and p-Channel Metal-Cxide-Semiconductor Transsitors", Solid-State Electronics, pp. 461-474, 2010.

[10] J. Jang, J, Jung, h. Chang, Y. Kim, S. pARK, H. Kim, J. Kim and S. Yi, "Mobility Enhanced Pwer CMOS", Proceedings of The 25th International Symposium on Power Semiconductor Devices & ICs, 2013.

978-1-5386-5284-8/18 $31.00 © 2018 IEEE

Modeling, Simulation and Optimization of 14nm High-K/Metal Gate NMOS with Taguchi Method

S.K.Mah
Department of Electrical Engineering
Faculty of Engineering and Technology
71800 Nilai, Negeri Sembilan, Malaysia
mahsiewkien@gmail.com

I.Ahmad, P.J. Ker, K.P. Tan, Noor Faizah Z. A.
Institute of Power Engineering
College of Engineering
Universiti Tenaga Nasional
43000 Kajang, Selangor, Malaysia
AIbrahim@uniten.edu.my

Abstract— The developments in electronics technology push the invention of Metal Oxide Semiconductor Field Effect Transistor (MOSFET) towards smaller physical dimension with improvements in both quality and performance. In this paper, design, fabrication and simulation of electrical characteristics of 14nm La_2O_3/WSi_2 NMOS is presented. The fabrication and simulation process of device were performed by using Virtual Wafer Fabrication (VWF) Silvaco TCAD Tools, which consists of ATHENA and ATLAS. The designed device was optimized using Taguchi Method that involves orthogonal arrays and analysis of variance (ANOVA). The original results before optimization process for V_{TH} is 0.212648V (7.5% lower than the targeted value) and I_{OFF} is 3.73851×10^{-9} A/μm while the optimized results for V_{TH} is 0.233321 V (1.44 % higher than the targeted value) and I_{OFF} is 4.732375×10^{-11} A/μm which fulfilled the targets based on International Technology Roadmap for Semiconductors (ITRS) 2013. The Taguchi optimization method yields a significantly lower I_{OFF} with an improved I_{ON}/I_{OFF} ratio by a factor of 25.

Keywords—MOSFET, Taguchi, ATHENA, ATLAS

I. INTRODUCTION

In current fast paced technological era, the development of Si-based ultra large scale integrated circuits (ULSIs) have grown sharply for the last 40 years. The advancement of digitization and Internet of Things (IoT) will further increase the demand for semiconductors such as in smartphones, wearable electronics and IoT devices. The advancements in computer and software technologies have driven the invention of decreasing physical dimension, faster processing speed and reduce power consumption MOSFETs [1]. Thus, technological advancements on Si-ULSIs have provided tremendous computing power in ever device dimensions. Moreover, civilization is very much affected by nanotechnology and much of advancement of nanotechnology is embedded in conventional products of everyday life.

The increasing number of transistor count on ULSIs in the electronic devices has prompted the continued physical feature size scaling of MOSFETs by researchers. MOSFET in smaller physical dimension have higher efficiency, faster switching times, and lower production cost. The size of the MOSFET has to be scaled down to nanometer regime in order to compete in this challenging microelectronics industry while maintaining excellent functionality and quality.

Through scaling down the physical dimensions of MOSFET, the switching frequency increased with higher transistor drive current. Besides that, higher integration density could be achieved. However, negative effects such as Drain Induced Barrier Lowering (DIBL), and Hot Carrier Effect can affect performance and reliability [2]. All these effects will have impact on the leakage current (I_{OFF}) and will result in increased power consumption. Furthermore, the phenomenon of Short Channel Effect (SCE) occurs when the transistor device reaches nanometer regime. The charge distribution will be influenced due to the changes in electric field at the drain and source area. The SCE is attributed to the electron drift characteristics in the channel which results in modification of the threshold voltage (V_{TH}) [3].

High-k dielectric materials are being introduced as one of the control methods for gate oxide to resolve the problem of excess I_{OFF}. Furthermore, higher gate capacitance can be obtained at greater thickness by using high-k dielectric. Considerations on dielectric constant, large band gap and high band offset to silicon, stability, interface quality, process compatibility and reliability should be taken care of, when deciding for an alternative high-k dielectric material. Moreover, the selection of metal gate material with the considerations of matching work function and process integration paired with high-k dielectric material is critical for better performance [4].

In this paper, WSi_2 has been chosen as the metal gate material and paired with La_2O_3 dielectric in a NMOS due to its high thermal stability, high conductivity and high work function [4]. La_2O_3 gate dielectric has relative high permittivity and large band gap. Thus, it is able to reduce I_{OFF} and also enable the further scaling of gate oxide [5]. However, the use of high-k dielectric is quite challenging and difficult due to several issues such as interface control, unsuitable charge trapping causes V_{TH} instability, limiting scaling potential below 2 nm, thermal instability, significant mobility degradation and an unclear integration path [6].

978-1-5386-5284-8/18 $31.00 © 2018 IEEE

Taguchi experimental design is used in this research as statistical optimization method with the aim of identifying controllable factors which give significant impacts on the performance of NMOS. Orthogonal arrays (OAs) design is the statistical method proposed by Dr. Genichi Taguchi [7] to study the parameters with the least number of experiments. The fabrication process of NMOS transistor with the selected high-k material and metal gate was carried out virtually using ATHENA module while the device's electrical characteristics were characterized through ATLAS module [4]. Both modules are obtainable from the SILVACO TCAD simulator. The V_{TH} and I_{OFF} are the target values of the critical design characteristics that must be achieved.

II. PROCESS AND DEVICE STRUCTURE

A. Fabrication Process

The device was fabricated using ATHENA module. All the fabrication processes were based on similar research on the technology of High-K/Metal Gate [2]. However, design parameters were different since the doping concentrations were optimized according to the specifications of high-k dielectric material and metal gate that were used [8]. The design parameters of 14 nm NMOS is shown in Table I.

TABLE I.   FABRICATION RECIPE OF NMOS

Process Step	Parameters
Silicon substrate	<100> orientation
Retrograde well implantation	200Å oxide layer by 970 °C, 20 min of dry oxygen.
	3.75x10^{12} atom/cm^2
	30 min, 900°C diffused in nitrogen
STI isolation	130Å stress buffer by 900°C, 25 min of dry oxygen
Gate oxide	diffused dry oxygen for 0.1 min, 815°C
V$_{TH}$ adjust implant	2.7x10^{12} atom/cm^2 Boron difluoride
	5keV implant energy, 7° tilt
	20 min annealing at 800 °C
High-K/Metal gate deposition	0.001um La$_2$O$_3$
	0.038um WSi$_2$
	30 min, 800°C annealing
Halo implantation	6.75x10^{13} atom/cm^2 Indium, 30° tilt
Sidewall spacer deposition	0.047um SiN
S/D implantation	0.98x10^{14} atom/cm^2 Arsenic
	12KeV energy dopant, 7° tilt
PMD deposition	20 min, 852°C annealing
Compensation implantation	0.60x10^{14} atom/cm^2 phosphor
	60KeV energy dopant, 7° tilt
Metal 1	0.04 um Aluminum
IMD deposition	0.05 um BPSG
	15 min, 950°C annealing
Metal 2	0.12 um Aluminum

After going through various processing steps in device fabrication, the cross section of the NMOS is depicted in Fig. 1. The NMOS has metal layer, oxide layer and semiconductor layer.

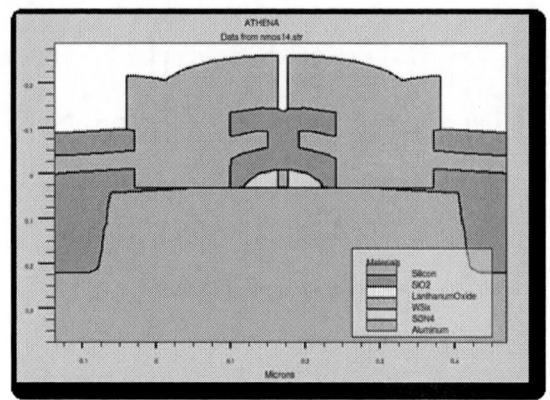

Fig. 1.   Cross section of NMOS.

B. Taguchi L9 (3⁴) Orthogonal Array Method

After the designed NMOS has been successfully fabricated and simulated, process parameters were optimized through Taguchi method by using experimental layout of L9 OA [9]. The values of the different levels of control factors are shown in Table II. Two noise factors (NF) were evaluated in the experiments as shown in Table III.

TABLE II.   CONTROL FACTORS AND THEIR LEVELS

Control Factor	Unit	Level 1	Level 2	Level 3
Halo Implantation	atom/cm^2	6.65x10^{13}	6.75x10^{13}	6.80x10^{13}
Halo Tilt Angle	°	29	30	31
S/D Implantation	atom/cm^2	0.97x10^{14}	0.98x10^{14}	1.00x10^{14}
Compensation Implantation	atom/cm$_2$	0.59x10^{14}	0.60x10^{14}	0.61x10^{14}

TABLE III.   NOISE FACTORS AND THEIR LEVELS

Noise Factor (NF)	Unit	Level 1	Level 2
Sacrificial Oxide Layer Temperature	°C	910 (N1)	915 (N2)
Annealing Temperature	°C	950 (M1)	949 (M2)

III. RESULTS AND DISCUSSION

A. Simulation of Electrical Characteristics

The designed NMOS was simulated in Silvaco TCAD in the ATLAS module. The simulation results for electrical characteristics of the NMOS are shown in Fig. 2 and Fig. 3.

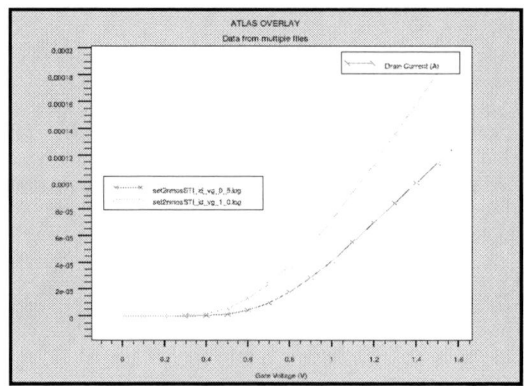

Fig. 2. $I_D - V_{GS}$ Curve

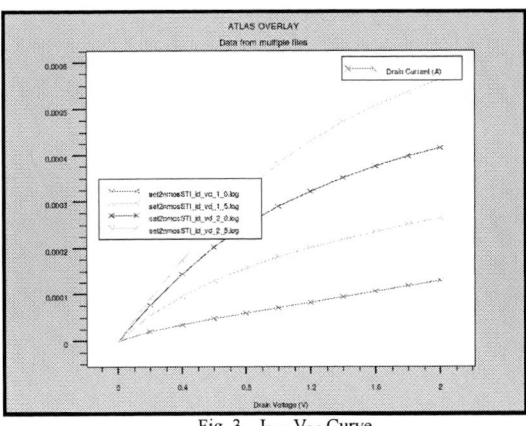

Fig. 3. $I_D - V_{DS}$ Curve

B. Optimization using Taguchi L9

Nine sets of different experiments for NMOS with 4 noise factors N1, N2, M1 and M2 were simulated according to the process parameter combinations as specified in the L9 OA. The simulation results for V_{TH} and I_{OFF} are shown in Table IV and Table V respectively.

TABLE IV.   RESULTS OF VTH (V)

Exp. No	Threshold Voltage (V_{TH})			
	N1M1	*N1M2*	*N2M1*	*N2M1*
1	0.287137	0.323786	0.287239	0.323885
2	0.0824418	0.120627	0.0825364	0.12072
3	-0.1388	-0.10289	-0.138714	-0.10281
4	0.122842	0.158831	0.122936	0.158924
5	0.291788	0.329504	0.291888	0.329601
6	0.197709	0.237258	0.197802	0.237361
7	0.263194	0.303888	0.263295	0.303987
8	0.176829	0.212477	0.176922	0.212569
9	0.375447	0.412502	0.375544	0.412596

TABLE V.   RESULTS OF IOFF (A/µM)

Exp No	Threshold Voltage (V_{TH})			
	N1M1	*N1M2*	*N2M1*	*N2M1*
1	1.29416×10^{-9}	7.83379×10^{-10}	1.29224×10^{-9}	7.82228×0^{-10}
2	2.88000×10^{-8}	1.63584×10^{-8}	2.87581×10^{-8}	1.63347×10^{-8}
3	8.07082×10^{-7}	4.89497×10^{-7}	8.06113×10^{-7}	4.88885×10^{-7}
4	1.54360×10^{-8}	9.04752×10^{-9}	1.54135×10^{-8}	9.03434×10^{-9}
5	1.22881×10^{-9}	7.3029×10^{-10}	1.22701×10^{-9}	7.29227×10^{-10}
6	4.63438×10^{-9}	2.64795×10^{-9}	4.62573×10^{-9}	2.64406×10^{-9}
7	1.87076×10^{-9}	1.07614×10^{-9}	1.86802×10^{-9}	1.07458×10^{-9}
8	6.48804×10^{-9}	3.80172×10^{-9}	6.47847×10^{-9}	3.79616×10^{-9}
9	3.41084×10^{-10}	2.00191×10^{-10}	3.40583×10^{-10}	1.99899×10^{-10}

Next was the process of finding the appropriate control factor levels to make the process less sensitive to variations in noise by studying the response variation using signal-to-noise (S/N) ratio. The optimal combinations were determined through signal-to-noise (S/N) ratio [4]. The level with the highest S/N ratio for each control factor was selected as to achieve better quality characteristic [10].

The quality characteristics in S/N ratio analysis are categorized into smaller-the-better (STB), larger-the-better (LTB) and nominal-the-best (NTB) performance characteristics [11]. V_{TH} analysis was determined by NTB characteristic while I_{OFF} analysis was determined by STB

characteristic [12]. The S/N ratio for V_{TH} and I_{OFF} are shown in Table VI and Table VII respectively.

TABLE VI.   S/N RATIO OF PROCESS PARAMETERS FOR VTH

Symbol	*Process Parameter*	S/N Ratio (Nominal-the-Best)			*Total Mean S/N*
		Level 1	*Level 2*	*Level 3*	
A	Halo Implantation	17.26	19.76	22.15	
B	Halo Tilt Angle	20.48	18.63	20.06	
C	S/D Implantation	20.76	18.40	20.01	19.72
D	Compensation Implantation	23.86	18.16	17.15	

TABLE VII.   S/N RATIO OF PROCESS PARAMETERS IOFF

Symbol	*Process Parameter*	S/N Ratio (Smaller-the-Better)			*Total Mean S/N*
		Level 1	*Level 2*	*Level 3*	
A	Halo Implantation	151.86	168.78	177.63	
B	Halo Tilt Angle	171.25	166.00	161.02	
C	S/D Implantation	171.13	167.22	159.92	166.09
D	Compensation Implantation	183.47	165.80	148.99	

C. Analysis of Variance (ANOVA)

Analysis of variance (ANOVA) is a collection of statistical techniques used to analyze the percentage contribution of each individual control factor on the entire process [9]. Thus, ANOVA was used to determine the ranking of impact of the control factors towards V_{TH} and I_{OFF}. The ANOVA results for V_{TH} and I_{OFF} are tabulated Table VIII and Table IX respectively. The higher percentage on factor effect indicates the particular control factor contributes larger effect to the NMOS in terms of V_{TH} and I_{OFF}.

TABLE VIII.   ANOVA ANALYSIS FOR VTH

Symbol	Process Parameter	Factor Effect	
		NTB	*Mean*
A	Halo Implantation	28	28.99
B	Halo Tilt Angle	4	2.40
C	S/D Implantation	7	3.53
D	Compensation Implantation	61	29.57

TABLE IX.   ANOVA ANALYSIS FOR IOFF

Symbol	Process Parameter	Factor Effect
		STB
A	Halo Implantation	33
B	Halo Tilt Angle	5
C	S/D Implantation	6
D	Compensation Implantation	56

According to Table VIII, Compensation Implantation (D) has the most dominant factor effect while Halo Implantation (A) was identified as the adjustment factor, while Halo Tilt Angle (B) and S/D Implantation (C) acted as pooled factors for V_{TH}. For I_{OFF}, ANOVA was used to find the highest factor effect contribution by a control factor, namely the dominant factor for I_{OFF}. Thus, Compensation Implant (D) which has the highest factor effect was defined as the dominant factor as shown in Table IX.

D. Confirmation Test

Confirmation tests for both V_{TH} and I_{OFF} were carried out in the final phase of this work. The objective of the confirmation test is to verify the final results after optimization using Taguchi method [9]. The best setting for V_{TH} was defined as A1B2C2D1 while for I_{OFF}, it was A3B1C1D1. Table X shows the best setting of process parameters for confirmation runs performed for V_{TH} and I_{OFF}. Table XI and Table XII show the final results of confirmation test for V_{TH} and I_{OFF}.

TABLE X. BEST SETTING OF PROCESS PARAMETERS

Symbol	Process Parameter	Unit	Best Value	
			V_{TH}	I_{OFF}
A	Halo Implantation	atom/cm^2	6.65×10^{13}	6.80×10^{13}
B	Halo Tilt Angle	°	30	29
C	S/D Implantation	atom/cm^2	0.98×10^{14}	0.97×10^{14}
D	Compensation Implantation	atom/cm^2	0.59×10^{14}	0.59×10^{14}

TABLE XI. CONFIRMATION TEST RESULT OF V_{TH} (V)

Noise Factor (N & M)				Best Value	
N1M1	N1M2	N2M1	N2M2	NTB	Mean
0.212935	0.253605	0.213038	0.253706	19.94	-12.64

According to ANOVA analysis and S/N ratio, the predicted S/N ratio for NTB falls between 14.27 dB to 22.01 dB. While the predicted S/N ratio for mean falls between -11.70 dB to -19.44 dB. This clearly shows that the obtained S/N ratio for NTB (19.94 dB) and mean (-12.64 dB) after optimization are in very good agreement with experimental. Furthermore, the average value of V_{TH} after optimization is 0.233321V (1.44 % higher than the targeted value).

TABLE XII. CONFIRMATION TEST RESULT OF I_{OFF} (A/μM)

Noise Factor (N & M)				Best Value
N1M1	N1M2	N2M1	N2M2	STB
5.84952×10^{-11}	3.62233×10^{-11}	5.84065×10^{-11}	3.617×10^{-11}	206.5

According to Taguchi method, the predicted S/N ratio for STB falls between 179.78 dB to 220.34 dB. This shows that the obtained S/N ratio for I_{OFF} (206.5 dB) after optimization is within the predicted range. Besides, the average value of I_{OFF} after optimization is 4.732375×10^{-11} A/μm.

Through optimization process, the original result has been optimized from 0.212648 V (7.5% lower than the targeted value) to 0.233321 V (1.44 % higher than the targeted value) for V_{TH} and I_{OFF} from 3.73851×10^{-9} A/μm to 4.732375×10^{-11} A/μm. The average results after optimization for V_{TH} and I_{OFF} are shown in Table XIII and are compared with ITRS 2013 prediction.

TABLE XIII. AVERAGE RESULT OF CONFIRMATION TEST

Controlled Parameter	ITRS 2013 Prediction	Average Result
V_{TH} (V)	0.230±12.7%	0.233321 (+1.44%)
I_{OFF} (A/μm)	100×10^{-9}	4.732375×10^{-11}

IV. CONCLUSION

This work has fulfilled the objectives and the simulation results have proven the possibility of fabricating 14nm La_2O_3/WSi_2 NMOS. Original results for the designed NMOS are 0.212648V for V_{TH} and 3.73851×10^{-9} A/μm for I_{OFF}. Then, Taguchi optimization was performed to optimize the process factors. The results obtained are compared with the predicted values of S/N ratio and ANOVA analysis. By using Taguchi method, the average V_{TH} of 0.233321V was observed with the control factors of A1B2C2D1. For I_{OFF}, the average value of 4.732375×10^{-11} A/μm was obtained. Therefore, this work has successfully demonstrated that La_2O_3/WSi_2 NMOS can be fabricated and optimized with both V_{TH} and I_{OFF} values fall within ITRS 2013 prediction.

ACKNOWLEDGMENT

The authors gratefully acknowledge the Tenaga Nasional Berhad (TNB) Seeding fund (Project code: U-TG-RD-18-04) for the access to the simulation software.

REFERENCES

[1] N. B. Atan, I. B. Ahmad, and B. B. Y. Majlis, "Effects of high-K dielectrics with metal gate for electrical characteristics of 18nm NMOS device," in 2014 IEEE International Conference on Semiconductor Electronics (ICSE2014), 2014, pp. 56-59.

[2] Z. A. N. Faizah, I. Ahmad, P. J. Ker, P. S. A. Roslan, and A. H. A. Maheran, "Modeling of 14 nm gate length n-Type MOSFET," in 2015 IEEE Regional Symposium on Micro and Nanoelectronics (RSM), 2015, pp. 1-4.

[3] A. H. A. Maheran, P. S. Menon, I. Ahmad, H. A. Elgomati, B. Y. Majlis, and F. Salehuddin, "Scaling down of the 32 nm to 22 nm gate length NMOS transistor," in 2012 10th IEEE International Conference on Semiconductor Electronics (ICSE), 2012, pp. 173-176.

[4] N. F. Zainul Abidin, "Statistical Modeling for 32 nm High-k Metal Gate Technology of Silicon MOSFET Device," Master, College of Engineering, Universiti Tenaga Nasional, Kajang, Malaysia, 2014.

[5] T. Koyanagi, K. Tachi, K. Okamoto, K. Kakushima, P. Ahmet, K. Tsutsui, et al., "Electrical characterization of La2O3-gated metal oxide semiconductor field effect transistor with Mg incorporation," Japanese Journal of Applied Physics, vol. 48, p. 05DC02, 2009.

[6] S.-H. Lo, D. Buchanan, Y. Taur, and W. Wang, "Quantum-mechanical modeling of electron tunneling current from the inversion layer of ultra-thin-oxide nMOSFET's," IEEE Electron Device Letters, vol. 18, pp. 209-211, 1997.

[7] K.-L. Tsui, "An overview of Taguchi method and newly developed statistical methods for robust design," Iie Transactions, vol. 24, pp. 44-57, 1992.

[8] S.K. Mah, I. Ahmad, P.J. Ker, Noor Faizah Z.A., "Modelling of 14NM Gate Length La2O3 -based n-Type MOSFET," Journal of Telecommunication, Electronic and Computer Engineering, pp. 107-110, 2016.

[9] H. Ramakrishnan, S. Shedabale, G. Russell, and A. Yakovlev, "Analysing the effect of process variation to reduce parametric yield loss," in Integrated Circuit Design and Technology and Tutorial, 2008. ICICDT 2008. IEEE International Conference on, 2008, pp. 171-176.

[10] N. Naidu, "Mathematical model for quality cost optimization," Robotics and Computer-Integrated Manufacturing, vol. 24, pp. 811-815, 2008.

[11] P. J. Ross, Taguchi Techniques for Quality Engineering: Loss Function, Orthogonal Experiments, Parameter and Tolerance Design., 2nd Edition ed.: McGraw-Hill, New York, 1996.

[12] S. Fauziyah, I. Ahmad, F. Azlee Hamid, A. Zaharim, H. Elgomati, B. Yeop Majlis, et al., "Optimization of HALO structure effects in 45nm p-type MOSFETs device using Taguchi Method," International Journal of Engineering and Applied Sciences, pp. 80-86, 2011.

List of Reviewers

Name	Affiliation	Country
Ab Rahim, Rosminazuin	International Islamic University Malaysia	Malaysia
Abd Aziz, Norazreen	Universiti Kebangsaan Malaysia	Malaysia
Abd. Rahman, Mohd. Yusri	IMEN, UKM, Bangi, Selangor	Malaysia
Abdul Aziz, Azlan	Universiti Sains Malaysia	Malaysia
Abdul Rashid, Affa Rozana	USIM	Malaysia
Abdullah, Mohammad Faiz Liew	Universiti Tun Hussein Onn Malaysia (UTHM)	Malaysia
Abu Bakar, Norhayati	Institute of Microengineering and Nanoelectronics (IMEN)	Malaysia
Ahmad, Badrul Hisham	Universiti Teknikal Malaysia Melaka	Malaysia
Ahmad, Ibrahim	Universiti Tenaga Nasional	Malaysia
Akhtaruzzaman, Md. Akhtaruzzaman	Universiti Kebangsaan Malaysia	Malaysia
Alam, Ahm Zahirul	International Islamic University Malaysia	Malaysia
Ali Umar, Akrajas	Universiti Kebangsaan Malaysia	Malaysia
Aliza Aini, Md Ralib	International Islamic University Malaysia	Malaysia
Al-Khashab, Yaareb	Ministry of Water Resources/Badush Dam	Iraq
Amin, Nowshad	Universiti Kebangsaan Malaysia	Malaysia
Azizan, Muhammad Mokhzaini	Universiti Malaysia Perlis	Malaysia
Babulak, Eduard	Fort Hays State University	USA
Bais, Badariah	Universiti Kebangsaan Malaysia	Malaysia
Battula, Krishna	Jawaharlal Nehru Technological University Kakinada	India
bin Mohd Saad, Wira Hidayat	Universiti Teknikal Malaysia Melaka	Malaysia
Burhanudin, Zainal	Universiti Teknologi PETRONAS	Malaysia
Buyong, Muhamad Ramdzan	UKM	Malaysia
Chang, I-Cheng	National DongHwa University	Taiwan
Chik, Mohd Azizi	UNIMAP	Malaysia
Chuah, Joon Huang	University of Malaya	Malaysia
Ciulin, Dan	E-I-A	Switzerland
Dee, Chang Fu	Universiti Kebangsaan Malaysia (UKM)	Malaysia
Dennis, John	Universiti Teknologi PETRONAS	Malaysia
Duryha, Dilla	IMEN, UKM	Malaysia
Ehsan, Abang Annuar	Universiti Kebangsaan Malaysia	Malaysia
Forsyth, David	UTM	United Kingdom
Ghasempour, Alireza	ICT Faculty	USA
Goh, Boon Tong	University of Malaya	Malaysia
Hamidon, Mohd Nizar	Universiti Putra Malaysia	Malaysia
Hamzah, Azlan	University Kebangsan Malaysia	Malaysia
Haroon, Hazura	Universiti Teknikal Malaysia Melaka	Malaysia
Harun, Sulaiman Wadi	Uni Malaya	Malaysia

Name	Affiliation	Country
Ibrahim, Siti Azlida	Multimedia University	Malaysia
Islam, Shabiul	Multimedia University (MMU)	Malaysia
Ismail, Ahmad Ghadafi	Universiti Kebangsaan Malaysia	Malaysia
Kamsani, Noor Ain	Universiti Putra Malaysia	Malaysia
Kerdvibulvech, Chutisant	National Institute of Development Administration	Thailand
Latif, Rhonira	Universiti Kebangsaan Malaysia	Malaysia
Lee, Poming	NCTU	Taiwan
Ma' Radzi, Ahmad Alabqari	Universiti Tun Hussein Onn Malaysia	Malaysia
Mahmood Zuhdi, Ahmad Wafi	Universiti Tenaga Nasional	Malaysia
Malaysia, EDS Malaysia	Universiti Kebangsaan Malaysia	Malaysia
Mamat, Hazian	Mimos Berhad	Malaysia
Manut, Azrif	UiTM	Malaysia
Md Arshad, Mohd Khairuddin	Universiti Malaysia Perlis	Malaysia
Menon, P. Susthitha	Universiti Kebangsaan Malaysia	Malaysia
Mohamad Saad, Puteri Sarah	Universiti Teknologi MARA	Malaysia
Mohamed, Mohd Ambri	Universiti Kebangsaan Malaysia (UKM)	Malaysia
Mohd Hassan, Siti Maisurah	TM Research & Development Sdn. Bhd.	Malaysia
Mohd razip wee, Farhanulhakim	IMEN	Malaysia
Mohd Tawil, Siti Nooraya	Universiti Pertahanan Nasional Malaysia	Malaysia
Mohmad, Abdul Rahman	Universiti Kebangsaan Malaysia	Malaysia
Mustafa, Mohd Amrallah	Universiti Putra Malaysia	Malaysia
Nabil, Md	UNITEN	Malaysia
Nasir, Haidawati	Universiti Kuala Lumpur	Malaysia
Nayan, Nafarizal	Universiti Tun Hussein Onn Malaysia	Malaysia
Ngajikin, Nor Hafizah	Universiti Tun Hussein Onn Malaysia	Malaysia
Palma-Orozco, Rosaura	Instituto Politécnico Nacional	Mexico
Sampe, Jahariah	Universiti Kebangsaan Malaysia (UKM)	Malaysia
Siow, Kim	Universiti Kebangsaan Malaysia	Malaysia
Soin, Norhayati	University of Malaya	Malaysia
Sudirman, Rubita	Universiti Teknologi Malaysia	Malaysia
Tengku Aziz, Tengku Hasnan	Universiti Kebangsaan Malaysia	Malaysia
Thong, Li Wah	Multimedia University	Malaysia
Wan Muhamad Hatta, Sharifah Fatmadiana	University of Malaya	Malaysia
Wei, Wei	Xi'an University of Technology	P.R. China
Yahya, Iskandar	Universiti Kebangsaan Malaysia	Malaysia
Yunas, Jumril	Universiti Kebangsaan Malaysia	Malaysia
Yunas, Jumril	Universiti Kebangsaan Malaysia	Malaysia
Yusoff, Zubaida	Multimedia University	Malaysia
Zolkapli, Maizatul	Universiti Teknologi MARA	Malaysia
Zoolfakar, Ahmad Sabirin	Universiti Teknologi MARA	Malaysia

Index

A

A. A. Hamzah	169
A. Arbae	128
A. Ibrahim	128
A. Manut	164
A. Sabirin Zoolfakar	164
A. S. Azlan	128
A.A. Zulkefle	205
A.I.A Rahman	205
A.S. Ismail	77
A.S. Zoolfakar	73, 77, 136
Abdelkader Hassein-Bey	29
Abdul Hafiz Mat Sulaiman	148
Abdul Rahman Mohmad	177
Affa Rozana Abdul Rashid	152
Ahmad Afif Safwan Mohd Radzi	184
Ahmad Rifqi Md Zain	180, 188
Ahmad Sabirin Zoolfakar	1, 21, 89
Ahmad Wafi Mahmood Zuhdi	192, 201, 209, 214
Aidil Aizat Mohd Hamir	21
Ales Hamacek	C4
Alipah Pawi	238
Aliza Aini Md Ralib	97
Anim Arifah Ahmad	242, 246
Asmaa Leila Hassein-Bey	29
Asnida Asli	53
Azam Mohamad	57
Azhan Hashim	53
Azlan Hamzah	117
Azri Husni Hasani	192, 201, 209, 214
Azrif Manut	21
Azrul Azlan Hamzah	5, 13, 85, 104, 124, 132, 156, 267

B

B. Y. Majlis	169
Badariah Bais	65, 85
Basuki Rachmatul Alam	218
Burhanuddin Yeop Majlis	5, 13, 49, 61, 65, 69, 85, 104, 117, 121, 124, 132, 140, 180, 188, 267

C

C.A.N. Fernando	93
Chang Fu Dee	121, 140
Chao-Hung Song	17
Chiew Ching Tan	271
Chiew Ching Tan	230
Ching-Wen Hsu	45

D

Dilla Duryha Berhanuddin	180, 188
Duu Sheng Ong	173

E

Edward Yi Chang	B2, 121

F

Faisal Mohd-Yasin	104
Farahdiana Wan Yunus	5, 13, 124
Farrah Masyitah Mohd Shuib	148
Fatin Nurdini Omar	1
Fauzan Ahmad	197
Fazlena Hamzah	53, 160
Fazliyana Za'abar	192, 201, 209, 214
Franky Lumbantoruan	121

H

H. Hashim	144
Hakim Tahi	29
Hamzah Azrul Azlan	148
Harzawardi Hasim	57
Haseeb A. Khan	1, 53, 89, 136
Herman D. Mendoza	254
Hira Shumail	9

I

I.Ahmad	275
Ibrahim Ahmad	192, 275
Intan Helina Hasan	25
Iskandar Yahya	250
Ismayadi Ismail	25

J

Jackie Y. Ying	B1
Jahariah Sampe	226, 238
Jan Reboun	C4
Jang Kyoo Shin	C3
John P. R. David	177
Jumril Yunas	5, 101, 117, 124, 238, 267

K

K.P. Tan	275
Kamarul 'Asyikin Mustafa	117, 267
Kan Yeep Choo	173
Karsono Ahmad Dasuki	152

Katrul Nadia Basri	21
Ke Kian Seng Joseph	234
Keith Newman	259
Kevin Homewood	C2
Kim Shyong Siow	13, 132
Kong Eng Ng	140
Kuei-Ann Wen	17, 33, 45

L

L. Hasanah	169
Lee Li Theng	250
Lee Wen Zhao	173
Luca Marchetti	221

M

M. A. Zakariya	81
M. H. Md. Khir	81
M. Kashif	169
M. M. Ramli	128
M. Najib Harif	192, 201, 209, 214
M. Robaiah	160
M. Rusop	77, 136, 160
M. S. Jaafar	128
M.S.P Sarah	144
M. Z. Jamaludin	109
M. Zainon	205
M.A. Sepee	205
M.A.M. Hanafiah	205
M. Anas A. Basir	164
M.F. Malek	77, 73, 136
M.H. Mamat	73, 77
M. Hafiz Mamat	164
M.J.Salifairus	73
M.K. Md Arshad	271
M.K. Radhakrishnan	A2
M.N. Afnan Uda	93
M.Rusop	21, 73
Mahamad Fariz Mohamad Taib	250
Mahdi All Khamis	109
Mahzaton Aqma Abu Talip	21
Manolo G. Mena	254
Mas Syarafina Norzin	5, 124
Masuri Othman	242, 246
Mazin Malik	192
Md. Shuhazlly Mamat	37
Mehdi Azadmehr	221
Mehmet Ertugrul	37
Minoru Watanabe	263
Mohamad Fauzee Mohamad Ryeeshyam	89
Mohamad Hafiz Mamat	1, 21, 89

Mohamad Marzuki Bin Mohd Fauzi	230
Mohamad Rusop	53, 164, 184
Mohamad Rusop Mahmood	1, 89
Mohamed El-Amine Benamar	29
Mohammad Faseehuddin	226
Mohd Adzir Mahdi	180, 188
Mohd Ambri Mohamed	61, 250
Mohd Husairi Fadzilah Suhaimi	1
Mohd Nizar Hamidon	25, 37
Mohd Nuriman Nawi	148, 180
Mohd Shaparuddin Bahrudin	192, 201, 209, 214
MohdAmbri Mohamed	140
Muhamad Ramdzan Buyong	13, 69, 85, 132
Muhammad Asnawi Mohd Kusaimi	25
Muhammad Fahmi Jaafar	49
Muhammad Izzuddin Abd Samad	13, 69
Muhammad Khairulanwar Abdul Rahim	85, 132
Muhammad Quisar Lokman	197
Muhammad Ramdzan Buyong	5
Munirah Safiay	53

N

N. A. A. Halim	169
N. A. M. Asib	160
N. A. Rahman	109
N.E.A. Azhar	77, 136
N.H. Sulimai	73, 77
N. Norizan	144
Najwa Ezira Ahmed Azhar	89
Navakantha Bhat	C1
Niew Soon Huat	234
Noor Fadzilah M. Sharif	197
Noor Faizah Z. A.	275
Noor Hidayah Mohd Yunus	238
Noor Sulaiman	101
Noorfazila Kamal	242, 246
Norazreen Abd Aziz	69
Norhayati Soin	41, 113
Nur Akmar Jamil	61
Nur Alin Mohd Azhari	25
Nur Athirah Mohd Taib	152
Nur Fatin binti Muhammad Razali	97
Nur Mas Ayu Jamaludin	85, 132
Nur Rabiatul Adawiyah Tajul Othamany	69
Nurul Azzyaty Jayah	65
Nurulhani Diana Rashid	180, 188

P

P. Susthitha Menon	61

P.J. Ker	275
Pao-Min Chu	33
Petr Kaspar	C4
Philip Beow Yew Tan	230, 271
Pin Jern Ker	109

R

R. Abdul Rani	77, 136
R.D.A.A Rajapaksha	93
Radek Soukup	C4
Razman Mohd Halim	152
Reena Sri Selvarajan	104, 156
Rhonira Latif	49
Robaiah Hj Mamat	53
Rosmalini Ab Kadir	164
Rosminazuin Ab Rahim	97
Rozina Abdul Rani	1, 21, 73, 89,160, 164

S

S. Ishak	128
S. Johari	128
S. K. Sahari	169
S. Marini	169
S. Munirah	160
S.Abdullah	73
S.C.B. Gopinath	93
S.K.Mah	275
S.S. Shariffudin	136, 144
Sadia Muniza Faraz	9
Sahbudin Shaari	61
Saifollah Abdullah	53, 160, 184
Salifairuz Mohamad Jafar	53
Salman Alrokayan	1, 73, 89, 136, 160
Salman.A.H.Alrokayan	53
Saman Azhari	25
Samar K. Saha	A1
Samira Abdelli-Messaci	29
Sawal Hamid Md Ali	148, 226, 242, 246
Shafii A. Wahab	148
Sharidya Rahman	41
Sharifah Fatmadiana Wan Muhamad Hatta	113
Sharifah Wan Muhamad Hatta	41
Shaw Fa Lim	259
Shinya Fujisaki	263
Siti Aisyah Zawawi	104, 156
Siti Amaniah Mohd Chachuli	25, 37
Siti Fazlili Abdullah	192, 201, 209, 214
Siti Raudzah Abdul Rahman	242, 246
Slimane Lafane	29
Suhaidi Shafie	197

Suhana Mohamed Sultan	57
Suraya Shaban	197

T

Tan Chan Lik	234
Teck Yaw Tiong	140
Terence Lucero F. Menor	254
Tineesha Naidu	41
Tomas Blecha	C4

U

U. Hashim	93

W

W. Emilin Rashid	109
Wan Ammar Fikri Wan Ali	117, 267
Wan Maisarah Mukhtar	152
Wei Sea Chang	140

Y

Yasmin Abdul Wahab	41
Yngvar Berg	221
Yuan Lin	121
Yueh Chin Lin	121
Yusran Sulaiman	197

Z

Z. H. A. Rahman	81
Z. Khusaimi	160
Z. Zakaria	205
Z.A. Baharudin	205
Z.Khusaimi	73
Zulfikar Zulfikar	113
Zuraida Khusaimi	53

IEEE
445 Hoes Lane
Piscataway, NJ 08854-4141

ISBN 978-1-5386-5284-8